CLIMATE CHANGE AND ITS CAUSES, EFFECTS AND PREDICTION

LAKE WATER

PROPERTIES AND USES

(CASE STUDIES OF HYDROCHEMISTRY AND HYDROBIOLOGY OF LAKES IN NORTHWEST RUSSIA)

Climate Change and its Causes, Effects and Prediction

Additional books and e-books in this series can be found on Nova's website under the Series tab.

CLIMATE CHANGE AND ITS CAUSES, EFFECTS AND PREDICTION

LAKE WATER

PROPERTIES AND USES

(CASE STUDIES OF HYDROCHEMISTRY AND HYDROBIOLOGY OF LAKES IN NORTHWEST RUSSIA)

OLEG S. POKROVSKY
YULIA BESPALAYA
LIUDMILA S. SHIROKOVA
TAISIA Y. VOROBYEVA
EDITORS

nova
science publishers
New York

Library of Congress Cataloging-in-Publication Data

ISBN: 978-1-53619-275-9

Published by Nova Science Publishers, Inc. † New York

CONTENTS

PREFACE

LAKES AS SENTINELS FOR CURRENT CLIMATE CHANGES AND ANTHROPOGENIC PRESSURE

Lake ecosystems are known to be valid sentinels for current climate changes and anthropogenic pressure because they provide indicators of these impacts either directly or indirectly through the influence of climate and human activity on their catchments. Among these indicators, to name just a few, are water temperature, dissolved organic carbon, nutrients and metals, phyto- and zooplankton composition as well as population and biodiversity of crustacea, mollusks and fish. The advantages of using lakes as tracers of climatic changes and anthropogenic impacts on aquatic ecosystems are multiple. Lake ecosystems are well constrained, confined and are studied in a sustained fashion; lakes respond directly to climate change and local and global pollution via incorporating the effects of these impacts occurring within the catchment; lakes integrate responses over time, and thus allow to avoid the random or unique single-time effects. Finally lakes of various sizes are distributed worldwide and, as such, can act as sentinels across various climatic conditions while exhibiting different degree of vulnerability to external pressure depending on their size and specific location capturing different aspects of climate change (e.g., changing precipitation regime, heat waves, permafrost thaw, invasion of new species, local and global (dispersed) pollution).

However, the majority of published studies on lakes in the boreal and subarctic zone deal with Western and Northern Europe and Northern America, with quite limited information on lakes in the NW Russia. This book is intended to partially filling this gap by presenting 13 chapters describing the hydrology, hydrochemistry and hydrobiology of various lakes located in the NW European Russia, from the Finland border in the west to the Ural Mountains in the East. The thirteen chapters of the book, written by the experts in the field of biogeochemistry, limnology and zoology cover full limnetic ecosystems, from lake physical characteristics to lake water chemistry, microbiology, phytoplankton and zooplankton population, Crustacea, mollusks and fish. A multidisciplinary approach across wide geographical zones, comprising both small and large lakes of the Russian Subarctic, presented in this book, will be interesting for a large community of scholars, students, and researchers from academic and private organizations.

The first six chapters describe multidisciplinary long-term (2006-2017) research on small and medium-size lakes of Kenozersky National Park (South of Arkhangelsk Region, NW Russia). The authors analyze climatic records over past decades and current water temperature, describe the hydrology and bathymetry of lakes and present thorough investigation of hydrochemical parameter of the water column, assessed during different seasons, typically all year around, and in different sites of the lake, including shallow and deep water stations. In each lake, the authors present thorough characterization of phytoplankton and zooplankton. In chapter 4, a unique event, massive death of European cisco *Coregonus albula* in mesotrophic Lake Lekshmozero during the summer of 1996 is considered. It is demonstrated that under conditions of sustained prolonged stratification, elevated water temperatures, and an increase in the active reaction of the pH of the medium during the phytoplankton bloom, the most likely scenario was the transformation of ammonium ions to non-ionized ammonia (under elevated pH) which caused poisoning of fish in the surface and deep lake horizons.

Chapter 7 presents the features of phytoplankton structure and dynamics under the influence of natural and anthropogenic factors such as eutrophication, pollution with heavy metals, acidification, thermal pollution. The species composition and quantitative characteristics of algae in large lakes in the North of the European Region, small Arctic lakes and lakes in the Arctic regions are described and analyzed in the context of environmental impact.

In Chapter 8, the authors describes geographical distribution and biology of representatives of the families Polyphemidae and Cercopagidae (Onychopoda, Cladocera) in lakes of the European North and the Arctic.

Chapters 9 deals with the distribution and biology of the northern lakes dominant species of amphipods and narrow-clawed crayfish, which have recently spread to the north, the features of their life cycles, metabolism and nutrition in cold habitats. The authors revealed patterns of energy metabolism, feeding in amphipods under changing climate, belonging to different ecological and biographical groups from northern latitudes.

The next two chapters (10 and 11) present comprehensive analyses of environmental factors control on the diversity and distribution of freshwater mollusks in 9 lakes on the Solovetsky Archipelago, White Sea and the relict Vashutkiny Lakes System (NE European tundra).

The two final chapters deal with ecosystem analyses of lakes in Arkhangelsk region and Bolshezemelskaya Tundra with special focus on fish stock and biodiversity. Altogether, the 13 chapters represent hierarchical analyses of poorly known Russian Ssbarctic lake ecosystems, from the physical and chemical parameters of the lake water column across the food web (bacteria - phytoplankton - zooplankton - Crustacea - mollusks - fish).

ACKNOWLEDGEMENTS

Recommended for publication by the Academic Council of N. Laverov Federal Center for Integrated Arctic Research of the Ural Branch of Russian Academy of Sciences 08/02/2021, No 2

In: Lake Water
Editors: Oleg S. Pokrovsky, Yulia Bespalaya et al.
ISBN: 978-1-53619-275-9

Chapter 1

MULTIDISCIPLINARY RESEARCH OF DIFFERENT TYPES OF BOREAL LAKES IN NORTH-WEST OF RUSSIA

T. Ya. Vorobyeva*, S. I. Klimov, S. A. Zabelina, L. S. Shirokova, O. S. Pokrovsky, E. I. Sobko, O. Y. Moreva, A. A. Chupakova, A. V. Chupakov, N. V. Shorina, N. V. Neverova and O. D. Kovalev
N. Laverov Federal Center for Integrated Arctic Research of the Ural Branch of the Russian Academy of Sciences, Arkhangelsk, Russia

ABSTRACT

This chapter presents the main characteristics of five lakes of the Kenozersky National Park, located in the north-west of Russia. An analysis of the climatic conditions of the study area is carried out. The temperature and precipitation records for the last 135 years (1883-2017) were retrieved from the meteorological station closest to the study area (Kargopol). A trend of increasing in average annual air temperatures is demonstrated. Average annual air temperatures for the periods 1961-1980, 1981-2000 and 2001-2017 were 1.5°C, 2.1°C and 3.1°C, respectively. During the observation period (2006-2017), the spring, summer and autumn seasons were within the normal range and generally warm. Winters were characterized as cold or very warm. We also provide the analytical methods used in Chapters 2-6.

Keywords: multidisciplinary monitoring of boreal lake ecosystems, Kenozersky National Park, climatic conditions.

* Corresponding Author's Email: vtais345@gmail.com.

INTRODUCTION

Climate change is a recognized threat to lake ecosystems, and although there are likely to be geographical differences in the sensitivity of lakes to climate, large-scale, long-term studies are needed to understand these changes. In addition, the potential role of the local and regional environmental context in the response of aquatic ecosystems is not well studied (McCullough et al., 2019). The boreal region is important reservoir of lakes, contributing 27% to the global area of lakes, despite the fact that it makes up only 14% of the total land surface (Hastie et al., 2019).

One of the most effective traditional methods for assessing the impact of environmental changes is the long-term comprehensive monitoring of lake ecosystems, which is usually carried out in various boreal and temperate regions (Zhang et al., 2010; Grönlund et al., 2012; Filella et al., 2014; Yang et al., 2014; Adamovich et al., 2017; Brömssen et al., 2018; Zhou et al., 2018; Warner et al., 2019).

Climate change can have a strong impact on the nutrients cycle and, as a consequence, on primary production. An analysis of the literature shows that long-term trends in the physical, chemical and biological parameters of boreal lakes have become a subject of increased interest in the context of the reaction of carbon biogeochemistry to climate change in high-latitude aquatic ecosystems, and increasing soil and water temperature is one of the key issues in predicting the evolution of ecosystems under ongoing environmental changes (Algesten et al., 2003; Worrall et al., 2004; Grönlund et al., 2012; Kothawala et al., 2014; Pagano et al., 2014).

Multidisciplinary monitoring of lake ecosystems, including physical (climatic), hydrochemical and hydrobiological parameters, allows us to assess not only interannual variability, but also the complexity of environmental factors, lake evolution under various scenarios of global changes and local impacts (Shirokova et al., 2016).

Previously, studies were conducted mainly on individual or small groups of ecosystems, so it is difficult to predict how future climate change will affect water quality (Collins et al., 2019). This forecast is substantially hindered by the lack of knowledge about the reaction of aquatic ecosystems to one of the main consequences of climate warming - an increase in surface water temperature.

In Russia, long-term monitoring is carried out mainly for large lakes, and observations of small lakes are scarce. Nevertheless, small lakes are the most sensitive indicators of current environmental changes, while being very vulnerable to both local anthropogenic impacts and global climate change. In this regard, the small boreal lakes of the European part of Russia deserve special attention. Here, the most urgent task is to assess changes in the ecosystems of lakes in relatively undisturbed territories such as national parks, under the influence of anthropogenic and climatic factors.

The main natural factors that influence the course of evolution of lakes (climate, river flow, morphometry of lake basins, the nature of vegetation in the lake catchment basin, etc.) are much less dynamic than the anthropogenic ones (plowing, deforestation, constructions, pollution). The complexity of natural relationships, on the one hand, and the insufficient knowledge of the quantitative interactions and relationships between hydrological, hydrochemical, and biological processes, on the other hand, make it difficult to predict the development of lake ecosystems. Moreover, a study of individual components of the ecosystem does not give a conception of the integrated picture of all the occurring processes. Only multidisciplinary integrated research approaches, covering both abiotic and biotic components, allow us to trace the transformation mechanisms of freshwater ecosystems.

It must be noted that even the most complete and accurate analysis provides information about the composition of water only at the point and at the time of sampling and in the absence of long series of observations. As a result, it is difficult to draw conclusions about long-term trends in the transformation of water quality and ecosystems in general. To predict natural processes, a thorough analysis of long-term observation series and the identification of ecosystem development trends are required.

As such, there is a need for comprehensive, interdisciplinary research of lakes in the boreal zone, since these "model" lakes can serve as important evidence of the ongoing environmental changes in the water ecosystems of the North-West of Russia. Long-term monitoring helps assess the importance of various external parameters and predict the overall response of aquatic ecosystems to environmental factors (Trifonova, 1998; Sobek et al., 2003, 2006; Xie et al., 2012; Adamczuk et al., 2019; Rimet et al., 2019; Schiaffino et al., 2019). However, most of the information obtained in the works cited above is geographically biased, since large areas of Northern Eurasia outside Scandinavia are still insufficiently studied. For example, very poorly information about lakes in the taiga zone of the North-West of Russia (Drabkova, 1974; Shirokova et al, 2010, 2013. Vorobyeva et al., 2012; Kokryatskaya et al., 2012, Pokrovsky, 2012).

In European Russia, such long-term monitoring is known only for large lakes (Kvasov et al., 1990; Rumyantsev et al., 2002, 2015); observations of small lakes were practically not conducted. It is known that with an increase in the discharge of pollutants into natural and man-made ecosystems, local and regional biogeochemical cycles of chemical elements are disturbed, which can lead to unpredictable reactions of organisms, in particular, depletion of biomass and biodiversity. In this regard, in recent years, attention has been paid to integrated studies of the northern lakes ecosystems of the European part of Russia, while similar studies of other regions are still at the initial stage (Moiseenko et al., 1997, 2006, 2009, 2010).

In this study, we chose representative lakes in the southern part of the Kenozersky National Park as model reservoirs for predicting the state of boreal lakes under changing natural environment. According to hydrochemical and hydrobiological indicators, the selected lakes are typical reservoirs of the North-West of Russia. Furthermore, the lakes of the Kenozersky National Park are representatives of the relatively undisturbed ecosystems of the boreal zone, and thus they can serve as indicatory water bodies to trace the influence of climate change and anthropogenic pressure.

Kenozersky National Park was established in December 1991. Kenozersky National Park is located in southwestern part of the Arkhangelsk region, the middle taiga subzone and it covers territories that are unique in their location and natural characteristics. Here lies the border of the Baltic Crystal Shield and the Russian Platform, marking the watershed divide between the basins of the White and Baltic Seas. Its vast lake-river system in terms of aesthetic and recreational potential has few analogues in the European part of Russia. The park has about 300 lakes with a total water area of more than 20 thousand hectares, which is almost 14.5% of its land area. In 2004, Kenozersky National Park received the status of a biosphere reserve and was included in the UNESCO List of Biosphere Reserves. Since 2014, it has been a candidate for inclusion in the list of World Heritage Sites.

Due to the minimal anthropogenic impact of the catchment area, the lakes of the KNP can serve as highly pristine model reservoirs. Studies of lake ecosystems may provide a reliable time benchmark for foreseeing possible future changes in these ecosystems, as well as providing background for analyzing community dynamics under the influence of changing environmental conditions.

The lakes under study are located in the zone of excessive precipitation, which, with a small catchment area, makes them low-mineralized and, therefore, causes a low buffer capacity of their waters with respect to acidification (lowering pH).

The purpose of this study was to assess changes in the physicochemical and hydrobiological parameters of lakes in response to climatic fluctuations and changes in the recreational load due to the development of ecotourism and the Park infrastructure. For this, we assessed the ecological state of lakes in the boreal zone in specially protected natural areas and identified factors affecting the stability of ecosystems in a changing climate and increasing anthropogenic impact, based on our own data obtained during 27 field studies in different seasons, from 2006 to 2017.

General Characteristics of Lakes

In five chapters below, 5 lakes of the Kenozersky National Park are described (Figure 1).

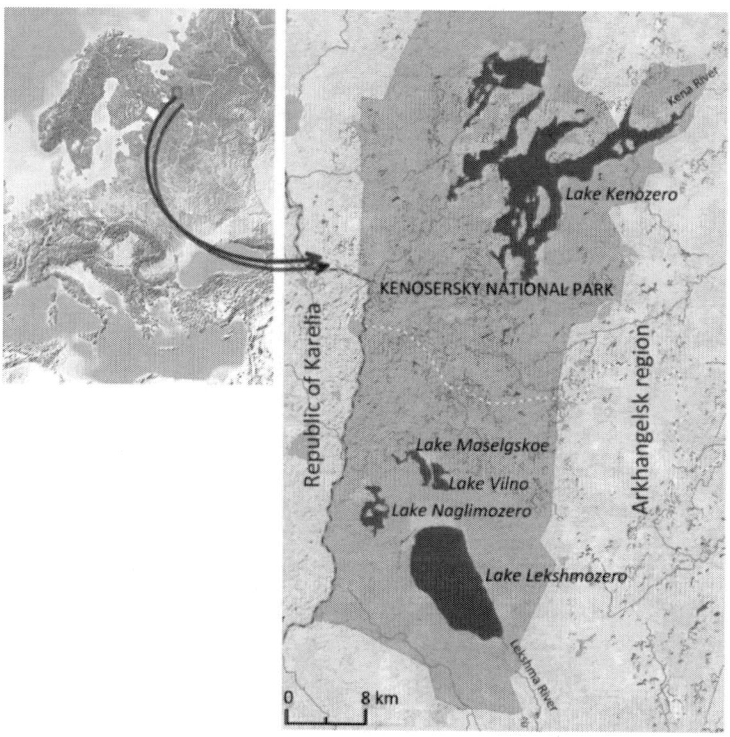

Figure 1. A schematic map of the location of research objects - 5 lakes of the Kenozersky National Park (large and medium lakes Kenozero and Lekshmozero; small lakes Naglimozero, Maselgskoe and Vilno).

Lakes Kenozero and Lekshmozero are among the largest lakes of national economic importance, Kenozersky National Park, located in the boreal zone in the north-west of the Russian Federation (Arkhangelsk region). Both lakes are open basins. The Kena River flows from Kenozero, the Lekshma River originates from Lekshmozero. Lake Lekshmozero is located south of Lake Kenozero at a relatively short distance - about 25 km, but the catchment areas of these lakes vary greatly in the composition of the constituent rocks, which significantly affects the chemical composition of their waters. The catchment of Lake Lekshmozero is entirely located on the Russian platform, where bedrock is represented by limestone. The catchment of Lake Kenozero is located on the border of the Baltic Crystal Shield and the Russian platform is composed mainly of crystalline rocks.

Kenozero Lake is deep, with high flow in the northern part, a steep decline in coasts and a narrow deep basin. Kenozero Lake is characterized by a complex hydrological regime. The northern part of the lake is highly flowing. The waters of a powerful tributary - the Pochi River, which determines the entire hydrochemical regime and the formation of biota, come here. The southern part is slow-flowing, however, it also experiences the influence of supply water, especially in winter. The lake is formed as a result of glacial activity. The coast is very indented by narrow bays. The relief is hilly and ridged. Water belongs to the bicarbonate-calcium class with a salinity of 37-67 mg/L (Kalinkina et al., 2006; Gordeeva & Novoseltsev, 1983).

Lake Lekshmozero has a smooth decrease in the depth and very low water flow rate. Lekshmozero Lake is a reservoir of a pronounced accumulative type. The lake is inhabited by commercial fish species, the most valuable of which is European cisco *Coregonus albula*. In the 70s of the last century, the lake was oligotrophic. By the beginning of the 90s, the reservoir lost oligotrophic status and became mesotrophic. Both lakes - Kenozero and Lekshmozero - in the 70-80s experienced a strong impact from agricultural activities on the catchment. On Kenozero, in addition, in its northern part, for 70 years (until 1993), mole rafting was carried out and deforestation was carried out in the northern part of the catchment. (Kalinkina et al., 2006).

From 1996 to 2004 the state of these lakes was monitored by the Institute for Water Problems of the North (Petrozavodsk, Karelia). In work (Kalinkina et al., 2006) based on literature data and our own research conducted in 1996, 1997 at Lekshmozero and in 1999-at Kenozero, it was concluded that over the 30-year period, the ecosystems of the lakes underwent noticeable changes, which, as a result, affected fishing. The main reason for these changes was the anthropogenic factor - active agricultural activity along the shores of lakes and deforestation.

The input of nutrients from the catchment (due to agricultural runoff), as well as humic substances from the catchment, which is prone to deforestation, led to a significant restructuring of the chemical regime, and, consequently, of the biological communities of water bodies. The chemical composition of the lake's water, and first of all, the increased content of organic matter and the high color of the water of Lake Kenozero, affect the quantitative development of biological communities and their structural composition. The water color in Kenozero in summer increased from 66 degrees in 1963 to 129 degrees in 1982, in winter - from 24 in 1962 to 150 degrees in 1999. High water color reduces light penetration into water, which is naturally accompanied by a decrease in the thickness of the euphotic layer and autogenous synthesis of organic matter, which is the basis of the bio-productivity of water bodies.

The staff of the Laboratory of Freshwater and Marine Ecosystems of the (N. Laverov Federal Center for Integrated Arctic Research of the Ural Branch of the Russian Academy of

Sciences) FECIAR UrB RAS carried out multidisciplinary studies on the lakes of the southern part of the Kenozersky National Park from 2006 to 2017. The ecological state of ecosystems was assessed by hydrological, hydrochemical, and hydrobiological characteristics. For the analysis, we used our own data obtained during 27 field studies from 2006 to 2017. As part of our research, we have been updating the morphometric and hydrophysical characteristics of Lekshmozero Lake for 12 years, including more than 60,000 depth measurements. In July 2015, comprehensive studies were conducted in the northern part of KNP on Kenozero Lake.

Lakes Naglimozero, Maselgskoe and Vilno belong to small reservoirs of the Kenozersky National Park, located in the boreal zone in the north-west of the Russian Federation (Arkhangelsk Region). Situated within the border of the Baltic crystalline shield and the Russian platform, the watershed divide between the basins of the White and Baltic seas passes through the territory of the Kenozersky National Park. Lakes Naglimozero and Maselgskoe belong to the beginning of the catchment basin of the Baltic Sea (Atlantic Ocean), whereas Lake Vilno belongs to the White Sea (Arctic Ocean).

Climatic Conditions of the Region: Analysis of Long-Term Trends

The climate of the Arkhangelsk region is determined by its geographical position in the north of the European territory of Russia, the transfer of moist air masses from the Atlantic and the influence of the Arctic Ocean waters. The western districts of the region are subject to intense impact of cyclones (especially in autumn and winter) and, consequently, of marine (Atlantic) air masses, while in the east the influence of continental air predominates. The entire territory is subject to frequent invasions of Arctic air masses. Frequent change of various air masses determines the great variability of the weather.

Features of the wind regime affect the annual distribution of air temperature. In the cold season, most of the territory is dominated by winds with a southern component (due to cyclonic activity), while in the warm season, north and north-east winds prevail due to the invasion of cold Arctic air on the heated mainland, where atmospheric pressure is reduced at that time. The air temperature in summer is determined by the magnitude of solar radiation, so it naturally rises from north to south, revealing a clear zonality. In the cold half of the year, the main factor in the temperature regime is heat transfer from the Atlantic, which is why the decrease in temperature from west to east is clearly pronounced.

The territory of the region is located in the zone of sufficient and excessive moisture. The annual rainfall in the Arkhangelsk region varies from 350-400 mm in the north to 550-600 mm in the southern and western regions. The zone with increased precipitation is the Konosha-Nyandoma Upland - 730 mm. The minimum rainfall in most of the territory is observed in February, the maximum in July-August. At least 30% of precipitation falls in the form of snow (Vasiliev, 2006).

The most intense increase in snow depth occurs from November to January, when the greatest frequency of cyclonic weather is observed. It reaches its maximum value in the second and third decades of March in most of the territory. The average duration of snow cover is 170-190 days. From March to April, under the influence of an intense influx of solar radiation and an increase in the length of the day, there is a sharp decrease in the height of the snow cover. Most of the region is freed from snow in late April - early May.

The annual runoff values are subordinated to latitudinal zonality, as well as to the influence of topography and rocks. The largest runoff layer is observed on the slopes of the hills. The main source of nourishment for the rivers is snowmelt, therefore the main share of the runoff falls on the period of spring flood. The lowest runoff values are observed in the winter low water. The lakes under study are located in the Onega river basin for which runoff characteristics are given in Table 1. The hydrochemical regime of the rivers is characterized by low mineralization (with a minimum during the spring flood and a maximum during the winter low water) and low water hardness.

Kenozersky National Park (KNP) is located in the southwestern, warmest part of the region in the mid-taiga geographical zone, where the climate is temperate continental. The coldest month is January, the warmest is July. According to the Kargopol hydrometeorological station (KHS), located at a distance of about 60 km from the studied KNI lakes, the meteorological spring begins from April 10 and lasts until May 25 (average dates of a stable transition of average daily temperatures, respectively, through 0°C and +10°C). The summer period with an average daily temperature above +10°C lasts during104 days (from May 25 to September 7). The warmest period is summer with temperatures above 15°C is 45 days (from June 23 to August 8). The onset of meteorological autumn is a steady transition of average daily air temperature through 10°C to lower values. With the transition of temperature through 0°C, the autumn period ends on October 22. The pre-winter period with unstable variable weather ends on average on November 19 after the air temperature decreases below -5°C.

According to the Kargopol HS, snow cover appears in the second or third decade of October (the average multi-year date is October 20). The first snow usually disappears under the influence of a short-term thaw. Only after two or three weeks does a stable snow cover form (November 12), its duration on average 175 days. The maximum height of the snow cover is observed in mid-March and amounts to 49 cm. The destruction and melting of the snow cover occurs in a shorter time than its formation. The average date of the destruction of stable snow cover is April 23, and the disappearance of snow cover is April 30 (Small Cities of Russia 2010-2011 [Electronic resource]/Access mode: http://www.small-cities.ru/kargopol/kargopol2_1.html). In the cold period (from October to March) the south and south-west winds prevail, in the warm (from April to September) - north and north-east. The average monthly wind speeds are small - 2-3 m/s, the minimum wind speeds are observed in July, and the maximum in December.

Table 1. Hydrological characteristics of the Onega River, the village of Porog (1943-1993 yr)

Characteristic	Jan	Feb	March	Apr	May	June	July	Aug	Sept	Oct	Nov	Dec	Annual
river discharge, m^3/c	178	145	133	480	1937	819	430	318	376	484	399	253	496
river flow, km^3	0.48	0.35	0.36	1.25	5.19	2.12	1.15	0.85	0.97	1.3	1.03	0.68	15.7
runoff rate, $L/c/km^2$	3.19	2.61	2.38	8.62	34.8	14.7	7.72	5.7	6.75	8.69	7.17	4.55	8.9
share of runoff, %	3.0	2.4	2.2	8.1	32.5	13.8	7.2	5.3	6.3	8.1	6.7	4.3	100

The air temperature for the Kargopol weather station was used from the Weather and climate website: http://www.pogodaiklimat.ru/ to describe the weather conditions. Average annual values were calculated from December of the previous year to November of the current one.

As can be seen from Figure 2, most of research on the KNP lakes from 2006 to 2017 occurred during the warm period. This decade differed from previous 20-year periods by positive temperature anomalies in all seasons, especially in spring and winter. At the same time warming began earlier and is most pronounced in winter and spring. The hydrological and hydrochemical regime of the lakes, especially in the meta- and hypolimnion, is influenced by weather conditions during the period of spring heating and autumn cooling. In climatology, the qualitative thermal state of the atmosphere is considered normal - with a deviation of the monthly average air temperature from the normal within 1°C, warm/cold - more than 1°C and less than 4°C and anomalously warm/cold - equal to and more than 4°C (http://meteopathy.ru/meteofaktory/klimaticheskaya-norma-temperatureury-vozduxa/). Figures 3 and 4 show that during the period of research, the thermal state of the atmosphere in the spring-summer and autumn periods is mainly characterized as normal and warm.

The maximum of water density reaches at a temperature of about 4 °C, so periods of prolonged cold spring and lingering warm autumn contribute to better mixing of deep waters. Rapid heating of water in the spring leads to early stratification and the cessation of water mixing, and, consequently, to the cessation of oxygen transport to depths. To estimate the duration of the periods of convective mixing of water in spring and autumn, and the intensity of stratification of water in spring and summer, we used the dates of a stable transition of air temperature through gradations of 0°C, 4°C and 10°C, shown in Table 2. It can be seen from it that the spread between the dates of a stable transition of air temperature through gradations is from a month to one and a half to two months.

Table 2. Dates of a stable transition of air temperature through accepted gradations according to the Kargopol weather station

Season	Spring					Autumn		
Characteristic	Date			Period, day		Date		Period, day
Graduation	0°C	4°C	10°C	0-4°C	4-10°C	4°C	0°C	4-0°C
2006	1.04	25.04	20.05	24	25	14.10	17.12	64
2007	9.03	15.04	15.05	37	30	10.10	4.11	25
2008	26.03	28.04	9.06	33	43	2.11	17.11	15
2009	11.04	25.04	26.05	14	31	29.09	29.10	30
2010	28.03	27.04	9.05	30	12	11.10	17.11	37
2011	3.04	23.04	16.05	20	23	13.10	4.11	22
2012	12.04	13.04	10.05	1	27	21.10	26.11	39
2013	12.04	15.04	12.05	3	27	14.10	25.11	42
2014	11.04	13.04	10.05	2	27	4.10	14.10	10
2015	29.03	28.04	7.05	30	9	6.10	6.11	31
2016	27.03	25.04	26.04	29	1	6.10	24.10	18
2017	22.04	16.05	5.06	24	20	19.10	20.11	32
Days	44	33	44	36	42	33	64	54

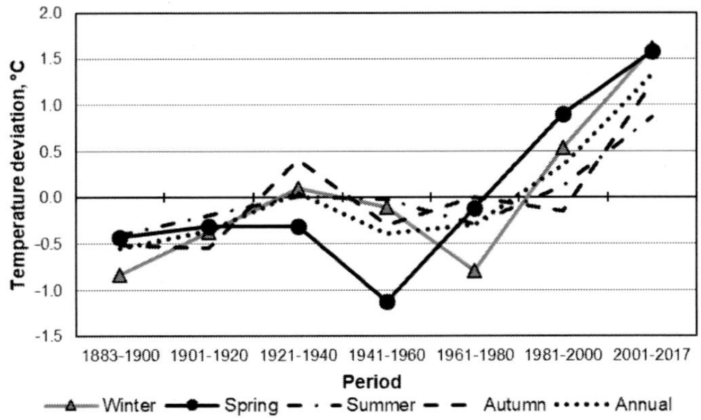

Figure 2. Deviation of air temperature standards for twenty-year periods at the Kargopol weather station relative to its 'normal' values for the observation period 1883-2017.

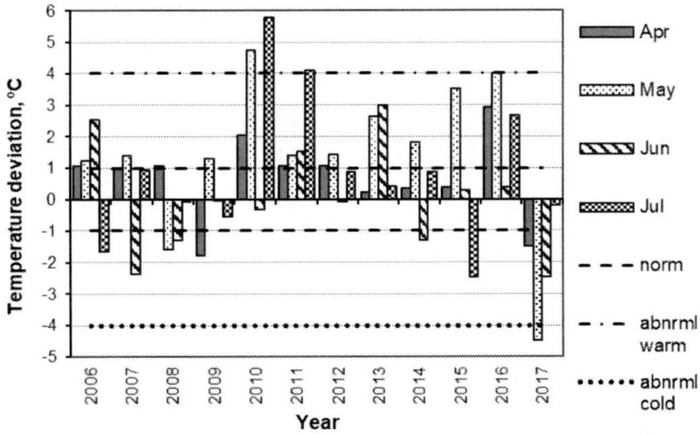

Figure 3. Deviations of average monthly air temperatures in the spring-summer period by years according to the Kargopol weather station relative to their normal values during 1961-2010.

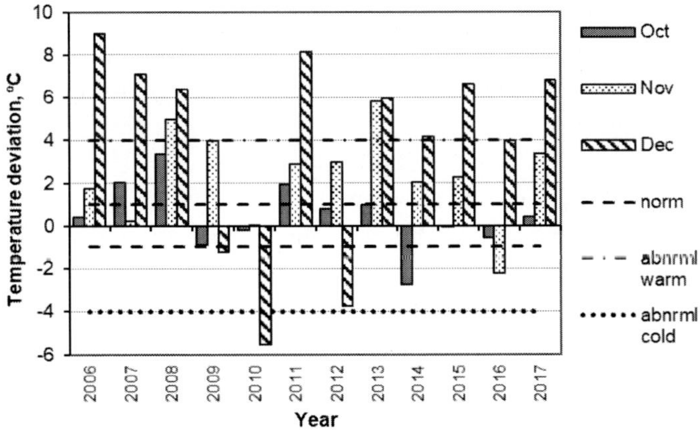

Figure 4. Deviations of average monthly air temperatures in the autumn period for years according to the Kargopol weather station relative to their normal values during 1961-2010.

Thus, the studies of lakes KNP were carried out in the warmest, since 1883, 20-year period, which differs from others in positive temperature anomalies in all seasons, while their growth was most pronounced in the winter and spring seasons. Average annual air temperature for the periods 1961-1980, 1981-2000 and 2001-2017 was, respectively, 1.5°C, 2.1°C and 3.1°C. During the study period, monthly air temperature anomalies in the spring-summer and autumn periods, influencing the formation of the thermal regime of the lakes, are mainly characterized as normal and warm.

METHODS

In the chapters 2 to 6 below, the following analytical methods were employed. The selection of sampling stations took into account the hydrological features of the water bodies and the anthropogenic influence on the catchment area. The depth of sampling was determined based on the results of preliminary probing. Temperature and conductivity were measured using the Multiparameter Mini Sonde Hydrolab MS5 hydrological probe (the accuracy of the sensor depth ±0.05 m). Dissolved O_2 level and temperature (used if there is no Hydrolab MS5) data were measured in situ using the Oxi 197i oximeter (WTW®, Weilheim, Germany) with a Cellox 325 submersible sensor (WTW®, Weilheim, Germany; ±0.5% and ±0.2°C uncertainty). The Winkler titration method was also used for O_2 determination. Conductivity was also measured in the samples (used if there is no Hydrolab MS5) measured on-site using a Hanna HI991300 conductivity meter. pH was measured in the field using a combined electrode calibrated against NIST (National Institute of Standards and Technology, Gaithersburg, MD, USA) buffer solutions (pH = 4.00 and 6.86 at 25°C). The accuracy of pH measurements was ±0.02 pH units.

The water samples were taken using a PVC boat in the summer time and from ice cover during the winter time from selected stations of the lake using a pre-cleaned polycarbonate horizontal water sampler (Aquatic Research Instruments Inc., Hope, ID, USA); they were immediately filtered through sterile, single use Minisart® filter units (acetate cellulose filter, Sartorius®) with a pore size of 0.45 μm. Filtration was performed directly on the boat in summer or on the ice in winter.

Filtered water samples for total forms nutrient analyses were frozen (-20°C) within 1 to 3 h after collection and analyzed either on-site or within 1 week after sampling.

The major anion concentrations (Cl^-, SO_4^{2-}) in the < 0.45 μm fraction were measured using ion chromatography (HPLC, Dionex ICS 2000i), with an uncertainty of 2%, estimated from the replicate analyses of PERADE and RAIN international certified materials.

The DOC and DIC were analyzed by high-temperature catalytic oxidation using TOC-VCSN, Shimadzu (uncertainty ± 2%, 0.1 mg/L detection limit). The DIC was measured after sample acidification with HCl and DOC was analyzed in acidified samples after sparging it with C-free air for 3 min at 100 mL/min as non-purgable organic carbon (NPOC). The nutrient (Si, P-PO_4, P_{tot}, N-NO_2, N-NO_3, N-NH_4, N_{tot}) determinations were based on colorimetric assays (Koroleff, 1983a, b; Chupakov et al., 2017).

The content of boron and iron was obtained as a result of determining the elemental composition of water by the ICP-MS method. ICP-MS Agilent 7500ce with In and Re as internal standards and 3 external in-house standards, placed each 10 samples in a series of lake water. The SLRS-4 and SLRS-5 (Riverine Water Reference Material for Trace Metals certified by the National Research Council of Canada) was used to check the accuracy and reproducibility of analyses (Yeghicheyan et al., 2013). The uncertainty of TE analyses were of 5% and detection limits of 0.01–0.02 μg/L.

Samples for microbiological analyses were collected in sterile 250-mL flacons and stored for less than 2 h in a portable cooler at 5 °C. Active bacteria number count (Colony forming units, CFU/mL) was done using Petri dishes inoculation (0.1 to 1.0 mL of lake water in three replicates) performed in a specially organized microbiological field laboratory (Shirokova et al., 2009, Shirokova et al., 2012, Shirokova et al., 2010).

All manipulations were performed in the vicinity of open flame on thoroughly alcohol- and detergent-treated and UV-sterilized microbiological table. Samples were inoculated on Nutrient Agar (5 g/L beef extract, 5 g/L gelatin peptone, 15 g/L bacteriological agar, pH = 6.8 ± 0.2 at 25°C) to determine the total number of heterotrophic bacteria. Difco™ agar (granulated powder) inoculation was used to assess the number of oligotrophic bacteria. Inoculation of blanks was routinely performed to assure the absence of contamination from external environments. Two duplicates of different volumes of lake water (normally 0.5, 1, 5 mL) were inoculated on three types of agar nutrient media (nutrient Bacto agar, 1:10 diluted nutrient agar, and Difco nutrient-poor agar, for enumeration of eutrophic, facultatively oligotrophic and oligotrophic bacteria, respectively). The number of colony-forming units (CFU) was evaluated by visual counting of colonies on the agar plate after 5, 10, and 30 days of incubation at 25°C in the dark for eutrophic, facultatively oligotrophic, and oligotrophic type bacteria, respectively.

For phytoplankton analyses, the lake water samples were stored in the refrigerator, concentrated by sedimentation, and counted in a Najotte chamber of 0.05 mL volume (Guseva, 1959, Trifonova, 1998). The biomass was determined as algal volume for each lake and converted to wet weight assuming a density of 1 g/ml after approximating the geometry of the cell (Makarova et al., 1970). The microscopes AxioLab A1 and Axiovert CFL-40 (Carl Zeiss, GmbH, Oberkochen, Germany) were used to assess the cell number and morphology. Species were classified into functional groups for the statistical analyses (Reynolds et al., 2002; Padisák et al., 2009).

Zooplankton was collected from the surface horizon by filtering 50 liters of water through the plankton net (64 μm), at deep-water stations, the selection was made by a barometer (volume 5 liters) or by vertical pulling the grid from the bottom to the surface. Samples were fixed with 4% formalin. Conventional methods were used for hydrobiological processing of samples (Methods for studying zooplankton, 1983; Methodical recommendations, 1984).

Samples were analyzed to determine the species composition of zooplankton, dominant complexes and the ratio of taxonomic groups, the number and biomass of organisms. The number of organisms per 1 m^3, the average data was taken from two samples, and the number of organisms in the sample was calculated for the entire sample volume. In the whole sample were considered rare and minor taxa. The zooplankton biomass was calculated using tables of average weights by F. D. Mordukhai-Boltovsky and S.N. Ulomsky (Mordukhai-Boltovsky, 1954; Ulomsky, 1958, 1961). Appropriate determinants were used to identify zooplankton species (Kutikova, 1970; Borutsky et al., 1991; Alekseyev et al., 1995, 2010).

Water quality was assessed by the Pantle and Bukk saprobity index in the Sladechek modification (Sladecek, 1973, 1983; Unified methods of water quality research, 1975). Species diversity was assessed by the Shannon-Weaver index (Shannon et al., 1949). To determine the trophic type of the reservoir by the amount of zooplankton biomass, the classification of S. P. Kitaev was used (Kitaev, 2007).

Main Research Results

In the chapters 2-6 of this monograph, for the first time for the limnic systems of the middle taiga subzone of the North-West of the European part of Russia, the results of long-term multidisciplinary studies are presented, including the simultaneous study of hydrological, hydrochemical and hydrobiological parameters in the same water samples in different hydrological seasons of the year.

The purpose of this study was to assess changes in the physicochemical and hydrobiological parameters of lakes in response to climatic fluctuations and changes in the recreational load due to the development of ecotourism and the Park infrastructure. This allowed to describe the ecological state of lakes in the boreal zone in specially protected natural areas and to identify factors affecting the stability of ecosystems in a changing climate and increasing anthropogenic impact.

For the analysis, we used our own data obtained during expeditions from 2006 to 2017. A retrospective analysis of the knowledge of the lakes of the Kenozero National Park for the period from 1982 to 2004 is also carried out. As a result of multidisciplinary integrated research, including field observations and stationary work on the KNP lakes, an assessment of the ecological state of aquatic ecosystems of five lakes of the park at the present stage was performed.

First of all, the choice of representative lakes in the southern part of the Kenozero National Park as model water bodies for predicting the state of boreal lakes in a changing natural environment was substantiated. In terms of hydrochemical and hydrobiological indicators, the studied lakes are typical water bodies of the North-West of Russia, but at the same time, the lakes of the Kenozero National Park are representatives of relatively undisturbed ecosystems of the Arkhangelsk Region, which can serve as indicators of long-term changes in water bodies under the influence of climate change and anthropogenic load. Kenozersky National Park (southwestern part of the Arkhangelsk region, the middle taiga subzone) covers territories that are unique in their location and natural characteristics. Here lies the border of the Baltic Crystal Shield and the Russian Platform, the watershed between the basins of the White and Baltic Seas. Its vast lake-river system in terms of aesthetic and recreational potential has few analogues in the European part of Russia. In view of the minimal influence of the drainage basin of the lake, KNP can serve as the most convenient model water bodies for studying the ecological state of lakes in the boreal zone.

Analysis of meteorological data showed that research work on the lakes of the Kenozero National Park from 2006 to 2016 took place in a warm period, which differs from the previous 20-year periods of the 20th century by positive anomalies in all seasons.

The lakes under study are located in the zone of excessive moisture and the predominant role in their nutrition belongs to atmospheric precipitation, which, with a small catchment area, makes them low-mineralized and, therefore, causes a low buffer capacity of their waters with a predisposition to acidification (decrease in pH value). Water belongs to the bicarbonate class of the calcium group. The active reaction (pH) of the environment of aquatic ecosystems during the periods of research was characterized mainly as neutral.

According to the nature of the vertical temperature distribution in the summer period, the studied lakes are represented by 3 types: epithermal (Vilno), metathermal (Lekshmozero, Maselgskoe, Naglimozero) and hypothermic (Pezhikherye is a relatively isolated northern part of Lake Maselgskoe).

The analysis of long-term data obtained in the course of complex seasonal studies of lake ecosystems, including the study of the spatio-temporal dynamics of hydrological and hydrochemical parameters, qualitative and quantitative characteristics of biological communities, showed that their current ecological state can be assessed as stable. However, there is a potential vulnerability of lake ecosystems due to the low buffer capacity of waters.

In the studied lakes, mineralization processes are actively taking place, among which aerobic processes dominate, contributing to the natural self-purification of waters. However, there are local zones in the bottom layers of the deep parts of Lake Maselgskoe and in the relatively isolated bays of Lake Naglimozero, where areas of oxygen deficiency are formed during periods of stratification due to its consumption for destruction processes, which favors the inflow of mineral compounds of phosphorus, nitrogen and ferrous forms of iron from bottom sediments.

The trophic status of the lakes Maselgskoe, Lekshmozero, Vilno, according to the main hydrochemical and hydrobiological indicators, is defined as mesotrophic, lakes Naglimozero - oligotrophic, Kenozero - oligotrophic with features of a dystrophic.

The study of the quantitative content of various ecological-trophic groups of heterotrophic bacteria in the water of lakes as a whole showed their low abundance, which is due to the low concentration of organic substances and nutrients. The high variability of the content of heterotrophic microorganisms in the water of lakes is mainly associated with seasonal processes. In general, microbiological indicators in the lake are typical for clean waters.

The status of lake ecosystems is mainly controlled by intra-basin processes. Biological communities are the main factor in the sustainability of ecosystems. Hydrobionts of the studied ecosystems at the present stage are in a stable and balanced state, reacting to changes in biotic and abiotic conditions, thereby maintaining ecosystems in a state of ecological balance under conditions of a changing climate and increasing anthropogenic influence.

However, for mesotrophic Lake Lekshmozero it has been established that the consequences of the bloom of cyanobacterium *Gloeotrichia echinulata* can be very significant. According to the testimony of local residents, the death of fish in Lake Lekshmozero in 1996 is a rare, but not a unique event. This is alarming, especially in light of current climate change, which may cause the likelihood of the risk of recurrence of large-scale environmental disasters. Therefore, there is a potential threat of high vulnerability of the mesotrophic lake Lekshmozero in the context of global climatic changes. If the unfavorable scenarios described above are realized, the situation with a sharp increase in the trophicity of the lake and the emergence of subsequent environmental problems are possible.

As we previously noted for three lakes in the boreal zone (Shirokova et al, 2020), as well as for lakes KNP, we did not find any significant increase in the concentration of DOC in lake waters over the past 10–12 years, which contradicts the widespread "browning" of boreal surface waters in Western Europe and Northern America. During the study period, there were also no significant changes in the hydrochemical and hydrobiological state of the lakes. The response of lake ecosystems to hot summers was expressed only on the scale of the surface layers of lakes.

Our results are generally consistent with the findings of other researchers. The authors of (Oliver et al, 2017) quantified trends in lake water quality using nutrient and chlorophyll (Chl) observations from the Lakes Database of the Northeastern United States (LAGOS-NE). LAGOS-NE was used to quantify changes in lake water quality from 1990 to 2013. The authors concluded that, despite significant changes in the environment, the water quality in lakes in the Midwest and Northeast United States did not change.

Thus, it can be summarized that studies of the ecological state of the lakes of the Kenozero National Park and forecasting their changes are of great importance, both in terms of studying the current state and the problem of a comprehensive ecological assessment of boreal water bodies. Eventually, this should allow to determine the degree of a particular load on a particular reservoir; predict further functioning and determine the restoration potential in the studied ecosystems.

The study of the structure, diversity and functional characteristics of communities of lake ecosystems helps to improve the understanding of the processes occurring in limnic systems, and will serve to forecasting the state of boreal lakes under the pressure of natural and anthropogenic factors.

ACKNOWLEDGMENTS

This work was supported by the State Task AAAA-A18-118012390200-5. We express our gratitude to the administration of the Kenozersky National Park and personally to the director, Shatkovskaya E. F., for the comprehensive support of this work.

REFERENCES

Adamczuk, M., Ferencz, B., Mieczan, T., Dawidek J. 2019. Allochthonous subsidies as driving forces for development of plankton in an autotrophic, temperate, and small lake. *Hydrobiologia,* 846: 59-73, doi.org/10.1007/s10750-019-04052-9.

Adamovich, B. V., Zhukova, T. V., Mikheeva, T. M., Kovalevskaya, R. Z., Makarevich, T. A., Zhukova, A. A. 2017. Eutrophication, oligotrophication, and benthiphication in naroch lakes: 40 years of monitoring. *Journal of Siberian Federal University. Biology*, 10 (4): 379-394. (In Russian).

Alekseyev, V. R., Tsalolikhin, S. Ya. Determinant of freshwater invertebrates in Russia and neighboring territories. Vol. 2. *Crustaceans,* 1995: 628. (In Russian).

Alekseyev, V. R., Tsalolikhin, S. Ya. Determinant of zooplankton and zoobenthos of fresh waters of European Russia. Vol. 1. *Zooplankton,* 2010: 495 p. (In Russian).

Algesten, G., Sobek, S., Bergstro, A. K., Ågren, A., Tranvik, L. J., Jansson, M. 2003. Role of lakes for organic carbon cycling in the boreal zone. *Global Change Biology,* 10: 141-147. doi.org/10.1111/j.1365-2486.2003.00721.x.

Borutsky, E. V., Stepanova, L. A., Kos, M. S. *Determinant of Calanoida of fresh waters of the USSR*, 1991: 504. (In Russian).

Brömssen, C., Fölster, J., Futter, M., McEwan K. 2018. Statistical models for evaluating suspected artefacts in long-term environmental monitoring data. *Environmental Monitoring and Assessment*, 190: 558.

Chupakov, A. V., Chupakova A. A., Moreva O Yu, Shirokova L. S., Zabelina, S. A., Vorobieva, T. Y., Klimov, S. I., Brovko, O. S., Pokrovsky, O. S. 2017. Seasonal dynamics of dissolved carbon in contrasting stratified lakes of the subarctic. *Boreal Environment Research,* 22: 213–230.

Drabkova, V. G. Production processes in lakes of Kola Peninsula. Lakes of Different Landscapes of Kola Peninsula, *Part 2. Hydrochemistry and Hydrobiology*, 1974: 63–77. (In Russian).

Filella, M., Rodrigues-Murillo, J. C. 2014. Long-term trends of organic carbon concentrations in freshwaters: Strengths and weaknesses of existing evidence. *Water*, 6 (5): 1360-1418.

Grönlund, E. 2012. The recovery of two polluted subarctic lakes-towards nutrient management or a pristine state? *Water*, 4: 793-814.

Guseva, K. A. 1959. Methods of phytoplankton counting. *Trudy Inst. Biol. Vodochr*. 2: 44–51. (In Russian).

Hastie, A., Lauerwald, R., Weyhenmeyer, G. 2017. CO_2 evasion from boreal lakes: Revised estimate, drivers of spatial variability, and future projections. *Global Change Biology*. 24(2): 711-728. doi: 10.1111/gcb.13902.

Kalinkina, N. M., Timakova, T. M., Lozovik, P. A., Ryzhakov, A. V., Kulikova, T. P., Ryabinkin, A. V., Chekryzheva, T. A. 2006. Research of transformation of water ecosystems of Kenozersky national Park (Arkhangelsk region) under the influence of agriculture and forestry//*Water resources of the European North of Russia: results and prospects of research*. Materials of the anniversary conference dedicated to the 15th anniversary of the Institute of water problems of the North. Petrozavodsk: 312-322. (In Russian).

Kitaev, S. P. *Fundamentals of Limnology for hydrobiologists and ichthyologists*. Petrozavodsk: Karelian scientific center of the Russian Academy of Sciences. 2007: 395. (In Russian).

Kokryatskaya, N. M., Zabelina S. A., Savvichev A. S., Moreva, O. Yu, Vorobjeva, T. Ya. 2012. Seasonal biogeochemical and microbiological studies of small lakes in taiga zone of North-western Russia (Arkhangelsk Province). *Water Resources*, 39: 105–117.

Koroleff, F. Determination of phosphorus. In Grasshoff, K., M. Ehrhardt & K. Kremling. *Methods for Seawater Analysis*. Verlag Chemie, Weinheim, 1983b: 125–136.

Koroleff, F. Total and organic nitrogen. In: Grasshoff, K., M. Ehrhardt & K. Kremling. *Methods for Seawater Analysis*. Verlag Chemie, Weinheim, 1983a: 162–168.

Kothawala, Dolly N., Stedmon, Colin A., Muller, Roger A., Weyhenmeyer, Gesa A., Kohler, Stephan, J., Tranvik, Lars J. 2014. Controls of dissolved organic matter quality: Evidence from a large-scale boreal lake suvery. *Global Change Biology*, 20: 1101-1114.

Kutikova, L. A. *Rotifers of the fauna of Russia*. 1970: 744. (In Russian).

Kvasov, D. D. Martinson, G. G., Raikas A. V. 1990. *History of Lake Ladoga, Lake Onega, Pskov-Peipsi Lake, Lake Baikal and Lake Khankey*, 280. (In Russian).

Makarova, I. V.; Pichkily, L. O. 1970. On some questions of the method of biomass of phytoplankton calculation. *Bot. J. (Botanicheskii Zhurnal)*, 55: 1488–1494. (In Russian).

McCullough, I. M., Cheruvelil, K. S., Collins, S., Soranno, P. A. 2019. Geographic patterns of the climate sensitivity of lakes. *Ecological Applications*, 29(2): 45-51 https://doi.org/10.1002/eap.1836.

Methodical recommendations on collection and processing of materials in hydrobiological studies on freshwater bodies. *Zooplankton and its products*. 1984: 33. (In Russian).

Methods for studying zooplankton. *Guide to methods of hydrobiological analysis of surface waters and bottom sediment*, 1983: 59-78. (In Russian).

Moiseenko, T.I., Dauvalter, V. A., Kagan, L. Y. 1997. Mountain lakes as air pollution indicators. *Water Resources*, 23 (5): 669-682.

Moiseenko, T. I., Gashkina, N. A., Bylinyak, Yu. A., Kudryavtseva, L. P., Sandimirov, S. S. 2006. Zonal features of the formation of water chemistry in small lakes in european Russia. *Water Resources*, 33 (2): 144-162.

Moiseenko, T. I., Gashkina, N. A., Sharov, A. N., Vandysh, O. I., Kudryavtseva, L. P. 2009. Anthropogenic transformations of the arctic ecosystem of Lake Imandra: tendencies for recovery after long period of pollution. *Water Resources*, 36(3): 296-309.

Mordukhai-Boltovsky, F. D. 1954. Materials on the average weight of aquatic invertebrates of the don basin. *Tr. problemy. and themed slots. no. ZIN an SSSR.* 2: 233-241. (In Russian).

Oliver, S., Collins, S., Soranno, P. A., Wagner, T., Stanley, E. H.; Jones, J. R., Stow, C. A., Lottig, N. R. 2017. Unexpected stasis in a changing world: Lake nutrient and chlorophyll trends since. *Global Change Biology*, 23(12): 5455-5467. doi: 10.1111/gcb.13810.

Padisák, J., Crossetti, L. O., Naselli-Flores, L. 2009. Use and misuse in the application of the phytoplankton functional classification: a critical review with updates. *Hydrobiologia*, 621: 1–19.

Pagano, T., Bida, M., Kenny, J. F. 2014. Trends in levels of allochthonous dissolved organic carbon in natural water: A review of potential mechanisms under a changing climate. *Water*, 6: 2862-2897.

Pokrovsky, O. S., Shirokova, L., S. Zabelina, S. A., Vorobieva, T. Ya., Moreva, O. Yu., Klimov S. I.,. Chupakov, A. V., Shorina, N. V., Kokryatskaya, N. M., Audry, S., Viers, J., Zoutien, C. 2012. *Size Fractionation of Trace Elements in a Seasonally Stratified Boreal Lake: Control of Organic Matter and Iron Colloids Aquatic Geochemistry*. - Berlin, New York: Springer, 2: 115–139. http://dx.doi.org/10.1007/s10498-011-9154-z.

Reynolds, C. S., Huszar, V., Kruk, C., Naselli-Flores, L., Melo, S. 2002. Towards a functional classification of the freshwater phytoplankton. *Journal of Plankton Research*, 24: 417–482.

Rimet, F., Feret, L., Bouchez, A. 2019. Factors influencing the heterogeneity of benthic diatom communities along the shoreline of natural alpine lakes. *Hydrobiologia,* 839 (1): 103-118 doi: 10.1007/s10750-019-03999-z.

Rumyantsev, V. A., Drabkov, V. G.. *Lake Ladoga: past, present, future*, 2002: 326. (In Russian).

Rumyantsev, V. A., Drabkova, V. G., Izmailova, A.V. *Lakes of the European part of Russia*, 2015: 392. (In Russian).

Schiaffino, M. R., Diovisalvi, N., Molina, D. M., Fermani, P., Li Puma, C., Lagomarsino, L., Quiroga, M. V., Perez, G. L. 2019. Microbial food-web components in two hypertrophic human-impacted Pampean shallow lakes: interactive effects of environmental, hydrological, and temporal drivers. *Hydrobiologia,* 830 (1): 255-276. doi: 10.1007/s10750-018-3874-7.

Shannon, C. E., Weaver W. *The Mathematical Theory of Communication*, 1949: 117. (In Russian).

Shirokova, L. S., Chupakov, A. V., Kokryatskaya, N. M., Titova, K. V., Klimov, S. I., Zabelina, S. A., Moreva, O. Y., Vorobieva, T. Y., Pokrovsky, O. S. 2020. Variability of hydrochemical parameters of small boreal lakes under natural and anthropogenic factors (case study of NW Russia)//*Hydrobiologia*.. https://doi.org/10.1007/s10750-020-04193-2.

Shirokova, L. S., Pokrovsky, O. S., Kirpotin, S. N., Desmukh, C., Pokrovsky, B. G., Audry, S., Viers, J. 2012. Biogeochemistry of organic carbon, CO_2, CH_4, and trace elements in thermokarst water bodies in discontinuous permafrost zones of Western Siberia. *Biogeochemistry*, 113: 573–593.

Shirokova, L. S., Pokrovsky, O. S., Kirpotin, S. N., Dupre, B. 2009. Heterotrophic bacterioplankton in thawed lakes of northern part of Western Siberia controls the CO^2 flux to the atmosphere. *Int. J. Environ. Stud.*, 66: 433–445.

Shirokova, L. S., Pokrovsky, O. S., Moreva, O. Yu., Chupakov, A. V., Zabelina, S. A., Klimov, S. I., Vorobieva, T. Ya. 2013. Decrease of concentration and colloidal fraction of organic carbon and trace elements in response to the anomalously hot summer 2010 in a humic boreal lake. *Science of the Total Environment*, 463: 78-90. http://dx.doi.org/10.1016/j.scitotenv.2013.05.088.

Shirokova, L.S., Pokrovsky, O. S., Viers, J., Klimov, S. I., Moreva, O. Yu. Zabelina, S. A., Vorobieva, T. Ya., Dupré, B. 2010. Diurnal variations of trace elements and heterotrophic bacterioplankton concentration in a small boreal lake of the White Sea basin. *International Journal of Limnology*, 46: 67–75.

Shirokova, L. S., Vorobieva, T. Y. Zabelina, S. A., Klimov, S. I., Moreva, O. Y., Chupakov, A. V., Makhnovich, N. M., Gogolitsyn, Y. A., Sobko, E. I., Shorina, N. V., Kokryatskay,a N. M., Ershova, A. A., Pokrovsky, O. S. 2016. Small boreal lake ecosystem evolution under the influence of natural and anthropogenic factors: results of a multidisciplinary longterm study. *Water*, 8: 316. doi: 10.3390/w8080316.

Sladecek V. 1983. Rotifer as indicators of water quality. *Hydrobiologia*, 100(2): 169 – 201.

Sladecek, V. 1973. System of water quality from biological point of view. *Arch. Hydrobiol.* Ergebnisse der Limnologie, 7: 218. (In Russian).

Small Cities of Russia 2010-2011 [Electronic resource]/Access mode: http://www.small-cities.ru/kargopol/kargopol2_1.html. (In Russian).

Sobek, S., Algesten, G., Bergstrom, A. K., Jansson, M., Tranvik, L. J. 2003. The catchment and climate regulation of p CO_2 in boreal lake. *Global Change Biology*, 9: 630-641. (In Russian).

The meteorological conditions characteristics of the study area were evaluated according to the weather archive [Weather of Russia [Electronic resource]/access mode http://meteo.infospace.ru.] According to the HMS Kargopol. (In Russian).

Trifonova I. S. 1998. Phytoplankton composition and biomass structure in relation to trophic gradient in some temperate and subarctic lakes of northwestern Russia and the Prebaltic. *Hydrobiologia*, 369/370: 99-108. (In Russian).

Ulomsky S. N. 1958. Materials on raw weight of lower crustaceans from reservoirs of the *Urals Sci. - tech*. Byul. UNIOR. 6: 81-89. (In Russian).

Ulomsky S. N. 1961. Raw weight of mass forms of lower crustaceans of the Kama reservoir and some lakes of the Urals and TRANS-Urals. *Tr. Ural. otd. UNIOR,* 5: 200-210. (In Russian).

Unified methods of water quality research. *Part III. Methods of biological analysis of water*, 1975: 176. (In Russian).

Vasiliev L. Yu. 2006. Climatic zoning of the Arkhangelsk region: Diss... Ph.D. S-Pb: 148. (In Russian).

Warner K. A., Saros J. 2019 .Variable responses of dissolved organic carbon to precipitation events in boreal drinking water lakes. *Water Research,* 156. doi: 10.1016/j.watres.2019.03.036.

Worrall, F., Harriman, R.; Evans, C. D. Evans, C. D.; Watts, C. D.; Adamson, J.; Neal, C.; Tipping, E.; Burt, T.; Grieve, I.; Monteith, D., Naden, P. S., Nisbet, T., Reynolds, B., Stevens, P. 2004. Trends in dissolved organic carbon in UK rivers and lakes. *Biogeochemistry*, 70: 369-402.

Xie, L., Rediske, R. R., Hong, Y., O'Keefe, J., Gillett, N. D., Dyble, J., Steinman, A. D. 2012. The role of environmental parameters in the structure of phytoplankton assemblages and cyanobacteria toxins in two hypereutrophic lakes. *Hydrobiologia*, 691(1): 255-268. doi: 10.1007/s10750-012-1077-1.

Yang, W., Yang, Z. F. 2014. Effects of long-term environmental flow releases on the restoration and preservation of Baiyangdian Lake, a regulated Chinese freshwater lake. *Hydrobiologia*, 730(1): 79-91 doi: 10.1007/s10750-014-1823-7.

Yeghicheyan, D., Bossy C., Bouhnik, Le Coz M., Douchet, C., Granier, G., Heimburger, A., F Lacan, F., Lanzanova, A., Tristan, C., Rousseau, C., Seidel, J. V, Tharaud, M., Candaudap, F., Chmeleff, J., Cloquet, C., Delpoux, S., Labatut, M., L., Pradoux, C., Sivry, Y., Sonke, E. J. 2013. A compilation of silicon, rare earth element and twenty-one other trace element concentrations in the natural river water reference material SLRS-5 (NRC-CNRC). *Geostandards and Geoanalytical Research,* 37: 449–467.

Zhang, J., Hudson, J., Neal, R., Sereda, J. Clair, T., Turner, M., Jeffries, D, Dillon, P., Molot, L., Somers, K., Hesslein, R. 2010. Long-term patterns of dissolved organic carbon in lakes across eastern Canada: Evidence of a pronounced climate effect. *Limnology and Oceanography*, 55(1): 30-42.

Zhou, Y., Davidson, T. A., Yao, X. Zhang, Y. L., Jeppesen, E., de Souza, J. G., Wu, H. W., Shi, K., Qin, B. Q. 2018. How autochthonous dissolved organic matter responds to eutrophication and climate warming: Evidence from a cross-continental data analysis and experiments. *Earth-Science Reviews*, 185: 928-937.

In: Lake Water
Editors: Oleg S. Pokrovsky, Yulia Bespalaya et al.
ISBN: 978-1-53619-275-9

Chapter 2

MULTIDISCIPLINARY RESEARCH OF THE LAKE KENOZERO (NORTHERN PART OF THE KENOZERSKY NATIONAL PARK) IN THE SUMMER

S. A. Zabelina*, S. I. Klimov, T. Ya. Vorobyeva, N. V. Shorina, E. I. Sobko, O. Y. Moreva, A. A. Chupakova, A. V. Chupakov, L. S. Shirokova, O. S. Pokrovsky and N. V. Neverova

N. Laverov Federal Center for Integrated Arctic Research of the Ural Branch of the Russian Academy of Sciences, Arkhangelsk, Russia

ABSTRACT

The results of a retrospective analysis of the data obtained in the course of comprehensive research at the deepwater station of Lake Kenozero, and literature data for the period 1962-2003. (with an interval of 2-20 years) are presented. Lake Kenozero is the deepest lake in the Arkhangelsk region with high flow in the northern part, a steep decline in coasts and a narrow deep basin. Lake Kenozero has a complex hydrological regime. The features of the formation of the hydrological and hydrochemical regimes of the lake are due to factors associated with their morphometric and hydrographic characteristics. Analysis of the dynamics of changes in the water color index over a 50-year period revealed an increase of 1.5-2 times in the 80s, which remains relatively stable over the past 30 years. The cause of this phenomenon is deforestation on the catchment of the of the main lake's tributary river Pocha, soil erosion and leaching of humic substances bound to complexes with iron from a wetland catchment. An increased iron content throughout the water column was recorded.

Given the level of dissolved oxygen, the prevailing amount of hypolimnion over the epilimnion thickness, poorly developed littoral, quantitative indicators of the main nutrients (phosphorus, nitrogen), and hydrobiological (quantitative and qualitative indicators of phyto-, zooplankton, heterotrophic bacteria) parameters of deep lake Kenozero at the present stage in the research area can be attributed to the oligotrophic type.

* Corresponding Author's Email: sv.zabelina@fciarctic.ru.

Some property of a dystrophic waterbodies are caused by the geochemical features of the catchment area (wetlands of about 14%), namely, the influx of a large amount of dissolved organic matter (humic substances) via the main tributary of the lake.

Keywords: lake, nutrients, plankton, hydrology, diversity, environmental factors

PHYSIO-GEOGRAPHICAL AND HYDROLOGICAL PARAMETERS OF THE LAKE

According to the State Water Cadastre, Kenozero Lake has an area of 68.6 km^2 and a catchment area of 5450 km^2. The lake area with islands is 99.4 km^2. Kenozero is the deepest lake in the Arkhangelsk region with a lobed-type basin and a very rugged coastline with average indicators of conditional water exchange and the ratio of lake area to that of the watershed. After analyzing the available publications, we presented the most reliable values of morphometric and hydrophysical parameters in Table 1.

The latest comprehensive studies of Lake Kenozero were carried out at the beginning of 2000, there are still no exact morphometric measurements, the maximum depth and areas of maximum depth are unknown, the hydrographic parameters are not specified.

According to Kalinkina et al. (2000, 2006), the lake is a large oligotrophic deep-water reservoir with an average depth of 30 m, the maximum depth according to various estimates is from 63 to 90 m (Kalinkina et al. 2000, 2006).

In order to study the current state of Kenozero Lake, we carried out comprehensive research in the northern part of Kenozero Lake on July 13 and 14, 2015 in the Glushevaya Lakhta Bay, Figure 1. On July 13, an echo-sounding of the depths was carried out in order to choose the place of the station. The area location with deepest depth was indicated by the locals. The Figure shows the measurements in the deepest area. The maximum measured depth was 65.6 meters, where the sampling station was established.

The hydrological station was carried out on July 14, 2015 using a Hydrolab MS5 probe and an oximeter. Figure 1 shows the vertical distribution of hydrological parameters. The depth of the station was 62 m.

Table 1. The Main morphometric and hydrographic characteristics of the Lake Kenozero

Length, km	28	Volume, km^3	1.34
The greatest width, km	5.5	Average annual runoff from the catchment, km^3	1.80
Average depth, m	20.3	The specific catchment area	79.4
Maximum depth, m	68	The coefficient of conditional water exchange, year^{-1}	1.1
Watershed area, km^2	5450	Water residence time in the lake, year	0.91
Area, km^2	68.6		

Notes. [1] Data on Lake Kenozero are given according to (Kalinkina et al. 2006; Surface Water Resources…, 1965).

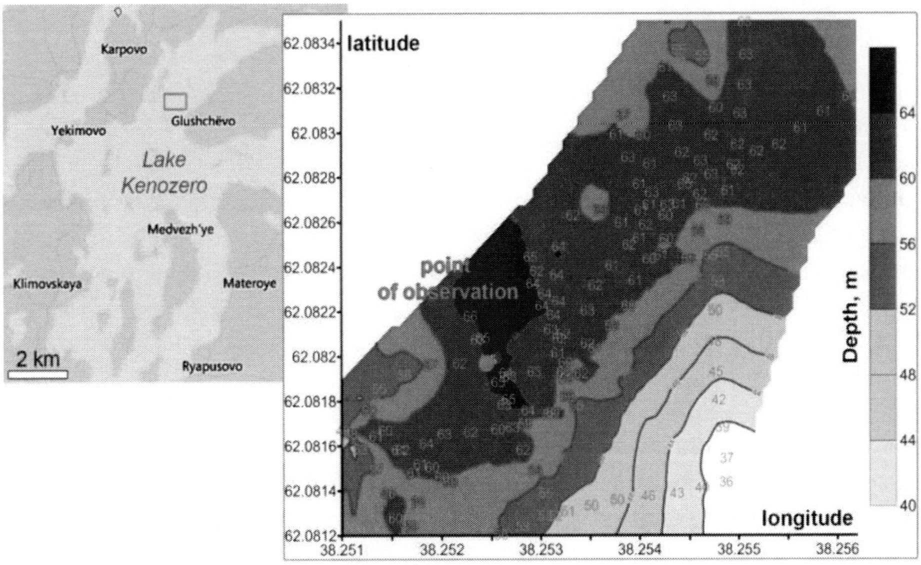

Figure 1. Location of the integrated sampling station on Lake Kenozero in July 2015.

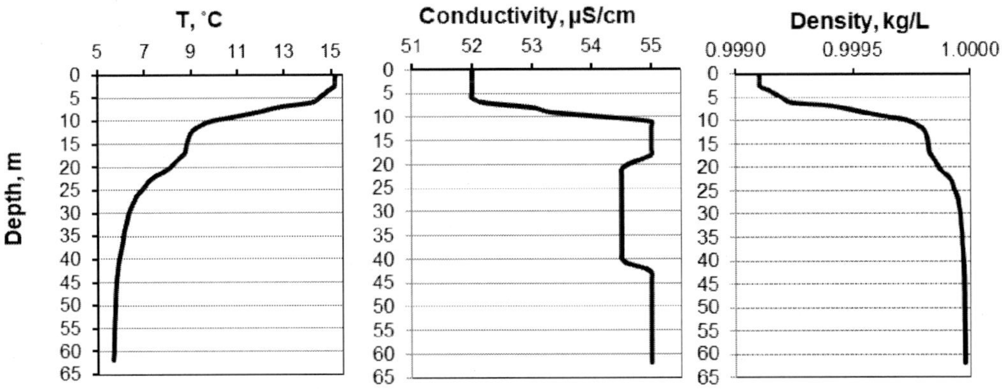

Figure 2. The vertical distribution of temperature, specific conductivity and water density at the integrated station in Glushevaya Lakhta Bay of Lake Kenozero July 14, 2015.

In the upper 3-meter layer, homothermia was observed with a water temperature of 15.2°C. In the layer from 6 to 10 m there was a thermocline with an average gradient of 1.1 °C/m with a maximum of 1.4°C/m in the layer of 6-7 m. At a depth of 12-17 m, there was a quasihomogeneous layer with a temperature of 7.3°C. Below in the layer of 18-26 m, the second thermocline was noted with an average vertical gradient of 0.23°C/m and a maximum of 0.40 18°C/m at 21-22 m. Then the temperature slowly decreased to 5.80°C at 45 m. below which the layer was quasi-homothermal. The water temperature in the bottom layers was 5.66°C.

In the distribution of conductivity, homohalinity up to 6 m was observed at 52 µS/cm. A homogeneous layer with the conductivity of 55 µS/cm to 18 m was observed below the halocline from 11 m. Then, the conductivity decreased to values of about 54.5 µS/cm, which remained constant to a depth of 40 m, below which there was a slow increase in electrical conductivity to about 55 µS/cm in the bottom layers. The resolution of the conductivity sensor is 1 µS/cm.

As can be seen in Figure 2, the density stratification was not violated, and with small variability of the conductivity was determined almost exclusively by the temperature factor, which is confirmed by the shape of the density graph, which mirrors the graph of the vertical temperature distribution.

Thus, very small variability in the vertical distribution of electrical conductivity can be noted in Lake Kenozero. Summer stratification is formed mainly due to the temperature factor. The homogeneity of the layer below 20 m in electrical conductivity is due to spring mixing of the waters, which caused the heating of deep waters to 5.7 °C. The reduced electrical conductivity in the layer of 20-40 m is due to the fact that melt spring waters with less mineralization were involved in the mixing. With increasing temperature stratification, these waters were blocked from mixing, and in the upper layer due to ground runoff, mineralization began to increase. Mineralization in the bottom layers increased due to runoff to lower bottom waters, in which, upon contact with the bottom, there was an increase in mineralization due to mineral enrichment from sediments and an increase in density due to lower temperatures.

The homogeneous layer at a depth of 12-17 m is probably due to storm wind mixing. According to the Kargopol meteorological station, south-west winds were observed from noon on June 16, coinciding with the direction of optimal acceleration for the bay, with an average speed of 3-5 m s^{-1} and gusts of up to 10-13 m s^{-1}, on June 17, the winds continued for 12 hours the same direction with an average speed of 4-6 m s^{-1} and gusts up to 10-12 m s^{-1}.

NUTRIENTS

The results of a study of the nutrients concentration showed that, despite the great depth, the concentrations of nutrients are quasi-homogeneous distributed over the water column.

Water is saturated with oxygen evenly throughout the water column, the pH of the water is neutral, Figure 3.

Concentrations of dissolved nitrogen and phosphorus are almost constant throughout the water column and have low values (Figure 4). The total phosphorus (16.0-24.0 μgP/L), and total nitrogen (<500.0 μgN\L) contents are at the level reported 20 years ago (Kalinkina et al. 2000, 2006).

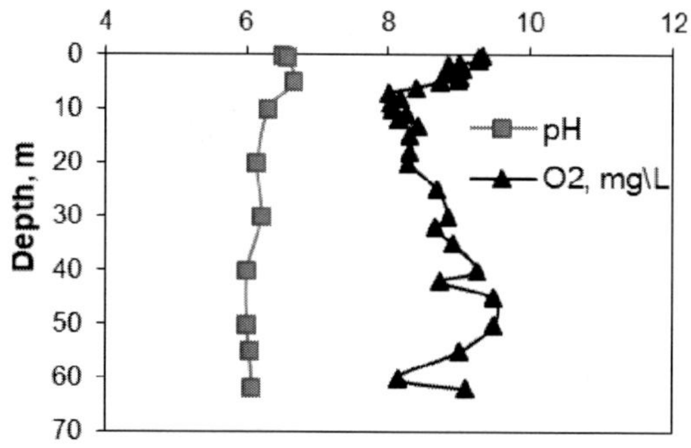

Figure 3. Profiles pH and dissolved oxygen concentration in the Kenozero Lake.

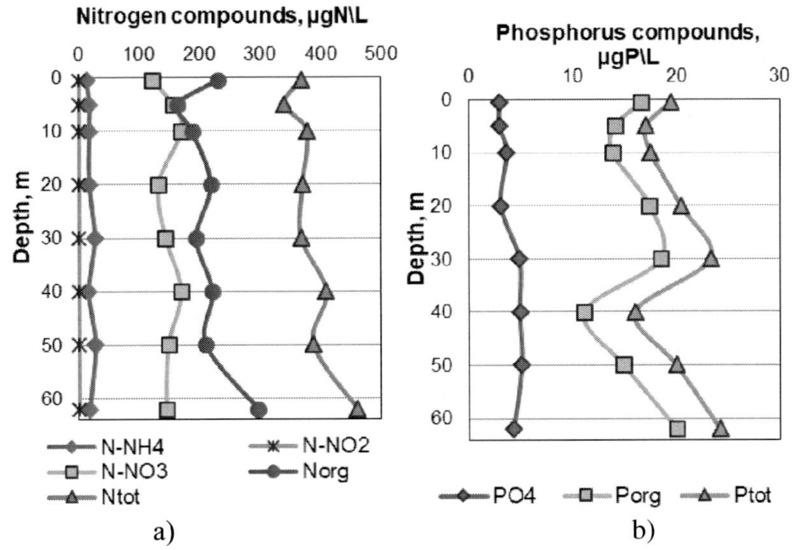

Figure 4. Concentration profiles of nitrogen (a) and phosphorus (b) compounds in the Kenozero Lake.

Table 2. Variation in nutrient concentration in the Kenozero Lake

Parameters	1999 [1]	2003 [2]	2015
Mineralization, mg\L	37-67	32-43	44-48
Phosphate-phosphorus, μgP\L	-	-	2.86-5.05
Total phosphorus, μg P\L	14.0-20.0	10.0-20.0	16.0-24.0
Ammonia-nitrogen, μgN\L	-	-	14.0-28.0
Nitrate-nitrogen, μgN\L	-	40.0-60.0	120.0-170.0
Organic nitrogen, μgN\L	-	400.0-560.0	160.0-300.0
Total nitrogen, μgN\L	≤ 500.0	460.0-610.0	340.0-460.0

Note [1]Kalinkina et al. 2000, 2006; [2]Kalinkina et al. 2006.

The silicon concentration insignificantly decreases in the surface ten-meter layer from 2579 μg/L to 2321 μg/L, and after 10 meters the concentration values remain quasihomogeneous (within 2579-2643 μg/ L to the bottom layer. A decrease in the concentration in the upper layer indicates an absorption of silicon by phytoplankton during the growing season.

Our results are comparable with those obtained in 1999, in 2003 by the results of other authors (Kalinkina et al. 2000, 2006), as can be seen from Table 2.

According to the concentration of the main nutrients in Kenozero water (total, mineral and organic nitrogen, total phosphorus), and following the classification of Kitaev (Kitaev, 1984), the trophic status of the lake during the study period can be characterized as oligotrophic. To assess the ecological state of Kenozero, the ratio of the main nutrients - total nitrogen and total phosphorus - was determined to be equal to 20. This value is located on the border between the moderate and middle classes and is optimal for the development of phytoplankton. At the same time, this value indicates the limiting role of phosphorus in the development of hydrobionts in this water body (Kitaev, 1984).

Overall, a study of the distribution of Kenozero's nutrients in the summer showed that the lake's ecosystem is in a stable and balanced state according to hydrochemical parameters.

IRON

Kenozero Lake is a narrow deep basin with a steep recession and. As noted above, the lake has a complex hydrological regime. The northern part of the lake receives the waters of a powerful tributary - the Pochi River, which determines the entire hydrochemical regime and the formation of biota. The southern part of the lake is weakly flowing, however, it also experiences the influence of supply water, especially in winter. The main reason for the impoverishment of the fish population in the lake was a violation of the food supply, associated with a strong increase in the color of water. This was preceded by deforestation in the northern part of the catchment, which drains the river Pocha.

Recent studies (Kritzberg et al. 2014. Ekvall et al. 2013; Grubisic et al. 2012; Weyhenmeyer, 2004) also indicate the effect of increased color on the phytoplankton community composition, the food web structure (Hansson et al. 2013; Wissel et al. 2003), and the biogeochemical role of fresh water in the global cycle. Several previous studies (Kalinkina et al. 2000, 2006; Hansson et al. 2013; Wissel et al. 2003) showed that for 40 years since 1962 the color of the Lake Kenozero water is gradually increasing, which is associated with forest cut on the catchment. Deforestation leads to soil erosion, increasing the flow of humic substances from soils and wetlands.

In the work of Sushchenya L.M. (Sushchenya et al. 1986) it was shown that with an impact of humified waters, a sharp decrease in the number of relic crustaceans occurs. A large number of organic substances in the water of the lake Kenozero coming from the catchment, as confirmed by the color index of river waters. According to (Kalinkina et al. 2000, 2006), this indicator is 125° Cr-Co. The water color of the lake Kenozero according to the same authors varies in the range of 107-110° Cr-Co, and the water color at the river is 150 degrees.

According to the classification of Kitaev, these indicators reflect the meso-poly-humous type of lake. Our data on the Kenozero deepwater station in the northeastern part of the lake opposite the village of Gory showed a color value of up to 120° Cr-Co at the surface and up to 117° Cr-Co at the bottom. A comparative analysis of the data on the color index over a 20-year period revealed an increase in this indicator in the near-bottom horizon, which may indicate a trend in the accumulation of organic substances in the lake's ecosystem (Table 3).

Table 3. Change in the color of water (° Cr-Co) in the surface and bottom water layer of lake Kenozero (W - winter, S – summer).

Layer	Years												
	1962[1]	1963[1]		1964[1]		1968[1]	1982[1]		1983[1]	1999[1]	2001[1]	2003[1]	2015[2]
	W	W	S	W	S	S	W	S	W	W	S	S	S
Surface	24	24	66	80	78	91	122	129	156	150	90	130	120
Bottom	66	70	-	-	84	91	106	130	125	90	109	95	117

Note [1] Kalinkina et al. 2006. [2] This study.

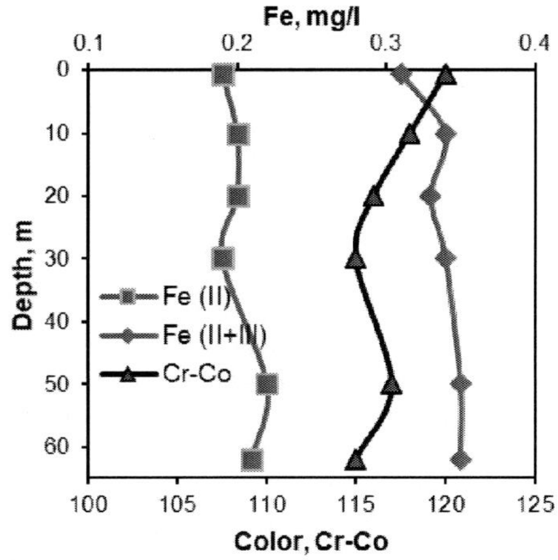

Figure 5. Distribution of the content of forms of iron and color in the water of Lake Kenozero.

The contribution of iron to increasing the color of water has recently been recognized (Björnerås et al. 2017), although earlier the increase in color was thought to be associated with DOM (dissolved organic matter). This phenomenon has long-term environmental and social consequences, as it affects the composition and functioning of aquatic ecosystems by reducing light transmission in water, which is naturally accompanied by a decrease in the thickness of the photic layer and autogenous synthesis of organic matter. This further decrease the bio-productivity of water bodies and their recreational and fishery values as well as their potential use as a source of drinking water.

The Fe concentration in the Lake Kenozero is comparable to its maximum permissible concentrations. The main source of iron in the lake's ecosystem is the Pocha River, which contains up to 0.99 mg/L of total dissolved iron (Kalinkina et al. 2006). Our research on the deep-sea reference station of Lake Kenozero also indicate an increased value of the total dissolved iron, where its content is evenly distributed from the surface to the bottom and ranges from 0.31 to 0.35 mg/L, respectively (Figure 5). Figure 5 shows that the main predominant form is the total dissolved iron, the bulk of which is in the form of Fe (II), whereas suspended iron could not be detected in the entire range of studied depths.

Neal et al. (2008) and Kritzberg et al. (2012) showed a connection between an increase in the iron content and an increase in DOM, which may indicate that the export of iron to aquatic ecosystems by river runoff is controlled by the presence of DOM as a transporting agent. Therefore, an increase in the flux of DOM to surface waters can also have an indirect effect on the change in the concentration of iron due to the maintenance of its compounds in soluble form, which can increase the iron transfer rate.

According to our data, the oxygen content at the deepwater station of Lake Kenozero accounts for 93-71% of saturation from the surface to the bottom or 9.33-9.1 mg/L. The active reaction of the medium is from slightly acidic to close to neutral (pH 6.66-6.07). In oxygen-saturated waters with a neutral pH, the interaction of iron and DOM is a necessary condition for iron to remain in a colloidal state. Consequently, the fact that peatlands and coniferous forests act as important sources of iron are associated with their relatively high concentrations of DOM compared with other types of vegetation cover. Peat and forest soils often produce more dissolved iron than mineral soils, although the latter type of soil contains more bound iron (Björnerås et al. 2017).

Thus, the analysis of data for a 50-year period on the color index revealed its increase by 1.5-2 times, over the 30-year period. There was no significant change in water color in the surface, but there was an increase in water color in the bottom layers, which indicates an increase in the humic acid content in Lake Kenozero. The reason for this phenomenon is deforestation on the watershed of the Pocha River, soil erosion and leaching of humic substances in complexes with iron from the boggy watershed. The main predominant form of iron in water is the total dissolved form. An increased content of iron in the water column was established over the entire range of studied depths.

ZOOPLANKTON

Zoocenosis of the Lake Kenozero is represented by three groups of organisms (Rotifera, Cladocera, and Copepoda) and consist of species common for lakes in the northwest (Kulikova, 2007; Kulikova, Ryabinkin, 2015). The Kenozero zooplankton community consisted of 18 species of aquatic organisms, of which 7 species were Cladocera, 4 were Copepoda, and 7 species of Rotifera.

Cladocera and Rotifera dominated in the species composition of zooplankton (its accounted for 39% of the total species composition), whereas Copepoda accounted for 22%. The contribution of various hydrobiont groups to the species structure of the lake's zoocenosis is illustrated in Figure 6.

The dominant core of zoocenosis is represented by 4-5 species. The structure-forming complex consisted of the following species: *Mesocyclops leuckarti, Bosmina longirostris, Asplanchna priodonta, Bipalpus hudsoni, Polyarthra dolichoptera.*

The dominant complex of planktonic fauna included indicator species, among which oligosaprobic and oligo-β-mesosaprobic predominated (83% of the total number of planktonic organisms).

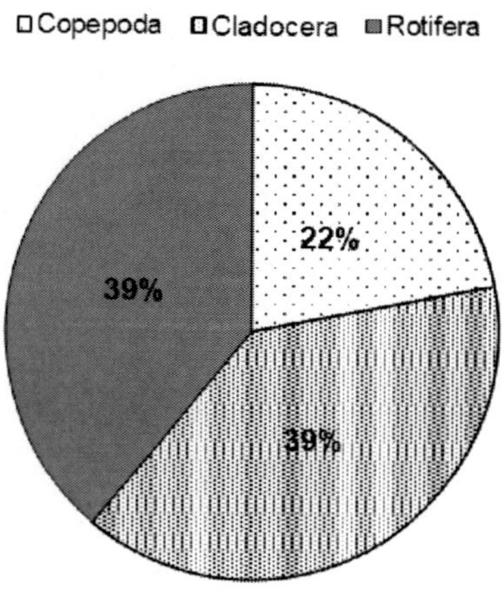

Figure 6. Species composition zoocenosis of Kenozero Lake.

Table 4. The structural characteristics of the Kenozero Lake zoocenosis (July 2015)

Parameter	**Parameter** value
Number of species	18
Density, thous. ind./m³	32.4
Biomass, g/m³	0.18
% Rotifera:Cladocera:Copepoda*	18:14:68 (contribution to the total density) 15:39:46 (contribution to the total biomass)
Shannon index by number, bits ind^{-1}	3.09
Saprobity index	1.33

The average abundance (density) and biomass of zooplankton during the observation period were 32,400 ind./m³ and 0.18 g/m³, respectively. The copepods dominated in numbers. Copepoda accounted for more than 68% of the total plankton abundance. In July, there was a massive development of Copepoda, the number of nauplii was 76% of the total number of Copepoda. The contribution of Cladocera to the total number of zooplankton was 14%, the share of Rotifera in the total number was 18%.

Copepoda also dominated by biomass (46% of the total biomass). The contribution of Cladocera and Rotifera to the total biomass of zooplankton was 39% and 15%, respectively. The bulk of the biomass was dominated by Copepoda *Mesocyclops leuckarti* (33%), *Eudiaptomus gracilis* (8%), Cladocera *Holopedium gibberum* (24%), and Rotifera *Asplanchna priodonta* (10%). The structural characteristics of Kenozero zooplankton are presented in Table 3.

To assess the general level of water body pollution and the ecological status of zooplanktocenoses, the Pantle - Buck saprobity index in the Sladechek modification and the ecological Shannon-Weaver index species diversity were calculated.

According to the state of plankton communities, the water quality of the lake in July 2015 corresponded to the α-oligosaprobic class (saprobity index 1.33; class II - pure waters). The Shannon index was 3.09 bit/ind, which corresponds to an oligotrophic water bodies (Table 3). The values of the index indicate a uniform distribution of species in the community and on the stability of zoocenosis.

A study of the zooplankton communities of Kenozero in the summer showed that zoocenosis is in a stable and balanced state. Species in the community are evenly distributed. Cladocera and Rotifera dominated in the species composition of zooplankton. The abundance and biomass in the community was dominated by Copepoda. The water quality of the lake according to the saprobity index during the research period corresponded to clean waters.

PHYTOPLANKTON

The Kenozero summer phytoplankton was studied at a depth of 0.5 and 5 m. Due to the increased color of water in Lake Kenozero (Table 4), the water transparency during the period of the comprehensive studies was low to 1.9 m, therefore, the photic layer did not exceed 5.7 m. Water is saturated with oxygen evenly throughout the water column, the pH of the medium is characterized by slightly acid to neutral (Figure 3). The content of nutrients was optimal for the development of phytoplankton.

The abundance and biomass of phytoplankton in the surface horizon were 1,452 thousand cells/L and 0.533 g/m^3, respectively; on the horizon of 5 m - 861 thousand cells/L and 0.325 g/m^3, respectively. Figure 7 shows the taxonomic diversity of phytoplankton in the photic layer (0.5 and 5 m).

In the surface layer, Cryptophyta (66.9% and 67.3%, respectively) dominated in terms of abundance and biomass, being mainly represented by the genus *Cryptomonas*. At the depth of 5 m, blue-green algae and diatoms prevailed in numbers (48.5% and 28.7%, respectively). The blue-green were represented by the genus *Anabaena* and diatoms - *Aulacoseira*, whereas the diatoms accounted for main part of biomass. The diatoms (9 species) also exhibited for the greatest variety.

For summer phytoplankton of the lake Kenozero species diversity is characteristic of the North-West of Russia. There are species from the genera *Aulacoseira (A. ambigua, A. subarctica A. tenella), Tabeliaria (T. fenestrata, T. flocculosa), Cyclotella (C. bodanica, C. stelligera), Asterionella formosa, Fragilaria ulna* from diatoms in Kenozero. Cyanophyta were represented by the genus *Anabaena (A. flos-aquae, A. lemmermanii)* and the species *Aphanizomenon flos-aquae*. Chrysophyta was represented by the genus *Mallomonas sp.*

Cryptophyta are represented by species of *Chroomonas acuta, Cryptomonas cf. reflexa,* small forms of the genus *Cryptomonas sp, Dinophyta - Gymnodinium cf. fuscum, Gymnodinium sp.* The genus Gonyostomum (G. semen and G. cf. latum) was found from the raphidophyte; and one species *Monoraphidium komarkovii* was found from the Chlorophyta.

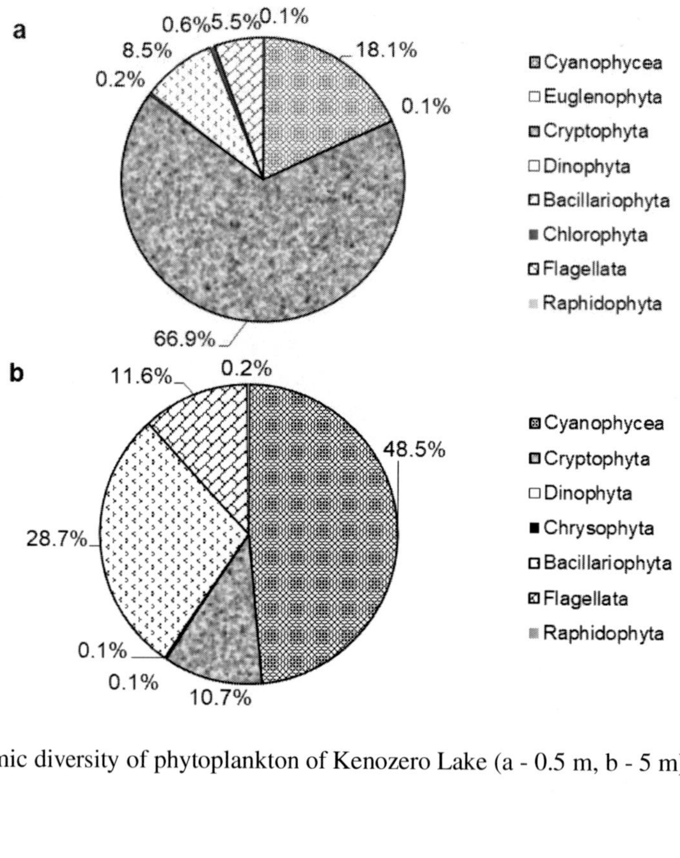

Figure 7. Taxonomic diversity of phytoplankton of Kenozero Lake (a - 0.5 m, b - 5 m).

Detected taxa are predominantly cosmopolitan, most species are oligo-β and α-β-mesosaprobes. A significant proportion of cryptophytic algae is explained by high concentration of organic matter (OM) in water (DOC 17 ± 0.2 mg / L), since this taxon is an indicator of OM.

In comparison with other algae, Cryptophyta are extremely sensitive to excessive illumination (Kitaev, 1984; Hansson et al., 2013; Wissel et al. 2003). However, in Kenozero, their maximum number is noted in the surface layer (0.5 m). This is due to the reduced light transmission through high-color water. Raphidophyta, in particular, the species *Gonyostomum semen*, began to occur more and more often in water bodies of the North-West of Russia. They were rare in northern water bodies until the end of the last century, but were quite widespread in the water bodies of the temperate zone, mainly in lakes with pH from 5.5 to 6.8 (Kitaev, 1984). The biomass of Raphidophyta amounted to 15.7% of the total biomass in the lake water.

Thus, mainly for the summer phytoplankton of Lake Kenozero species diversity is characteristic of the North-West of Russia. Dominant taxa are characteristic of waters with a high concentration of organic substances.

BACTERIOPLANKTON

The number of heterotrophic bacterioplankton (HB) in the water of Lake Kenozero in July 2015 was low and ranged from 54 to 600 CFU/ml for eutrophic bacterioplankton and from 130 to 490 CFU/ml for oligotrophic bacterioplankton. Such a low development of heterotrophic bacterioplankton is probably due to the low availability for microbial mineralization of OM present in the lake, primarily the humic component.

As can be seen from Figure 8, the distribution of HB in the water column was uneven.

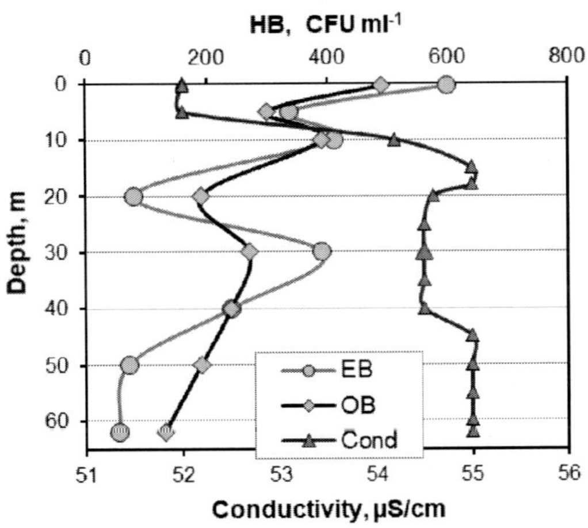

Figure 8. Vertical distribution of heterotrophic bacterioplankton (HB) (EB, eutrophic bacteria; OB, oligotrophic bacteria) at the deep station of Lake Kenozero in July 2015.

At the depth of 20 m and in the bottom layers, their numbers were minimal. As noted above, layers of 20–40 m in the spring were blocked from mixing, which affected the distribution of bacterioplankton. According to study of Kalinkina et al. (2006), conducted in 1999, in viex of lakes's bacterial inventory, Kenozero corresponded to the dystrophic type (a special type of oligotrophy). The quantitative content of eutrophic bacteria that we observed in July 2015 in Kenozero does not exceed similar indicators obtained in (Kalinkina et al. 2006), which indicates the stability of the state of the lake ecosystem. In general, microbiological indicators in the lake are typical for clean waters.

Conclusion

Comparative analysis of the data obtained during our studies at the deep water station of Lake Kenozero in July 2015 together with earlier results of Kalinkina et al. (2000, 2006) by the Institute of Water Problems of the Karelian Research Center RAS acquired in summer of 2003 demonstrated the following. At the beginning of 2000, Lake Kenozero was classified as an oligotrophic type with signs of dystrophy. Based on current state of the ecosystem of Lake Kenozero, in terms of oxygen content, the prevalence of hypolimnion volume over epilimnion volume, poorly developed littoral, quantitative indicators of the main biogenic elements (phosphorus, nitrogen), hydrobiological parameters, this deep lake can also be classified as an oligotrophic type. Partially dystrophic features of the lake are caused by specificity of the catchment area (wetland coverage of about 14%), and by the influx of a large amount of dissolved organic matter (humic substances) by the main lake inlet (the Pocha River).

Acknowledgments

This work was supported by the State Task AAAA-A18-118012390200-5. We are grateful to O. A. Pavlova for determining the quantitative and qualitative composition of phytoplankton.

References

Björnerås, C., Weyhenmeyer, G. A., Chris, D., Evans, C. D., Gessner, M. O., Grossart, H.- P., Kangur, K., Kokorite, I., Kortelainen, P., Lottig, N., Laudon, H., Lehtoranta, J., Monteith, D. T., Nõges, P., Nõges, T., Oulehle, F., Räike, A., Riise, G., Rusak, J. A., Sire, J., Sterling, S., Kritzberg, E. S. 2017. Widespread increases in iron concentration in European and North American freshwaters // *Global Biogeochemical Cycles, Accepted manuscript online:* 5 October 2017. doi: 10.1002/2017GB00.

Ekvall, M. K., Martin, J. d. l. C., Faassen, E. J., Gustafsson, S., Lurling, M., Hansson, L. A. 2013. Synergistic and species-specific effects of climate change and water colour on cyanobacterial toxicity and bloom formation. *Freshwater Biology*, 58: 2414-2422. doi: 10. 1111/fwb.12220.

Gordeeva, L. I., Novoseltsev, G. E. 1983. Hydrobiological characteristics of the Kenozero group of lakes // *Modern quality of the waters of the Onega River and the reservoirs of its basin. Petrozavodsk: Karelian branch of the Academy of Sciences of the USSR*: 103-131. (In Russian).

Grubisic, L. M., Brutemark, A., Weyhenmeyer, G. A., Wikner, J, Bamstedt, U., Bertilsson, S. 2012. Effects of stratification depth and dissolved organic matter on brackish bacterioplankton communities. *Marine Ecology Progress Series*, 453: 37-48. doi: 10.3354/meps09634.

Hansson, L.-A., Nicolle, A., Graneli, W., Hallgren, P., Kritzberg, E., Persson, A., Björk, J., Nilsson, P. A., Brönmark, C. 2013. Food-chain length alters community responses to global change in aquatic systems. *Nature Climate Change*, 3: 228-233. doi: 10.1038/NCLIMATE1689.

Kritzberg, E. S., Ekström, S. M. 2012. Increasing iron concentrations in surface waters – A factor behind brownification? *Biogeosciences*, 9: 1465–1478. doi: 10.5194/bg-9-1465-2012.

Kalinkina, N. M., Timakova, T. M. 2000. Assessment of the current state of Lake. Kenozero (Arkhangelsk region) // Pomerania in the Barents region. Economics, ecology, culture. Materials of reports. *Arkhangelsk*: 101-102. (In Russian).

Kalinkina, N. M., Timakova, T. M., Lozovik, P. A., Ryzhakov, A. V., Kulikova, T. P., Ryabinkin, A. V., Chekryzheva, T. A. 2006. A study of the transformation of water ecosystems of the Kenozero National Park under the influence of agriculture and forestry // *Water Resources of the European North of Russia: Results and Prospects of Research: Proceedings of the anniversary conference dedicated to the 15th anniversary of the IVPS.* Petrozavodsk: KarRC RAS: 312–323. (In Russian).

Kitaev, S. P. 1984. Ecological basis of bioproductivity of lakes of different natural zones, M.: *Nauka*: 207. (In Russian).

Kritzberg, E. S., Graneli, W., Björk, J., Brönmark, C, Hallgren, P., Nicolle, A., Persson, A., Hansson, L.-A. 2014a. Warming and browning of lakes: consequences for pelagic carbon metabolism and sediment delivery. *Freshwater Biology*, 59: 325-336. doi.org/10.1111/fwb.12267.

Kulikova, T. P. 2007. *Zooplankton of water bodies of the Onega Lake basin*. Petrozavodsk: Karelian Research Center of the Russian Academy of Sciences: 223. (In Russian).

Kulikova, T. P., Ryabinkin, A. V. 2015. Zooplankton and macrozoobentos of small bodies of different types of landscapes of south Karelia. *Proceedings of the Karelian Scientific Center of the Russian Academy of Sciences*, 6: 47–63. doi: 10.17076 / bg25 (In Russian).

Neal, C., Lofts, S., Evans, C. D., Reynolds, B., Tipping, E., Neal, M. 2008. Increasing iron concentrations in UK upland waters. *Aquatic Geochemistry*, 14: 263–288. doi: 10.1007/s10498-008-9036-1.

Surface Water Resources of the USSR: Hydrological Understanding. 1965. L.: *Gidrometeoizdat*, Vol. 3. The Northern Territory / Ed. N. M.: 612. (In Russian).

Sushchenya, L. M., Semenchenko, V. P., Vezhnovets, V. V. 1986. *Biology and production of glacial relic crustaceans:* 160. (In Russian).

Weyhenmeyer, G. A., Willen, E., Sonesten, L. 2004. Effects of an extreme precipitation event on water chemistry and phytoplankton in the Swedish Lake Malaren. *Boreal Environment Research,* 9: 409-420.

Wissel, B., Boeing, W. J., Ramcharan, C. W. 2003. Effects of water color on predation regimes and zooplankton assemblages in freshwater lakes. *Limnology and Oceanography*, 48: 1965-1976. doi.org/10.4319/lo.2003.48.5.1965.

In: Lake Water
Editors: Oleg S. Pokrovsky, Yulia Bespalaya et al.
ISBN: 978-1-53619-275-9

Chapter 3

LONG-TERM STUDY OF THE ECOSYSTEM COMPONENTS VARIABILITY OF THE LAKE LEKSHMOZERO (THE SOUTHERN PART OF THE KENOZERSKY NATIONAL PARK)

S. A. Zabelina*, T. Ya. Vorobyeva, S. I. Klimov, N. V. Shorina, E. I. Sobko, O. Y. Moreva, A. A. Chupakova, A. V. Chupakov, L. S. Shirokova, O. S. Pokrovsky, N. V. Neverova and O. D. Kovalev
N. Laverov Federal Center for Integrated Arctic Research of the Ural Branch of the Russian Academy of Sciences, Arkhangelsk, Russia

ABSTRACT

The results of studies of the current ecological state of Lake Lekshmozero, which is of national economic importance, are presented. The lake is inhabited by commercial fish species, the most valuable of which is European cisco. Lekshmozero is a waterbody of a pronounced accumulative type with a small indicator of conditional water exchange and a small specific catchment. Lake Lekshmozero is a dimictic lake with an unstable metalimnion.

An analysis of long-term data (2006-2017) obtained in the course of multidisciplinary studies of the spatiotemporal dynamics of hydrological and hydrochemical parameters, qualitative and quantitative characteristics of the biological communities of the lake ecosystem, showed that the current ecological state of Lake Lekshmozero can be assessed as stable. It has been established that the state of the lake's ecosystem is mainly controlled by intrabasin processes. The hydrobionts of the lake's ecosystem at the present stage are in an active state, responding to changes in biotic and abiotic conditions, thereby maintaining the lake's ecosystem in a state of ecological balance in a changing climate. The quality of the water column of the lake during the study period corresponded to mesotrophic waters.

Keywords: lake, nutrients, plankton, hydrology, diversity, environmental factors, dissolved organic carbon

* Corresponding Author's E-mail: sv.zabelina@fciarctic.ru.

INTRODUCTION

The main task of the Kenozersky National Park (KNP) is to preserve the unique heritage of nature, incl. aquatic ecosystems and biological diversity of hydrobionts. Of the numerous lakes of the KNP, Lake Lekshmozero stands out, as it has an important recreational value, which is actively used for amateur and sport fishing. Lekshmozero is the second largest lake in the Kenozersky National Park, has important recreational importance, is actively used for amateur and sport fishing in the Park. The lake is inhabited by commercial fish species (bream, pike, burbot, perch, european cisco), the most valuable of which is European cisco *Coregonus albula* (Dvoryankin et al., 2016). Here, 70% of the entire European cisco of the region is caught annually.

Observations of the ecological state of the lake were carried out by us from 2006 to 2017 in order to study the response of the ecosystem of the medium-size lake to climate fluctuations and changes in the recreational load due to the development of ecotourism, and the infrastructure of the park. Climate-related factors can have a multilateral and prolonged impact on lakes. They affect not only the physical stimulation of primary production growth due to the availability of solar radiation, rainfall, wind and changes in the catchment, but also changes in the thermal stratification and regeneration of nutrients.

These combined effects can affect the phenology and succession of plankton in aquatic ecosystems (Adrian et al., 2006, Blenckner et al., 2007). A recent study of Dvoryankin et al. (2019) suggested that one of the reasons for the change in the structure of the ichthyofauna of KNP lakes is climate change. The statistics of amateur fishing over the past 20 years indicates a sharp increase in the number of heat-loving fish species in the lakes of the Pontocaspian faunistic complex - bream and zope.

Lakes are indicators of climate change, and at the same time, their responses to climate change depend on the particular lake, on its geographical location, the topography of the lake's catchment and the bathymetry of the lake basin in combination with the background of global, regional and local weather (Weyhenmeyer et al., 2008).

In this chapter, a retrospective analysis of the available information (1997-2004) and an assessment of the current ecological state of the lake ecosystem by hydrological, hydrochemical and hydrobiological indicators were carried out. For analysis, we used our own data obtained during 27 field studies conducted from 2006 to 2017. As part of our research, we have been updating the morphometric and hydrophysical characteristics of Lake Lekshmozero for 12 years, providing more than 60,000 depth measurements. The sampling procedure and analytical methods employed in this study are described in chapter 1.

MORPHOMETRIC AND HYDROGRAPHIC CHARACTERISTICS OF LEKSHMOZERO

Lake Lekshmozero is an oval-shaped body of water with a low coefficient of conditional water exchange and a small specific catchment. After analyzing the available publications, we presented the most reliable values of morphometric and hydrophysical parameters in Table 1.

The coastline is weakly indented (Figure 1). The bottom of the lake is flat without sharp drops in depth, with the exception of the fault zone. As can be seen from Figure 1b, the area of Lake Lekshmozero evenly decreases to a depth of 7 m, then to 9 m the bottom is shallow, then the area changes with depth more sharply. The share of the lake area is up to 7 m, from 7 to 10 m and from 10 m to the bottom, respectively, 39.7, 41.1 and 19.2%. About 73.0% of the lake volume is located prior to the 7 m isobath and the 86.2% is located prior 10 m isobath.

Table 1. The Main morphometric and hydrographic characteristics of Lekshmozero

Length, km	12.5	Volume, km^3	0.426
The greatest width, km	5.8	Average annual runoff from the catchment, km^3	**0.0461**
Average depth, m	8.0	The specific catchment area	**2.96**
Maximum depth, m	**29.0**	The coefficient of conditional water exchange, $year^{-1}$	**0.11**
Watershed area, km^2	**158.5**	Water residence time in the lake, year	**9.2**
Area, km^2	**53.5**		

Notes. [1] The literature gives values for the area of Lake Lekshmozero - 54.4 km^2, the catchment area - 197 km^2 (Surface Water Resources…, 1965). We have refined the morphometric and hydrophysical parameters for Lake Lekshmozero (in bold).

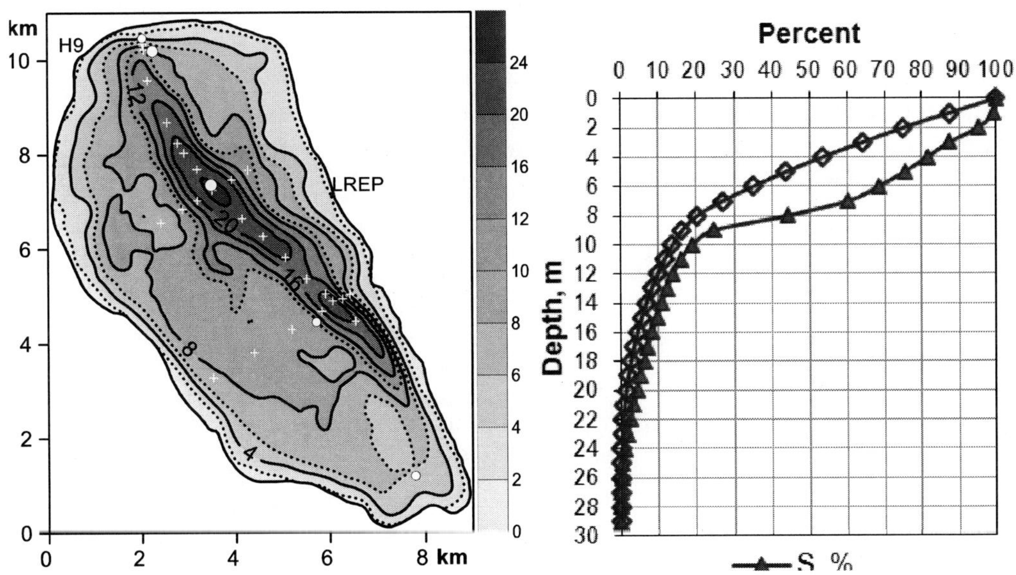

Figure 1. Bathymetric map and curves of changes in the area and volume of Lake Lekshmozero. The circle indicates the location of the reference stations, the cross mark the winter hydrological survey.

HYDROLOGICAL REGIME

Water Temperature

Lekshmozero belongs to a dimictic lake with an unstable metalimnion in summer. By the nature of the vertical distribution of water temperature in summer, Lake Lekshmozero belongs to the metathermal typc (Tikhomirov ct al., 1970). It is characterized by unstable thermal stratification of waters. The upper homogeneous mixed layer develops mainly to a depth of 4-11 m, but it may be practically absent, as it was in July 2012 and 2014. A pronounced layer of a temperature jump (with a gradient of more than 1°C/m) with a thickness of 2 to 4 m can be located in the depth range from 5 to 13 m with maximum gradients from 1.9 to 3.6°C/m. Sometimes, the second layers of temperature jumps are observed with a thickness of 1 to 2 m at depths of 13-15 m. In 2007 and 2015 there were no pronounced layers of the temperature jump.

Figure 2 shows the vertical distribution of water temperature in the form of quartile characteristics of a Box-Whisker plot. The central position of the data is expressed by the median, whereas the variability is represented by the interquartile range. In summer, the temperature variability in Lekshmozero is comparable across the entire layer; the largest variability was in the layer where the temperature jump is 3-7 m (since its position changes depending on meteorological conditions). In winter, as can be seen from Figure 2, in Lekshmozero, the vertical variability is comparable over the entire depth.

In order to study the processes of formation of the thermal state of the lake, a correlation analysis was carried out using Spearman's nonparametric rank correlation coefficient, which was chosen due to the short length of the rows. The robust correlation coefficient was also calculated by the regression coefficients determined by iterative linear least squares method with weighting by the modulus of the median deviations (Mosteller et al., 1982, Frederick Mosteller et al., 1977), Table 2.

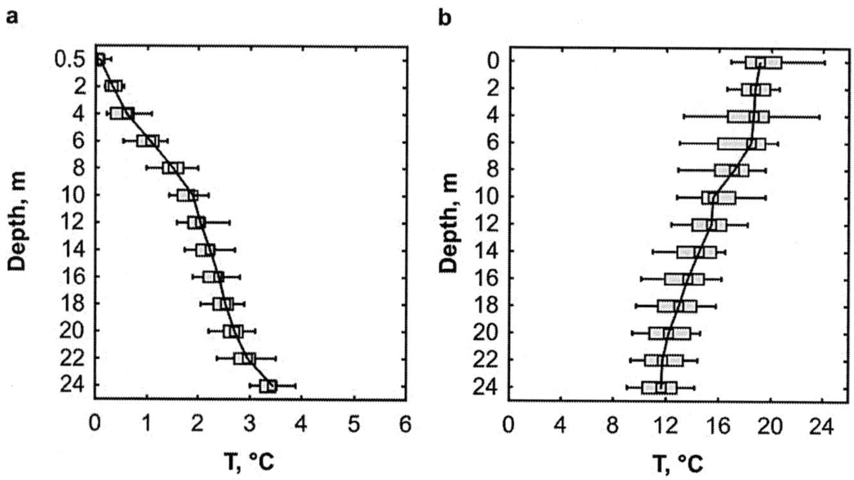

Figure 2. Vertical distribution of temperature at a deep-water station in winter (a) and summer (b).

Table 2. Correlation coefficients for the connection of thermal content (Q), weighted average water temperature (Tw) and anomalies of specific specific conductivity (dCond) from meteorological parameters

Variance	Layer, m	$SumT_{10}$ °C	$SumT_{all}$ °C	T_{may} °C	T_{jun} °C	T_{mj} °C	W_{may}	W_{jun}	W_{all}
N		11	11	11	11	11	7	7	7
Q, PJ	0-24	**0.61**	**0.61**	0.40	0.16	0.41	0.00	0.00	- 0.21
		0.59	0.59	0.35	0.17	0.17	- 0.23	- 0.23	- 0.27
	10-24	0.37	0.26	- 0.11	**- 0.66**	- 0.39	0.36	**0.86**	**0.89**
		0.32	0.27	- 0.16	**- 0.73**	- 0.43	0.26	**0.98**	**0.87**
T_w,°C	0-10	**0.66**	**0.70**	0.48	0.30	0.56	- 0.36	- 0.14	- 0.43
		0.63	**0.64**	0.21	0.21	0.35	- 0.31	- 0.13	- 0.30
	10-24	0.17	0.05	- 0.20	**- 0.72**	- 0.50	0.36	**0.79**	**0.89**
		0.14	0.10	- 0.19	**- 0.77**	- 0.48	0.32	**0.91**	**0.91**
	0-24	0.55	0.45	0.14	**- 0.64**	- 0.22	0.32	**0.89**	**0.86**
		0.65	**0.66**	- 0.09	**- 0.64**	- 0.33	0.25	**0.88**	**0.89**
dCond, μS/cm	10-24	- 0.28	- 0.02	**- 0.66**	- 0.59	**- 0.76**	**0.82**	0.43	**0.93**
		- 0.63	- 0.15	- 0.53	**- 0.65**	**- 0.71**	**0.76**	0.66	**0.99**

Note *In the top line, Spearman's correlation coefficients, in the bottom - robust values. Significant coefficients at $p < 0.05$ are marked in bold.

We considered the correlation between the thermal content of the lake (Q), the weighted average water temperature (T_w), and the weighted average deviations of the specific specific conductivity of water (dCond) with the following factors: dates of stable transition of air temperature through gradations of 4°C and 10°C and the period between them, average air temperatures for May (T_{may}) and June (T_{jun}), the values of heat degree-days (the sum of average daily air temperatures) from the dates of the transition through 4°C ($SumT_{all}$) and 10°C (SumT10) to the day of sampling, wind speeds in May, June and in total until the day of sampling. Weather data were taken from the Kargopol weather station (website:http://www. pogodaiklimat.ru).

The deviations of the specific electrical conductivity, determined for each station from its weighted average value, were used to assess its relationship with water stratification and to exclude its interannual variability. Since ice melt on Lake Lekshmozero occurs at the end of the first - beginning of the second decade of May, the wind conditions were considered from May 10. The average wind speed at the weather station is determined over a 10 minute interval. Winds were taken into account, with a speed equal to or greater than 3 m/s, not falling in a row at least 5 meteorological periods, i.e., more than half a day to exclude short-term and weak winds that do not significantly affect mixing. As indicators, we took the sum of the speeds of such winds from May 10 to the end of the month (Wmay), for June (Wjun) and the total amount from May 10 to the day of the station (Wall). (Data on wind speeds are available on the website since 2011). Table 2 shows the indicators for which significant correlation coefficients were obtained.

In general, it can be noted that the summer thermal content of the lake does not depend on air temperature. There was a weak relationship on the sum of heat degree-days, providing about 40% of the variability. This is due to the fact that in June-July, as a result of the increased length of the day, the lake water warms up due to their ability to effectively absorb solar radiation, regulated by cloudiness, and depends little on heat exchange processes with air.

The negative correlation of the June air temperature with the thermal content of the bottom layer is a consequence of the fact that when the cyclonic type of weather predominates, the probability of water mixing is higher, and therefore the thermal content of the bottom waters will increase. With anticyclonic weather, there is a sharp increase in water stratification and practically to the cessation of heat exchange processes in the lake between the surface and bottom layers of water. Due to the openness of the lake, the thermal content of the bottom waters is largely determined by wind conditions, as evidenced by the strongest correlations.

Specific Conductivity (Mineralization)

The investigated lake is located in a zone of excessive moisture and the predominant role in nutrition belongs to atmospheric precipitation, which, with a small catchment area, makes it low-mineralized. The water belongs to the bicarbonate class of the calcium group.

As shown in Figure 3, Lake Lekshmozero is characterized by low variability of specific conductivity. In winter, mineralization variability is greatest in the bottom layers.

Since 2013, winter hydrological sections have been carried out on Lake Lekshmozero using the Hydrolab MS5 hydrosonde. As can be seen from Figure 4, which shows a section along the fault, there are two basins of almost equal depth in the lake. According to echo sounding, the depth of the northern depression, located near the reference station of the Hydrometeorological Service, was 28 m, the depth of the southern depression was 29 m. The following features of the distribution of specific conductivity can be noted: in the central part of the lake, in a layer of 2-7 m, there is the least mineralized water; the water with the highest mineralization is concentrated at the bottom of the northern depression.

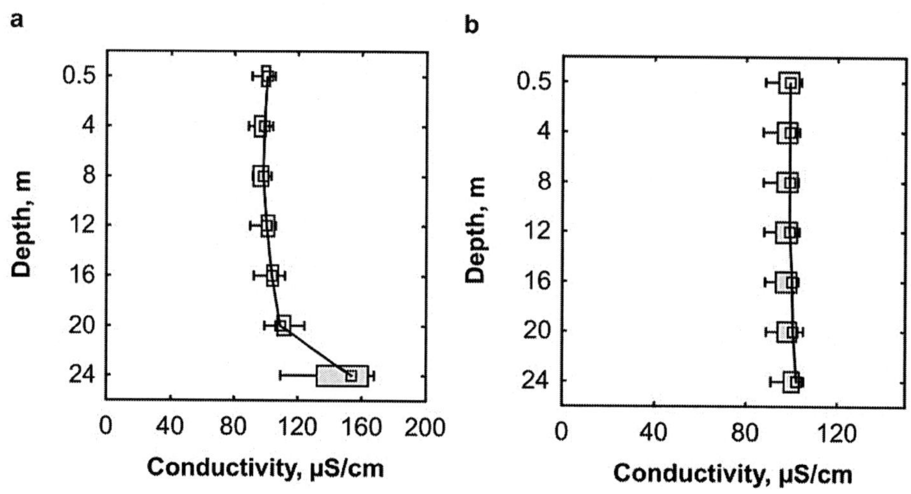

Figure 3. Vertical distribution of specific specific conductivity of water in March (a) and July (b) and its variability for the period 2006-2017.

In the southern depression, separated from the nothern by a depth of up to 16 m, the mineralization is lower, which is explained by the smaller volume of bottom waters enriched with salts flowing into the depression. The bottom temperature is higher in the nothern depression; for 4 years the difference was from 0.2 to 0.7°C. In 2015, the maximum value of the temperature of bottom waters in winter was noted for Lake Lekshmozero, which amounted to 4.70°C in the northern depression (4.29 in the second depression). The difference in specific conductivity of waters between depressions below 24 m varied from 70-80 to 110-150 μS/cm; the maximum values in 2015 were 470 and 330 μS/cm, respectively, in the northern and southern depression.

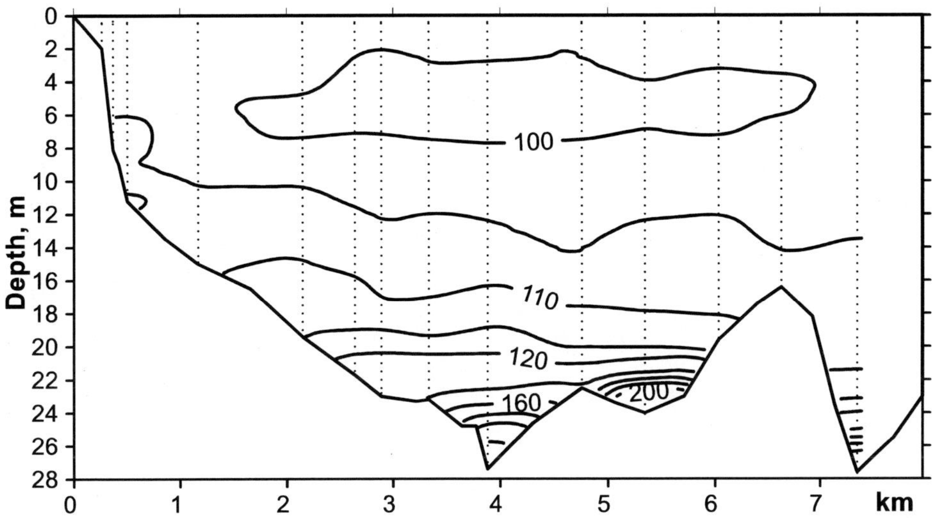

Figure 4. The vertical distribution of the specific conductivity of water in the hydrological section along in the fault Lake Lekshmozero in March 2016.

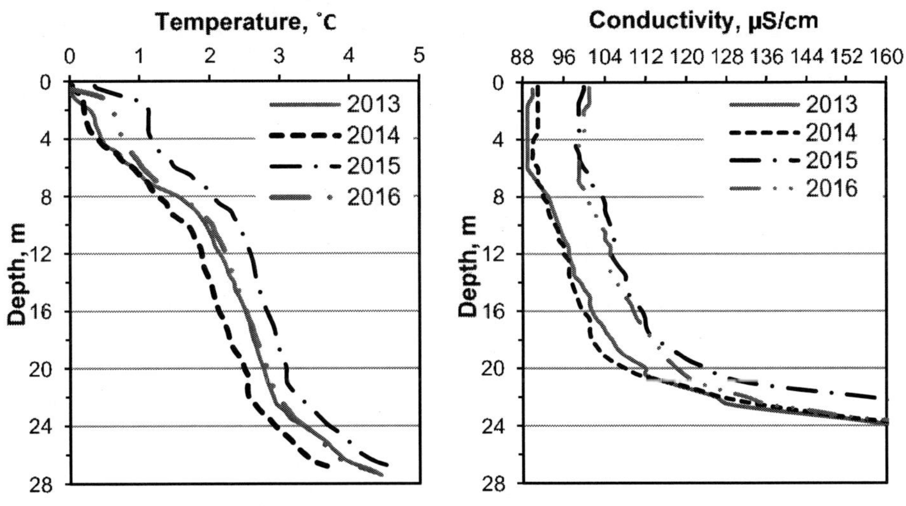

Figure 5. Vertical distribution of temperature (a) and (b) specific conductivity of water in Lake Lekshmozero in winter.

Figure 5 shows the vertical distribution of the temperature and specific conductivity of water at the station of the first basin. It can be noted that at higher water temperatures, the water mineralization is also higher. This is due to the conditions of autumn mixing, with a prolonged autumn, more cooling of the waters and the bottom occurs, and the incoming atmospheric precipitation contributes to the demineralization of water, and the processes of mineralization of organic substances occurring under anaerobic conditions are also delayed. As shown in Table 2, a steady transition of air temperature occurred on November 26 and 25, October 14 and November 6, respectively, in 2012, 2013, 2014 and 2015. The graphs show that the temperature of water and mineralization in winter were lower in 2013 and 2014, and their maximum values were observed in 2015.

The strong correlations between the specific conductivity anomalies and wind conditions in the summer period, shown in Table 1 can be explained as follows. Figure 6 shows the vertical distributions of hydrological parameters in July at the reference stations of Lake Lekshmozero performed at 5-day intervals. From the evening of July 11, northern winds were observed with speeds of 3-4 m/s with gusts up to 11 m/s which lasted more than a day. The second station was studied 2 days after the winds subsided. From the temperature graphs it can be seen that the destruction of the bottom temperature jumps and a layer with a uniform temperature (20-24 m) occurred, i.e., mixing took place down to a depth of 20 m.

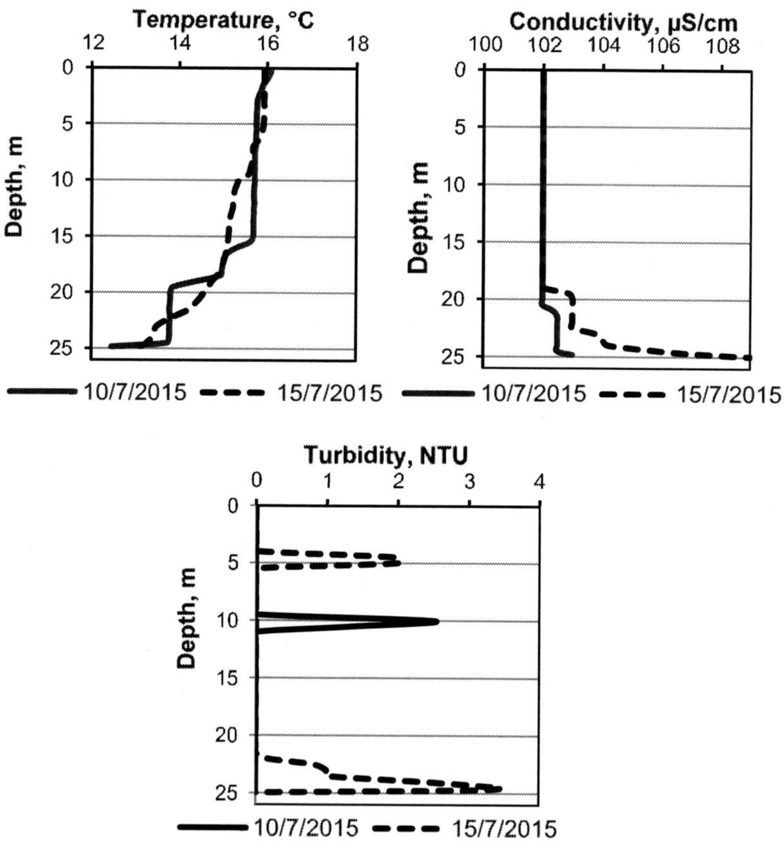

Figure 6. Vertical distribution of water temperature, electrical conductivity and turbidity in Lake Lekshmozero in July 2015.

There was also an increase in specific conductivity in the bottom layer. Probably, wave mixing facilitated the flow of dissolved matter from bottom sediments due to sediment detachment, which could lead to an increase in the density of bottom waters and their sliding into the basin. At thc layers of 19-22 m, the calculated water density decreased due to an increase in water temperature (up to 0.000005 g/cm^3) and the main contribution to the increase in density was most likely due to the presence of suspended matter in water. The content of suspended matter in the bottom layer was recorded by a sensor turbidity, as can be seen from the graph. It was absent at the previous station. Thus, presumably strong winds contribute to an increase in mineralization in the bottom layers.

Dissolved Oxygen

In summer, as shown in Figure 7, in Lake Lekshmozero, the greatest variability of the dissolved oxygen concentration occurs in the subsurface layer (as a result of photosynthesis) and in the supra-bottom layers as a result of variability in interannual meteorological conditions, which cause mixing of lake waters. In Lekshmozero, the formation of intrawater oxygen maxima is observed as a result of the development of phytoplankton. No significant relationships associated with air temperature and oxygen content in the lake were manifested, since the processes of spring mixing are multivariate.

As shown in Figure 7, in winter, the greatest variability of the oxygen content in Lake Lekshmozero is observed in the intermediate layers, the smallest - in the surface, bottom and at a depth of 10 m. In the upper layer, the variability was due to the conditions of autumn mixing, in the deep layer, it was a consequence of a combination of factors such as the heating of bottom sediments in the summer.

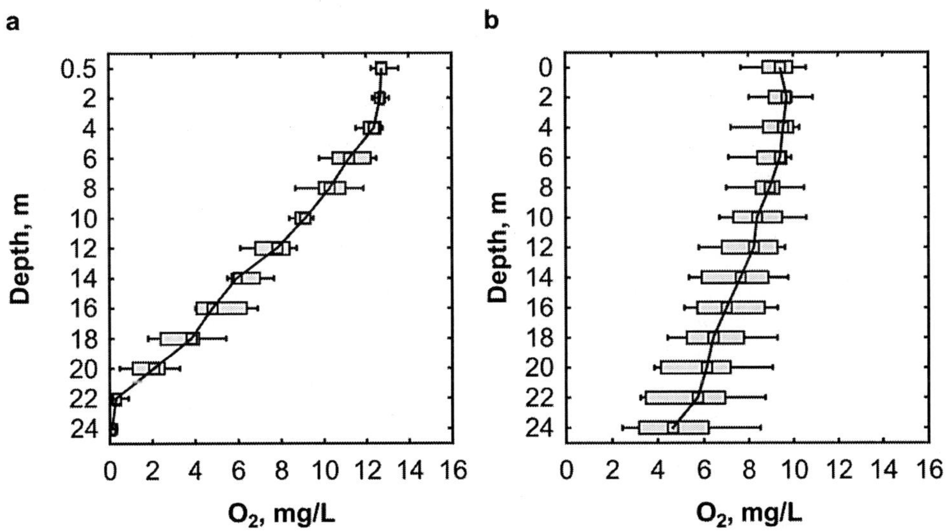

Figure 7. Vertical distribution of dissolved oxygen [O_2] in water in March (a) and in July (b) and its variability for the period 2006-2017 in Lake Lekshmozero.

The coefficients of linear regression for the dependence of the average concentration of dissolved oxygen [O_2] in March for the surface 8-m layer during the period between the dates of stable transition of air temperature in the fall of the previous year through gradations of 4°C and 0°C showed that the change in [O_2] by 60-80% is due to the duration of the period of convective mixing of water in the fall before the formation of the ice cover.

Thus, the studied lake is located in the zone of excessive moisture and the predominant role in water input belongs to atmospheric precipitation. This, with a small catchment area, makes it low-mineralized and leads to a low buffer capacity of lake waters. Water belongs to the bicarbonate class of the calcium group. The active reaction (pH) of the lake water environment during the study periods was characterized mainly as neutral. By the nature of the vertical temperature distribution in the summer period, the lake is metathermal.

In the studied periods, in terms of [O_2] for the littoral of Lake Lekshmozero, a homogeneous distribution of temperature is observed (Vorobieva et al., 2011, 2012, 2013). At deep stations of Lake Lekshmozero, the [O_2] is similar to the vertical profile of water temperature and the formation of a density jump. In general, Lake Lekshmozero is characterized by a favorable oxygen regime in all seasons. Only in the winter under-ice period, there was a weak stratification in the bottom layers. The [O_2] does not fall below 20-50% within the upper 20 m and remains within 1.2 to 10.0% in the bottom layers. In the spring and autumn periods of intensive mixing of waters, a homogeneous vertical distribution of hydrological and hydrochemical characteristics is generally noted.

Hydrochemical Characteristic

Dynamics of Main Nutrients in Lekshmozero

The nutrient content is one of the important factors for assessing the state and functioning of aquatic ecosystems. Biogenic elements are mineral substances that are most actively involved in the life of aquatic organisms. These include compounds of phosphorus, nitrogen, silicon and iron. Biogenic elements, being the basis of the biological productivity of water bodies, in most cases determine the quality of water used for economic purposes. Thus, the study of the regime and dynamics of the transformation of nutrients in water bodies is an important hydrochemical and environmental issue.

This section summarizes the content of the main nutrients for the period 2009-2017 for Lekshmozero. Sampling at the Lake Lekshmozero was carried out in various hydrological seasons, over discrete water column ~~column~~ horizons. Figures 8 - 13 show the vertical distribution of nutrients in the lake water in the form of a box-Whisker plot. The central position of the data is expressed by the median, the variability is reflected by the interquartile range.

The distribution of nutrients in the water column is affected by the degree of stratification. The results showed that in the water of Lake Lekshmozero in spring and autumn, during periods of water mixing, mainly the quantitative indicators of various forms of silicon, nitrogen and phosphorus are aligned according to the depth profile. In periods of water stagnation, an uneven distribution of silicon, various forms of nitrogen and phosphorus over the thickness of the water column is characteristic with a general tendency to increase concentrations from the surface to the bottom.

The silicon content in the periods under study was as following: in winter - from 385 μg/L in the surface horizons to 5158 μg/L in the near-bottom (median is 1165 μg/L); in summer - from 418 μg/L in the surface horizons to 2930 μg/L in the bottom horizons (median is 1215 μg/L), Figure 8. In spring and autumn, due to mixing, the concentration was almost constant: 1035-1580 μg/L (median is 1076 μg/L) and 1114-1193 μg/L (median is 1158 μg/L), respectively. In epilimnion in summer, spring and autumn, the silicon concentration does not exceed 1800 μg/L (median is 1175 μg/L).

In winter, during a period of stable stratification, due to the uptake of silicon by psychrophilic diatoms, Si concentration in the photic layer decreased to 385-2060 μg/L (median is 797 μg/L).

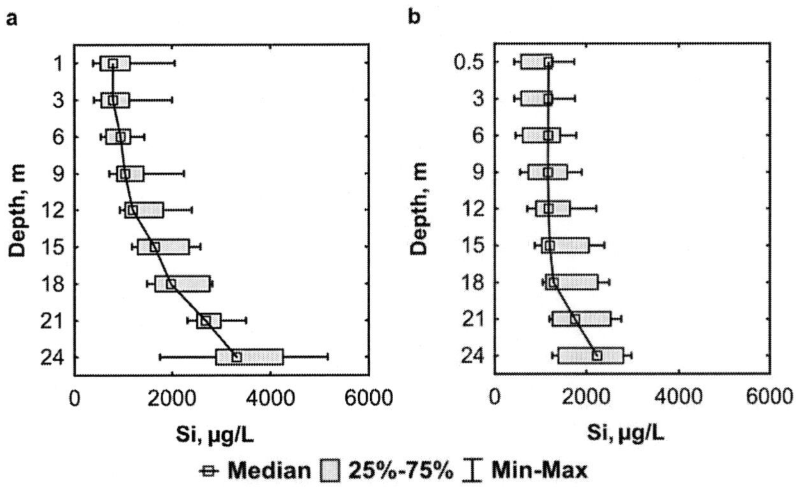

Figure 8. The vertical Silicon Variability deep station (a – winter; b – summer) in the water of Lekshmozero Lake.

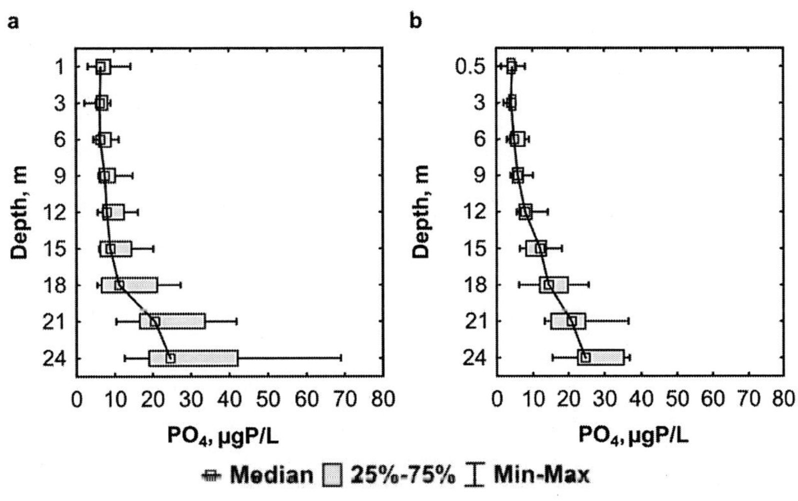

Figure 9. The vertical profile of P-PO_4 in winter (a) and summer (b) in the Lekshmozero Lake.

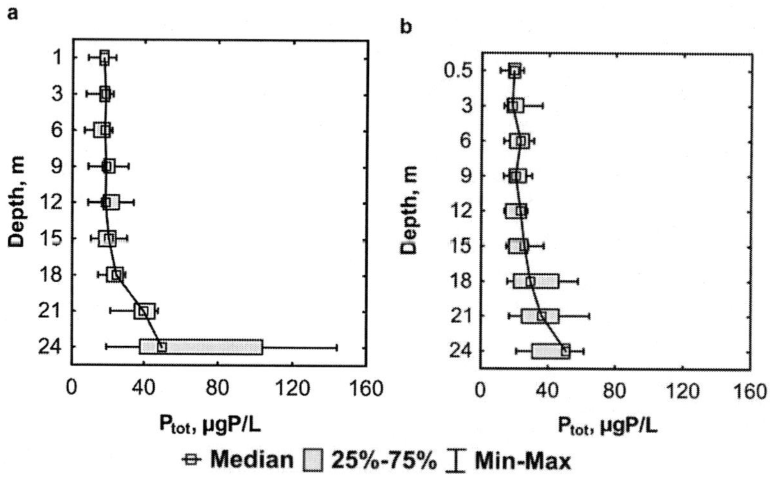

Figure 10. The vertical of Total Phosphorus in winter (a) and summer (b) in the Lekshmozero Lake.

The concentration of mineral phosphorus in the water of Lake Lekshmozero was generally low in the epilimnion and did not exceed 16.0 µgP/L, only in the bottom layer it increased to 21.0-63.7 µgP/L (Figure 9).

The total phosphorus content in the photic layer is 17.0-22.2 µgP/L (median values), gradually increasing to 52-62 µgP/L in the bottom layer (Figure 10). As can be seen in Figure 9 and 10, the low variability of phosphates and total phosphorus is characteristic of Lake Lekshmozero due to the sufficient conditions of water mixing, both in spring and summer. In winter, as shown in Figure 9 and 10, the variability of phosphorus compounds is greatest in the bottom layers. The concentration of Ptot indicates its sufficient amount for primary production processes in Lake Lekshmozero. However, in the epilimnion, the proportion of PO_4^{3-} is not more than 29% of total P. The low concentration of phosphates at the end of the winter low-water season, as well as the predominance of the organic form of phosphorus indicate the incompleteness of the processes of mineralization of organic substances, which can lead to their accumulation in sediments.

In spring, the content of phosphates and total phosphorus in the water column was almost quasi-uniform and amounted to: for phosphates in 2009 and 2010 - from 2.3 to 6.9 µgP/L, for total phosphorus from 16 to 50 µgP/L in the bottom horizon in 2009 and from 11 to 15 µgP/L in 2010. In autumn, the content of phosphates and total phosphorus was also quasi-uniform and amounted to: phosphates in 2009 from 6.2 to 10.6 µgP/L and in 2012 from 2.09 to 3.10 µgP/L; total phosphorus in 2009 from 24 to 35 µgP/L and in 2012 from 19 to 30 µgP/L.

The vertical distribution of the mineral forms of nitrogen shows that the NO_3 form always prevails throughout the Lekshmozero water column, and an increase in NH_4 content is observed from a depth of 21-22 meters, with its subsequent predominance in the bottom layer. The NO_2 content in all seasons is very low, fluctuating around hundredths and thousandths of µgN/L in the photic layer, up to 11.4 µgN/L in the bottom layer, and makes up 0.1–4% of the total inorganic forms of nitrogen. Figures 11-13 show the distribution of nitrates and ammonia nitrogen at the deep-water station Lekshmozero in different seasons over a long-term observation period.

The nitrate content in the water column is characterized by the greatest variability in all seasons. The concentration of nitrates is most intensively affected by biological processes in aquatic ecosystems. On the one hand, the concentration of nitrates depends on the intensity of aerobic destruction processes, on the other hand, nitrates are most actively absorbed by phytoplankton. In Lake Lekshmozero, the intensity of respiration processes prevails over the primary production.

The ratio of the intensity of primary production to aerobic destruction is typically below 1, and varies depending on the season, which indicates that the ecosystems are in a relatively stable state. Aerobic destruction processes can exceed up to 70 times the amount of organic matter synthesized by phytoplankton. The most active OM destruction processes take place in the thermocline zone (Timakova, Kalinkina et al., 2000, Kalinkina et al., 2006, Vorobieva et al., 2012, 2013).

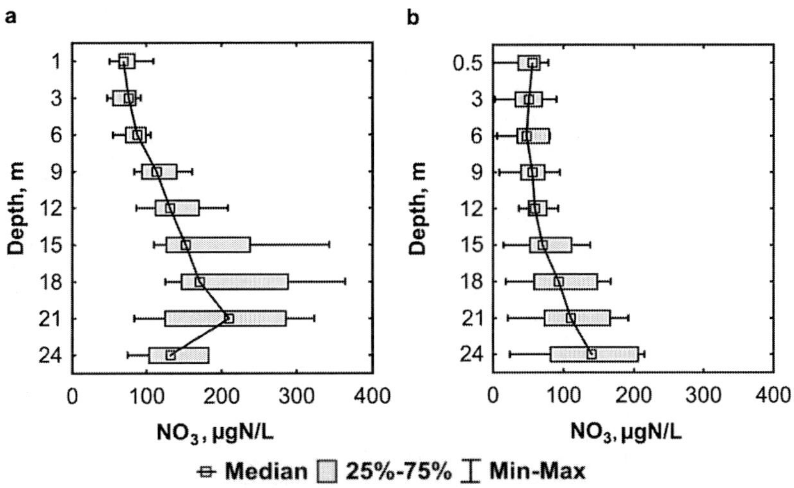

Figure 11. The vertical distribution variability of dissolved nitrates (2009-2017) in winter (a) and summer (b) in Lake Lekshmozero.

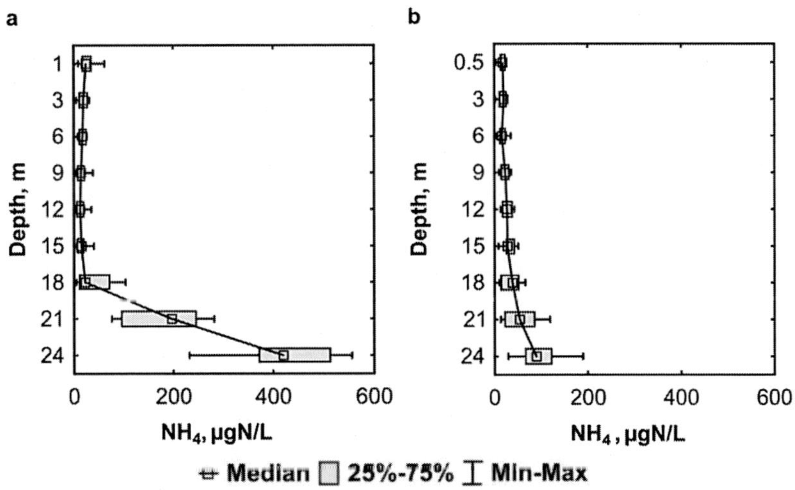

Figure 12. The vertical distribution of dissolved ammonium (2009-2017) in winter (a) and summer (b) in Lake Lekshmozero.

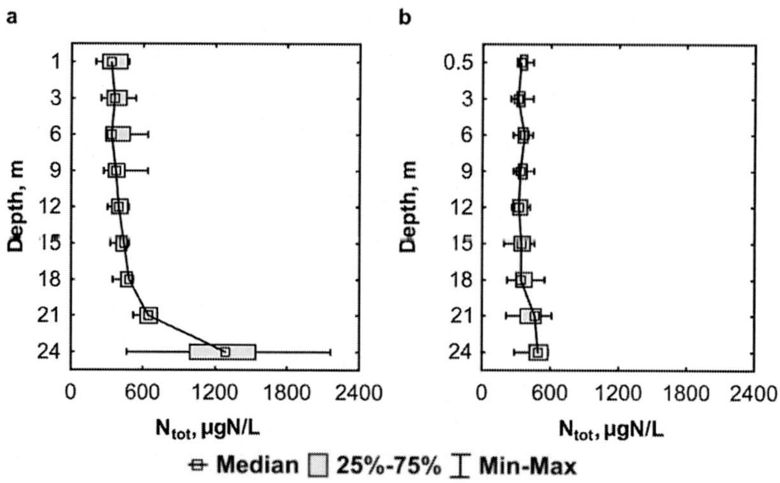

Figure 13. Variability of the total nitrogen vertical distribution in winter (a) and summer (b) in Lake Lekshmozero (2009-2017).

Studies of the biogeochemical nitrogen cycle in Lake Lekshmozero indicate that the processes of ammonification and nitrification are intensively taking place over almost the entire water column (Timakova, Kalinkina et. al., 2000, Kalinkina et al., 2006, Vorobieva et al., 2012).

The predominance of $N\text{-}NO_3^-$ over other mineral forms of nitrogen is most likely associated with active nitrification processes under favorable oxygen conditions, and this, in turn, leads to a decrease in the concentration of ammonium nitrogen. In winter, under the conditions of anaerobic zones, the number of nitrifying bacteria decreases, which leads to the accumulation of ammonia nitrogen (Vorobieva et al., 2012). Analysis of the total nitrogen content in water column indicates its accumulation by the end of the winter. In summer, there is a uniform nitrogen content distribution across the water column (Figure 13).

As can be seen from Figure 13, Lake Lekshmozero is characterized by a low variability of total nitrogen throughout the water column. In addition to the bottom layers, especially in the winter low water. In summer, in the epilimnion, the N_{org} content is up to 80-83% of N_{tot} (even in the bottom layer up to 51%), in the winter, the N_{org} content reaches 70% of N_{tot} in the epilimnion and 49% in the hypolimnion. Our results are comparable with those obtained in 1997 and 2004 by other authors, as can be seen from Table 3.

Table 3. Variation in physico-chemical parameters in Lekshmozero Lake in different years of research

PARAMETERS	08.1997 [1]	08.2004 [2]	07.2007-2017
pH	-	7.10-7.15	6.74-8.63
$P\text{-}PO_4$, µg P\L	1,4	2.0-6.0	2.86-5.05
Total phosphorus, µg P\L	8.0-22.0	4.0-14.0	16.0-24.0
$N\text{-}NO_2$, µgN\L	-	≤ 2.0	≤ 4.0
$N\text{-}NH_4$, µgN\L	6.0-121.0	30.0-90.0	14.0-28.0
$N\text{-}NO_3$, µgN\L	-	20.0-90.0	120.0-170.0

Note: [1] Kalinkina et al., 2006; [2]. Kalinkina et al., 2005; dashes - no data available.

Table 4. Median values concentrations of Nitrogen and Phosphorus compounds and their ratios

Observation period	Ptot µgP\L	N_{min}, µgN\L	N_{tot} µgN\L	N_{org} µgN\L	N_{min}/PO_4	N_{tot}/P_{tot}
March	20.5	123.8	435.5	249.7	17.7	19.8
May	17.2	94.4	352.6	263.0	27.9	21.1
July	24.4	88.3	362.9	255.9	12.5	15.1
November	26.8	74.1	427.3	346.1	9.4	15.8
Entire observation period	23.3	99.7	392.0	264.1	15.9	17.2

We determined the average values of phosphorus and nitrogen during different hydrological seasons for the entire observation period. The median concentrations of total nitrogen and phosphorus fluctuate within the limits characteristic of mesotrophic water bodies and are presented in Table 4. The ratio of the main nutrients used by primary producers is essential for assessing the ecological state. (Bogdanovskaya et al., 2001., Samarina et al., 2008., Shaporenko et al., 2006., Redfild et al., 1963).

Table 4 shows the average N_{tot}/P_{tot} ratios for Lekshmozero in different hydrological seasons. These numbers demonstrate limiting role of P in phytoplankton development, although in summer and sometimes in autumn, a deficit of N can be encountered in the epilimnion.

Overall, the ratio of the main nutrients shows that they are optimal for the development of phytoplankton. Thus, analysis of the dynamics of the concentrations of nutrients during different seasons showed that the maximum concentrations of many mineral forms of nutrients occurs during winter period. For nitrates, this can be seen throughout the water column, and for phosphates only in the photic layer. The maximal content of silicon, total nitrogen and total phosphorus falls on the summer period. For total phosphorus, this happens over full depth of the water column, whereas for total nitrogen and silicon, it occurs only in the photic layer. For ammonium, the interannual variability is insignificant.

The values of the ratio of the main biogenic elements are optimal for the development of phytoplankton. Phosphorus is mainly the limiting compound for the development of aquatic organisms in the studied lake. Overall, results of spatio-temporal changes in nutrients can be used to assess changes in the lake ecosystem under the influence of anthropogenic factors and a changing climate.

Forms of Iron and Color

Iron distribution at shallow stations of Lake Lekshmozero (H9 and littoral, with depths of up to 9 m and 2 m, respectively) are characterized by low concentrations of total dissolved (Fe (II + III)) iron throughout the water column with a predominance of total dissolved iron (up to 0.06 mg/L) in winter, and particulate iron (Fe part), to 0.03 mg/L, in summer (Figure 14).

These numbers are almost 10 times lower than the maximum permissible concentration (MPC) of water bodies for drinking purposes.

In summer, the deep reference station is characterized by sufficient oxygen saturation down to the bottom (up to 20% O_2 saturation, pH = 7.13-7.2) due to good wind mixing and active development of phytoplankton. The main form of iron in the lake during this period is the particulate form, Figure 14a. Waters of the Lake Lekshmozero belong to low-colored waters. An increase in the color index (up to 13°Cr-Co) is indicative of the tendency for dissolved organic matter to accumulate in the near-bottom layer at the deep reference station (Figure 15a). In winter, the deep-water reference station undergoes an intensive process of decomposition of organic matter, forming an oxygen deficit in the bottom layers (1-2 meters above the bottom) (0.3-4% O_2 saturation, pH 6.5-6.9).

This creates favorable conditions for the release of phosphorus and iron from bottom sediments and the prevalence of dissolved forms of iron over suspended ones (Figure 15). In addition, it was noted that the conditions of oxygen deficiency and additional release of iron from bottom sediments lead to an increase in the concentration of iron in the winter period compared to the summer period. As noted (Kalinkina et al., 2006), at the present time, so much organic matter, phosphorus, and other elements have been accumulated in bottom sediments that, with the slightest shifts in the hydrological, temperature, and hydrochemical regimes, the sediments are capable releasing substantial amount of phosphorus, iron, and other elements to water column.

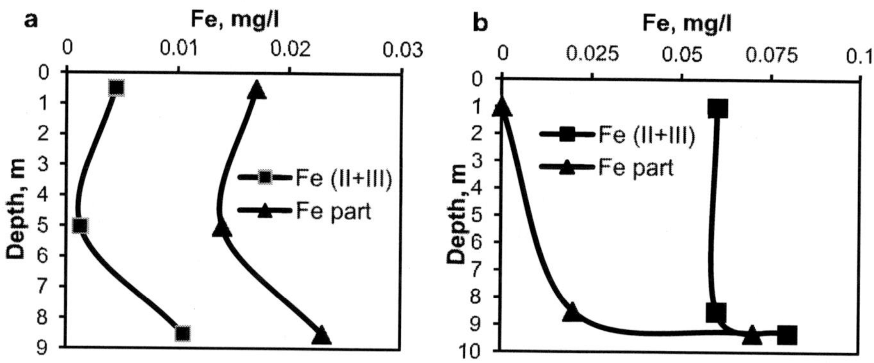

Figure 14. The vertical distribution of iron forms in a shallow water station H9 in summer (a) and winter (b) in the water of Lake Lekshmozero, 2016.

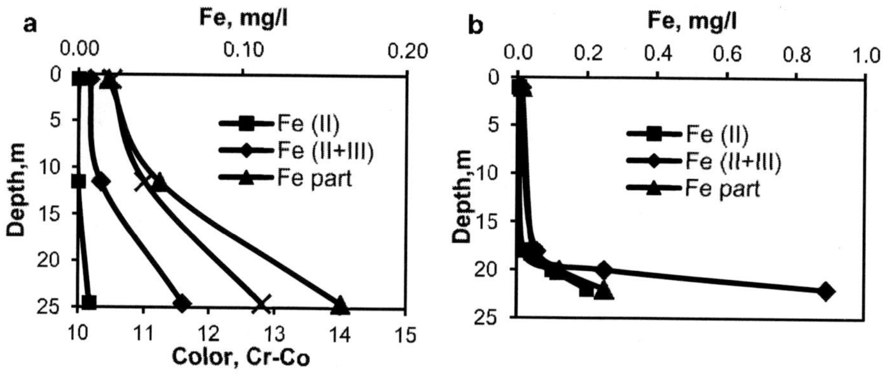

Figure 15. The vertical distribution of iron forms and water color in deep sea station H9 – summer (a) and winter (b) in the Lekshmozero Lake.

After the ice cover has been established and as a result of heating of bottom waters from sediments, there is an increase in the water. At the same time, water is saturated with mineral compounds from sediments, which additionally increases the density of bottom waters and facilitates their transfer towards depressions. Thus, in the deep area of the lake, where the reference station is located, various nutrients including iron accumulate in sediments. The result of these processes was a 5-fold increase in the iron content in the bottom layer in the winter compared with the summer, which is confirmed by long-term observations. Presumably, the bottom sediments of Lake Lekshmozero act as the main regulators of nutrients in water, including iron (Figure 16).

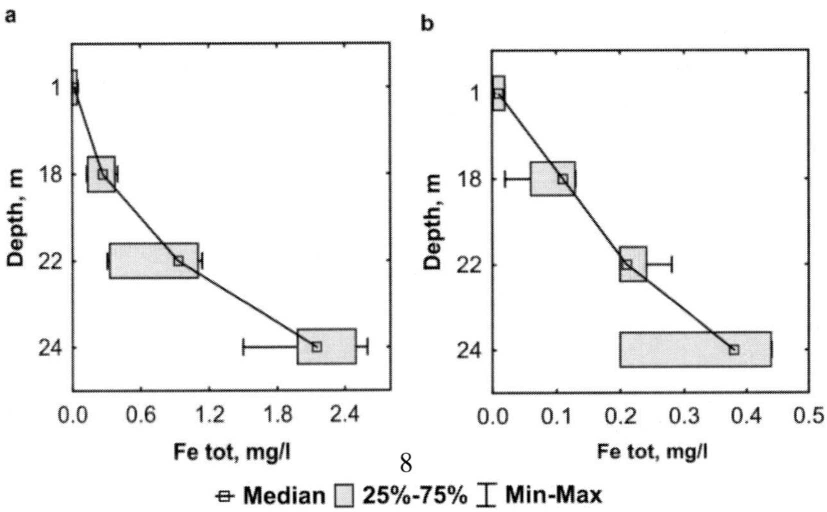

Figure 16. The vertical distribution of total iron and water color in deep station in winter (a) and summer (b) in the water of Lake Lekshmozero.

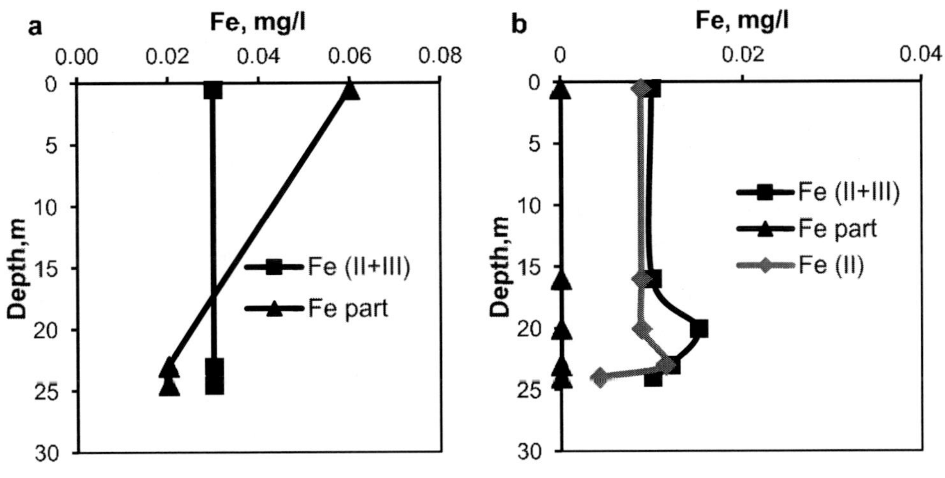

Figure 17. Seasonal dynamics of the distribution of forms of iron at the reference deep station of Lake Lekshmozero (a –spring turnover, 2010 and b –autumn turnover 2012).

In 2010 and 2012, a study of the distribution of iron during homothermia (lake overturn) was carried out. At the deep station during the spring homothermy, the iron content was 0.02–0.06 mg/L from the surface to the bottom, with a predominant of particulate iron, and from 16 m to the surface - total dissolved iron. Hydrological and hydrochemical parameters of the water column showed a low variability of the main indicators at the surface and at the bottom, respectively: pH = 5.9-6.6; O_2 from 118% to 86.7%, T from 14°C to 4°C. Iron (II) in water during the spring homothermy was not detected (Figure 17). In November 2012, the period of autumn homothermy, with a general tendency to equalize the hydrological and hydrochemical parameters of the environment, the dissolved forms of iron dominated, whereas suspended or particulate iron was not detected in the entire depth range (Figure 17).

During the period of homothermia, the concentration of iron decreases sharply, and the distribution pattern becomes more uniform from the surface to the bottom. Suspended iron (Fe part) was not detected at shallow-water stations during the period of autumn homothermy. The concentration of iron in water samples taken during the period of off-season homotherms is significantly lower compared to periods of stratification, especially compared to winter.

In 2009, in November and July, water samples were taken from streams flowing into Lake Lekshmozero. Analysis of iron delivered by streams to the lakes showed the following: the main form of iron is the dissolved form, the highest concentrations of iron in the streams were recorded in the summer period (Figure 18). It should be noted that the streams flowing into Lake Lekshmozero contain 3-5 times more iron than the waters of the lake. The iron content in the Lekshma River, flowing out of Lake Lekshmozero, is significantly lower than that in small inlet streams (Figure 19). The incoming iron is actively involved in the biological cycle inside the reservoir itself, and partially accumulates in bottom sediments.

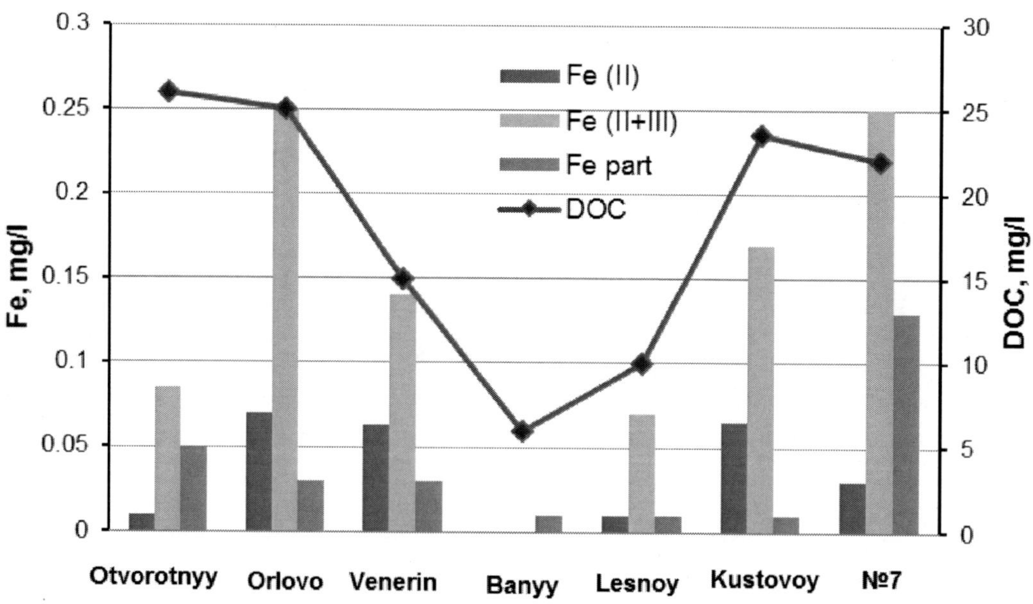

Figure 18. Distribution of forms of iron and dissolved organic carbon (DOC) in inlet streams of Lake Lekshmozero, November 2009.

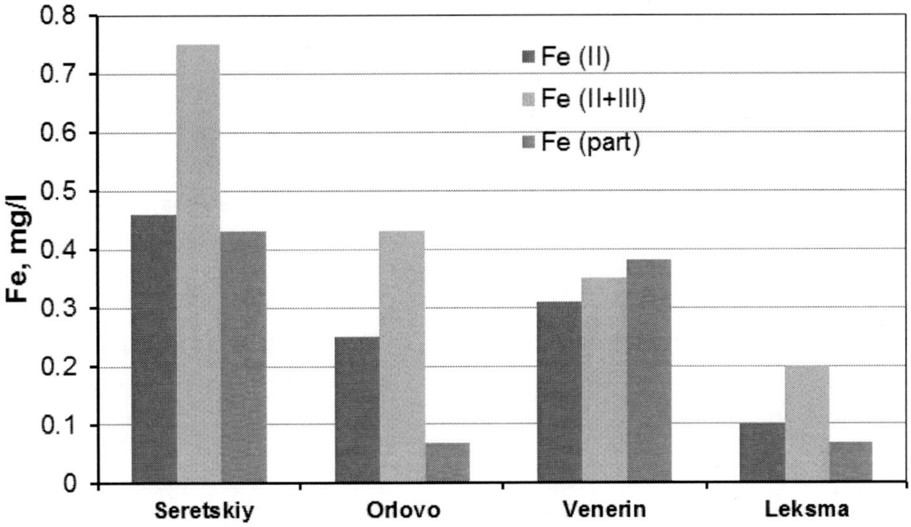

Figure 19. The distribution of forms of iron in the inlet streams of Lekshmozero and the only outlet stream (river Lekshma), July 2009.

The predominance of the dissolved form of iron in most streams in the summer can be explained by the transport of iron in complexes with dissolved organic matter (DOM), primarily with humic and fulvic acids (Levshina et al., 2006., Shirokova et al., 2012). Iron is highly abundant in soils and peat of raised bogs (Dobrovolsky et al., 2003). The ability of humic substances to form stable compounds with metals is one of the reasons for the delivery of iron into surface water bodies. In watersheds enriched in organic matter, iron mainly enters streams, lakes, and rivers in a form associated with organic complexes. Its content increases especially significantly in the waters of forest and tundra streams, rivers and lakes.

Eleven small tributaries flow into Lake Lekshmozero, which do not have a significant impact on the formation of the hydrological regime of the lake, and, consequently, the hydrochemical one. One river, Lekshma, flows out of the lake. The flow rate of the lake is quite low, the conditional water exchange coefficient is 0.11, that is, the full water exchange occurs once every 9.2 years. Lake Lekshmozero has the lowest watershed slope among all the studied lakes (0.008), this circumstance, and the insignificant runoff of the lake itself relative to its volume led to the fact that the increased color of 70-80° Cr-Co in streams does not affect the color of the waters of Lake Lekshmozero (10-15° Cr-Co).

The low flow rate and morphometric features of the reservoir (a smooth decrease in depth to the central part) cause the accumulation of almost all nutrients and organic substances coming from the catchment area, including iron, which is reflected in an increase in all forms of iron and color from the surface to the bottom. The main predominant form of iron during the period of winter stratification is the dissolved form, whereas in summer, the suspended Fe dominates. Iron concentration in winter, under conditions of oxygen deficiency, is an order of magnitude higher than iron concentration in summer. It is shown that the main form of transport of iron in the streams flowing into Lake Lekshmozero is the dissolved form, which is caused by the presence of iron in the organic complexes with humic acids.

Dissolved Organic and Inorganic Carbon Concentrations

In recent years, comprehensive fundamental research on the ecosystems of northern lakes has become a priority. Interest in studies of the carbon cycle is growing, primarily due to climate change. Comparison of the processes of the biogeochemical carbon cycle in various freshwater limnic ecosystems makes it possible to study it in the natural environment under different conditions of the uptake of allochthonous organic matter (OM), production of autochthonous OM, mineralization of dissolved and suspended OM by a heterotrophic bacterial unit, Red/Ox conditions and residence time in the system (lake water exchange time). We used the concentration of dissolved organic carbon (DOC), the concentration of dissolved inorganic carbon (DIC) and their ratio as the main indicators of the course of these processes. The ratio of DOC and organic nitrogen was used as a characteristic of the origin of DOC.

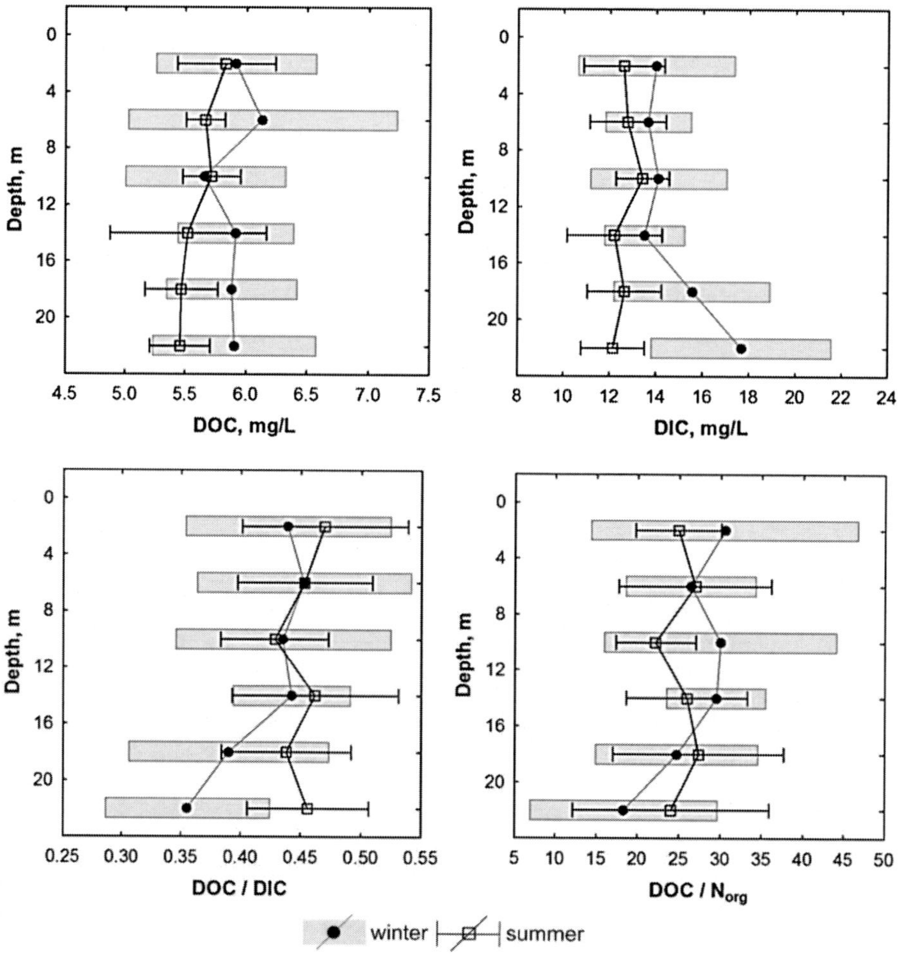

Figure 20. The vertical distribution of DOC and DIC in Lake Lekshmozero (point - mean; box of whisker – standard deviation).

The average concentration of dissolved organic carbon (DOC) for the period of research (2007-2015) in the water of Lekshmozero lakes was 5.58 ± 0.55 mg/L. Due to hydrological features, Lekshmozero Lake is characterized by a relatively uniform distribution of DOC with a pronounced tendency to decrease in concentration with depth, which may be due to the photosynthetic activity of phytoplankton and the destruction activity of bacterioplankton. During the periods of spring and autumn homothermy in Lake Lekshmozero, the concentrations of DOC are similar along the water column due to the mixing of water column.

For Lake Lekshmozero (Figure 20), the vertical distribution of dissolved inorganic carbon shows dominance and a slight increase with depth in winter, which in turn indicates the prevailing processes of mineralization of dissolved and suspended organic matter under ice conditions.

In the spring, with the input of allochthonous organic matter into the lakes, an increase in the concentration of inorganic carbon was observed, which can be explained by the acceleration of destruction processes, in accordance with direct field observations (Shirokova et al., 2016).

An analysis of the ratio of seasonally averaged values of organic and inorganic dissolved carbon (DOC/DIC) in Lake Lekshmozero (Figure 21; Table 5) revealed the prevailing inorganic component in the total balance of dissolved carbon (the average ratio is 0.43). The ratio of organic and inorganic carbon in turn confirms the data on the intensity of production-destruction processes in their seasonal aspect and, in general, indicate the heterotrophic status of the Lekshmozero ecosystems (Vorobieva et al., 2013).

The C_{org}/N_{org} ratio for Lake Lekshmozero was mainly in the range of 5.75 to 41.3 (the average is 25.2). These values are characteristic of aquatic phytoplankton, macrophytes, and their humification products (~ 20) (Chupakov et al., 2017). During periods of stratification, the C/N ratio in hypolimnion decreases. A probable reason for this decrease is the mineralization of organic material in sedimentary bottom layers.

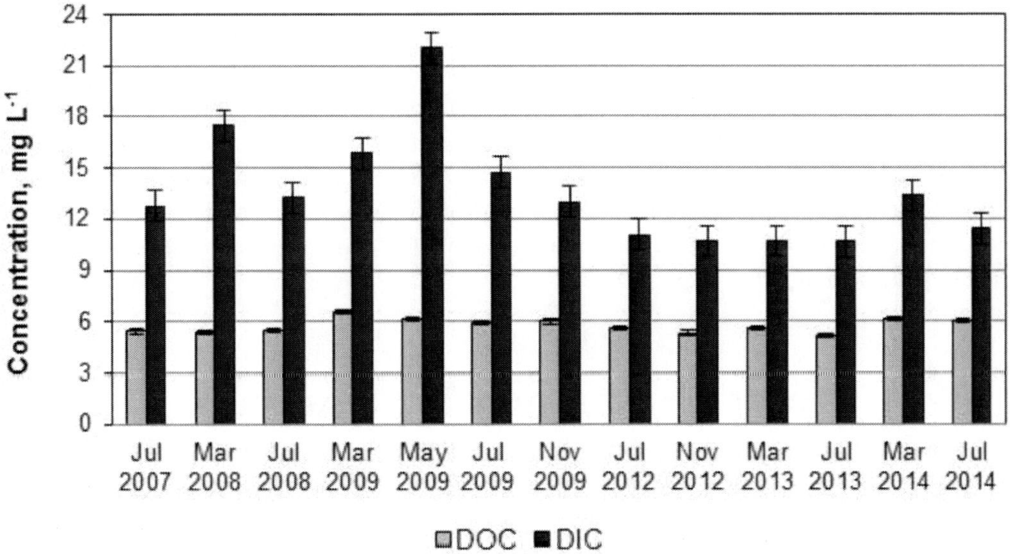

Figure 21. Seasonal and interannual dynamics of DOC and DIC in Lake Lekshmozero.

Table 5. The values of the DOC, DIC, DOC/DIC and DOC/N_{org} are averaged layer by layer. Statistical characteristics

Interval depth	N	Mean	Min	Max	SD	N	Mean	Min	Max	SD
	DOC, mg/L					DIC, mg/L				
(0;4]	33	5.9	5.2	7.7	0.6	35	14	11	24	4
(4;8]	19	5.9	5.1	7.9	0.7	19	14	11	23	3
(8;12]	15	5.7	5.2	6.9	0.5	15	14	10	22	3
(12;16]	17	5.6	4.2	6.5	0.6	18	13	8	21	3
(16;20]	21	5.7	5.1	7.6	0.6	22	14	10	23	3
(20; bottom]	28	5.6	5.0	7.3	0.5	29	15	10	24	4
	DOC/DIC					DOC/N_{org}				
(0;4]	33	0.45	0.24	0.60	0.09	28	26	10	63	11
(4;8]	18	0.44	0.26	0.56	0.08	13	26	16	38	8
(8;12]	14	0.42	0.27	0.53	0.08	11	26	12	55	12
(12;16]	17	0.45	0.29	0.56	0.07	14	26	14	37	7
(16;20]	21	0.42	0.24	0.53	0.07	19	26	10	48	10
(20;bottom]	27	0.41	0.23	0.53	0.08	23	22	6	51	11

In Lake Lekshmozero, low-molecular forms of DOC (less than 1 kDa or 1 nm) account for 20-30% of the total dissolved concentration in March and May, while the role of suspended forms and large sub-colloidal particles (0.025 μm - 20 μm) is insignificant. The relative amount of organic colloids (1 kDa - 0.45 μm) decreases 2-3 times during the summer period (July), while the low molecular weight fraction (< 1 kDa) makes up to 80% of the total DOC in the water column.

Table 5. Statistical characteristics of DOC, DIC and ratio of lake water for 4-m layers for the LREP reference point of Lake Lekshmozero for 2007-2015.

Thus, Lake Lekshmozero is characterized by low concentrations of dissolved organic carbon with a relatively uniform distribution and a tendency toward a decrease in concentrations with depth, which may be due to the processes of photosynthetic activity of phytoplankton and respiration of bacterioplankton. The Corg/Norg ratio for Lake Lekshmozero (5.8-41.3 (average is 25.2) and the predominance of potentially "bioavailable" (low molecular weight) forms of DOC and a low proportion of colloidal forms indicate essentially autochthonous nature of organic matter.

Zooplankton Community Composition

During the period from 2009 to 2017, 60 forms of zooplankton were noted in the Lekshmozero Lake zooplankton community: Copepoda - 12 species (Calanoida - 3 and Cyclopoida - 9), Cladocera - 33 species and Rotifera - 15 species. The planktonic fauna is represented by species that are usual for water bodies in the northwest. The majority of zooplankton (30 species, or 50% of their total number) were representatives of planktonic forms; the share of literal (16 species) and eurytopic (14 species) forms was 27% and 23%, respectively. In the zooplankton community of the studied lake, oligo- and oligo-β-mesosaprobes predominated (68% of the total number of planktonic organisms).

Species and trophic structures, and dynamics of the quantitative development of planktonic organisms in the lake are analyzed in a seasonal aspect. In winter, a decrease in species diversity was observed, since most species of zooplankton organisms were not present in the community. In zoocenosis, herbivorous and predatory Copepoda and Rotifera prevailed (up to 94% of the total zooplankton abundance). Analysis of the vertical distribution of aquatic organisms in the winter showed their confinement to the lower layers of water (12-20 m).

The quantitative characteristics of the development of zooplankton in this period are lower than in summer. The contribution of Copepoda to the total number of zooplankton in different years of research ranged from 59 to 70%, Rotifera - from 19 to 24%. *Cyclops scutifer*, *Cyclops vicinus*, *Eudiaptomus gracilis* dominated in numbers from Copepoda. Of the Rotifera, the dominant species were *Asplanchna priodonta*, *Kellicottia longispina* and *Keratella quadrata*. The average values of quantitative indicators of winter zooplankton varied over the years. Average abundances varied from 130 to 31800 ind./m^3 and the biomass ranged from 0.0001 to 0.21 g/m^3. The interannual dynamics of quantitative indicators in the winter period is presented in Figures 22.

The summer is characterized by maximum heating of the water and the formation of stable thermal stratification. At the beginning of summer, cold-water species move from the water column to the bottom layers. The planktonic fauna is enriched in Cladocera (*Daphnia*, *Bosmina*, *Chydorus*, *Polyphemus*, *Ceriodaphnia*, *Diaphanosoma)*. Cladocera determined the species diversity of the zooplankton community. The Lekshmozero Lake, in comparison with other KNP lakes, is characterized by a later date of the onset of Cladocera development. This is due to the morphometric characteristics of the lake, namely, the area and depth of the lake. The zooplankton in the summer is dominated by the following species: *Mesocyclops leuckarti*, *Eudiaptomus gracilis*, *Thermocyclops crassus*, *Bosmina coregoni*, *Bosmina longirostris*, *Daphnia cucullata*, *Daphnia cristata*, *Kellicottia longispina*, *Keratella cochlearis*. A significant proportion of the zooplankton community of Lake Lekshmozero in the summer was primary filter-feeders and Rotifera (56% of the total number) among which representatives of the genera *Kellicottia*, *Conochilus*, *Keratella*, *Daphnia*, *Bosmina*, *Eudiaptomus* preyed in the water column.

The next numerous trophic group of organisms (swimming/active capture) are optional and obligate predators (25% of the total number). This group included representatives of the families Asplanchnidae, Polyphemidae, Leptodoridae, Cercopagidae, Cyclopidae. The third largest trophic group (19% of the total number) consisted of floating-crawling secondary filter-feeders that extract food from the surface of the substrate. This group includes representatives of the family Chydoridae. The dominance of large filter-feeders and predators in the food networks of the zooplankton communities of the lake indicates the oligotrophic status of the studied lake (Derevenskaya et al., 2002)

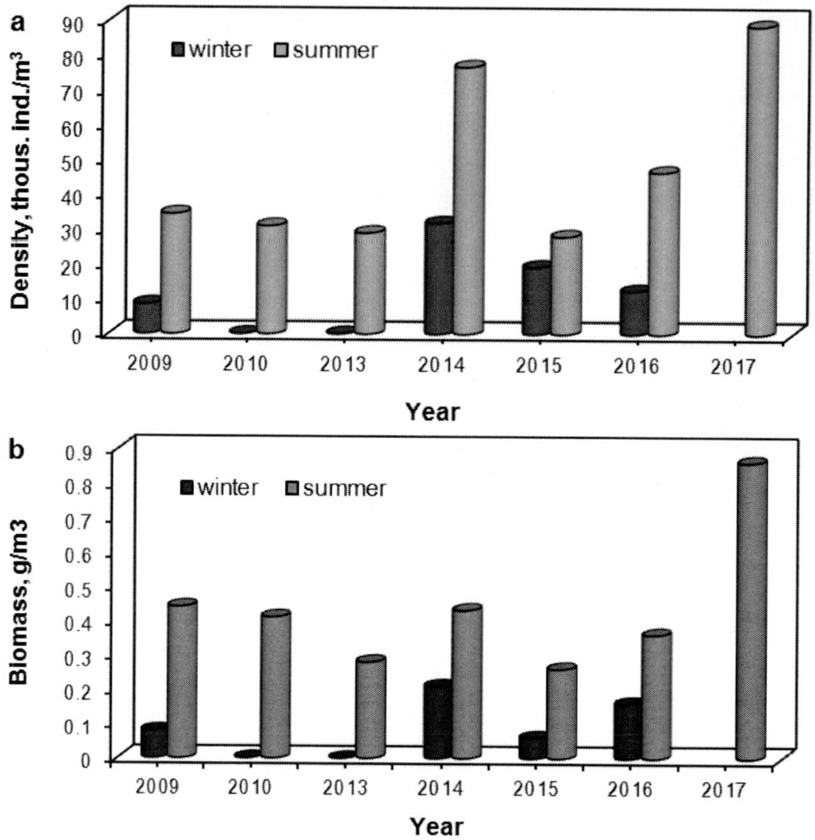

Figure 22. Interannual dynamics of zooplankton abundance (a) and biomass (b) in winter and summer in the lake Lekshmozero (2009-2017).

In the littoral zone overgrown with macrophytes, a huge amount of phytophilic zooplankton develops in summer, which allows us to consider this zone as the most populated area of the lake. Mass development in the littoral zones reached Cladocera (40 - 61% of the total number). In the thicket of zone of Lake Lekshmozero phytophilic, littoral, and sublittoral species of Cladocera (*Sida crystallina*, *Ceriodaphnia quadrangula*, *Acroperus harpae*, *Pleuroxus truncatus*, *Polyphemus pediculus*, *Simocephalus vetulus*, *Chydorus sphaericus*, *Diaproma pomera*, etc.) are widespread. Among Copepoda, typical representatives of plankton from macrophytes zone are *Eucyclops macrurus*, *E. serrulatus*, *Macrocyclops albidus*). Rotifera represented *Asplanchna*, *Kellicottia*, *Polyarthra*, *Keratella*. Typical planktonic forms (*Daphnia cristata*, *Bosmina coregoni*, *Eudiaptomus gracilis*, *Mesocyclops leuckarti)* are introduced from the pelagic zone of the lake into macrophyte thickets.

The listed zooplankton forms were structure-forming in the zoocenoses of the thicket zone. Quantitative indicators of zooplankton development in the littoral zone are the highest. The abundance of zooplankton in the thicket zone in different years of research varied from 23,000 to 181,000 ind./m^3, and the biomass varied from 0.17 to 1.45 g/m^3. Littoral ecosystems are the most productive areas of the lake.

Quantitative characteristics of the zooplankton in the summer season varied among different years. A significant contribution to the total abundance and biomass was provided by Copepoda (from 44 to 54% of the total zooplankton abundance and from 41 to 50% of the total biomass). The contribution of the Cladocera to the formation of the total abundance and biomass was from 36 to 46% and from 46 to 53%, respectively.

Among Cladocera, structure-forming species were representatives of the genera *Daphnia*, *Bosmina*. Among Rotifera, *Keratella cochlearis* and *Kellicottia longispina* dominated. *Asplanchna priodonta* contributed up to 95% to Rotifera biomass. On average, the zooplankton development level of Lake Lekshmozero in summer was characterized by the following quantitative indicators – the average abundance depending on the year of study ranged from 28,000 to 88,400 ind./m3, biomass from 0.26 to 0.86 g/m^3.

The lake is characterized by high values of the average body weight of planktonic organisms, which indicates the great value of zooplankton as a food for fish. The interannual dynamics of quantitative indicators in the summer period is presented in Figure 22. Analysis of interannual dynamics of abundance and biomass of zooplankton of Lake Lekshmozero during the research period, as well as the analysis of literature data showed an increase in these parameters of zooplankton (Balushkina et al., 1997, Kulikova et al., 2008.)

The average seasonal indicators of zooplankton total abundance almost doubled between 2014 and 2017 (Figure 22). In general, the average seasonal values of the total zooplankton biomass for the period of this study were at the level of long-term average indicators (Table 6).

Analysis of zooplankton vertical distribution showed its confinement to the upper layer of water (0-4 m). Representatives of the genera *Thermocyclops*, *Eudiaptomus*, *Mesocyclops*, *Daphnia*, *Chydorus*, *Keratella*, *Kellicottia*, *Diaphanosoma* most intensively developed in the surface layers. With depth, the role of Rotifera decreased, and the role of representatives of Cyclopoida and Calanoida increased. At depths of 12-18 m, clusters of Daphnia cristata were observed. In water layers below 10 m, a relic crustacean *Limnocalanus macrurus* is noted.

To assess the state of the lake's zooplankton communities, the Shannon species diversity index (H) was calculated based on the number of hydrobionts. The zoocenoses of Lake Lekshmozero were characterized by significant species diversity and complex structure. The Shannon Species Diversity Index varied from 2.58 (2017) to 3.36 bit/specimen (2010), depending on the year of research. The index values indicate a relatively stable and balanced status of the lake's zooplankton community. According to the state of zooplankton communities and following the classification of Kitaev et al. (1984), the lake can be characterized as oligotrophic, although in terms of biomass, the coastal areas in the northern and southern regions of the reservoir can be classified as α-mesotrophic. The most favorable feeding conditions for planktophagous fish have developed here.

Table 6. General characteristics of the zoocenosis of Lake Lekshmozero in various years of research

Parameters	1991[1]	1997[2]	2003[3]	2004[2]	2009-2017
Study period, month	**VII-VIII**	**VIII**	**VII**	**VII**	**VII-VIII**
Number of species	19	21	25	23	60
Density, thous. ind./m³	47.5	13.9	14.8	27.4	28.0-88.4
Biomass, g/m³	0.42	0.60	0.74	0.42	0.26-0.86
Saprobity index	-	1.67-1.83	1.75-1.92	1.70-1.89	1.20-1.54

Note [1]Balushkina et al., 1997; [2]Kulikova et al., 2008; [3]Kalinkina et al., 2006.

The quality of the lake water masses corresponded to α-oligotrophic waters (clean waters) throughout the study period. The saprobity index ranged from 1.20 (2017) to 1.54 (2014). According to literature data, the trophic status of Lake Lekshmozero in the period from 1977 to 2004 defined as mesotrophic. In terms of quality, the water masses of the lake belonged to the β-mesosaprobic class (saprobic index 1.67-1.92) (Kalinkina et al., 2006, Balushkina et al., 1997., Kulikova et al., 2008., Kalinkina et al., 2004.)

In general, the low indices of water saprobity, the average values of the Shannon-Weaver coefficient, and the significant quantitative development of organisms indicate the relative stability and favorable state of zooplanktocenoses in Lake Lekshmozero. The taxonomic composition of zooplankton is typical for cold-water oligotrophic water bodies of the boreal zone. The trophic status of the lake during the study period was determined as oligotrophic with signs of mesotrophy.

PHYTOPLANKTON COMMUNITY COMPOSITION

The development of phytoplankton communities in aquatic ecosystems is controlled by the interaction of various environmental factors. Physical processes, in particular the formation of thermal stratification, is considered the most important factor for the regulation of the composition and abundance of phytoplankton in the summer (Carey et al., 2012; Yang, Y. et al., 2016a; Yang et al., 2016b). By the nature of the vertical distribution of temperature in summer, Lake Lekshmozero is metathermal. In summer, due to the openness of the lake and wind mixing, the stratification is unstable. The study of the ratio of the main nutrients showed that they are optimal for the development of phytoplankton. These values also indicate mainly the limiting role of phosphorus in the development of phytoplankton in the lake, but in summer, and sometimes in autumn, nitrogen deficiency can be created in the epilimnion (<16). The water transparency in Lake Lekshmozero in the deep part was within 3.0-5.7 m (average 4.48 ± 0.95 m), in the littoral - 2.5-4.8 m (average value 3.69 ± 0.77 m). The study of the qualitative and quantitative content of phytoplankton communities in Lake Lekshmozero was carried out from 2007 to 2017. Seasonal variations in phytoplankton in Lake Lekshmozero mainly indicate its greatest activity in spring and summer.

The number of phytoplankton in summer reached 3913 thousand cells/L. The phytoplankton biomass during the biological summer in the photic layer for Lekshmozero was in the range of 0.235-2.32 mg/L. Quite high values of biomass (0,464-1,839 mg/L) were observed in the lake in November 2009 before freezing. In winter, the number and biomass of phytoplankton in the photic layer of water were low - 16.5-40 thousand cells/L and 0.051-0.163 mg/L, respectively.

When describing the phytoplankton composition of Lake Lekshmozero, we used the functional characteristic proposed in literature (Reynolds et al., 2002, Padisák et al., 2009). The results of the analysis showed that representatives of Dinophyta, Cryptophyta, Bacillariophyta predominate in winter phytoplankton, when species from Euglenophyta, Chrysophyta are found. Functional groups **X2** (chrysoflagellates, small *Cryptomonas sp.*, *Chroomonas acuta*) and **X2** (small *Gymnodinium sp*, *Cryptomonas ovata*) have the largest share in the community in terms of number and biomass. The representative Dinophyta (*Peridinium inconspicuum*) belongs to the Lo group. From Bacillariophyta, it is represented by functional groups **B** (*Aulacoseira islandica*), **C** (*Asterionella formosa*), $\mathbf{T_B}$ (*Melosira varians*), **MP** (*Eunotia sp.*).

In spring and autumn, they are also represented by groups **X2, Y**, the proportion of the functional group Lo increases significantly, due to mainly representatives of Dinophyta (*Peridinium aciculiferum, P. inconspicuum, Gymnodinium inversum*), species from groups **E, F, J.** Lake Lekshmozero is similar to Lake Erken (Sweden), which is also unstably stratified during the open water period (Yang et al., 2016a). The absence of islands, frequent winds, a high proportion of the littoral part (up to the 7 m isobath is 73.0% and up to 10 m - 86.2% of the lake volume) affect variability hydrodynamic conditions, which affects the distribution of the phytoplankton community.

In summer, when the water warms up to 18-22°C, Cyanobacteria dominate in the epilimnion mainly in terms of numbers (code **S1** - *Planktolyngbya limnetica*, code **K** - *Aphanocapsa delicatissima, Cyanodictyon sp.*, code **H1**- *Anabaena circinalis, A. flos- aquae, Aphanizomenon flos-aquae*, code **H2** - *Anabaena lemmermanii, Gloeotrichia echinulata*), the biomass is dominated by the complex of diatom taxa (code B - centric diatoms, code C - *Asterionella formosa, Aulacoseira ambigua*, code P - *Aulacoseira granulate, Fragilaria crotonensis*, code **D** - - *Fragilaria ulna*), Cryptophyta (code **X2, Y**) and Chlorophyta (code **F** - *Botryococcus braunii*, code **N** - *Cosmarium sp.*). The vertical distribution of various phytoplankton taxa by abundance (a) and biomass (b) in Lake Lekshmozero is shown in Figure 23, for example on July 10, 2015.

In Lekshmozero in the second half of summer, the intensive development of nitrogen-fixing Cyanobacteria *Gloeotrichia echinulata* (J.E. Smith et Sowerby) P. Richt is possible. As a result, firstly, there is a sharp increase in the productivity of the lake, and secondly, there is a change in the active reaction of the environment towards an increase in water pH, which can cause secondary pollution of the aquatic ecosystem (see Chapter 4). *Gloeotrichia echinulata* prefers a stratified water column with an increased temperature and transparency of the water, and the need for prior intensive mixing of the water column (Barbiero, 1993, Hyenstrand et al., 2001, Carey et al., 2009). For example, on Lake Erken (Sweden), optimal conditions for the summer bloom of *Gloeotrichia echinulata* are created every year in mid-summer as more stable stratification occurs (Pettersson et al., 1993, Karlsson-Elfgren et al., 2003, Carey et al., 2008, Yang et al., 2016b), and in Lake Lekshmozero, similar in morphometric parameters, such conditions are not created every summer, due to more dynamic water column.

As an example of the phytoplankton community composition on 10 and 15 July 2015, it is clearly seen how intensive wind mixing of water for 1 day between sampling dates contributes to a change in the composition of dominant species (Figure 6). On 10th July, nitrogen-fixing cyanobacteria from the functional group H1 (11% of the total phytoplankton number), represented by the species *Anabaena flos-aquae*, and H_2 (45%) - the dominant *Anabaena lemmermanii*, dominated in epilimnion. Relative biomass H_2 was only 6.48%, of which the biomass of *G. echinulata* was only 0.52%. The ratio of nutrients Ntot/Ptot in the epilimnion at this time indicated nitrogen limitation (9.75-15.8), together with low concentrations of phosphates (4-8 μg P/L), nitrates (44.3-63.7 μg N/L) and ammonium (27.3-37.9 μg N/L. On July 15, a further decrease in the concentrations of mineral forms of phosphorus and nitrogen was observed, and the concentration of phosphates decreased to the minimum values - 1.2-2.3 μg P/L. On July 15, the relative biomass of nitrogen-fixing cyanobacteria of the H2 functional group (*Anabaena lemmermanii, Gloeotrichia echinulata*) increased to 45%, of which

Gloeotrichia echinulata was 41%, Figure 24. At the same time, the total biomass increased from 0.362 mg/L to 1.33 mg/L in the surface layer.

Thus, the phytoplankton community of Lake Lekshmozero is represented by functional groups, mainly characteristic of well-mixed mesotrophic water bodies with a transparent epilimnion and a low content of nutrients (codons **Lo, H1, X2, Y, C, B, F, N, A**). For the summer phytoplankton in Lake Lekshmozero, the species diversity is typical for the North-West of Russia. The presented taxa are predominantly cosmopolitan forms, most of the species are oligo-β and β-mesosaprobes. The biomass of summer phytoplankton makes it possible to determine the trophic status of the studied lakes as mesotrophic.

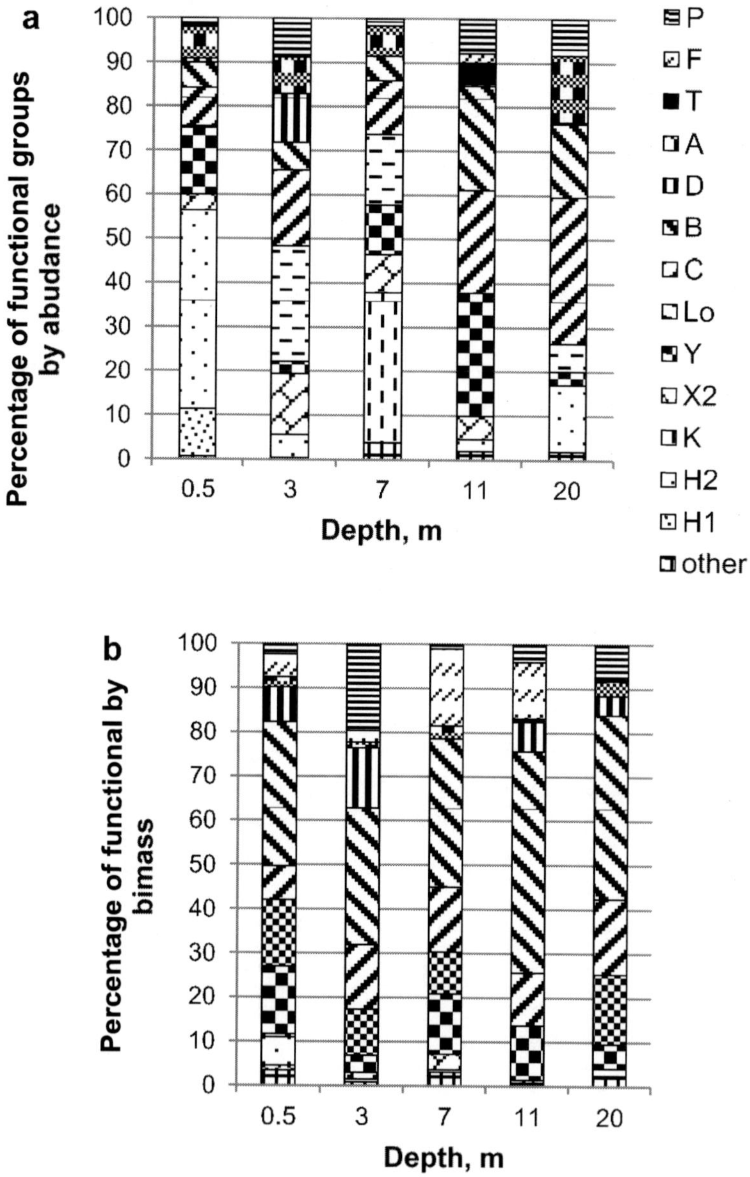

Figure 23. Vertical distribution of various phytoplankton taxa by abundance (a) and biomass (b) in Lake Lekshmozero (July 10, 2015).

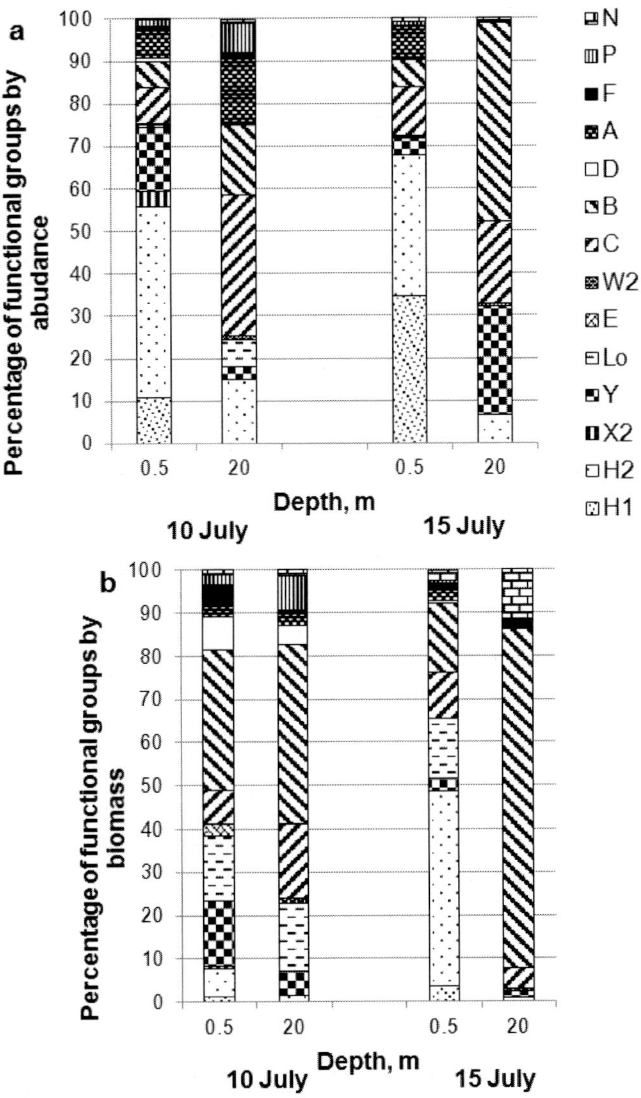

Figure 24. Vertical distribution of various phytoplankton taxa by abundance (a) and biomass (b) in Lake Lekshmozero at the surface and bottom layers (July 10 and 15, 2015).

MICROBIOLOGICAL CHARACTERISTICS

The status of lake ecosystem is the result of a complex interaction of processes determined by both natural and anthropogenic factors. Almost all components of the water body biocenoses are involved in the processes of self-purification and formation of water quality in aquatic ecosystems. Bacterioplankton plays an important role in the functioning of ecosystems of water bodies: in the processes of decomposition of autochthonous and allochthonous organic matter, the cycle of biogenic elements, decomposition and removal of pollutants from the cycle, i.e., in the process of self-purification of water bodies, and, ultimately, in maintaining a stable status of aquatic ecosystems. Heterotrophic bacteria are a key component of the self-purification of aquatic ecosystems. They are the most active participant in the biotransformation of dissolved and suspended organic matter, which serves as a food source for this group of microorganisms. Besides, heterotrophic bacteria are only organisms that return dissolved organic matter to the carbon cycle (Tranvik et al., 1992., Del Giorgio et al., 1994). In addition, the high rate of bacterial reproduction makes it possible to synthesize a significant biomass during the year, often commensurate with the primary production of OM in the process of photosynthesis, which is a constant source of food for zooplankton and zoobenthos.

Nevertheless, bacterioplankton remains one of the least studied components of aquatic ecosystems, including its trophic relationships and factors influencing its development. Bacterial respiration is the main cause of pure heterotrophy (the ratio of photosynthesis to respiration <1) in oligo- and mesotrophic lakes (del Giorgio et al., 1994., Biddanda et al., 2002). The destruction of dissolved organic matter by aerobic heterotrophic bacterioplankton in surface waters is one of the key factors that regulate the exchange of CO2 in lakes and rivers with the atmosphere in the arctic and subarctic zones (Shirokova et al., 2010, 2017; Tranvik et al., 1992). Heterotrophic bacteria quickly recycle nutrients and keep them within the plankton community, which allows phytoplankton to reuse nutrients. It is well known that, despite their heterotrophy, bacteria absorb a significant part of inorganic nutrients: on average, 60% of phosphate (PO_4) and 30% of ammonium (NH_4) are absorbed by heterotrophic bacteria (Kirchman et al., 1994.). This creates competition between bacteria and phytoplankton for growth-limiting nutrients (Caron et al., 1994.)

Microorganisms are the most sensitive indicators of changes in the ecological state of aquatic ecosystems and, therefore, are widely used in environmental studies. Obtaining such characteristics of the microbial community as the abundance and seasonal dynamics of ecological and trophic groups of microorganisms is an important tool for understanding and explaining the processes occurring in a water body.

This section presents the results of the seasonal and interannual dynamics studies of the abundance and ecological-trophic structure of heterotrophic bacterioplankton in Lake Lekshmozero (White Sea basin) of the Kenozersky National Park. To generalize the material, we used the data obtained by the authors in 2006-2016, as well as literature data.

The study of the quantitative content dynamics of various ecological and trophic groups of heterotrophic bacteria in the lake water showed their high seasonal and interannual variability. The content of heterotrophic bacteria in Lake Lekshmozero varied within several orders of magnitude, but, in general, according to the median values, their numbers were low, which is probably due to the low DOC values in the lake water. According to the average DOC concentrations for the period of research in Lake Lekshmozero, the lowest concentrations (~ 6 mg/L) were observed among all the KNP lakes studied by our group (Shirokova et al., 2010)

Figure 25 shows the interannual dynamics of the weighted average values for the water column of the HB population. The lowest concentrations of HB were observed during the ice period. By the end of the winter stratification period, fluctuations in abundance values for the period 2007-2014 (n = 63) were: EB, 1 - 360 CFU/ml (median 11 CFU/ml); FOB, 2-450 CFU/ml (median 22 CFU/ml); and OB, 1-430 (median 15 CFU/ml)); however, the median values of the abundance of all groups of bacteria were units and tens of CFU/ml, which indicates the completion of the transformation of bioavailable organic matter by the end of the winter stratification period.

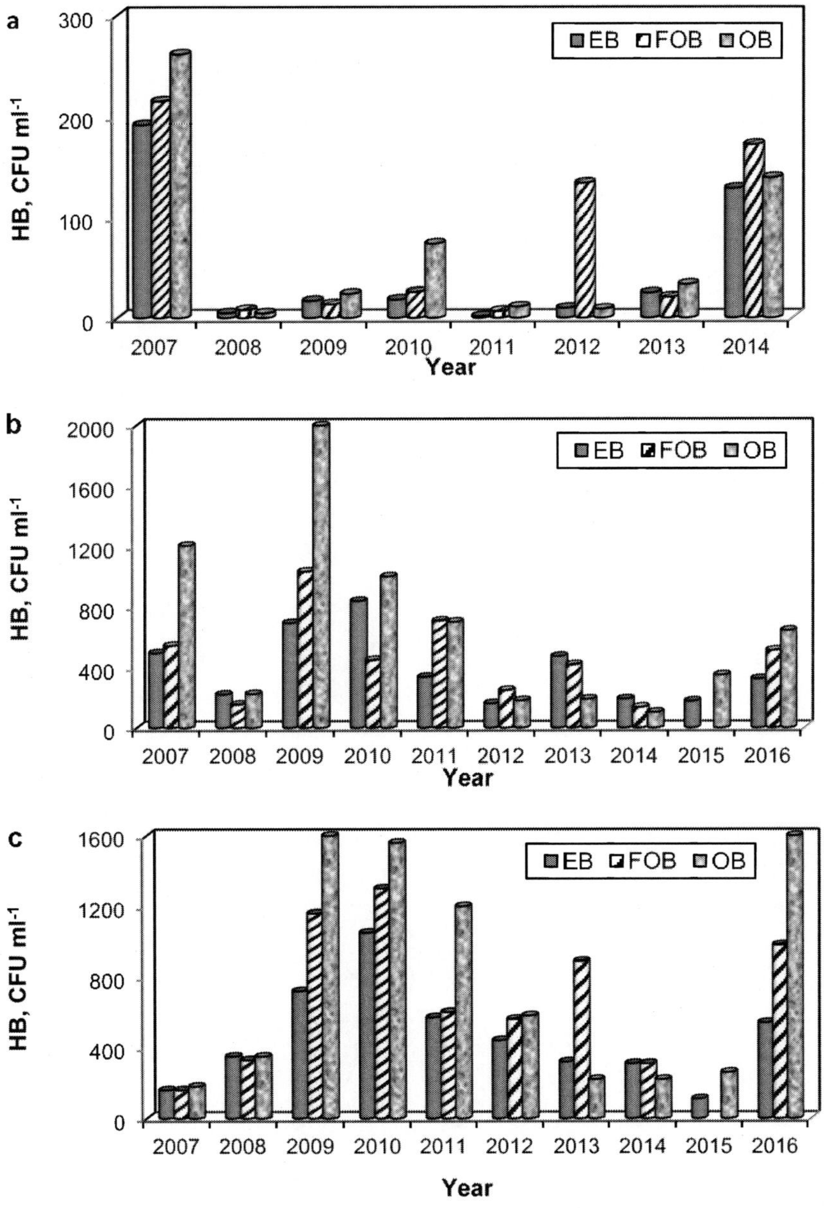

Figure 25. Interannual dynamics of quantitative indicators (weighted average values (0-24 m)) of various ecological-trophic groups of HB (CFU/ml) in Lake Lekshmozero during winter (a) and summer (b), (c - surface layer (0.5 m)) in 2007-2016.

In summer, the number of all HB groups increases by several orders of magnitude, which is a reflection of an increase in the intensity of destruction processes of organic matter newly formed during photosynthesis, as well as a favorable temperature regime. The highest concentrations of eutrophic bacteria were observed in the water of the littoral zone of Lake Lekshmozero, where macrophytes contribute to the DOM content. Content of EB, FOB, OB for the summer periods 2007-2016 (n = 66) varied, respectively, from 40 to 2070 CFU/ml (median 320 CFU/ml), from 10 to 1420 CFU/ml (median 420 CFU/ml), from 20 to 4000

CFU/ml (median 340 CFU/ml). In the ecological-trophic structure of HB in different periods, FOB and OB prevailed; however, in general, the concentrations of various ecological-trophic groups differed insignificantly, maximum 2 times. The data obtained indicate the active self-purification activity of bacterioplankton, as well as the development of the microbial community most adapted to the habitat conditions in the lake ecosystem.

The maximum concentrations of heterotrophic bacterioplankton in the surface water (Figure 25), as the most dynamic part of the lake, were observed in July 2010 and 2016, and were associated with the abnormally hot summer periods observed in the European part of Russia. The ecosystem of the lake reacted to the increase in water temperature by increasing the intensity of production and destruction processes, as evidenced by a sharp increase in the concentrations of all groups of heterotrophic bacterioplankton, which is clearly demonstrated in Figure 25.

However, the weighted mean values of the abundance by the water column in 2010 showed a maximum of only eutrophic bacteria, as the most sensitive group of HB to changes in the content of organic matter, which indicates the stability of the main water column of the lake. At the same time, there were no significant changes in the content of nutrients during abnormally warm periods (Zabelina et al., 2014). These facts indicate that heterotrophic bacterioplankton maintains the equilibrium of lake ecosystems, carrying out a rapid recycling of nutrients, reducing their flux into the bottom layers and bottom sediments of lakes and preventing the lake eutrophication under the current anthropogenic load.

The minimum HB concentrations were recorded during the under-ice periods of 2008, 2009 and 2011, when the abundance of heterotrophic bacterioplankton was 1-30 CFU/ml. An earlier transition of the air temperature through zero leads to an earlier freeze-up and reduced heat loss by water bodies, which led to higher water temperatures values of the bottom layers and weighted mean value in the lake in March, which allowed the microbocenosis of lakes to basically complete by the end of the stratification period the mineralization formed in summer period OM.

As noted above, research work on the KNP lakes from 2006 to 2016 took place in a warm period, which differs from the previous 20-year periods by positive anomalies in all seasons. Periods of warm autumn promote better mixing of deep waters, saturation with oxygen and, as a consequence, more active aerobic mineralization of organic matter thus increasing the self-purification capacity of the lake ecosystem.

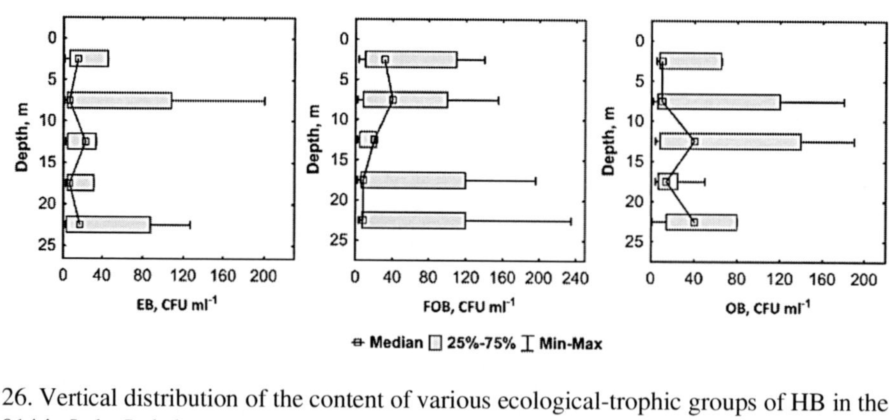

Figure 26. Vertical distribution of the content of various ecological-trophic groups of HB in the period 2008-2014 in Lake Lekshmozero.

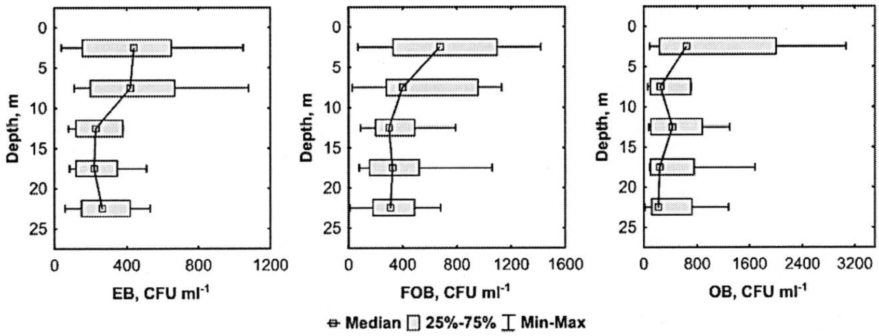

Figure 27. Vertical distribution of the content of various ecological-trophic groups of HB in the summer period 2007-2016 in Lake Lekshmozero.

The data on the vertical distribution of HBs in the lake water column in winter and summer are presented in Figures 26 and 27, respectively.

In the water column, the distribution of HBs was uneven, with a general tendency towards a decrease in the abundance of HB towards the bottom layer in summer. In the summer, higher indicators and the greatest fluctuations in the number of bacterioplankton are characteristic of the surface horizons, as the most dynamic part of the water column. In winter, a more complex dynamics of the vertical distribution was observed, both in the number and structure of the HB.

It is necessary to note the high variability of the content of heterotrophic microorganisms in the water of the lake, which is mainly associated with seasonal processes. The number of eutrophic bacteria observed in Lekshmozero for the period 2007-2016 does not exceed similar indicators obtained in 1997 (Timakova et al., 2000), which indicates the stability of the lake ecosystem. In general, microbiological indicators in the lake are typical for clean waters.

Thus, the study of the quantitative content of various ecological-trophic groups of heterotrophic bacteria in the water of Lake Lekshmozero as a whole, with a large range of fluctuations, showed their low abundance, which is presumably due to the low concentration of organic matter and nutrients.

Heterotrophic bacterioplankton, responsive to changes in natural and anthropogenic factors, maintains the ecosystems of lakes in a stable state. On the one hand, via participating in the cycle of organic matter, creating biomass for zooplankton, recycling biogenic elements, determines the level of primary production, contributes to the self-purification processes and, ultimately, regulates the quality of the aquatic environment. As such, heterotrophic bacteria are one of the key factors determining the sustainability of lake ecosystems.

CONCLUSION

Lake Lekshmozero is located in a zone of excessive moisture and the predominant role in nutrition belongs to atmospheric precipitation, which, with a small catchment area, makes it low-mineralized and, therefore, causes a low buffer capacity of their waters with a predisposition to acidification (decrease in pH value). Water belongs to the bicarbonate class of the calcium group. The active reaction (pH) of the environment of aquatic ecosystems during the periods of research was characterized mainly as neutral. By the nature of the vertical distribution of temperature in summer, Lake Lekshmozero is metathermal.

Research on lakes KNP from 2006 to 2016 was carried out in a warm period, which differs from the previous 20-year periods by positive anomalies in all seasons. During the periods under study, the content of dissolved oxygen for the littoral of Lake Lekshmozero is characterized by homothermy, the content of dissolved oxygen changes insignificantly from the surface to the bottom layer (Vorobieva et al., 2013, 2017)

At the deep zone of Lake Lekshmozero, the vertical profile of the dissolved oxygen content is similar to that of water temperature. In general, Lake Lekshmozero is characterized by a favorable oxygen regime in all seasons. Only in winter, weak stratification in the bottom layers is observed. The content of dissolved oxygen does not fall below 20-48% oxygen saturation in the water column to a depth of 20 meters and decreases to 1.2-10.0% in the bottom layers.

Analysis of the dynamics of nutrient content by seasons showed that the maximum concentration of many mineral forms of nutrients occurs in the winter period. For nitrates, this is seen throughout the water column, and for phosphates only in the photic layer. The maximum of Si, N and P is observed in summer throughout the water column (P), and in the photic layer (N and Si). For ammonium nitrogen, interannual variability is insignificant. The values of the ratio of the main nutrients are optimal for the development of phytoplankton. The limiting component for the development of phytoplankton in the studied lake is mainly phosphorus.

Study of the space-time distribution of iron forms in the water column of Lake Lekshmozero, including the inlet streams, revealed a relationship between the distribution of iron and the color index. The low flow rate and morphometric features of the lake (smooth lowering of depths towards the central part) cause the accumulation of almost all biogenic substances and organic substances coming from the catchment area, including iron, which is reflected in an increase in all forms of iron and color from the surface to the bottom. Predominant form of iron in the winter period of stratification is the dissolved form, and in summer it is the particulate form. Iron concentration in winter, under conditions of oxygen deficiency, is an order of magnitude higher than iron concentration in summer. The main form of iron in the inlet streams is dissolved organic complexes with humic acids.

Lake Lekshmozero is characterized by low concentrations of dissolved organic carbon with a relatively uniform distribution and a tendency toward a decrease in concentrations with depth, which may be due to the processes of photosynthetic activity of phytoplankton and respiration of bacterioplankton. The Corg/Norg ratio for Lake Lekshmozero (5.8–41.3 (average - 25.2) and the predominance of potentially "bioavailable" (low molecular weight) forms of DOC and a low proportion of colloidal forms indicate the autochthonous nature of dissolved organic matter.

To assess the state of the lake's zooplankton communities, the Shannon species diversity index (H) was calculated based on the number of hydrobionts. The zoocenoses of Lake Lekshmozero were characterized by significant species diversity and complex structure. The Shannon Species Diversity Index varied from 2.58 (2017) to 3.36 bit/specimen (2010), depending on the year of research. The index values indicate a relatively stable and balanced state of the lake's zooplankton community. According to the state of zooplankton communities, the lake can be characterized as oligotrophic, although in terms of biomass, the coastal areas in the northern and southern regions of the reservoir can be classified as α-mesotrophic. The most favorable feeding conditions for planktophagous fish have developed here.

In general, the low indices of water saprobity, the average values of the Shannon-Weaver coefficient, and the significant quantitative development of organisms indicate the relative stability and favorable state of zooplanktocenoses in Lake Lekshmozero. The taxonomic composition of zooplankton is typical for cold-water oligotrophic water bodies of the boreal zone. The trophic status of the lake during the study period was determined as oligotrophic with signs of mesotrophy.

The phytoplankton community of Lake Lekshmozero is represented by functional groups, mainly characteristic of well-mixed mesotrophic water bodies with a transparent epilimnion and a low content of nutrients. For the summer phytoplankton in Lake Lekshmozero, the species diversity is typical for the North-West of Russia. The presented taxa are predominantly cosmopolitan forms, most of the species are oligo-β and β-mesosaprobes. The biomass of summer phytoplankton makes it possible to determine the trophic status of the studied lakes as mesotrophic.

Study of the quantitative content of various ecological and trophic groups of heterotrophic bacteria (HB) in the water of Lake Lekshmozero generally showed their low abundance, which is due to the low concentration of organic matter and biogenic elements. It is necessary to note the high variability of the content of heterotrophic microorganisms in the water of Lake Lekshmozero, associated mainly with seasonal processes. In general, microbiological indicators in the lake are typical for clean waters. The number of eutrophic bacteria observed in Lekshmozero for the period 2007-2016 does not exceed similar indicators obtained in 1997 (Timakova et al., 2000), which indicates the stability of the lake ecosystem. In general, microbiological studies of KNP lakes have shown that heterotrophic bacterioplankton is responsive to changes in natural and anthropogenic factors and maintains lake ecosystems in a stable state. Based on the comprehensive seasonal studies of the ecosystem of Lake Lekshmozero in the southern part of the Kenozero National Park, which include the assessment of the spatio-temporal dynamics of hydrological and hydrochemical indicators, qualitative and quantitative characteristics of biological communities, we demonstrate that the current ecological state of Lake Lekshmozero can be assessed as stable. Overall, the status the Lekshmozero Lake ecosystem is mainly controlled by intrabasin processes. Biological communities are the main factor in the lake ecosystem sustainability. Hydrobionts of the studied ecosystem are in active state at the present stage, responding to changes in biotic and abiotic conditions, thereby maintaining the ecosystem of the lake in a state of ecological equilibrium.

ACKNOWLEDGMENTS

This work was supported by the State Task AAAA-A18-118012390200-5. We are grateful to O.A. Pavlova for determining the quantitative and qualitative composition of phytoplankton.

REFERENCES

Adrian, R., Wilhem, S. & Gerten, D. (2006). Life-history traits of lake plankton species may govern their phonological response to climate warming. *Global Change Biology*, 12, 652–661. Doi.org/10.1111/j.1365-2486.2006.01125.x.

Balushkina, E. V., Golubkov, S. M., Ivanova, M. B., Nikulina, V. N., Umnov, A. A. & Umnova, L. P. (1997). Prediction of eutrophication consequences in Lake Lekshmozero based on ecosystem functioning patterns: Response of lake ecosystems to changes in biotic and abiotic conditions. *Proceedings of the Zoological Institute of the Russian Academy of Sciences*. Saint- Petersburg, 272, 228-265. (In Russian).

Barbiero, R. P. (1993). A contribution to the life history of the planktonic cyanophyte, *Gloeotrichia echinulata*. *Archiv fur Hydrobiologie*, 127 (1), 87–100.

Blenckner, T., Adrian, R., Livingstone, D. M., Jennings, E., Weyhenmeyer, G. A., George, D. G., Jankowski, T., Järvinen, M., Aonghusa, C. N., Nõges, T., Straile, D. & Teubner, K. (2007). Large-scale climatic signatures in lakes across Europe: a meta-analysis. *Global Change Biology*, 13, 1314–1325. doi: 10.1111/j.1365-2486.2007.01364.x.

Bogdanovskaya, V. V. (2001). The ratio of nutrients in the bottom and silt water of coastal areas. *Oceanology*, 41(1), 61-68. (In Russian).

Caron, D. A. (1994). Inorganic nutrients, bacteria and the microbial loop. *Microbial Ecology*, 28, 295-298.

Carey, C. C., Ibelings, B. W., Hoffmann, E. P., Hamilton, D. P. & Brookes, J. D. (2012). Eco-physiological adaptations that favour freshwater cyanobacteria in a changing climate. *Water research*, 46 (5), 1394-1407. doi: 10.1016/j.watres.2011.12.016.

Carey, C. C., Weathers, K. C. & Cottingham, K. L. (2008). Gloeotrichia echinulata blooms in an oligotrophic lake: helpful insights from eutrophic lakes. *Journal of Plankton Research*, 30 (8), 893-904. Doi.org/10.1093/plankt/fbn055.

Carey, C. C., Weathers, K. C. & Cottingham, K. L. (2009). Increases in phosphorus at the sediment-water interface may influence the initiation of cyanobacterial blooms in an oligotrophic lake. *Proceedings of the International Association of Theoretical and Applied Limnology*, 30 (8), 1185-1188. doi: 10.1080/03680770.2009.11923908.

Chupakov, A. V., Chupakova, A. A., Moreva, O. Yu., Shirokova, L. S., Zabelina, S. A., Vorobieva, T. Y., Klimov, S. I., Brovko, O. S. & Pokrovsky, O. S. (2017). Allochthonous and autochthonous carbon in deep, organic-rich and organic-poor lakes of the European Russian subarctic. *Boreal Environment Research*, 22, 213–230.

Cotner, J. B. & Biddanda, B. A. (2002). Small players, large role: Microbial influence on biogeochemical processes in pelagic aquatic ecosystems. *Ecosystems*, 5, 105-121. doi: 10. 1007/s10021-001-0059-3.

Del Giorgio, P. A. & Peters, R. H. (1994). Patterns in planktonic P: R ratios in lakes: Influence of lake trophy and dissolved organic carbon. *Limnology Oceanography*, 39, 772-787. Doi.org/10.4319/lo.1994.39.4.0772.

Derevenskaya, O. Yu. (2002). Trophic structure of zooplankton in the lakes of the Middle Volga region. *Biology of internal waters*, 2, 46-50. (In Russian).

Dobrovolsky, V. V. (2003). *Fundamentals of biogeochemistry: Textbook for students of higher educational institutions*. Moscow: Publishing Center "Academy", 400. (In Russian).

Dvoryankin, G. A. (2019). Current status of the ichthyofauna of the lakes of Kenozersky National Park and the influence of natural and anthropogenic factors on it. In collection: *Lakes of Eurasia: problems and solutions*. Materials of the II International Conference: 240-244. (In Russian).

Dvoryankin, G. A. (2016). *Fish of the Kenozersky National Park*. Arkhangelsk: Ministry of Natural Resources and Ecology of the Russian Federation: 100. (In Russian).

Hyenstrand, P., Rydin, E., Gunnerhed, M., Linder, J. & Blomqvist, P. (2001). Response of the cyanobacterium Gloeotrichia echinulata to iron and boron additions - an experiment from Lake Erken. *Freshwater biology*, 46 (6), 735-741. doi: 10.1046/j.1365-2427.2001.00710.x.

Kalinkina, N. M., Timakova, T. M., Lozovik, P. A., Ryzhakov, A. V., Kulikova, T. P., Ryabinkin, A. V. & Chekryzheva, T. A. (2006). Research of transformation of water ecosystems of Kenozersky national Park (Arkhangelsk region) under the influence of agriculture and forestry // *Water resources of the European North of Russia: results and prospects of research.* Materials of the anniversary conference dedicated to the 15th anniversary of the Institute of water problems of the North. Petrozavodsk: 312-322. (In Russian).

Karlsson-Elfgren, I., Rydin, E., Hyenstrand, P. & Pettersson, K. (2003). Recruitment and pelagic growth of Gloeotrichia echinulata (Cyanophyceae) in Lake Erken. *Journal of Phycology*, 39 (6), 1050–1056. doi: 10.1111/j.0022-3646.2003.03-030.x.

Kirchman, D. L. (1994). The uptake of inorganic nutrients by heterotrophic bacteria. *Microbial Ecology*, 28, 255-271.

Kitaev, S. P. (1984). *Ecological foundations of the biological productivity of lakes in different natural zones*. Moscow: Nauka, 207. (In Russian).

Kulikova, T. P. (2008). Zooplankton of the water bodies in Kenozero national park (Arkhangelsk region). *Proceedings of the Karelian scientific center of the Russian Academy of Sciences.* Series: biogeography, issue 12, 103-110. (In Russian).

Levshina, S. I. (2006). Humic acids in the river waters of the Amur region. *Geography and natural resource*, 2, 101-105. (In Russian).

Mosteller, F. & Tukey, J. W. (1977). *Data analysis and regression.* Addison-Wesley, 588.

Pettersson, K., Herlitz, E. & Istvanovics, V. (1993). The role of Gloeotrichia echinulata in the transfer of phosphorus from sediments to water in Lake Erken. *Hydrobiologia*, 253, 123 – 129.

Redfild, A. C., Ketchum, B. H. & Richards, F. A. (1963). The influence of organisms on the composition of sea water // The sea. New York – London, *John Wiley and Sons*, 2, 26-49.

Samarina, V. P. (2008). Spatial and temporal variability of biogenic substances in the Oskol river water. *Water Resources*, 35(3), 346-352. (In Russian).

Shaporenko, S. I. & Shilkrot, G. S. (2006). Stability and variability of hydrochemical parameters of Lake Glubokoe in the conditions of a natural reserve. *Water Resources*, 33 (4), 459-474. (in Russian).

Shirokova, L. S., Bredoire, R., Rols, J. L. & Pokrovsky, O. S. (2017). Moss and Peat Leachate Degradability by Heterotrophic Bacteria: The Fate of Organic Carbon and Trace Metals. *Geomicrobiology journal*, 34 (8), 641-655. Doi.org/10.1080/ 01490451.2015.1111470.

Shirokova, L. S., Pokrovsky, O. S., Viers, J., Klimov, S. I., Moreva, O. Yu., Zabelina, S. A., Vorobieva, T. Y. & Dupré, B. (2010). Diurnal variations of trace elements and heterotrophic bacterioplankton concentration in a small boreal lake of the White Sea basin. *Annales de Limnologie*, 46, 67–75. doi: 10.1051/limn/2010011.

Shirokova, L. S., Pokrovsky, O. S. & Zabelina, S. A. (2010). Dissolved organic carbon and production and destruction processes in the lakes of the southern part of the Arkhangelsk region: Materials of the international symposium *"Ecology of the Arctic and practical territories"*, Arkhangelsk, June 6-10, 143-145. (In Russian).

Shirokova, L., Vorobyeva, T., Zabelina, S., Klimov, S., Moreva, O., Chupakov, A., Makhnovitch, N., Gogolitsyn, V., Sobko, E., Shorina, N., Kokryatskaya, N., Ershova, A. & Pokrovsky, O. (2016). Small boreal lake ecosystem evolution under the influence of natural and anthropogenic factors: results of a multidisciplinary long-term study. *Water*, 8 (8), 316. doi: 10.3390/w8080316.

Shirokova, L. S., Vorobieva, T. Ya., Zabelina, S. A., Klimov, S. I., Moreva, O. Yu., Pokrovsky, O. S., Chupakov, A. V., Shorina, N. V. & Sobko, E. (2012). The current ecological state of the lakes in the south of the Arkhangelsk region: Ecology and geological changes in the environment of the northern regions. *Materials of the All-Russian Conference with international participation.* Arkhangelsk, September 24-28, 253-256. (In Russian).

Tikhomirov, I. A. (1982). *Thermal classification of lakes in the temperate zone: Proceedings of the All Union Symposium on the main problems of freshwater lakes.* Vol. 1. Lakes mode. Vilnius, 1970, 174-185. Tikhomirov, I.A. Thermic of large lakes. L.: 232. (In Russian).

Timakova, T. M. & Kalinkina, N. M. (2000). Ecological problems of Lake Lekshmozero (Arkhangelsk region): *Pomorie in the Barents region. Economy, ecology, culture.* Materials of reports. Arkhangelsk, 228-229. (In Russian).

Tranvik, L. J. (1992). Allochthonous dissolved organic matter as an energy source for pelagic bacteria and the concept of microbial loop. *Hydrobiologia*, 229, 107-114.

Vorobieva, T. Y., Moreva, O. Y., Sobko, E. I., Shirokova, L. S., Zabelina, S. A., Klimov, S. I., Shorina, N. V., Pokrovsky, O. S., Ershova, A. A. & Chupakov, A. V. (2013). Assessment the ecological state of Kenozero National Park lakes (Arkhangelsk oblast) *Izvestia of Samara Scientific Center of the RAS*, 15 (3), 825-83. (In Russian).

Vorobieva, T. Ya., Shirokova, L. S. & Moreva, O. Yu. (2017). Comprehensive studies of the current state of lakes in the Kargopol sector of the Kenozero National Park: Abstracts of the international conference *"Wildlife of the Arctic: conservation of biodiversity, assessment of the state of ecosystems."* Arkhangelsk, October 30 - November 3: 54-56. (In Russian).

Vorobieva, T. Ya., Sobko, E. I. & Zabelina, S. A. (2012). Spatio-temporal dynamics of plankton communities and components of their habitat in the lakes of the southern part of the Kenozero National Park // Kenozerskie Reading - 2011. *Man and Environment: Harmony and Contradictions*: Collection of Materials of the V All-Russian Scientific and Practical Conference, Arkhangelsk, 53-61. (In Russian).

Yang, Y., Colom, W., Pierson, D. & Pettersson, K. (2016a). Water column stability and summer phytoplankton dynamics in a temperate lake (Lake Erken, Sweden). *Inland Waters*, 6(4), 499-508. Doi.org/10.1080/IW-6.4.874.

Yang, Y., Pettersson, K. & Padisak, J. (2016b). Repetitive baselines of phytoplankton succession in an unstably stratified temperate lake (Lake Erken, Sweden): a long-term analysis. *Hydrobiologia*, 764, 211–227. doi: 10.1007/s10750-015-2314-1.

Zabelina, S. A., Moreva, O. Y., Vorobieva, T. Ya. & Shirokova, L. S. (2014). The role of heterotrophic bacterioplankton in maintaining the stability of lake ecosystems in the southern part of the Kenozero National Park // *Biodiagnostics of the state of natural and natural-technogenic systems:* Materials of the XII All-Russian Scientific and Practical Conference with International Participation. Book 1. (Kirov, 2–3 December 2014). Kirov: Publishing house of OOO Vesi, 296-299. (In Russian).

In: Lake Water
Editors: Oleg S. Pokrovsky, Yulia Bespalaya et al.
ISBN: 978-1-53619-275-9

Chapter 4

FISH DEATH IN THE MESOTROPHIC LAKE LEKSHMOZERO: POSSIBLE CAUSES

T. Ya. Vorobyeva*[*], *S. I. Klimov*, *A. V. Chupakov* and *O. S. Pokrovsky
N. Laverov Federal Center for Integrated Arctic Research of the Ural Branch of the Russian Academy of Sciences, Arkhangelsk, Russia

ABSTRACT

In this chapter, the most probable causes in the death of European cisco *Coregonus albula* (Linnaeus, 1758) in mesotrophic Lake Lekshmozero during the biological summer of 1996 are considered. In July-September 1996, there was a sharp increase in the trophicity of the water column, due to the unusual extensive development of phytoplankton. In the second half of summer, after the phosphorus and nitrogen compounds were reduced to the minimum values in the photic aqueous layer, an extensive development of the nitrogen-fixing cyanobacteria *Gloeotrichia echinulata* (J.E. Smith) P. Richter 1894 occurred.

As a result of the eutrophication of the water column, a large amount of ammonium nitrogen was formed during the destruction of organic matter. Under conditions of sustained prolonged stratification, elevated water temperatures, and an increase in the active reaction of the pH of the medium during the phytoplankton bloom, the most likely scenario was the transformation of ammonium ions to non-ionized ammonia (under elevated pH). These hydrochemical conditions could turn out to be unfavorable and even catastrophic for ichthyofauna. We made the assumption that the main reason for the death of the fish was poisoning by ammonia.

Keywords: lake, *Gloeotrichia echinulata*, ammonium, pH and temperature, fish death

[*] Corresponding Author's E-mail: vtais345@gmail.com.

INTRODUCTION

At the end of the XX - beginning of the XXI centuries, the world scientific community came to almost a unanimous opinion that significant climatic changes are taking place on the Earth (Report ... 2017, Summary for Policymakers, 2018, 2019). It was revealed that in temperate and high-latitude these changes are most pronounced, and their consequences are significant for the balance of the natural environment at the planetary level, as well as ensuring environmental safety in the region (Summary for Policymakers, 2018, 2019, McGuire et al., 2006).

In a changing climate, aquatic ecosystems, incl. lakes will be affected by increased temperatures, more intense and longer periods of thermal stratification, altered hydrology (increased frequency, intensity and duration of storms), and increased precipitation and altered nutrient loading (McGuire et al., 2006, Carey et al., 2012a, Reichwaldt, Ghadouani, 2012, Stockwell et al., 2020). In response to changes in many physicochemical parameters of lakes, changes will also occur in phytoplankton populations, because the most favorable conditions may occur for cyanobacteria (Carey et al., 2012a, Reichwaldt, Ghadouani, 2012, Stockwell et al., 2020), and this causes a certain concern. Blooms are known to be a major factor in water quality deterioration, including unpleasant odors and tastes, deoxygenation of bottom waters (hypoxia and anoxia), toxicity, fish mortality, and alteration of the food web (Ibelings et al., 2005, Carey et al., 2012a).

In addition, toxins resulting from cyanobacterial blooms can adversely affect the health of animals (including humans) when used for recreational and drinking purposes. It is believed that cyanobacterial toxins can kill fish and waterfowl (Rodger, 1994, Matsunaga et al., 1999). However, most studies on these events lack direct factual evidence for the role of toxic cyanobacteria (Pietsch et al., 2001). The specific contribution of cyanobacterial toxins to the death of fish or birds in natural waters remains largely unclear (Pietsch et al., 2001). Relatively little is known about the concentration of microcystins and their toxicological effects in the aquatic ecosystem. For example, high concentrations of MC-LR were found in the cyanobacteria *Microcystis aeruginosa*, *Radiocystis fernandoi*, while the lowest concentrations were measured in *Gloeotrichia echinulata* (Vezie et al., 1997, Dos S Vieira et al., 2005, Carey et al., 2007). However, the harmful sublethal effect of microcystins seems to be very likely for many species of biota in the lake. Examples of such effects can be found in a number of publications (Bury, 1995, Baganz, 1998, Ibelings et al., 2005, Hinojosa et al., 2019), which describes the chronic effect of microcystin on fish, leading to decreased growth, impaired reproduction, damage to the nervous systems, etc.

During periods of intense cyanobacterial bloom, fish mortality can be affected by other harmful factors, such as extreme temperature or pH, low oxygen levels, high ammonia concentrations, or a combination of these factors (Paerl et al., 2001, Sun et al., 2012). It was found that the greatest amount of toxins is released from cells during degradation, and at the same time, increased concentrations of ammonium/ammonia are formed, which can persist for several days (Yang et al., 2010). It was revealed that high concentrations of ammonia can be toxic to a number of organisms, including fish (Kopp et al., 2000, Ip et al., 2001, Randall, Tsui, 2002, McKenzie et al., 2003, Sun et al., 2012).

Cyanobacteria possess a number of unique and highly adaptive ecophysiological traits (Lichman et al., 2010, Carey et al., 2012a, Elliott, 2012), such as: the ability to grow at warmer temperatures; buoyancy due to the production of gas bubbles; high affinity for phosphorus and the ability to store it; nitrogen fixation; production of akinet (dormant cells at the bottom), etc. Many of the above physiological properties are possessed by the cyanobacterium *G. echinulata*, which is capable to affect the functioning of ecosystems in various water bodies, and even lakes with low nutrient content (Carey et al., 2008, 2012b). Studies show that in water bodies, even in oligotrophic and mesotrophic ones, where *G. echinulata* is widespread, cases of a sharp increase in trophicity and the occurrence of ecological disasters are possible (Barbiero, Welch, 1992; Karlsson-Elfgren et al., 2003; Carey et al., 2009, 2012b, etc.) associated with sharp outbreaks of mass "blooming" of this cyanobacteria. Modern scientific research is still unable to determine exactly which complex of factors preceding such outbreaks of "bloom" is at play (Pettersson et al., 1993, Hyenstrand et al., 2001, Carey et al., 2008, 2012).

The life cycle of *G. echinulata* is rather complex; it has both benthic and pelagic stages in its development. Most of its life cycle takes place on sediments, either in the form of dormant cells (akinet), or as "germinant" growing on the sediment before migrating into the water column. Germination is an important component of the life cycle of this organism, and its germination and subsequent growth depend on combination of light, temperature, nutrient conditions, and mixing of waters (Barbiero, 1993, Karlsson-Elfgren et al., 2003, 2004). In summer, such a complex of optimal conditions for the germination of dormant cells is created in relatively shallow lakes down to depths of 5-8 m (Hyenstrand et al. 2001, Carey et al., 2008, 2012). In response to increased light and temperature, akinetes germinate and grow on the lake sediment, take up excessive concentrations of phosphorus (P) from the pore water, and then migrate into the water column using gas vesicles (Istvánovics et al., 1993, Pettersson et al., 1993). During the planktonic phase, colonies of *G. echinulata* are predominantly positively buoyant, and therefore the mixing of water by the wind largely controls their distribution (Istvánovics et al., 1993). Via acquiring excess of phosphorus, the colony retains this P stock for subsequent growth in the water column. These reserves allow *G. echinulata* to develop intensively in the P-limited epilimnion unlike the other phytoplankton. In addition, *G. echinulata* is a nitrogen-fixing cyanobacteria, and thus it can use molecular nitrogen for life, when most plankton requires ammonium and nitrogen oxides (Roelofs and Oglesby 1970, Barbiero, Welch, 1992, Pettersson et al., 1993, Karlsson-Elfgren et al., 2003). Thus, the minimal concentrations, and even a complete absence of main nutrients (phosphorus and nitrogen) in water are not a limiting factor for the development of *G. echinulata* (Barbiero and Welch, 1992).

Nutritional restrictions on the development of pelagic *G. echinulata* are also interesting from an ecological point of view, since the question of how the population behaves in the lake is still open. The nutritional requirements of *G. echinulata* are not yet fully understood and may include both inorganic and organic factors, as well as direct and indirect effects (Karlsson-Elfgren et al., 2005). Hyenstrand et al. (2000, 2001) found that the development of *G. echinulata* depends not only on phosphates, inorganic nitrogen, but also on iron and boron. In a photosynthetic cell, there is a need of iron for photosynthetic and respiratory electron transport, nitrate assimilation and nitrogen fixation. The latter process has particularly high demand for iron (Raven et al., 1999, Hyenstrand et al., 2000), and most likely for boron (Cakmak and RoEmheld, 1997, Hyenstrand et al., 2001), which play an important role in protecting heterocysts from oxygen. In freshwater environments, nitrogen fixation is usually accomplished by heterocysts, specialized cells that prevent oxygen from entering the surrounding water and neighboring vegetative cells, which would otherwise inactivate nitrogenase activity (Oliver and Ganf, 2000).

The aim of the present study is to analyze the reasons that caused the death of the European cisco *Coregonus albula* (Linnaeus, 1758) in 1996 during the summer period of anomalously prolonged phytoplankton bloom in the mesotrophic lake Lekshmozero. This raises 2 questions: 1) what factors could provoke a prolonged blooming of phytoplankton communities; 2) what are the possible reasons for the mass death of fish.

FEATURES OF LAKE LEKSHMOZERO

Lake Lekshmozero has a rich diversity of ichthyofauna, which is represented by cisco, vendace, pike, roach, perch, burbot, bream, and ruff (Dvoryankin, 2016). The most valuable fish in the Lake Lekshmozero is the European cisco *Coregonus albula* (Linnaeus, 1758), which serves as an object of lake fishing, including amateur.

Coregonus albula (Linnaeus, 1758), known as the vendace or as the European cisco, is a species of freshwater whitefish in the family *Salmonidae*. It is found in lakes in northern Europe, especially northern Russia, Finland, Sweden, and Estonia, and in some lakes of Norway, the United Kingdom, northern Germany, and Poland. It is also found in diluted brackish water in the Gulfs of Finland and Bothnia, both of which are in the Baltic Sea (Froese and Pauly, 2015). In Lake Lekshmozero (Arkhangelsk region) it is the main commercial fish, represented by a small lake form. In the commercial catches, there are individuals from 10 to 15 cm in length, the age of which varies from 2 to 5 years. Fish aged 1+-3+ years constitute the bulk of the catches. Studies have shown that the vendace population in the Lake Lekshmozero is in a satisfactory state. With annual monitoring and management of the fishery in the Kenozersky National Park, it is possible to guarantee the average annual catches of vendace in Lake Lekshmozero at the level of 20-22 tons (Dvoryankin et al., 2016).

Coregonus albula (Linnaeus, 1758) is a species demanding abiotic environmental factors (Vavilkin et al., 1985, Potapova, 1978). It belongs to the psychrophilic species, so the temperature of the water is of high importance for its life. The temperature conditions of the water body affect the European cisco both indirectly, through the provision of food (qualitative and quantitative development of crustacean plankton), and directly, by changing the level of metabolic processes in fish organisms. The optimum nutrition and development occur at 12-14°C. At the same time, on hot summer days, when the water near the surface warms up to 20°C, the food activity of European cisco decreases, and it migrates to the bottom layers (Potapova, 1978). According to our observations, the main limiting factors in the summer period may be water temperature and oxygen content. The positions of the horizons and the corresponding volume of lake waters with unfavorable conditions are shown in Table 1. Thus, with the accepted critical values in the years under study, during the period of summer stratification in July, the volume of lake waters with unfavorable conditions varied significantly in different years - from 0 to 89%. According to 2017 data, it can be seen that the volume of the zone with optimal conditions can vary greatly during the summer period.

Table 1. Layers (m) of critical values and the share of lake volume (%) with unfavorable conditions for *Coregonus albula* in Lekshmozero in summer by years

Year	T_{water} > 18°C	O_2 < 3.5 mg/L	Volume
2007	8		80
2008	8		80
2009	0		0
2010	9	21	84
2011	6.5		61
2012	0		0
2013	8.5	-	(82)
2014	2.5		31
2015	0		0
2016	11	22	89
2017 (July 2)	0		0
2017 (August 12)	12	22	91

Long-term comprehensive studies of Lake Lekshmozero by the laboratory of freshwater and marine ecosystems of the IEPN FECIAR UrB RAS (2006-2017) at the present stage indicate a stable mesotrophic state of the lake (Vorobyeva et al., 2012, 2013; Chapter 2). Aerobic bacterial destruction prevails in the lake, the processes of ammonification and nitrification are intensive, bacterioplankton is in an active state. In terms of primary production, abundance and biomass of summer phytoplankton, the trophic status of the lake corresponds to the mesotrophic one.

The watershed area of Lake Lekshmozero is 158.5 km^2, the water area is 53.5 km^2, the maximum depth is 29 m, and the average depth is 8 m). The share of the lake area up to 7 m, from 7 to 10 m is 39.7 and 41.1%, respectively. Up to the 7 m isobath there is 73.0% and up to 10 m – 86.2% of the lake volume. Ten small tributaries flow into the Lekshmozero, which do not have a great influence on the formation of the hydrological regime of the lake; one river flows out Lekshma. The flow rate of the lake is low, the coefficient of conditional water exchange is 0.11, and that is, water exchange occurs once every 9.2 years. Water stratification is observed in winter and summer during long periods of stagnation. Due to the openness of the lake and wind mixing, summer water stratification is unstable. By the nature of the vertical distribution of temperature in summer, Lake Lekshmozero is metathermal. The transition of water temperature through 0°C occurs at the end of April beginning of May. In May, the water warms up to 3-10°C, in July up to 15-24°C. The content of dissolved oxygen varies insignificantly from the surface to the bottom, and only in the bottom waters of deep depression can it reach low values.

The active reaction (pH) of the lake environment is mainly characterized as neutral; in the summer months this indicator can reach values up to 88.2 in the epilimnion. The transparency of the water along the Secchi disk is 3-6 m (average 4.5 ± 1.2). According to the content of the main nutrients (phosphorus and nitrogen), the lake also belongs to the mesotrophic type (see Chapter 2.2). The limiting factor in the development of phytoplankton in Lake Lekshmozero is the phosphorus.

Table 2. Average values and statistical characteristics of dissolved (< 0.45 µm) concentrations of boron and iron in different horizons of the water column at the reference points of lakes (2013-2017 years, all seasons)

Interval depth, m	index	N	Mean	Median	Min	Max	25%	75%	SD
Lekshmozero									
(0;10]	B	63	7	7	5	13	6	8	2
(10;20]	B	19	7	7	5	8	6	8	1
(20; bottom)	B	15	8	8	5	13	6	10	2
(0;10]	Fe	29	14	11	3	102	9	13	17
(10;20]	Fe	19	36	18	4	196	12	29	49
(20; bottom)	Fe	15	595	47	13	5203	20	676	1326
Maselgskoe									
(0;5]	B	17	3	2	1	6	2	2	1
(5;15]	B	29	2	2	0.2	5	1	3	1
(15; bottom)	B	20	2	2	0.006	5	0.7	3	1
(0;5]	Fe	17	62	43	18	155	34	81	46
(5;15]	Fe	29	192	144	39	575	113.3	235	124
(15; bottom)	Fe	20	2732	1663	118	11970	392.2	3687	3176
Vilno									
(0;2]	B	12	2	2	0.09	3	1	3	1
(2;4]	B	8	2	2	0.2	3	0.7	3	1
(4; bottom)	B	8	2	2	0.1	5	0.6	2	1
(0;2]	Fe	12	57	53	25	117	35	74	28
(2;4]	Fe	8	59	60	29	119	36	67	29
(4; bottom)	Fe	8	340	76	24	2163	53	139	738

In summer, iron is mainly presented in suspended forms, and is characterized by low concentrations throughout the water column, increasing only in the bottom layers of the deep areas of the lake (Chapter 3). In summer, there is a low bioavailability of iron in the aerobic zone of the epilimnion, largely controlled by the low solubility of $Fe(OH)_3$. The main source of bioavailable dissolved iron can be streams flowing into the Lake Lekshmozero. However, the tributaries do not have a great influence on the formation of the hydrological regime of the lake due to hydrophysical conditions (the lake has a low slope – 0.008), and only during periods of intense rains the significance of the inflow can increase.

The boron concentration in Lake Lekshmozero was higher than in neighboring lakes Maselgskoe and Vilno in all studied periods (Table 2).

The differences in B and Fe concentrations among lakes are due to the influence of lithological (underlying rocks) and hydrological (underground runoff) factors. The minimum boron concentrations were observed in small lakes (Maselgskoe, Vilno), whose drainage basins are drained almost entirely by granite-gneisses. In the watershed of Lake Lekshmozero, the underlying rocks are limestone and dolomites. The underlying rocks are represented by limestones and dolomites of the Moscovian Stage of the middle section of the Carboniferous system. In the southeastern part of Lake Lekshmozero, the boundary between the Podolskian (C2mpd) and Myachkovskian (C2mmč) horizons of the Moscovian stage runs in the transverse direction. ~ 1/3 of the lake in the southeast, unlike all other water bodies, is located on the Myachkovskian Stage. In the process of vital activity of ancient marine organisms, boron accumulated in their skeleton, which led to the enrichment of these elements of limestone rocks.

In addition, in the same area there is an outflow of groundwater (springs), most likely confined to the boundary of the lake basin. In Lekshmozero, boron-enriched carbonate rocks contribute to groundwater supply at the lake bottom. In lakes Maselgskoe and Vilno, no clear contribution of groundwater input to lake is observed, and the carbonate rocks occur at much higher depth. It can be assumed that the presence of carbonate rocks contributes, via gradual diffusion or as a result of intense storms, to the delivery of boron into the water column of Lake Lekshmozero.

Analysis of the Situation

The events of June-October 1996 were described based on information from local residents and the results of hydrochemical and hydrobiological studies carried out by the Northern UGMS and the Institute of Water Problems of the North of the Karelian Scientific Center of the Russian Academy of Sciences (Timakova, Kalinkina, 2000, Kalinkina et al., 2006). June and July 1996 were rainy (the amount of precipitation was 191 and 176% of the norm, respectively), and were accompanied by strong winds, which contributed to the intensive mixing of the lake waters. In August, the weather was mostly calm, dry (the amount of precipitation was 39% of the norm) and hot. In July and August, the locals observed an unusually long and strong "bloom" of the lake. The mass death of *Coregonus albula* began at the end of August and continued throughout September and, in part, October. According to local residents, unusual behavior of the dying fish was observed: it rose to the surface, gasped for air, went deep into the depths, rose to the surface again; the dead fish had their mouths open. According to expert estimates, about 70 tons of *Coregonus albula* died (at least 90% of the population), although after two years the vendace fully recovered its numbers (Dvoryankin, 2016).

At the end of August, toxicological studies were carried out using biotesting methods on *Scenedesmus quadricauda* algae. Undiluted water from the surface layers (5 and 10 m) rendered acute toxicity to algae, but already twice diluting the water removed the toxic effect. Cyanobacteria *Gloeotrichia echinulata* were found in water samples taken on September 17-18, 1996 by the Northern Directorate of the HMS, but no massive "blooming" of the reservoir was observed. It was concluded that there is no relationship between cyanobacterial toxins and fish death in Lake Lekshmozero.

Hydrobiological and hydrochemical studies in October 1996, carried out by the Institute of Water Problems of the North of the Karelian Scientific Center of the Russian Academy of Sciences, Lake Lekshmozero, showed that the trophic status of the ecosystem corresponded to the highly eutrophic type during this period. The content of Chl *a* in the epilimnion was 9 μg/L, the concentration of total phosphorus was 31.5 μgP/L (Timakova, Kalinkina, 2000, Kalinkina et al., 2006). Unfortunately, there is no information in the available literature on the content of mineral forms and total nitrogen in water.

Hydrobiological and hydrochemical studies in the winter and summer of 1997 of the Institute of Water Problems of the North of the Karelian Scientific Center of the Russian

Academy of Sciences showed that the Lekshmozero ecosystem corresponded to a transitional mesotrophic-eutrophic type (Timakova, Kalinkina, 2000, Kalinkina et al., 2006). According to the results of studies, the toxicity of released toxins during the decay of *Gloeotrichia echinulata*, as well as anaerobic conditions associated with anthropogenic eutrophication of the lake, was named as the presumable cause of fish death in 1996 (Timakova and Kalinkina, 2000).

A retrospective analysis of data for 1977-2005 (Timakova, Kalinkina, 2000, Kalinkina et al., 2006), as well as our own studies, showed that over the past 20-25 years, only in a single period of the biological summer of 1996, the Lekshmosero ecosystem was characterized as highly eutrophic. We associate a possible reason for such a sharp increase in status and the ecological catastrophe that has occurred with the vital activity of the cyanobacterium *Gloeotrichia echinulata*, which has the ability to significantly affect the ecosystem via changing the phosphorus cycle and fixing atmospheric nitrogen.

In summer, *G. echinulata* in Lake Lekshmozero is always present in the phytoplankton complex, although in minor amount. The density of *G. echinulata* is usually low (<5 colonies L^{-1}), reaching 15 colonies in L^{-1} in August. Usually, the peak of the mass development of phytoplankton on the lake occurs in mid-to late July, when the biomass reaches 2000-2300 μg/L and pH increases to 8-8.2, as confirmed by our data (Vorobyeva et al., 2012; Chapter 3). Gradually, due to the complete consumption of biogenic elements, primarily the mineral forms of phosphorus and the peak of phytoplankton development falls.

Since the hydrochemical studies were not carried out in a timely manner in the summer of 1996, we can only use the fragmentary data and testimonies of local residents to present a possible scenario for the development of events. In the second half of the summer of 1996, almost immediately after the usual peak in the development of the phytoplankton community in August, when the amount of the main biogenic elements was already exhausted, the mass "blooming" of phytoplankton continued. At the same time, the nitrogen-fixing cyanobacterium *G. echinulata* has already become the dominant species, which entailed the formation of an abnormal amount of organic matter (OM) in the lake. Additionally, in the absence of available forms of nitrogen and phosphorus in the water column, some of the fixed and retained by the cyanobacterium *G. echinulata* P and N can be used by other phytoplankton (Pitois et al. 1997; Nõges et al. 2004; Fey et al. 2010, Carey et al., 2012b), which can also slightly increase the productivity of the lake. The pH of the lake water continues to increase, reaching values of 8.7-9.0, and at the same time, degradation of OM into mineral forms could release large amount of ammonium.

In natural waters, ammonium nitrogen is present mainly in the form of ammonium ions. But under certain conditions, it can turn into non-ionized ammonia (Camargo, Alonso, 2006). Ammonium ions and non-ionized ammonia are linked according to chemical equilibrium:

$$NH_3 + H_2O \underset{\text{pH} > 7, \text{high temperature}}{\overset{\text{pH} < 7, \text{low temperature}}{\rightleftarrows}} NH_4^+ + OH^- \quad (1)$$

The relative concentrations of ammonium ions and ammonia depend mainly on the pH and temperature of the water (Camargo, Alonso, 2006).

$$K_a = ([NH_3][H^+])/[NH_4^+] \quad (2)$$

$$NH_3\ (\%) = 100/(1 + 10^{(pKa-pH)}) \quad (3)$$

where, $pK_a = (0.09108+2729.92)/(273.2+T)$

Already at the water temperature of 17°C and pH of 8.5, the equilibrium shifts towards the dominance of ammonia. As an example, Figure 1 shows the dependence of the concentration of ammonia NH_3 in water on pH at constant water temperature (17°C) and ammonium NH_4^+ concentration (400 μgN/L). An increase in temperature from 20 to 25°C, together with an increase in pH of the medium from 7.5 to 8.5, can cause a 15-fold increase in the amount of NH_3 (Richardson, 1997). Elevated levels of ammonia in the environment are toxic. At the same time, water temperature has only an insignificant effect on the toxicity of ammonia, expressed in the total ammonia content in water (Randall, Tsui, 2002). The threshold concentration of total ammonia ($[NH_3] + [NH_4^+]$) leading to unacceptable biological effects in fresh water published by EPA (1998) is 3.48 mgN/L at pH 6.5 and 0.25 mgN/L at pH 9.0 (Ip et al., 2001).

Non-ionized ammonia is highly toxic to aquatic organisms, especially fish, while ammonium ions are non-toxic or noticeably less toxic. They can be partially toxic to *Nitrosomonas* and *Nitrobacter* bacteria, slowing down the nitrification process. This inhibition can lead to increased accumulation of ammonium ions in the aquatic environment, increasing toxicity to bacteria and aquatic organisms. The presence of ammonia in concentrations above 50 μg/L reduces the ability of fish hemoglobin to bind oxygen. Signs of severe intoxication are excitement, convulsions, the fish rushes in the water and jumps to the surface. The mechanism of toxic action is manifested in the excitation of the central nervous system, damage to the gill epithelium, stimulation of glycolysis and suppression of the Krebs cycle, causing progressive acidolysis and a decrease in the ability of blood to carry oxygen, hemolysis (rupture) of erythrocytes (Randall and Tsui, 2002). Modern criteria for the toxicity of ammonia for aquatic ecosystems are based on research on resting, well-fed fish that are not stressed. However, in natural conditions, the fish can experience stress from various negative environmental influences and be more sensitive to the action of ammonia. Consequently, in stressful situations, even lower concentrations can become fatal (Ip et al., 2001, Randall and Tsui, 2002). Elevated ammonia levels have many harmful effects. Ammonia in the blood damages the walls of the capillaries and other vessels of fish. As a result, a fish once poisoned with ammonia may suddenly die even after 2 weeks or more, without any external signs. Further, poisoning negatively affects immunity, making fish more susceptible to other infections (Thurston at al., 1984, Ip et al., 2001, Randall, Tsui, 2002).

Fish have various strategies to prevent ammonia toxicity. There are some species that are ammonia-tolerant, but most fish species cannot tolerate high levels of ammonia in the environment (Randall, Tsui, 2002). It must be assumed that the cold-water *Coregonus albula*, and especially its juveniles, which are demanding on abiotic environmental factors, are the least protected from the effects of ammonia. First, when the water temperature rises above the optimum (14–17°C), the food activity of vendace decreases (Potapova, 1978), and secondly, when a high concentration of ammonia occurs, the fish also stops feeding (Wicks and Randall, 2002). Such a stressful situation can be aggravated by eutrophication of the reservoir, with a decrease in dissolved oxygen and an increase in the active reaction of the environment. Ultimately, even lower concentrations of ammonia can be detrimental to sensitive fish. Studies carried out on cold-water fish – lake trout *Salvelinus namaycush* and whitefish *Coregonus artedi* have shown that temperature preferences are the main factors determining increased physiological stress or even their death (Ferguson, 1958, Magnuson et al., 1990, Rudstam and Trapp 1987, De Stasio et al., 1996).

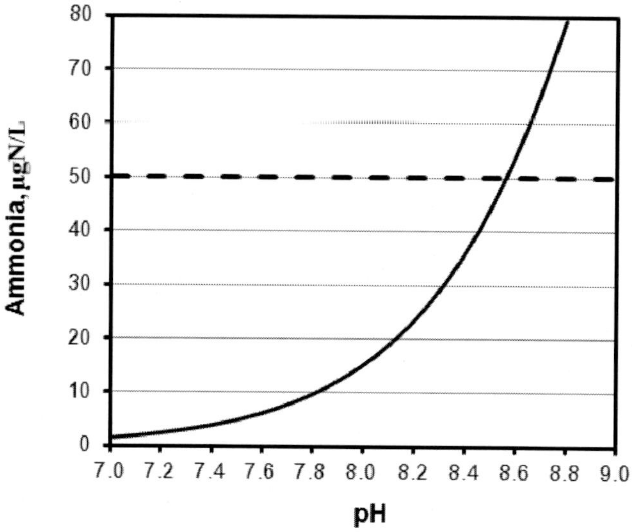

Figure 1. Dependence of the concentration of ammonia NH_3 in water on pH at water temperature (17.0°C) and ammonium NH_4^+ concentration (400 µgN/L).

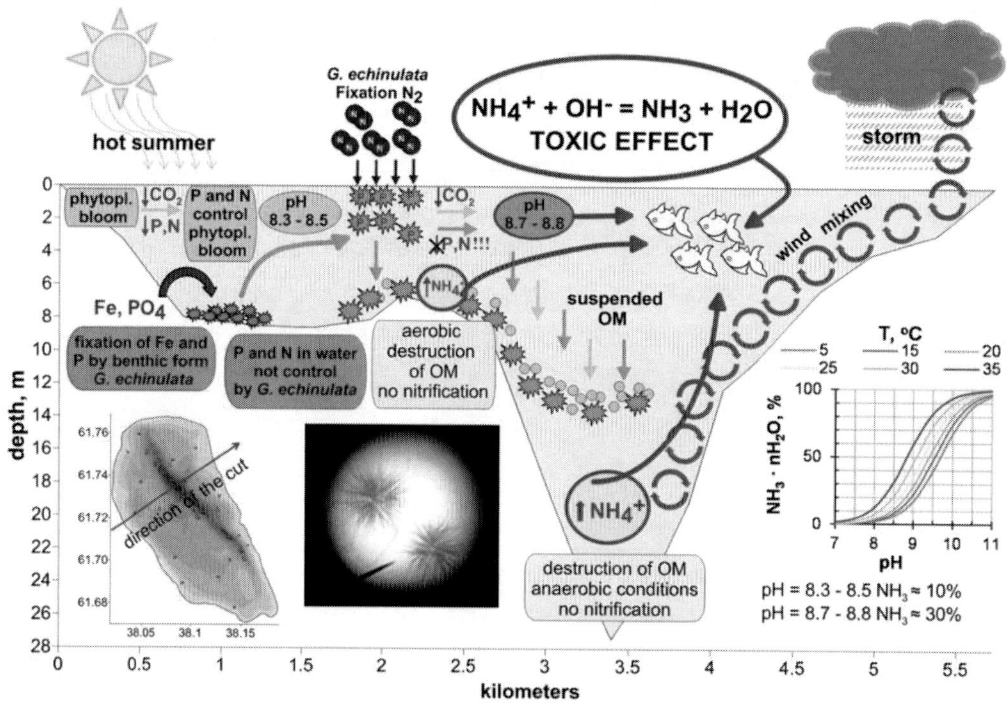

Figure 2. Schematic representation of a possible scenario of the development of events during fish death in Lekshmozero.

Discussion and Conclusion

1. The main factor in increasing the trophic status of the mesotrophic lake Lekshmozero to highly trophic during the biological summer (temperature above 10°C) in 1996 was the long-term mass bloom of phytoplankton communities. Even with a decrease to the minimum values of the main biogenic elements (phosphorus, nitrogen), during the stratification of the water column, the bloom of cyanobacteria continued with the dominance of nitrogen-fixing *Gloeotrichia echinulata*, which has the ability, in a resting state, to accumulate large amount of phosphorus from bottom sediments.
2. Which conditions could have caused the bloom of the cyanobacterium *G. echinulata*, which is so unusual for this mesotrophic lake? Studies show that in water bodies, even in oligotrophic and mesotrophic ones, where *G. echinulata* is widespread, cases of a sharp increase in trophicity are possible, especially during recent trends of climatic changes (Barbio, Welch, 1992, Karlsson-Elfgren et al., 2003, Carey et al., 2008, 2012b, etc.) associated with sharp outbreaks of mass "blooming" of this cyanobacteria. Modern scientific research is still unable to determine exactly what are the factors that precede and accompany such outbreaks of "bloom".
3. Perhaps one of the important reasons is the creation of optimal conditions for the deposition of a large amount of *G. echinulata* in the bottom layers in the dormant stage during the preceding autumn, which was possible under warm weather conditions. This prolonged the period of active growth of cyanobacteria, and then, with a cold snap, its buoyancy is lost, since akinets are usually formed under unfavorable growth conditions caused by changes in temperature, lighting, etc. (Karlan-L et al., 2010, Carey et al., 2012a). A decrease in water temperature in autumn is the main factor in the loss of buoyancy and, as a result, the settling of many cyanobacterial cells into bottom sediments (Visser et al., 2005). A warmer autumn can prolong the development of cyanobacteria and, in this case, with a sharp cooling of the water, a massive precipitation of cells will occur. In September 1995, the actual air temperature, as recorded by the HMS observations, was equal to 9.7°C, which is 1.4°C above the normal. Actual temperature in October, according to observations, was 4.3°C, which is even higher above the norm (+2.1°C). After October 20, there was a cold snap. Thus, in terms of air temperature, September and October 1995 were characterized as very warm months.
4. The next important factor is the percentage of germination and the growth rate of colonies of *G. echinulata*, depending on the interaction of light, temperature and nutrient conditions. Light conditions of Lake Lekshmozero are favorable for the development of phytoplankton, over the depth of 8 m. The share of the lake area up to 9 m depth is 75%, and, accordingly, the majority of the lake bottom is in the photic layer. The water temperature in the summer of 1996 was also favorable for the development of phytoplankton communities.
5. For further migration of colonies of *G. echinulata* into the water column, in addition to the increase in water temperature, which promotes the formation of gas bubbles, increasing their buoyancy, the wind also affects the water mixing. Strong winds were noted in June and July, which contributed to intensive mixing of the lake waters.

6. The low availability of iron in the oxygenic zone of the epilimnion in summer is the most likely limiting factor for the diazotrophic cyanobacterium *G. echinulata*. However, in the summer (June and July) of 1996, due to prolonged rains (the amount of precipitation amounted to 191 and 176% of the norm, respectively), the share of the input of the bioavailable dissolved form of iron with waters of streams flowing into Lake Lekshmozero has increased. During normal periods, the tributaries do not have a great influence on the formation of the hydrological regime of the lake, due to hydrophysical conditions (the lake has a low slope of 0.008), and only during periods of intense rains the significance of the inflow of streams water can increase. Further, the influx of iron and boron into the lake waters from bottom sediments could be facilitated by intense wind mixing of the water column during this period.
7. As a result of massive cyanobacteria bloom, the pH of the lake water could increase to a strongly alkaline (up to 9.0), with the formation of a high concentration of ammonium, and its subsequent transition to non-ionized ammonia.
8. Presumably, in August, *Coregonus albula*, which is demanding on abiotic environmental factors, found itself in a stressful situation. In the epilimnion, the water temperature was higher than the optimum for this fish (above 17°C), which caused a decrease in locomotor activity. The intensity of nutrition of vendace also decreased due to stressful pH of the environment (up to 9.0). Some toxicological effects on fish during this period of cyanobacterial toxin production cannot be ruled out. In the usual summer period, when conditions in the epilimnion become unfavorable, vendace goes to the bottom layers, where the temperature is lower (Potapova, 1978, Dvoryankin, 2016). However, in the summer of 1996, it found itself in a kind of "trap", since a high concentration of ammonium nitrogen accumulated in the bottom horizons with a transition to non-ionized ammonia.

The toxic effect of ammonia continued over the end of August, September and, partially, in October, which led to the mass death of fish. In the summer of 1996 in Lake Lekshmozero, mainly the *Coregonus albula* was affected. At the same time, there were also isolated cases of death of burbot Lota lota (Linnaeus, 1758) and juvenile perch *Perca fluviatilis* Linnaeus, 1758. At the end of October-November, as a result of intense cooling and convective mixing of waters, the hydrochemical conditions returned to their usual state and the trophic status of the lake restored back to the mesotrophic.

To summarize, it has been established that the consequences of the bloom of the cyanobacterium *Gloeotrichia echinulata* for a mesotrophic lake can be very significant. According to the testimony of local residents, the death of fish in Lake Lekshmozero is a rare, but not a unique phenomenon. Similar cases of fish deaths, albeit on a smaller scale, were observed in the last century. This is alarming, especially in view of on-going climate change, which may increase the likelihood of large-scale environmental disasters. Overall, there is a potential threat of high vulnerability of the mesotrophic Lake Lekshmozero and other similar lakes in the context of global climatic changes.

ACKNOWLEDGMENTS

We are grateful to A.P. Novoselov for valuable advice and constructive comments, O.A. Pavlova for determining the quantitative and qualitative composition of phytoplankton, and O. Yu. Moreva for the nutrient analyses. Project number State Task Number AAAA-A18-118012390200-5.

REFERENCES

Baganz, D., Staaks, G. & Steinberg, C. (1998). Impact of the cyanobacteria toxin, microcystin-LR on behaviour of zebrafish, Danio rerio. *Water research*, 32 (3), 948-952. doi: 10.1016/ S0043-1354(97)00207-8.

Barbiero, R. P. (1993). A contribution to the life history of the planktonic cyanophyte, *Gloeotrichia echinulata. Archiv fur Hydrobiologie*, 127 (1), 87–100.

Barbiero, R. P. & Welch, E. B. (1992). Contribution of benthic blue-green algal recruitment to lake population and phosphorus translocation. *Freshwater Biology*, 27, 249–260.

Bury, N. R., Eddy, F. B. & Codd, G. A. (1995). The effects of the cyanobacterium Microcystis aeruginosa the cyanobacterial hepatotoxin microcystin–LR, and ammonia on growth rate and ionic regulation of brown trout. *Journal of Fish Biology*, 46 (6), 1042-1054. doi: 10. 1111/j.1095-8649.1995.tb01408.x.

Cakmak, I. & Romheld, V. (1997). Boron deficiency-induced impairments of cellular functions in plants. *Plant and Soil*, 193, 71-83. doi: 10.1023/A:1004259808322.

Camargo, J. A. & Alonso, A. (2006). Ecological and toxicological effects of inorganic nitrogen pollution in aquatic ecosystems: A global assessment. *Environment International*, 32, 831-849. doi: 10.1016/j. envint.2006.05.002.

Carey, C. C., Ewing, H. A., Cottingham, K. L., Weathers, K. C., Thomas, R. Q. & Haney, J. F. (2012b). Occurrence and toxicity of the cyanobacterium *Gloeotrichia echinulata* in low-nutrient lakes in the northeastern United States. *Aquatic Ecology*, *46*(4), 395-409. doi: 10. 1007/s10452-012-9409-9.

Carey, C. C., Ibelings, B. W., Hoffmann, E. P., Hamilton, D. P. & Brookes, J. D. (2012a). Eco-physiological adaptations that favour freshwater cyanobacteria in a changing climate. *Water research*, 46 (5), 1394-1407. doi: 10.1016/j.watres.2011.12.016.

Carey, C. C., Weathers, K. C. & Cottingham, K. L. (2008). *Gloeotrichia echinulata* blooms in an oligotrophic lake: helpful insights from eutrophic lakes. *Journal of Plankton Research*, 30 (8), 893-904. doi: 10.1093/plankt/fbn055.

Carey, C. C., Haney, J. F. & Cottingham, K. L. (2007). First Report of Microcystin-LR in the Cyanobacterium *Gloeotrichia echinulata*. *Environmental toxicology*, 22(3), 337-339. doi: 10.1002/tox.20245.

Carey, C. C., Weathers, K. C. & Cottingham, K. L. (2009). Increases in phosphorus at the sediment-water interface may influence the initiation of cyanobacterial blooms in an oligotrophic lake. *Proceedings of the International Association of Theoretical and Applied Limnology*, 30 (8), 1185-1188. doi: 10.1080/03680770.2009.11923908.

De Stasio, B. T., Hill, D. K., Kleinhans, J. M., Nibbelink, N. P. & Magnuson, J. J. (1996). Potential effects of global climate change on small north-temperate lakes: Physics, fish, and plankton. *Limnology and Oceanography*, 41 (5), 1136-1149. doi: 10.4319/lo.1996.41.5.1136.

Dvoryankin, G. A. (2016). *Fish of the Kenozersky National Park*. Arkhangelsk: Ministry of Natural Resources and Ecology of the Russian Federation: 100. (In Russian).

Elliott, J. A. (2012). Is the future blue-green? A review of the current model predictions of how climate change could affect pelagic freshwater cyanobacteria. *Water Research*, 46, 1364-1371. doi: 10.1016/j.watrcs.2011.12.018.

Ferguson, R. G. (1958). The preferred temperature of fish and their midsummer distribution in temperate lakes and streams. *Journal of the Fisheries Research board of Canada*, 15 (4), 607-624.

Fey, S. B., Mayer, Z. A., Davis, S. C. & Cottingham, K. L. (2010). Zooplankton grazing of *Gloeotrichia echinulata* and associated life history consequences. *Journal of Plankton Research*, 32, 1337–1347. doi: 10.1093/plankt/fbq052.

Froese, R. & Pauly, D. (2015). "*Coregonus albula*" in FishBase. November 2015 version: 16.

Handbook of fish diseases. (1978). Eds. Osetrov, V. S., Moscow: Kolos: 351. (In Russian).

Hinojosa, M. G., Gutierrez-Praena, D., Prieto, A. I., Guzman-Guillen, R., Jos, A. & Camean, A. M. (2019). Neurotoxicity induced by microcystins and cylindrospermopsin: A review. *Science of The Total Environment*, 668, 547-565. doi: 10.1016/j.scitotenv.2019.02.426.

Hyenstrand, P., Rydin, E. & Gunnerhed, M. (2000). Response of pelagic cyanobacteria to iron additions—enclosure experiments from Lake Erken. *Journal of Plankton Research*, 22 (6), 1113-1126. doi: 10.1093/plankt/22.6.1113.

Hyenstrand, P., Rydin, E., Gunnerhed, M., Linder, J. & Blomqvist, P. (2001). Response of the cyanobacterium Gloeotrichia echinulata to iron and boron additions - an experiment from Lake Erken. *Freshwater Biology*, 46 (6), 735-741. doi: 10.1046/j.1365-2427.2001.00710.x.

Ibelings, B. W., Bruning, K., de Jonge, J., Wolfstein, K., Pires, L. M. D., Postma, J. & Burger, T. (2005). Distribution of microcystins in a lake foodweb: no evidence for biomagnification. *Microbial Ecology*, 49 (4), 487-500. doi: 10.1007/s00248-004-0014-x.

Ip, Y. K., Chew, S. F. & Randall, D. J. (2001). Ammonia toxicity, tolerance, and excretion. *Fish Physiology*, *20*, 109–148. doi: 10.1016/S1546-5098(01)20005-3.

Istvánovics, V., Pettersson, K., Rodrigo, M., Pierson, D., Padisák, J. & Colom, W. (1993). *Gloeotrichia echinulata*, a colonial cyanobacterium with a unique phosphorous uptake and life strategy. *Journal of Plankton Research*, 15, 531–552. doi: 10.1093/plankt/15.5.531.

Kalinkina, N. M., Timakova, T. M., Lozovik, P. A., Ryzhakov, A. V., Kulikova, T. P., Ryabinkin, A. V. & Chekryzheva, T. A. (2006). A study of the transformation of water ecosystems of the Kenozero National Park under the influence of agriculture and forestry. Water Resources of the European North of Russia: Results and Prospects of Research: Proceedings of the anniversary conference dedicated to the 15th anniversary of the IVPS. *Petrozavodsk: KarRC RAS*, 312–323. (In Russian).

Kaplan-Levy, R. N., Hadas, O., Summers, M. L., Rucker, J. & Sukenik, A. (2010). Akinetes: dormant cells of cyanobacteria. In: Lubzens, E., Cerda, J., Clark, M.S. (Eds.), *Dormancy and Resistance in Harsh Environments*. Springer, Heidelberg, Vol. 21, 5-27. doi: 10.1007/978-3-642-12422-8_2.

Karlsson-Elfgren, I., Rengefors, K. & Gustafsson, S. (2004). Factors regulating recruitment from the sediment to the water column in the bloom-forming cyanobacterium *Gloeotrichia echinulata*. *Freshwater Biology*, 49 (3), 265-273. doi: 10.1111/j.1365-2427.2004.01182.x.

Karlsson-Elfgren, I., Hyenstrand, P. & Riydin, E. (2005). Pelagic growth and colony division of *Gloeotrichia echinulata* in Lake Erken. *Journal of Plankton Research*, 27 (2), 145-151. doi: 10.1093/plankt/fbh165.

Karlsson-Elfgren, I., Rydin, E., Hyenstrand, P. & Pettersson, K. (2003). Recruitment and pelagic growth of Gloeotrichia echinulata (Cyanophyceae) in Lake Erken. *Journal of Phycology*, 39 (6), 1050–1056. doi: 10.1111/j.0022-3646.2003.03-030.x.

Kopp, R. & Heteša, J. (2000). Changes of haematological indices of juvenile carp (Cyprinus carpio L.) under the influence of natural populations of cyanobacterial water blooms. *Acta Veterinaria Brno*, 69 (2), 131–137. doi: 10.2754/avb200069020131.

Litchman, E., de Tezanos Pinto, P., Klausmeier, C. A., Thomas, M. K. & Yoshiyama, K. (2010). Linking traits to species diversity and community structure in phytoplankton. *Hydrobiologia*, *653*, 15-28. doi: 10.1007/s10750-010-0341-5.

Magnuson, J. J., Meisner, J. D. & Hill, D. R. (1990). Potential changes in the thermal habitat of Great Lakes fish after global climate warming. *Transactios of the American Fisheries Society*, 119 (2), 254-264. doi: 10.1577/1548-8659(1990)119<0254:PCITTH>2.3.CO; 2.

Matsunaga, H., Harada, K. I., Senma, M., Ito, Y., Yasuda, N., Ushida, S. & Kimura, Y. (1999). Possible cause of unnatural mass death of wild birds in a pond in Nishinomiya, Japan: sudden appearance of toxic cyanobacteria. *Journal of Natural Toxins*, 7(2), 81–84. doi: 10. 1002/(sici)/1522-7189(199903/04)7:2<81.

McGuire, A. D., Chapin, F. S., Walsh, J. E. & Wirth, C. (2006). Integrated regional changes in arctic climate feedbacks: implications for the global climate system. *Annual Review Environment and Resources*, 31, 61–91. doi: 10.1146/ annurev.energy.31.020105.100253.

McKenzie, D. J., Shingles, A. & Taylor, E. W. (2003). Sub-lethal plasma ammonia accumulation and the exercise performance of salmonids. *Comparative Biochemistry and Physiology Part A: Molecular & Integrative Physiology*, 135 (4), 515-526. doi: 10.1016/ S1095-6433(03)00116-8.

Nõges, T., Tonno, I., Laugaste, R., Loigu, E. & Skakalski, B. (2004). The impact of changes in nutrient loading on cyanobacterial dominance in Lake Peipsi (Estonia/Russia). *Archiv fur Hydrobiologie*, 160 (2), 261–279. doi: 10.1127/0003-9136/2004/0160-0261.

Oliver, R. L. & Ganf, G. G. (2000). Freshwater blooms. In: Whitton, B.A., Potts, M. (Eds.), *The Ecology of Cyanobacteria*. Kluwer Academic Publishers, Dordrecht: 149-194.

Paerl, H. W. & Paul, V. J. (2012). Climate change: Links to global expansion of harmful cyanobacteria. *Water Research*, 46 (5), 1349-1363. doi: 10.1016/j.watres.2011.08.002.

Paerl, H. W., Fulton, R. S., Moisander, P. H. & Dyble, J. (2001). Harmful freshwater algal blooms, with an emphasis on cyanobacteria. *The Scientific World Journal*, 1, 76–113. doi: 10.1100/tsw.2001.16.

Pettersson, K., Herlitz, E. & Istvanovics, V. (1993). The role of Gloeotrichia echinulata in the transfer of phosphorus from sediments to water in Lake Erken. *Hydrobiologia*, 253, 123 – 129.

Pietsch, C., Wiegand, C., Ame, M. V., Nicklisch, A., Wunderlin, D. & Pflugmacher, S. (2001). The effects of a cyanobacterial crude extract on different aquatic organisms: evidence for cyanobacterial toxin modulating factors. *Environmental Toxicology*, 16 (6), 535–542. doi: 10.1002/tox.10014 11769252.

Pitois, S. G., Jackson, M. H. & Wood, B. J. B. (1997). Summer bloom of Gloeotrichia echinulata and Aphanizomenon flos-aquae and phosphorus levels in Antermony Loch, central Scotland. *International Journal of Environmental Health Research*, 7 (2), 131–140. doi: 10.1080/09603129773922.

Potapova, O. I. (1978). *Big vendace*. Moscow: Agropromizdat., 114 p. (In Russian).

Randall, D. J. & Tsui, T. K. N. (2002). Ammonia toxicity in fish. *Marine Pollution Bulletin*, *45*, 17–23. doi: 10.1016/S0025-326X(02)00227-8.

Raven, J. A., Evans, M. C. W. & Korb, R. E. (1999). The role of trace metals in photosynthetic electron transport in O_2 - evolving organisms. *Photosynthesis Research*, 60, 111–149. doi: 10.1023/a:1006282714942.

Reichwaldt, E. S. & Ghadouani, A. (2012). Effects of rainfall patterns on toxic cyanobacterial blooms in a changing climate: Between simplistic scenarios and complex dynamics. *Water Research*, 46, 1372-1393. doi: 10.1016/j.watres.2011.11.052.

Report on climate risks in the Russian Federation. (2017). Saint-Petersburg: 106. (In Russian).

Richardson, J. (1997). Acute ammonia toxicity for eight New Zealand indigenous freshwater species. *New Zealand Journal of Marine and Freshwater Research*, 31, 185–190. doi: 10. 1080/00288330.1997.9516757.

Rodger, H. D., Turnbull, T., Edwards, C. & Codd, G. A. (1994). Cyanobacterial (blue-green-algal) bloom associated pathology in brown trout, Salmo trutta L., in Loch Leven, Scotland. *Journal of Fish Diseases*, 17 (2), 177–181. doi: 10.1111/j.1365-2761.1994.tb00211.x.

Roelofs, T. D. & Oglesby, R. T. (1970). Ecological observations on planktonic cyanophyte Gleotrichia echinulata. *Limnology and Oceanography*, 15, 224–229. doi: 10.4319/lo.1970. 15.2.0224.

Rudstam, L. G. & Trapp, T. W. (1987). Diel patterns of behavior and habitat utilization of cisco *Coregonus artedi* two Wisconsin USA lakes. *Transactions of the Wisconsin Academy of Sciences Arts and Letters*, 75, 70-78.

Stockwell, J. D., Doubek, J. P., Adrian, R., Anneville, O., Carey, C. C., Carvalho, L., De Senerpont Domis, L. N., Dur, G., Frassl, M. A., Grossart, H. P., Ibelings, B. W., Lajeunesse, M. J., Lewandowska, A. M., Llames, M. E., Matsuzak, S. I. S., Nodine, E. R., Nõges, P., Patil, V. P., Pomati, F., Rinke, K., Rudstam, L. G., Rusak, J. A., Salmaso, N., Seltmann, C. T., Straile, D., Thackeray, S. J., Thiery, W., Urrutia-Cordero, P., Venail, P., Verburg, P., Woolway, R. I., Zohary, T., Andersen, M. R., Bhattacharya, R., Hejzlar, J., Janatian, N., Kpodonu, A. T. N. K., Williamson, T. J. & Wilson, H. L. (2020). Storm impacts on phytoplankton community dynamics in lakes. *Global Change Biology*, 26 (5), 2756-2784. doi: 10.1111/gcb.15033.

Summary for Policymakers/Nerilie Abram, Carolina Adler, Nathaniel L. Bindoff et al. // *Special Report on the Ocean and Cryosphere in a Changing Climate: special report IPCC* /eds. Pörtner, H.-O., Roberts, D. C., Masson-Delmotte, V., Zhai, P., Tignor, M., Poloczanska, E., Mintenbeck, K., Alegría, A., Nicolai, M., Okem, A., Petzold, J., Rama, B., Weyer, N. M. 2019. Monaco, 36. https://www.ipcc.ch/srocc/chapter/summary-for-policymakers.

Summary for Policymakers. In: *Global warming of 1.5°C. An IPCC Special Report on the impacts of global warming of 1.5°C above pre-industrial levels and related global greenhouse gas emission pathways, in the context of strengthening the global response to the threat of climate change, sustainable development, and efforts to eradicate poverty/* Masson-Delmotte, V., Zhai, P., Pörtner, H.O., Roberts, D., Skea, J., Shukla, P.R., Pirani, A., Moufouma-Okia, W., Péan, C., Pidcock, R., Connors, S., Matthews, J.B.R., Chen, Y.,

Zhou, X., Gomis, M. I., Lonnoy, E., Maycock, T., Tignor, M., Waterfield, T. 2018. World Meteorological Organization. IPCC. Geneva, Switzerland, 32.

Sun, H. J., Lu, K., Minter, E. J. A., Chen, Y. F., Yang, Z. & Montagnes, D. J. S. (2012). Combined effects of ammonia and microcystin on survival, growth, antioxidant responses, and lipid peroxidation of bighead carp Hypophthalmythys nobilis larvae. *Journal of Hazardous Materials*, 221, 213-219. doi: 10.1016/j.jhazmat.2012.04.036.

Thurston, R. V., Russo, R. C., Luedtke, R. J., Smith, C. E., Meyn, E. L., Chakaoumakos, C., Wang, K. C. & Brown, C. J. D. (1984). Chronic toxicity of ammonia to rainbow trout. *Transactions of the American fisheries society*: Cleveland Rubber Company Press; 113: 56–73. doi: 10.1577/1548-8659(1984)113<56:CTOATR>2.0.CO;2.

Timakova, T. M. & Kalinkina, N. M. (2000). Environmental problems of the Lake Kenozero (Arkhangelsk region). *Pomerania in the Barents region. Economics, ecology, culture.* Materials of reports. Arkhangelsk: 228-229. (In Russian).

Vavilkin, A. S., Ivanov, A. P. & Kuranova, I. I. (1985). *Basics of ichthyology and fish farming.* Moscow: Agropromizdat, 185. (In Russian).

Vezie, C., Brient, L., Sivonen, K., Bertru, G, Lefeuvre, J. C. & Salkinoja-Salonen, M. (1997). Occurrence of microcystin-containing cyanobacterial blooms in freshwaters of Brittany (France). *Archiv fur Hydrobiologie*, 139 (3), 401-413. doi: 10.1127/archiv-hydrobiol/ 139/1997/401.

Vieira, J. M. D., Azevedo, M. T. D., Azevedo, S. M. F. D., Honda, R. Y. & Correa, B. (2005). Toxic cyanobacteria and microcystin concentrations in a public water supply reservoir in the Brazilian Amazonia region. *Toxicon.*, 45, 901–909. doi: 10.1016/j.toxicon.2005. 02.008.

Visser, P. M., Ibelings, B. W., Mur, L. R. & Walsby, A. E. (2005). The ecophysiology of the harmful cyanobacterium Microcystis. In: Huisman, J., Matthijs, H.C.P., Visser, P.M. *Harmful Cyanobacteria*. Springer, Dordrecht, 109-142.

Vorobyeva, T. Ya., Ershova, A. A., Moreva, O. Yu., Chupakov, A. V., Zabelina, S. A. & Klimov, S. I. (2012). Microbiological and hydrochemical aspects of nitrogen cycle in the lakes of Kenozero National park. *Arctic Environmental Research*, 4, 13-21. (In Russian).

Vorobyeva, T. Y., Moreva, O. Y., Sobko, E. I., Shirokova, L. S., Zabelina, S. A., Klimov, S. I., Shorina, N. V., Pokrovsky, O. S., Ershova, A. A. & Chupakov, A. V. (2013). Assessment the ecological state of Kenozero National Park lakes (Arkhangelsk oblast) *Izvestia of Samara Scientific Center of the RAS*, 15 (3), 825-831. (In Russian).

Wicks, B. J. & Randall, D. J. (2002). The effect of feeding and fasting on ammonia toxicity in juvenile rainbow trout, Oncorhynchus mykiss. *Aquatic Toxicology*, 59 (1), 71-82. doi: 10.1016/S0166-445X(01)00237-5.

Yang, W., Xiang, F. H., Liang, L. G. & Yang, Z. (2010). Toxicity of ammonia and its effects on oxidative stress mechanisms of juvenile crucian carp (Carassius auratus). *Journal of Freshwater Ecology*, *25* (2), 297-302. doi: 10.1080/02705060.2010.9665080.

Zabelina, S. A., Moreva, O. Yu., Vorobyeva, T. Ya. & Shirokova, L. S. (2014). The role of heterotrophic bacterioplankton in maintaining the stability of lake ecosystems in the southern part of the Kenozero National Park // *Biodiagnostics of the state of natural and natural-technogenic systems:* Materials of the XII All-Russian Scientific and Practical Conference with International Participation. Book 1. (Kirov, 2–3 December 2014). Kirov: Publishing house of OOO Vesi: 296-299. (In Russian).

In: Lake Water
Editors: Oleg S. Pokrovsky, Yulia Bespalaya et al.
ISBN: 978-1-53619-275-9

Chapter 5

OLIGOTROPHIC LAKE NAGLIMOZERO DURING SEASONAL STRATIFICATION

S. I. Klimov, T. Ya. Vorobyeva**, *S. A. Zabelina, E. I. Sobko, O. Yu. Moreva, A. A. Chupakova, L. S. Shirokova, O. S. Pokrovsky, A. V. Chupakov and N. V. Neverova

N. Laverov Federal Center for Integrated Arctic Research of the Ural Branch of the Russian Academy of Sciences, Arkhangelsk, Russia

ABSTRACT

The morphometric features and quantitative hydrochemical and hydrobiological parameters of the Naglimozero Lake observed during summer and winter stratification demonstrate a vulnerability of this limnic ecosystem even to small climatic fluctuations. This is primarily associated with a small buffer capacity of the water column. According to nutrient concentrations and hydrobiological indicators (quantity and species diversity of zoo-, phytoplankton), the Naglimozero Lake belongs to the oligotrophic type. The N_{tot}/P_{tot} ratios of the water column (13.4-33.1) in the summer and winter are optimal for the development of aquatic organisms and indicate the limiting role of phosphorus.

Due to the sufficient oxygen saturation of the lake water column, the oxidized forms of nitrogen prevailed. A decrease in the concentrations of PO_4^{3-}, NH_4^+ and Si in the summer indicates strong involvement of mineral forms of nutrients in the biological cycle. According to the concentration of mineral, organic and total nitrogen and phosphorus, the lake can be characterized as oligotrophic.

The indices of zooplankton diversity indicate a uniform distribution of species in the community and its high stability. On average, the level of development of zooplankton in Naglimozero was relatively low. The summer phytoplankton community was represented by species which are characteristic of clear waters with high transparency, and low content of nutrients, typical for oligotrophic and mesotrophic water bodies of North-West of Russia. The presented taxa are predominantly cosmopolitan forms; most of the species are oligo-, oligo-β and β-mesosaprobes. The biomass of summer phytoplankton is consistent with the trophic status of the studied lake as oligotrophic. The maximum of heterotrophic

* Corresponding Author's E-mail: vtais345@gmail.com.

bacterioplankton content was observed in the surface layers of the lake during the abnormally hot July of 2010, when the upper layer of the lake warmed up to 24-25°C.

Keywords: oligotrophic lake, nutrients, phytoplankton, zooplankton, bacterioplankton, trophic status

Introduction

Lakes of Kenozersky National Park with a small catchment area can serve as efficient indicators of the climate change. Indeed, the ecosystems of small lakes most clearly reflect the zonal, regional and local conditions together with global anthropogenic processes (Moiseenko et al., 2006). Therefore, it is essential to assess the changes in lacustrine ecosystems in relatively undisturbed territories under the influence of anthropogenic and climatic factors.

Lake Naglimozero belongs to small reservoirs of the Kenozersky National Park, located in the boreal zone in the north-west of the Russian Federation (Arkhangelsk Region). It is situated at the beginning of watershed, on the border of the Baltic crystalline shield and the Russian platform. This site represents a watershed divide between the basins of the White Sea and the Baltic Sea. Lake Naglimozero is suitable for studying numerous processes occurring in aquatic ecosystems. This lake is a heterogeneous limnic system with different biotopes. Being located at the beginning of the catchment, the lake is dominated by intra-basin processes with a minimum input of allochthonous organic matter. The sampling procedure and analytical methods employed in this study are described in chapter 1.

Hydrographic and Morphometric Characteristics

The least studied reservoir in the southern part of the Kenozersky Park is the Naglimozero Lake due to its complex morphometry and location in a fairly inaccessible area. The lake is a popular destination for tourists and fishing activity. In the summer of 2012, a detailed bathymetric survey of the lake was carried out for the first time (more than 26,500 depth measurements). This made it possible to construct a detailed bathymetric map (Figure 1a) and assess the main morphometric and hydrographic characteristics of the lake. In the summer of 2017, additional hydrological stations were completed in the northern part of the lake (Figure 1b).

The morphometric and hydrographic characteristics of the lake are presented in Table 1.

According to the classification of S.V. Grigoriev (Grigoriev, 1958) in terms of the specific catchment area (the ratio of the catchment area to the water area), lake Naglimozero belongs to the subgroup of lakes with a very small specific catchment area (< 5). According to the conditional water exchange indicator (the ratio of the volume of water entering the lake to the volume lakes) the lake has a small conditional water exchange (< 0.5). The full conditional water exchange of Naglimozero is 4.9 years.

Table 1. The main morphometric and landscape characteristics of the Naglimozero Lake

Specifications	Naglimozero
Height above sea level, m	163
Area, km^2	5.940
Shore line, km	34.66
Volume, km^3	0.03805
Average depth, m	6.4
Maximum depth, m	18.0
Length, km	5.36
The greatest width, km	2.66
Watershed area, km^2	27.58
The specific catchment area	4.64
Average annual runoff from the catchment, km^3	0.00777
Rate of replacement, year $^{-1}$	0.20
Water residence time in the lake, year	4.9

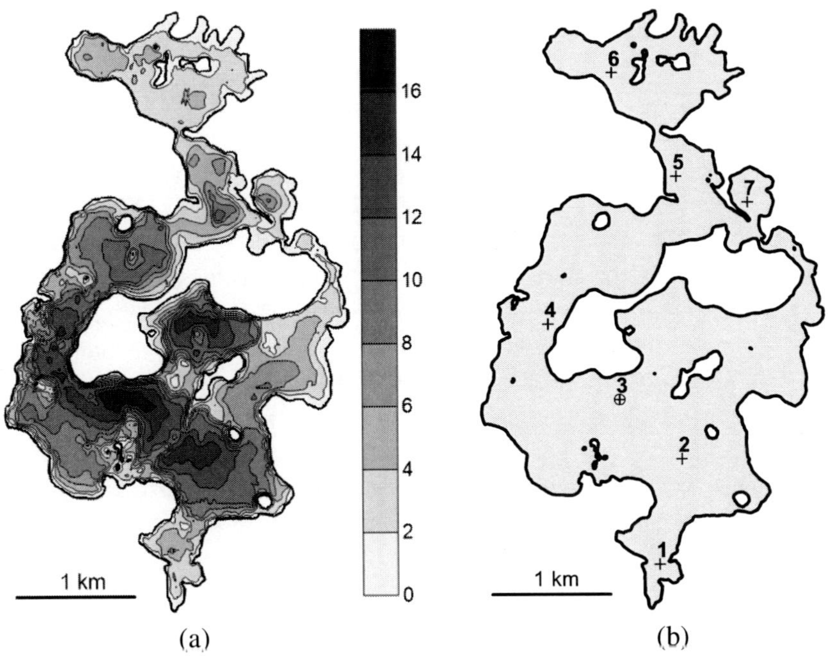

Figure 1. A bathymetric map of the lake Naglimozero and the scheme of stations carried out in 2017. A deep station is marked with a circle with a cross.

Lake Naglimozero is located 3 km northwest of Lake Lekshmozero. The lake has a rounded-lobed shape with a complex winding coastline with numerous capes and bays of varying degrees of openness. The structure of the lake basin is also complex, with a large number of small islands. The largest island, Bolshoi, is located in the central part of the lake. There are 21 islands with a total area of 1.17 km^2. The bottom relief is uneven, shallows alternate with sharp drops in depth. The deepest southern part of the lake had a maximum depth of 18.0 m. In the shallow part, the bottom is sandy, stony in places; at a depth of more than 5 m, brown and dark gray clayey silt prevails. The shores of the lake are steep, sloping in places, covered with forest. The Ustyanka River flows from the lake and continues into Lake Onega.

HYDROLOGICAL CHARACTERISTICS

Water Temperature

According to the vertical distribution of water temperature in the summer period, the lake belongs to the metathermal type (Tikhomirov I.A). In summer, at the deep-water station, the homothermy is noted in the upper 3-4 m layer. A sharply expressed thermocline is 3 to 4 m thick with a maximum gradient in the jump layer from 3.6 to 6.3°C/m (Figure 2). Further, there is a gradual decrease in temperature, reaching 6.5-7.3°C at the bottom. In winter, in the layer under the ice, there are small vertical temperature gradients of 0.5-0.7°C/m and a slight increase in temperature downwards, reaching 3.7 to 4.2°C at the bottom.

Water Specific Conductivity

In summer, at the deep station, the specific conductivity (SC) is practically uniform throughout the entire layer and remains very low (36-37 μS/cm), which corresponds to a mineralization of about 20-21 mg/L (Figure 2). In winter, the specific conductivity is slightly higher. The lowest values 40.0-40.5 μS/cm were observed at the depth of 8 to 10 m, and slightly increased towards the bottom to 44-46 μS/cm and towards the surface (42-44 μS/cm).

In the northern part of the lake, at stations 1-4, uniform values of SC from the surface to the bottom were observed (36-37 μS/cm) as shown in Figure 3. In the bay of the southern part of the lake at station 6, the specific conductivity was the lowest and amounted to 32 μS/cm. This can be explained by an inflow of low-mineralized waters from adjacent bogs. In the bottom layer, the specific conductivity increased to 46 μS/cm, while the absence of oxygen was observed from a depth of 3.5 m. The maximum specific conductivity was observed at station 7 in a relatively closed bay (38 μS/cm at the surface and 84 μS/cm at the bottom at a depth of 8.5 m). Below 2.5 m, a chemocline was observed with maximum gradients of 14 μS/cm/m in a layer of 4-6 m. The water temperature at the bottom was 4.4°C. An oxygen-free zone was noted below 4.5 m.

Overall, in most parts of the Naglimozero Lake, low values of electrical conductivity and its homogeneous vertical distribution are noted. In the northern enclosed bay, a chemocline is formed in the bottom layer. In the bay where point 7 is located, meromictic conditions have been noted. They can be related to protection from the main lake basin by a narrow forested cape, and shallow water up to 1.5 m deep in front of the bay entrance, thus producing quite stagnant zone around point 7.

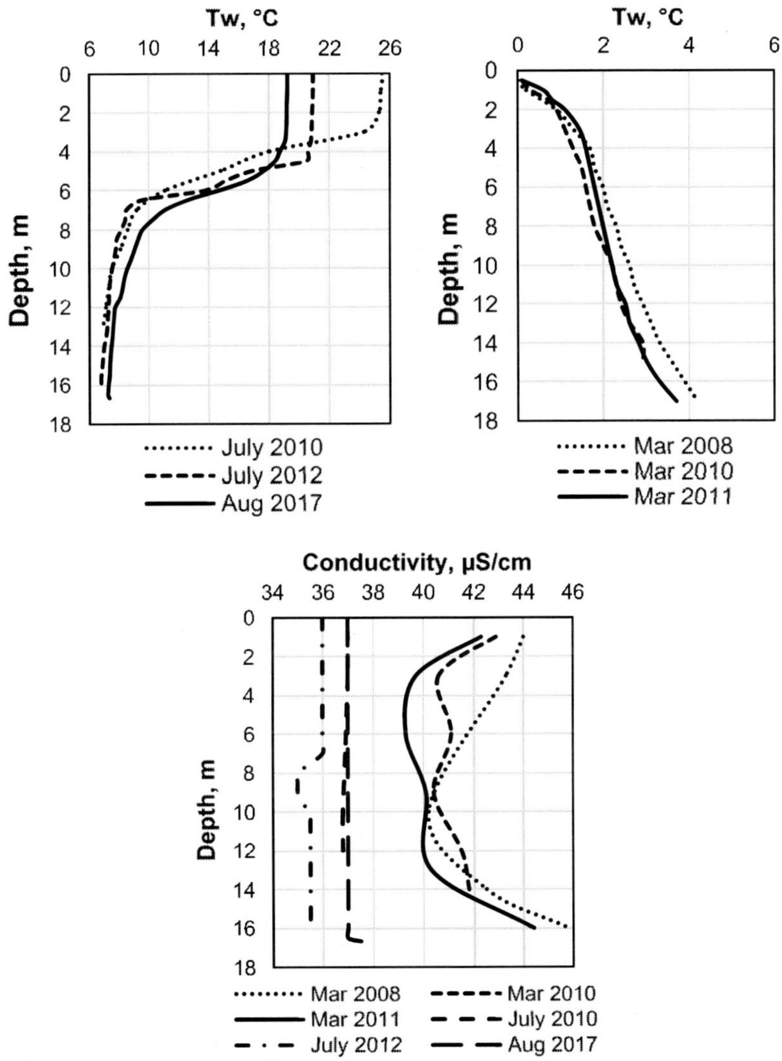

Figure 2. Vertical distribution of temperature and electrical conductivity at a deep-water station of the Naglimozero Lake in winter and summer.

The low specific conductivity of the waters is due to the fact that atmospheric precipitation plays a predominant role in nutrition, while the influences of the drainage basin and underlying rocks (granite moraine) are minimal. Low-mineralized water determines a low buffer capacity of the lake with respect to possible acidification.

pH

The pH value is characterized mainly as neutral, and ranges from 6.25 to 8.24 (Figure 4). The highest pH values in Naglimozero (8.09 to 8.24) were noted in the surface horizons during the peak of the phytoplankton development.

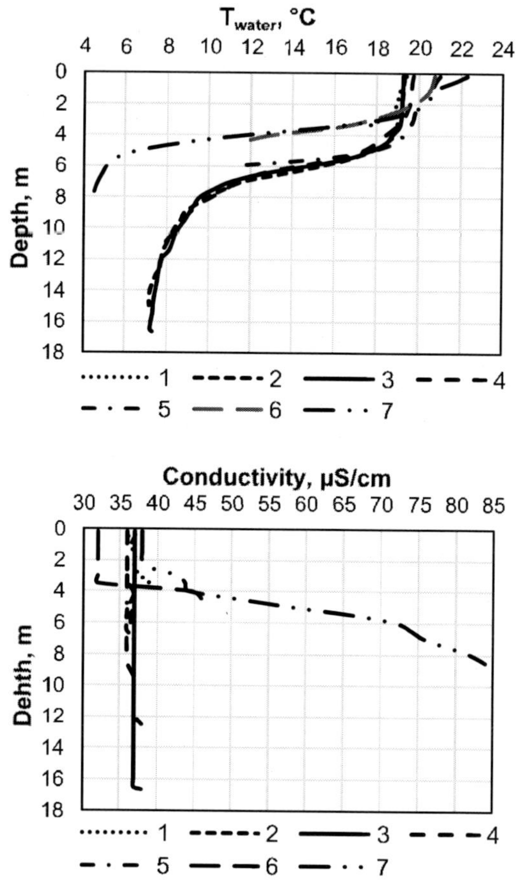

Figure 3. Vertical distribution of temperature and specific conductivity in the Naglimozero Lake in August 2017. The numbers correspond to stations and their location is shown in Figure 1.

Nutrients

The nutrients concentrations were measured during the stratification period of winter (2008, 2010, 2011) and summer (2010, 2012). One of the factors affecting the content of nutrients and their transformation between different forms is the content of dissolved oxygen (Filatov et al., 2003; Kuznetsov et al., 1952; Datsenko et al., 2007; Henderson-Sellers and Marland, 1990; Hutchinson et al., 1969). In Naglimozero, there was a linear decrease in the dissolved oxygen concentration over the water column during both winter and summer (Figure 4).

In winter and summer, approximately the same concentration of dissolved oxygen was noted: 80-110% in the upper layers, decreasing towards the bottom horizons to 16.7-1.0%. Phosphate concentrations in water of Naglimozero were very low, and did not exceed 2.0 µg P/L (Figure 5). In the bottom layers, especially in summer, the concentration reached 15.2 µgP/L, but most often did not exceed 5.0-7.0 µgP/L.

The P_{tot} content in summer was uniform throughout the water column and amounted to 23.5 µgP/L (weighted average). In winter, the P_{tot} concentration averaged 22 µg P/L, there was an increase in the concentration of P_{tot} from 14 meters to the bottom layer, reaching the values of 52.5-61.0 µg P/L (Figure 5).

The proportion of organic phosphorus in Naglimozero was higher than that of mineral phosphorus, exceeding 70% of P_{tot}, and even higher in March 2008, when PO_4^{3-} was not detected at all. This may indicate a photosynthetic activity of aquatic organisms, which leads to complete consumption of mineral phosphorus by phytoplankton in summer, and sometimes in winter (March 2008). In winter, accumulation of total phosphorus within the bottom layer was observed.

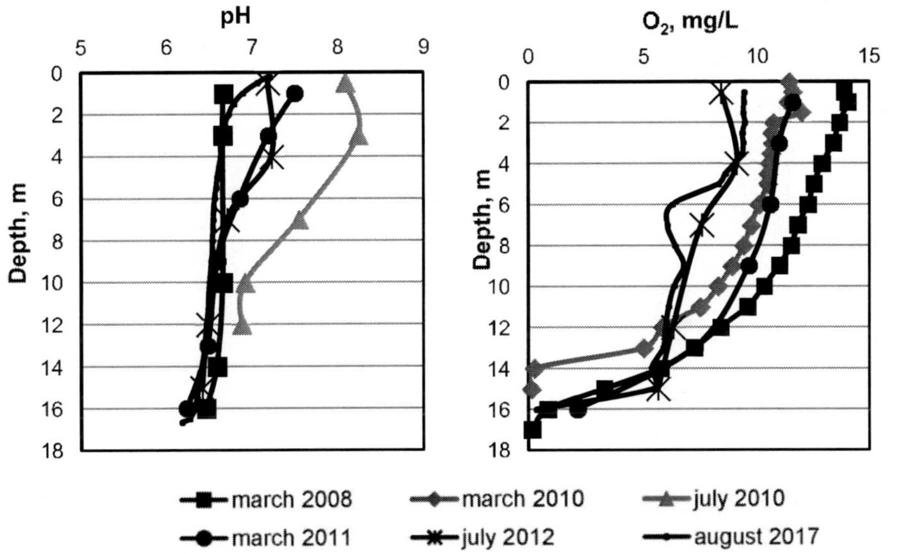

Figure 4. Vertical profiles of pH and dissolved oxygen concentration in the Naglimozero Lake.

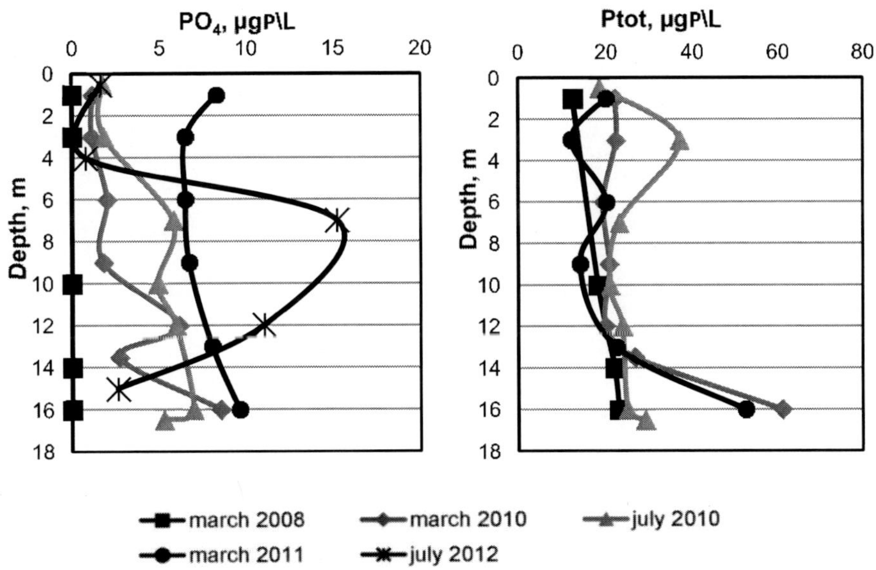

Figure 5. Vertical concentration profiles P-PO_4 and total phosphorus in the Naglimozero Lake.

The content of nitrites (NO_2^-) in Naglimozero was very low; the range of concentration is from the detection limit to 1.0 µg N/L. The mineral nitrogen was dominated by NO_3^-, its concentration was an order of magnitude higher than that of NH_4^+ and amounted to 50-170 µg N/L throughout the water column. The NH_4^+ concentration did not exceed 23.0 µg N/L in the water column. During certain periods (summer 2012), the NH_4^+ concentration remained below the detection limit (Figure 6).

Total nitrogen (N_{tot}) concentration was around 412 µg N/L (weighted average values) in the water column in winter and 368 µg N/L in summer (Figure 6). The proportion of organic forms of nitrogen was between 42 and 75% of N_{tot} in winter and between 58 and 72% in summer. The concentration of organic nitrogen was higher than that of mineral nitrogen, which indicates the predominance of primary production processes in the lake during the period of investigation.

The N_{tot}/P_{tot} ratio (13.4-33.1) in the summer and winter was optimal for the development of aquatic organisms and indicated the limiting role of phosphorus.

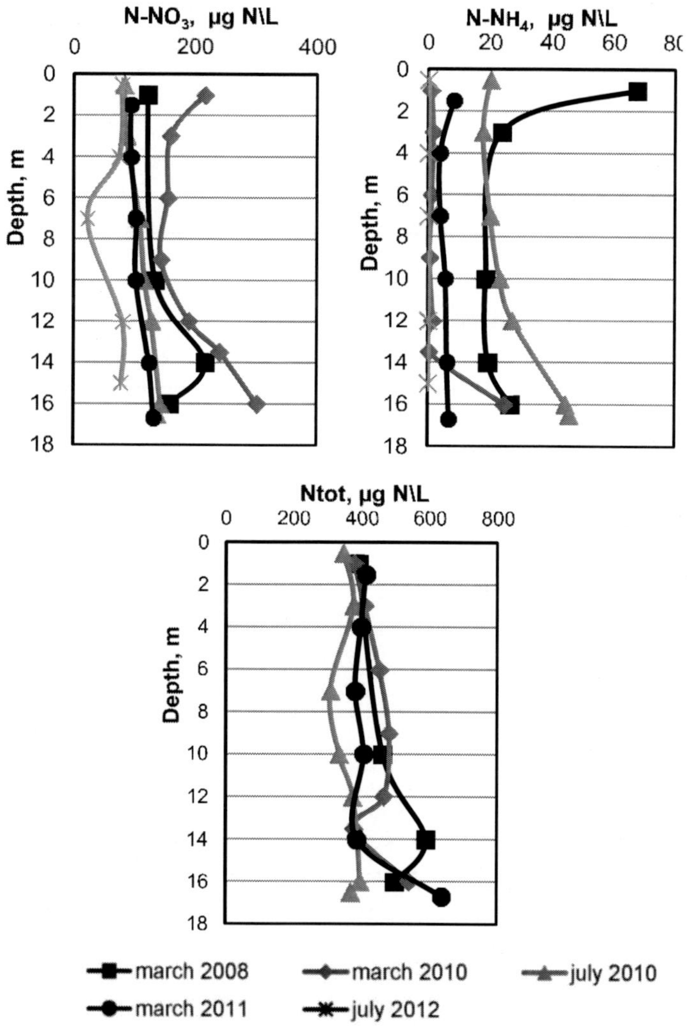

Figure 6. Concentration profiles of Nitrogen compounds in the Naglimozero Lake.

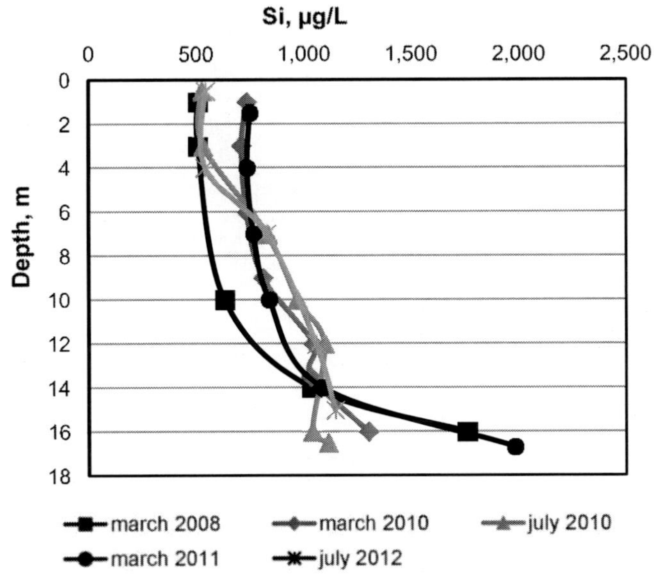

Figure 7. Concentration profile of Si in the Naglimozero Lake.

The dissolved Si concentration also depended on the hydrological season. An increase in the content of Si compounds was observed in winter and a decrease was observed in summer during their active consumption by living organisms, Figure 7. The Si concentration in Naglimozero ranged from 507 µg/L in the surface layers to 1980 µg/L in the bottom layer in the winter and from 526 µg/L in the surface layers to 1145 µg/L in the bottom layer in the summer.

To resume, due to the sufficient oxygen saturation of the lake water column, the oxidized forms of nitrogen prevailed. A decrease in the concentrations of PO_4, NH_4 and Si in the summer indicates intensive involvement of mineral forms of nutrients in the biological cycle. According to the concentration of mineral, organic and total nitrogen and total phosphorus, and following the classification of Kitaev (2007), the Naglimozero Lake can be classified as oligotrophic.

ZOOPLANKTON COMMUNITIES IN THE NAGLIMOZERO LAKE

The studies of zoocenoses in the Naglimozero Lake were carried out in July 2010 and 2012, when the species diversity of zooplankton reached its maximum values. The planktonic fauna of the reservoir was represented by three groups of organisms (Rotifera, Cladocera and Copepoda) and consisted of species common for lakes in the northwest of Russia. Copepoda and Cladocera played the leading role in the summer plankton. The species diversity was dominated mainly by Cladocera.

The zooplankton communities included 21 species. Cladocera were represented by 11 species, Copepoda by 7 species (Calanoida, 3 and Cyclopoida, 4), and Rotifera were represented by 3 species (Figure 8).

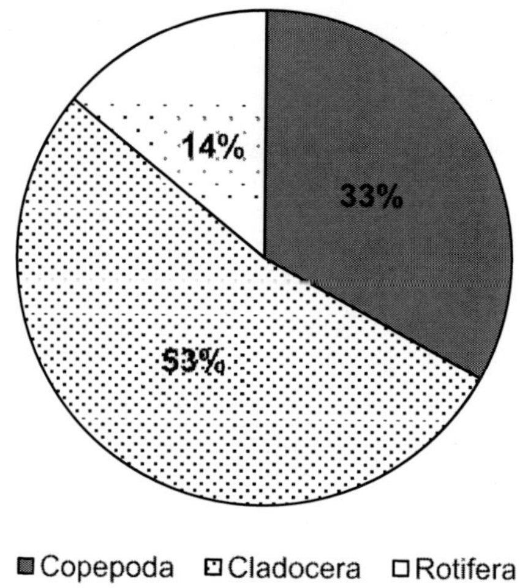

Figure 8. The species composition in the Naglimozero Lake.

The core of the planktonic complex consisted of 4-6 species in all areas of the lake. The structure-forming complex is represented by the following types: *Mesocyclops leuckarti*, *Eudiaptomus gracilis*, *Bosmina longispina*, *Daphnia cucullata*, *Kellicottia longispina*, *Conochilus unicornis.* Copepoda dominated in both abundance and biomass. Significant contribution to the total number was provided by *Mesocyclops leuckarti* (36%), *Eudiaptomus gracilis* (16%), *Daphnia cucullata* (16%), and *Conochilus unicornis* (20%).

The bulk of the biomass was represented by the Copepoda *Eudiaptomus gracilis* (35%) and *Mesocyclops leuckarti* (31%). The share of Cladocera in the total biomass ranged from 24 to 32%. In terms of biomass, representatives of the genus *Bosmina* (32% of the total biomass) and the genus *Daphnia* (41% of the total biomass) dominated among Cladocera. The bulk of the Rotifera biomass was dominated by large *Asplanchna priodonta*; its contribution to the Rotifera biomass achieved 85%.

The average abundance and biomass of zooplankton in the summer period amounted to 18000 ind./m^3 and 0.16 g/m^3, respectively (Table 2), which indicates the low feeding capacity of the reservoir for planktophagous fish.

Table 2. Structural characteristics of the zooplankton communities in the Naglimozero Lake (July 2010 and 2012)

Parameter	2010	2012
Number of species	20	21
Density, thous. ind./m^3	16.8	18
Biomass, g/m^3	0,16	0,16
% Rotifera:Cladocera:Copepoda*	29:35:36	13: 41:46
	17:22:61	2:28:70
Shannon index by number, bits ind^{-1}	3.71	3.52
Saprobity index	1.51	1.55

Note*– above the line is contribution to the total density (%), below the line is contribution to the total biomass (%)

To assess the ecological status of zooplanktocenoses, the coefficient of species diversity of Shannon-Weaver was calculated. The Shannon's index was 3.52 bits. The index values indicate a relatively stable and balanced state of the zooplankton community. According to the state of planktonic communities, and following the classification of S.P. Kitaev (2007), the Naglimozero Lake can be characterized as oligotrophic.

Analysis of the species composition and quantitative characteristics of the zooplankton of Naglimozero during two years did not reveal significant changes in the structural parameters of plankton communities. At the same time, certain differences can be noted in the ratio of the main groups of the community and the nature of their distribution (Table 2). The vertical heterogeneity of the distribution of different taxa was noted in the lake. Thermal stratification of the water column contributes to the formation of additional ecological niches in water bodies, allowing thermophilic and cold-loving zooplankton complexes to develop under the most optimal conditions. The epilimnion was dominated by the thermophilic species, Copepoda and *Cladocera nauplii.* More psychrophilic Copepoda species (representatives of the Cyclops genus) remained in the lower layers of the hypolimnion; this distribution was also typical for stenothermal Rotifera.

The values of the index of species diversity of zooplankton indicate a uniform distribution of species in the community and its high stability. On average, the level of development of zooplankton in the Naglimozero Lake was relatively low.

PHYTOPLANKTON COMPOSITION

The study of summer phytoplankton in the lake was carried out during July 2010. The water transparency during the periods of complex studies was 3.5 m by the Sekki disk. Water was saturated with oxygen evenly throughout the water column; the pH ranged from slightly acidic to neutral (Figure 4). The nutrient content was optimal for phytoplankton development. The N_{tot}/P_{tot} ratio for Lake Naglimozero (13 to 33) in the summer season indicates the limiting role of phosphorus in the development of phytoplankton communities.

Quantitative and qualitative composition of phytoplankton was studied in the photic layer (10.5 m) and at a depth of 12 m. The phytoplankton biomass was quite low, from 0.104 to 0.244 g/m^3. The number varied within 681-3436 thousand cells/L. In the summer, Cyanophyta dominated in abundance by the genera *Anabaena, Aphanocapsa (A. elachista and A.inserta), Cyanodiction reticulatum, Woronichinia compacta, Snowella lacustris*, reaching 81-87% at the depth of 10 m. Below the photic layer, cyanobacteria were not found, whereas at the depth of 12 m, the diatoms dominated.

The complex of taxa Bacillariophyta (*Aulacoseira ambigua, Cyclotella + Stephanodiscus spp., Tabellaria fenestrata, Aulacoseira subarctica*), Cryptophyta (*Cryptomonas marssonii*), and Chlorophyta dominated in the biomass. A dominant species at the depth of 3 m was *Botryococcus braunii* (56.4% of the total biomass). At the depth of 12 m, *Pleurotaenium ehrenbergii* dominated (62.3% of the total biomass)). The vertical distribution of various phytoplankton taxa by abundance and biomass is shown in Figure 9.

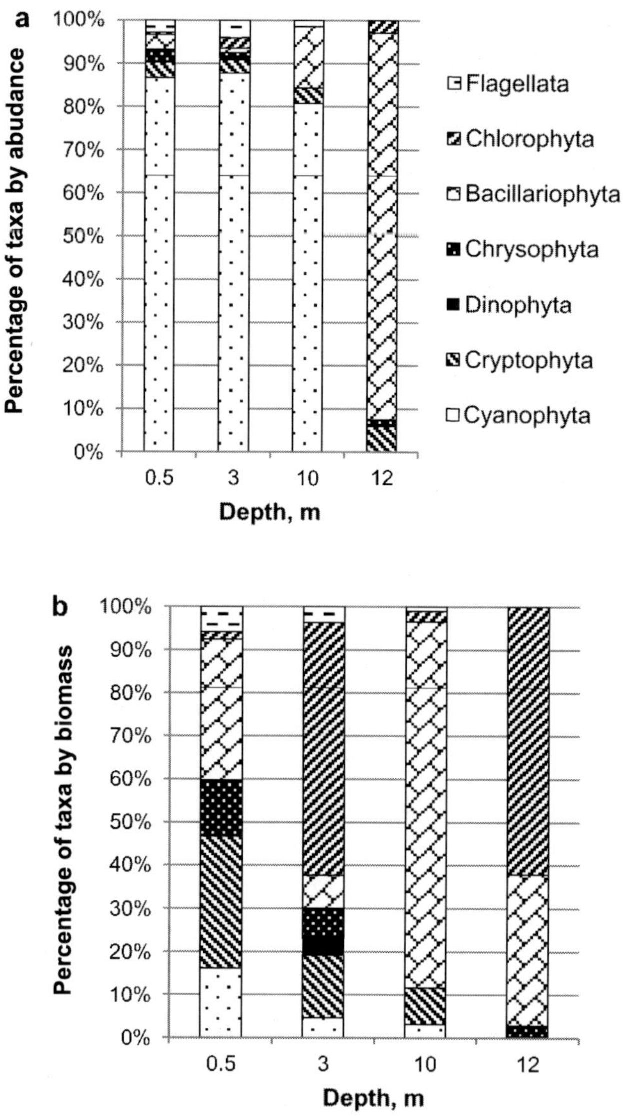

Figure 9. Vertical distribution taxonomic diversity of phytoplankton relative abundance (a) and biomass (b) in Lake Naglimozero (July 2010).

To describe the structure of the phytoplankton of Lake Naglimozero, we used a functional characteristic (Reynolds et al., 2002, Padisák et al., 2009); a total of 13 groups were identified. According to the dominant species (genera) in terms of abundance and biomass, the phytoplankton of the lake was represented by eight functional groups, among which seven predominate in numbers (**Lo, H1, C, B, Y, E, X2**), and eight in the biomass (**Lo, H1, Y, E, C, B, N, F, A**) as illustrated in Figure 10.

In the epilimnion, cyanobacteria and cryptophytes (groups **Lo, H1, Y**) predominated in the abundance; algae from groups (**Y, F, C, Lo, H1, A**) predominated in the biomass. Phytoplankton species of group **Y** are characterized by a wide range of habitats; they are able to inhabit all types of limnic aquatic ecosystems, where grazing pressure is low. The main dominant specie of this group is *Cryptomonas marssonii*. Group **F** includes species that prefer clean, deeply mixed meso-eutrophic lakes with a transparent epilimnion. At the lower boundary of the epilimnion, group **F** is represented by the species *Botryococcus braunii* with a 56% contribution to the total biomass. Group **J** unites algal species that live in shallow, well-mixed, highly nutrient-rich systems of lakes, ponds and rivers. As in the epilimnion, the **Lo** group, represented by small-sized colonial forms of cyanobacteria, dominated in abundance at the boundary of the photic layer (10 m). In general, representatives of the **Lo** group develop in deep and shallow, oligo- and eutrophic, medium and large lakes. During the summer, they can massively develop in the epilimnion of mesotrophic and oligotrophic lakes. These species are susceptible to prolonged or deep mixing of the water column (Padisák et al., 2009).

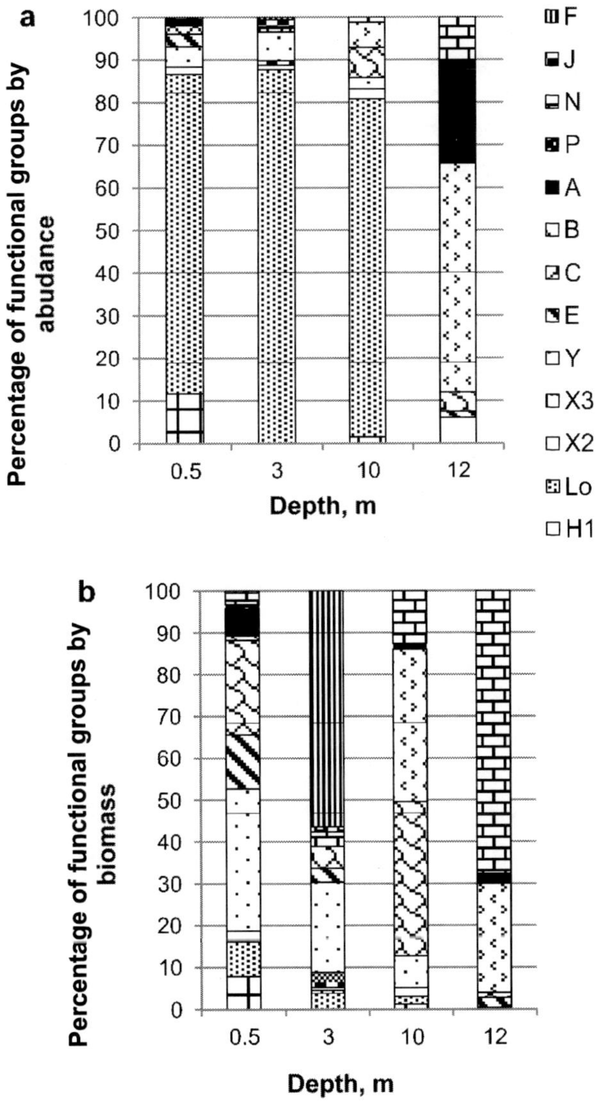

Figure 10. Vertical distribution functional groups of phytoplankton: relative abundance (a) and biomass (b) in Lake Naglimozero (July 2010).

When analyzing functional groups according to the classification (Reynolds et al., 2002, Padisák et al., 2009), some species of cyanobacteria (*Aphanocapsa elachista* and *A.inserta*, *Cyanodiction reticulatum*) were initially assigned to the **K** codon, but later we transferred *Aphanocapsa elachista* and *A.inserta* to the **Lo** group, as suggested by Padisák et al. (2009), since these species can occur in the epilimnion of oligotrophic lakes in the second half of summer. Most likely, it was also necessary to transfer from codon **K** to codon **Lo** the species *Cyanodiction reticulatum*, whose share of the total abundance in the surface layer reached 11.64%. An article on functional classification (Padisák et al., 2009) proposed to include Cyanoprokaryota of the genera Aphanocapsa, Aphanothece and Cyanodictyon to the **K** codon (Habitat template shallow, nutrient-rich water columns). However, *Cyanodiction reticulatum* in this case hardly fits this group, since this specie requires high purity and transparency of water; it is vulnerable and even included in the Red Data Book of the Leningrad Region (Red data…., 2000). This species is sensitive to pollution and increased trophy; it rapidly disappears from the phytoplankton composition when water bodies are polluted.

Below the epilimnion, the diatoms (centric algae more than 15 μm in size) and *Aulacoseira subarctica* (from the **B** + **C** functional group) and *Tabellaria fenestrata* (from codon **N**) dominated in terms of the biomass. Dysmidium green algae dominated from codon **N** at a depth of 12 m (63% of the total biomass, most represented by the species *Pleurotaenium ehrenbergii* 61.3%). Desmidia algae are known to be sensitive to environmental conditions. The main factor controlling their distribution and quantity is the geological context (Life of plants, 1977). Overall, they prefer granite rocks, which is consistent with lithology of the Naglimozero Lake (the underlying rocks are granito-gneisses).

To sum up, the summer phytoplankton community of Lake Naglimozero in July 2010 was represented by groups of species which are mainly characteristic of clear waters with high transparency, and low content of nutrients, typical for oligotrophic and mesotrophic water bodies (codons **Lo**, **H1**, **Y**, **E**, **C**, **B**, **N**, **F**, **A**). For the summer phytoplankton in Lake Naglimozero, the species diversity is typical for the North-West of Russia. The presented taxa are predominantly of cosmopolitan forms; most of the species are oligo-, oligo-β and β-mesosaprobes. The biomass values of summer phytoplankton make it possible to determine the trophic status of the studied lakes as oligotrophic.

HETEROTROPHIC BACTERIA DISTRIBUTION IN THE WATER OF LAKE NAGLIMOZERO

The study of the quantitative content of various ecologo-trophic groups of heterotrophic bacteria (HB) in the lake water showed their low abundance. This can be linked to rather low concentration of organic matter and nutrients, typical of an oligotrophic lake.

Figure 11 illustrates vertical distribution of heterotrophic bacteria in winter and summer. The lowest concentrations of HB were observed during the ice period. By the end of the period of winter stratification, the abundance of heterotrophic bacterioplankton in the water of Lake Naglimozero varied within two orders of magnitude and amounted to: Eutrophic bacterioplankton (EB), 2-190 CFU/ml (median 20 CFU/ml), Facultative oligotrophic bacterioplankton (FOB), 3-110 CFU/ml (median 25 CFU/ml), and Oligotrophic bacterioplankton (OB), 2-130 (median 10 CFU/ml). During the studied summer periods of 2010 and 2012 in the water column, the number of EB and FOB was on average 19 times, and the number of OB was 92 times higher than in winter and varied for EB within the range of 30-2200 CFU/ml (median 375 CFU/ml); FOB, 150-1980 CFU/ml (median 470 CFU/ml) and OB, 220-2600 (median 920 CFU/ml).

The maximum of heterotrophic bacterioplankton concentration was observed in the surface layers of the lake during the abnormally hot July of 2010, when the upper layer of the lake warmed up to 24-25°C.

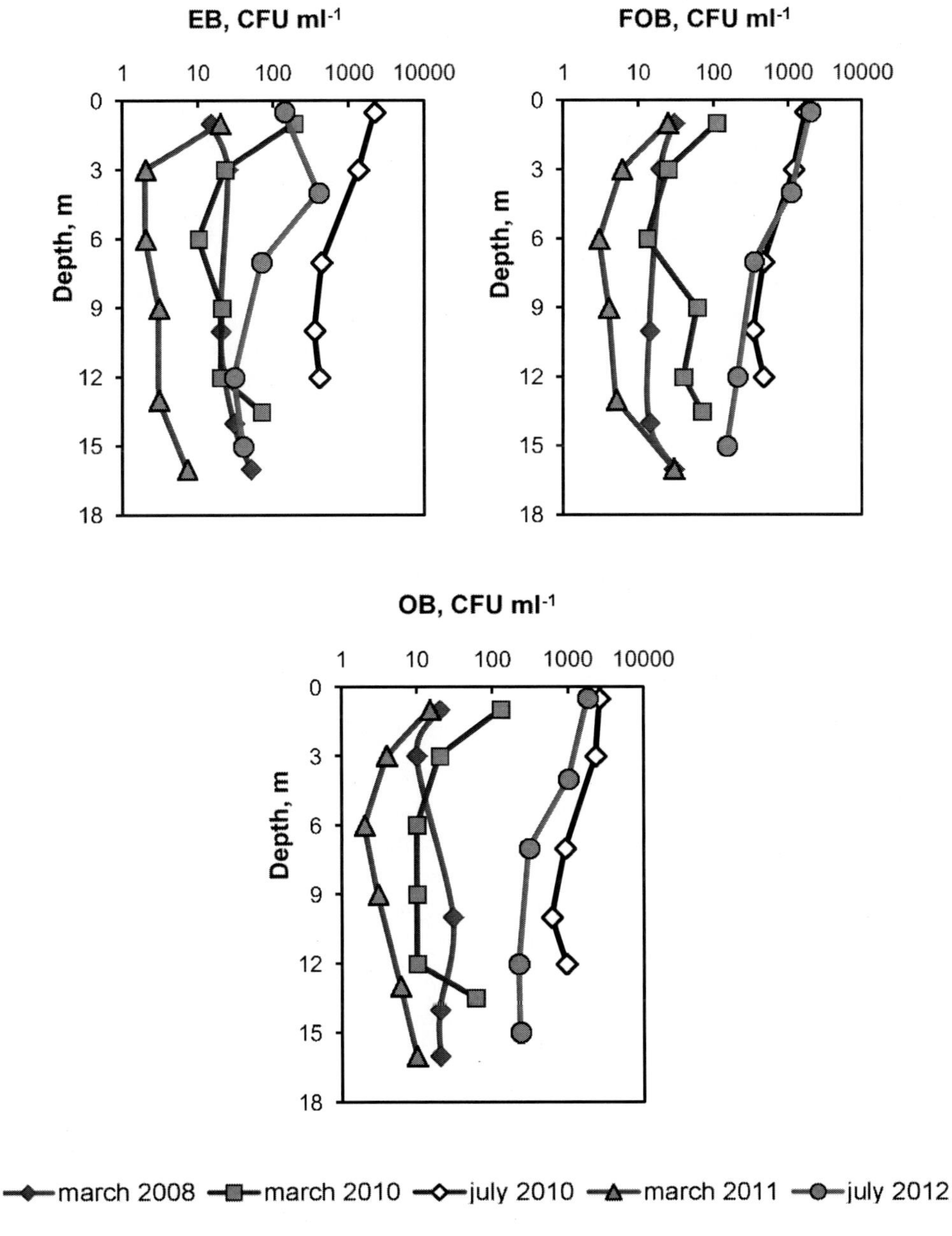

Figure 11. Vertical distribution of heterotrophic bacteria (EB - eutrophic bacteria, FOB - facultative oligotrophic bacteria, OB - oligotrophic bacteria) in the water column of the Naglimozero Lake.

CONCLUSION

The Naglimozero Lake is the least explored reservoir in the southern part of the Park due to its complex morphometry and location in a rather inaccessible area, but at the same time, it is a popular place for tourists and fishing enthusiasts. The lake is characterized by a complex relief with a large number of small islands and indented shores. The lake belongs to the meta-thermal type with a pronounced thermocline and a relatively uniform layer below the thermocline with temperatures ranging from 6.8 to 7.7°C. The lake is an ultra-fresh water body, which is mainly due to predominant role of atmospheric precipitation in the lake's nutrition. The influence of the catchment area is minimal, since the reservoir is located at the beginning of the catchment area.

Ultra-fresh water determines a low buffer capacity of the lake with respect to possible acidification. The active reaction of the environment is weakly acidic to neutral, and ranges from 6.25 to 8.24. The lake is classified as medium-transparent with a transparency of 3.5 to 4.0 m. In summer, the surface waters of the lake are well saturated with oxygen, whereas below the thermocline, the saturation is about 50%. Quite low oxygen content and its absence are noted at the bottom and in the bottom layers of isolated bays. In winter, water saturation in oxygen ranges from 50 to 80% except at the bottom, where quite low values were recorded.

The ratio of the main nutrients N_{tot}/P_{tot} for Lake Naglimozero ranges from 13 to 33 in the summer and winter periods, respectively. This is optimal for the development of aquatic organisms and indicates the limiting role of phosphorus. By the content of mineral, organic and gross nitrogen, as well as by the content of gross phosphorus, Naglimozero can be characterized as oligotrophic.

Analysis of the species composition and quantitative characteristics of the zooplankton of Naglimozero in different years did not reveal significant changes in the structural parameters of plankton communities. At the same time, certain differences can be noted in the ratio of the main groups of the community and the nature of their distribution. The vertical heterogeneity of the distribution of different taxa was also noted. Thermal stratification of the water column contributes to the formation of additional ecological niches in water bodies, allowing both thermophilic and psychrophilic zooplankton complexes to develop under the most optimal conditions. The epilimnion is dominated by the thermophilic species, Copepoda and *Cladocera nauplii*. The psychrophilic Copepoda species (representatives of the Cyclops genus) remains in the bottom layers of the hypolimnion; this distribution is also typical for stenothermal Rotifera. The Shannon Species Diversity Index (3.25 bit) indicates a stable and balanced state of the zooplankton community. On average, the level of development of zooplankton in Naglimozero was relatively low.

The summer phytoplankton of the lake is represented by groups which are mainly characteristic of clear oligotrophic and mesotrophic waters with high transparency and low content of nutrients (codons **Lo, H1, Y, E, C, B, N, F, A**). For the summer phytoplankton in the Naglimozero Lake, the species diversity is typical for the North-West of Russia. The specie *Cyanodiction reticulatum*, included in the Red Data Book of the Leningrad Region, has been identified as part of the summer phytoplankton. This cyanobacterium is very demanding on the purity and transparency of water. The presented taxa are predominantly cosmopolitan forms, most of the species are oligo-, oligo-β and β-mesosaprobes. According to the state of planktonic communities, Naglimozero can be characterized as oligotrophic.

The quantitative content of various ecologo-trophic groups of heterotrophic bacteria in the lake water exhibited quite low abundance, which is presumably due to the low concentration of organic matter and nutrients, characteristic of stable oligotrophic lake.

ACKNOWLEDGMENTS

This work was supported by the State Task AAAA-A18-118012390200-5. We also express gratitude to O.A. Pavlova for determining the quantitative and qualitative composition of phytoplankton.

REFERENCES

Datsenko, Yu. S. Eutrophication of reservoirs. *Hydrological and hydrochemical aspects*, 2007, 252. (In Russian).

Grigoriev, S. V. Experience of hydrological typology of lakes of the Latvian SSR // Fisheries of inland reservoirs of the Latvian SSR. - *II. Riga*, 1958, 245-248. (In Russian).

Henderson-Sellers, B; Marland, HR. Dying lakes. *Causes and control of anthropogenic eutrophication*, 1990, 275. (In Russian).

Hutchinson, D. *Limnology*, 1969, 592. (In Russian).

Kitaev, S. P. *Basics of limnology for hydrobiologists and ichthyologists*, 2007, 395. (In Russian).

Kuznetsov, S. I. *The role of microorganisms in the cycle of substances in lakes*, 1952, 300. (In Russian).

Moiseenko, T. I; Gashkina, N. A; Bylinyak, Yu. A; Kudryavtseva, L. P; Sandimirov, S. S. Zonal features of the formation of water chemistry in small lakes in European Russia. *Water Resources*, 33(2), 2006, 144-162.

Padisák, J; Crossetti, L. O; Naselli-Flores, L. Use and misuse in the application of the phytoplankton functional classification: a critical review with updates. *Hydrobiologia*, 621, 2009, 1–19. (In Russian).

Plant life. In 6 vols. Ed. CHL. - cor. AI of the USSR, prof. al. A. Fedorov. T. 3. Algae. Lichens. Edited by prof. M. M. Gollerbach. "*Enlightenment*", 1977, 487. (In Russian).

Pokrovsky, O. S; Shirokova, L. S; Zabelina, S. A.; Vorobieva, T. Ya.; Moreva, O. Yu.; Klimov, S. I., Chupakov, A. V., Shorina, N. V., Kokryatskaya, N. M., Audry, S., Viers, J.; Zoutien, C. Size fractionation of trace elements in a seasonally stratified boreal lake: control of organic matter and iron colloids. *Aquatic Geochemistry*, 2012, 18, 115–139. DOI: 10.1007/s10498-011-9154-z.

Red Data Book of Nature of the Leningrad region. (2000). Eds. Tzvelev N. N. St. Petersburg. *World & Family*, 2, 361, 672 p. (In Russian).

Reynolds, C. S; Huszar, V; Kruk, C; Naselli-Flores, L; Melo, S. Towards a functional classification of the freshwater phytoplankton. *Journal of Plankton Research*, 2002, 24, 417–482.

Guidelines for the chemical analysis of marine and fresh water in the environmental monitoring of fishery reservoirs and promising for the fishing areas of the World Ocean /Editor. V. V. Sapozhnikov. – Moscow: VNIRO, 2003, 202. (In Russian).

Tikhomirov, I. A. Thermal classification of lakes in the temperate zone. *Proceedings of the All-Union symposium on the main problems of freshwater lakes.*, Vol. 1. Lakes mode. - Vilnius, 1970, 174-185. (In Russian).

Yefremova, T. V.; Palshin, N. I.; Potakhin, M. S. Role of zonal factors and lake morphometry in forming thermal stratification // Water resources of the European North of Russia: results and prospects of research. Proc. anniversary confer. dedicated to the *15th anniversary of the IVPS – Petrozavodsk: KarRC RAS*, 2006, 377-391. (In Russian).

In: Lake Water
Editors: Oleg S. Pokrovsky, Yulia Bespalaya et al.

ISBN: 978-1-53619-275-9

Chapter 6

Complex Studies of Biotic and Abiotic Components of Boreal Lakes Located at the Beginning of the Catchment Basins of the Arctic (Lake Vilno) and Atlantic (Lake Maselgskoe) Oceans

T. Ya. Vorobyeva*, S. I. Klimov, L. S. Shirokova, O. S. Pokrovsky, A. V. Chupakov, O. Yu. Moreva, E. I. Sobko, S. A. Zabelina, N. V. Shorina, A. A. Chupakova, N. V. Neverova and O. D. Kovalev

N. Laverov Federal Center for Integrated Arctic Research of the Ural Branch of the Russian Academy of Sciences, Arkhangelsk, Russia

Abstract

Lakes Maselgskoe and Vilno are promising objects for studying the dynamics and variability of processes occurring in nature, since they represent a heterogeneous limnetic system with different biotopes. These lakes are located at the beginning of the catchment, therefore, they are dominated by intrabasin processes with a minimal input of allochthonous material. The ecosystems of the studied lakes are seasonally dynamic in terms of concentration of DOC, which is controlled mainly by autochthonous processes in the water column, such as primary production of phytoplankton, respiration of heterotrophic bacterioplankton, and anaerobic processes.

Comprehensive seasonal studies (2006 - 2017) of the ecosystems of Lake Maselgskoe and Lake Vilno in the southern part of Kenozero National Park included spatio-temporal dynamics of the qualitative and quantitative characteristics of phyto- and zooplankton communities. These studies showed that the ecological state of the studied lakes can be assessed as stable. The trophic status of the Maselgskoe and Vilno reservoirs is identified

* Corresponding Author's Email: vtais345@gmail.com.

as mesotrophic according to the main hydrobiological indicators and the content of nutrients.

However, there is a potential vulnerability of lake ecosystems due to the low buffer capacity of water. In the deep water areas of the lake Maselgskoe, where anaerobic layer is present during stratification periods, aerobic destruction plays a subordinary role (22-54%). During periods of stagnation in the bottom horizons and bottom sediments of Lake Maselgskoe, there are active processes of anaerobic degradation of organic matter (sulfate reduction and methanogenesis). In the bottom horizons of the deep areas of Lake Maselgskoe, during the periods of stratification, oxygen deficiency zones are formed. This creates favorable conditions for the input of mineral compounds of phosphorus, nitrogen and reduced forms of iron from bottom sediments into the water column.

Keywords: lake, hydrological characteristics, nutrients, phytoplankton, zooplankton, bacterioplankton, primary production, trophic status

INTRODUCTION

Boreal lakes represent an important and highly attractive subject of research in view of various environmental issues raised in the subarctic region, including, for example, pristine concentrations of trace metals that are most likely to be affected by the human impact (Roederer, 1991; de Caritat et al., 1996). Another important environmental issue is the fate of organic carbon in lakes that is likely to significantly control the CO_2 balance in high latitudes (Smith et al., 2004; Shirokova et al., 2009). High concentrations of Dissolved Organic Matter (DOM) and thus, the colloidal status of most trace elements (TE) is the most important characteristic feature of TE geochemistry in European Russian Arctic zone (Pokrovsky et al., 2002; 2005).

This chapter presents the results of an interdisciplinary study of two small (seasonally stratified and unstratified) lakes located at the beginning of catchments, and differing in morphometric, hydrological and hydrochemical characteristics, carried out from 2006 to 2017. We studied the hydrological, hydrochemical and hydrobiological characteristics of lake water during summer and winter periods to research the ecological state of lakes and the relationship between the evolution of their hydrochemical parameters with short-term weather events and climatic fluctuations. The seasonal dynamics of indicators was studied in 2009.

Lakes Maselgskoe and Vilno belong to small reservoirs of the Kenozersky National Park, located in the boreal zone in the north-west of the Russian Federation (Arkhangelsk Region).

The border of the Baltic crystalline shield and the Russian platform, the watershed between the basins of the White and Baltic seas passes through the territory of the Kenozersky National Park. Lake Maselgskoe belongs to the beginning of the catchment area of the Baltic Sea (Atlantic Ocean), whereas Lake Vilno belongs to the White Sea (Arctic Ocean). The watershed between the ocean basins runs along the western border of the Park and is most pronounced between lakes Maselgskoe and Vilno. Although the distance between the lakes is small - from 70 to 300 m, they are completely isolated by the lake ridge and have different water levels (Lake Vilno is 1 m lower, see Table 1) (http://www.kenozero.ru/o-parke/prirodnoe-nasledie/unikalnye-prirodnye-obekty/).

Lakes Maselgskoe and Vilno are highly suitable for studying the dynamics of processes occurring in aquatic ecosystems, since they are heterogeneous limnic systems with different biotopes. These lakes are located at the beginning of the catchment area; therefore, are essentially controlled by intrabasin processes with a minimum input of allochthonous organic matter. The sampling procedure and analytical methods employed in this study are described in chapter 1.

HYDROGRAPHIC AND MORPHOMETRIC CHARACTERISTICS OF SMALL LAKES MASELGSKOE AND VILNO (KENOZERSKY NATIONAL PARK)

As part of the study of lakes Maselgskoe and Vilno, we have refined the morphometric and hydrophysical characteristics for 12 years (2006-2017). The morphometric and hydrographic characteristics of the lakes are presented in Table 1. According to the classification of S.V. Grigoriev (Grigoriev, 1958) in terms of the specific catchment area (the ratio of the catchment area to the water area), lakes Maselgskoe and Vilno belong to the subgroup of lakes with a very small specific catchment area (< 5). According to the conditional water exchange indicator (the ratio of the volume of water entering the lake to the volume lakes) are lakes with a small conditional water exchange (< 0.5). The full conditional water exchange of lakes Maselgskoe and Vilno takes about 3.5 years.

Table 1. The main morphometric and landscape characteristics of the lakes

Specifications	Maselgskoe	Vilno
Height above sea level, m	166	167
Area, km^2	3.443	2.780
Shore line, km	21.44	12.13
Volume, km^3	0.0164	0.00947
Average depth, m	4.8	3.4
Maximum depth, m	20.2	5.9
Length, km	6.50	3.00
The greatest width, km	1.00	1.10
Watershed area, km^2	15.00	9.720
The specific catchment area	4.36	3.50
Average annual runoff from the catchment, km^3	0.00425	0.00273
Rate of replacement, year $^{-1}$	0.28	0.29
Water residence time in the lake, year	3.6	3.5

Lake Maselgskoe is characterized by a complex relief and relatively indented shores. The northwestern part of the lake is divided into a separate lake, Pezhikherye, relatively deep with a poorly developed littoral zone. The greatest depth of the lake noted here is 20.2 m. The second area of great depths is located directly in the northeastern part of Lake Maselgskoe, close to the coast near the village of Maselga, with a maximum depth of 19.5 m and in the form of a trench stretching from here along the eastern coast for a group of islands, as shown in Figure 1. Here it deepens to 11.5 m and turns to the west, between Bolshoy Island and the cape again reaches a depth of 11.5 m and ends south of the island near the western coast with the formation of a depression with a depth of up to 9.5 m.

Lake Vilno is elongated meridionally; the lengths of the major and minor axes are 2.9 and 1.4 km, respectively. The coastline is poorly developed. Lake Vilno is represented by two stretches separated by a promontory protruding into the lake, the deepest northern one. The slope of the depths is gradual, the bottom is even. Lake Vilno gives rise to the Vilenka River, has no tributaries.

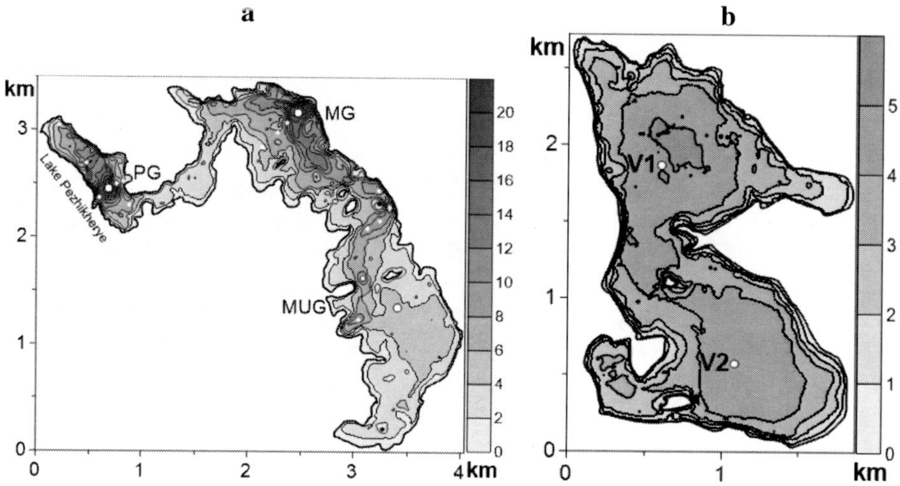

Figure 1. Bathymetric map of the Lake Maselgskoe (a) and of the Lake Vilno (b).

Hydrological Characteristics

Water Temperature

The vertical distribution of water temperature and its interannual variability are shown in Figures 2 and 3.

According to the thermal classification of Tikhomirov (1970), Lake Maselgskoe belongs to the metathermal type. In Lake Maselgskoe in summer, the upper layer is mixed to a depth of 4 m, but sometimes there are strong variations in water temperature due to sharp onset of hot weather. The maximum vertical water temperature gradients in the metalimnion ranged from 2.2 to 4.7ºC/m. These gradients were mainly located in the 5-6 m layer, but sometimes they are noted in the 9-10 m layer. The northwestern part of the Maselgskoe Lake, the so-called Lake Pezhikherye, belongs to the hypothermic type with a pronounced hypolimnion - a layer with a quasi-uniform distribution of water temperature, which is located below a layer of 10 m that does not heat up above 6˚C.

In Lake Pezhikherye, the upper homogeneous layer is thin and does not exceed 2 to 3 m. The thermocline extends to a depth of 6-7 m. The temperature jump layer is sharply expressed with maximum gradients from 3.8 to 7.5ºC/m and occurs at depths from 3 to 5 m. Deeper than 10 m, the vertical temperature gradients are less than 0.1ºC/m, and here the amplitude of interannual changes in water temperature is the lowest among all studied lakes.

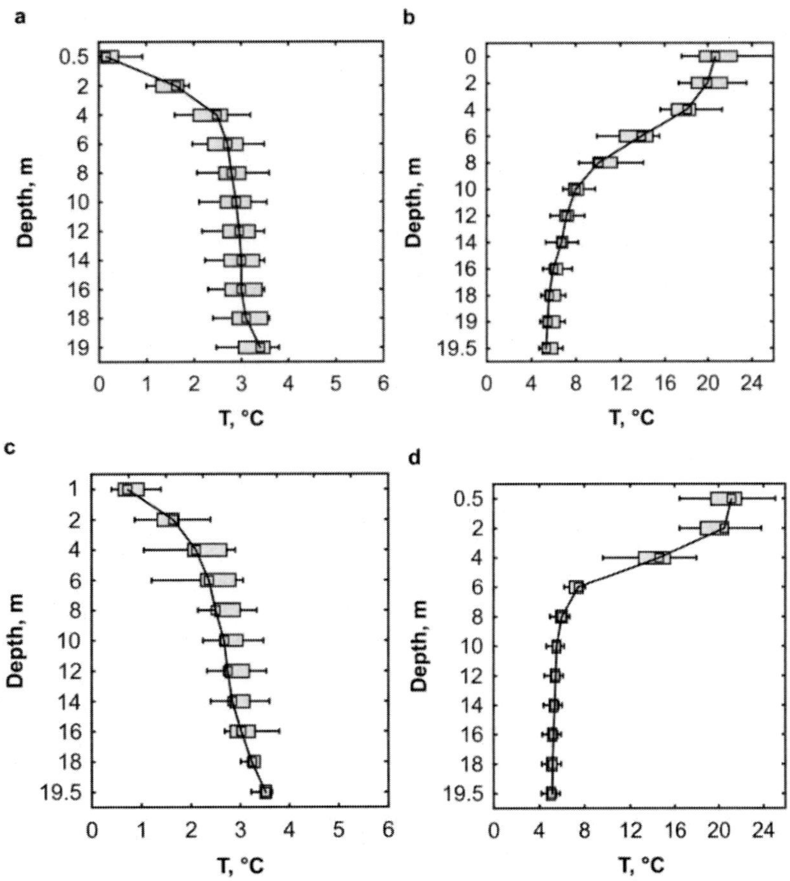

Figure 2. Vertical distribution and variability of water temperature in Lake Maselgskoe (2007-2017) in March (a) and July (b); Lake Pezhikherye, in March (c) and July (d).

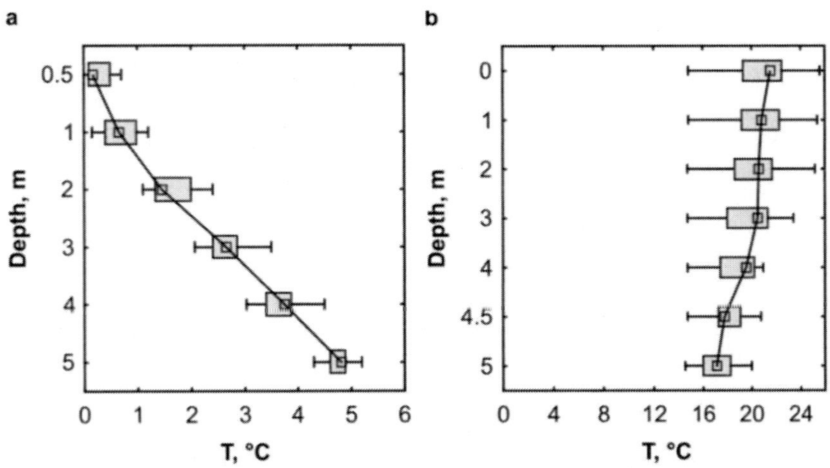

Figure 3. Vertical distribution and variability of water temperature in Lake Vilno (2007-2017) a) March; b) July.

The shallow lake Vilno belongs to the epithermal type. In Lake Vilno, the upper 3-meter layer is usually well mixed and warmed up. Mixing occurs to the bottom, but more often a temperature jump is formed in the bottom layer with a maximum vertical gradient of up to 5°C/m with a median value of 2.2°C/m.

In winter, the water temperature in lakes Maselgskoe and Pezhikherye is lower than the temperature of maximum water density (about 4°C), which indicates complete mixing of waters during the autumn cooling period. In Lake Vilno, the temperature linearly increases towards the bottom, reaching 5°C and more, which may be linked to warming up from the bottom: due to the shallowness of the lake in summer, the sediment stores a significant amount of heat relative to the volume of water.

Specific Conductivity

Similar to Lake Naglimozero (Chapter 5), the water of Lake Maselgskoe is characterized by low values of Specific Conductivity, which reflects low mineralization. In the upper 10-meter layer, in summer, the specific conductivity varied from 44 to 54 μS/cm, which corresponded to a mineralization of 25-31 mg/L. In winter, these variations were from 46 to 61 μS/cm or 27 to 35 mg/L (Figure 4). The largest changes in specific conductivity were observed in the bottom layers in summer (from 49 to 134 μS/cm) and in winter (from 74 to 178 μS/cm), which reflected particular conditions of spring and autumn turnover, respectively.

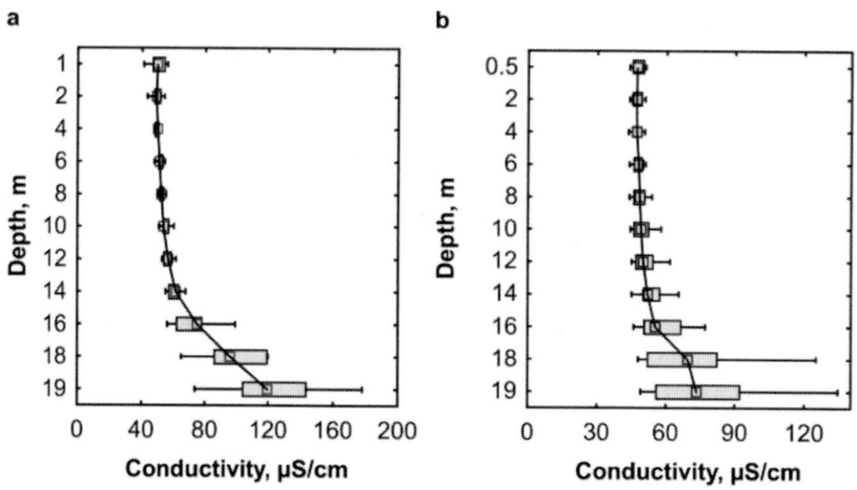

Figure 4. Vertical distribution of specific conductivity of Lake Maselgskoe water in March (a) and July (b) and its variability for the period 2006-2017.

Changes in specific conductivity in Lake Vilno ranged from 47 to 58 μS/cm in summer, which corresponded to mineralization from 27 to 33 mg/L, and from 47 to 85 μS/cm or from 27 to 50 mg/L in winter (Figure 5).

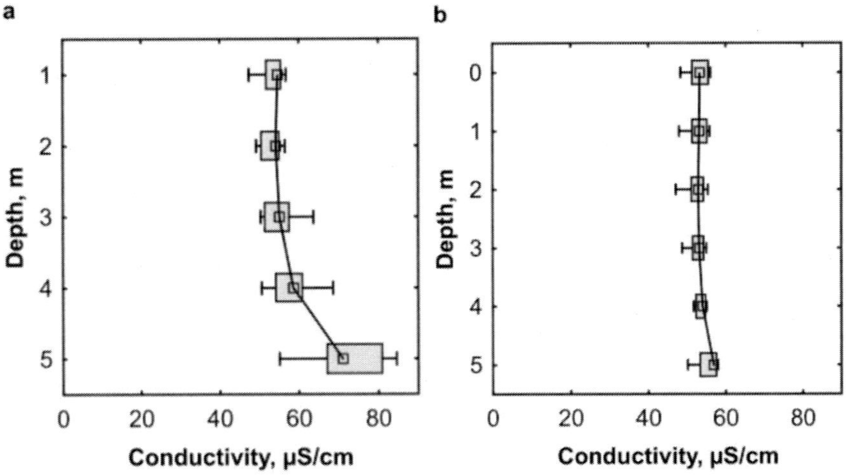

Figure 5. Vertical distribution of specific electrical conductivity of Lake Vilno water in March (a) and July (b) and its variability for the period 2006-2017.

Hydrochemical Characteristics

Dissolved Oxygen

As can be seen from Figures 6 and 7, in summer, the oxygen content in the surface water layer is practically the same in all lakes and exhibits low interannual variability. The greatest variability of the oxygen content is noted in the metalimnion at the reference station of Lake Maselgskoe and in the hypolimnion in Lake Pezhikherye. The conditions for oxygen saturation of deep waters are determined by spring mixing during relatively short period. Indeed, in May, the correlation coefficients of CO_2 and temperature in the first half of May are significant and amount to - 0.90 and - 0.62 for layers 4-12 and 12-19 m, respectively, in Lake Maselgskoe, and 0.94 and - 0.89 for layers 3-6 and 6-19 m, respectively in Lake Pezhikherye. Therefore, the lower the air temperature in early May, the longer the period of mixing and, thus the greater the water saturation with oxygen. Due to the greater openness, the mixing conditions in the area of the reference station of Lake Maselgskoe are better than in Pezhikherye, which is confirmed by the high oxygen content of deep waters in the spring. However, as a result of greater O_2 consumption in Maselgskoe in summer, the [O_2] content in July is 3-4 mg/L higher in Lake Pezhikherye compared to Maselgskoe. In Lake Vilno, low oxygen content is noted only in the bottom layer.

In winter, the oxygen content in the surface layer is higher than in summer due to better O_2 solubility at low temperatures. With depth, O_2 values decrease and oxygen-free zones can form at the bottom. Interannual changes in oxygen content are controlled by the intensity of autumn mixing of waters and the intensity of oxygen consumption under the ice, which, in turn, is determined by the time of the establishment of the ice cover. There is a significant relationship between the oxygen content and the duration of the period of stable transition of air temperature through gradations of 4 and 0ºC with a correlation coefficient of 0.75 in Lake Maselgskoe and 0.79 in Lake Vilno, i.e., the longer the period of water cooling, the higher oxygen content in winter.

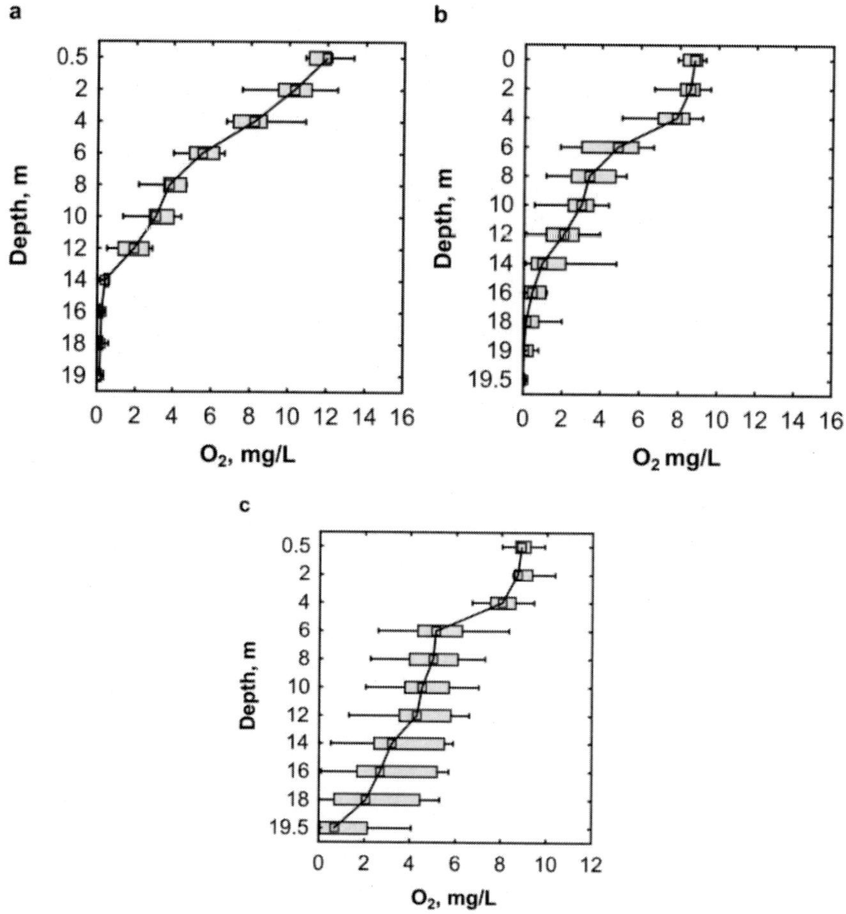

Figure 6. Vertical distribution of oxygen dissolved in water of Lake Maselgskoe during March (a) and July (b) and Lake Pezhikherye in July (c) over the period 2006-2017.

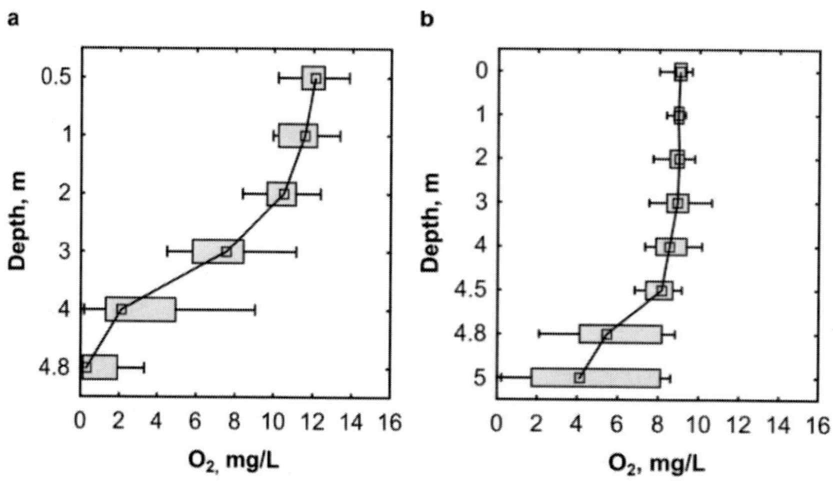

Figure 7. Vertical distribution of oxygen dissolved in water in March (a) and in July (b) and its variability for the period 2006-2017 in Lake Vilno.

Interannual Variability of Nutrients in Lakes Maselgskoe and Vilno

Figures 8-13 show the vertical distribution of nutrients in the lake water in the form of quartile characteristics in the form of a Box-Whisker plot, where the central rectangle corresponds to the median, and the variability is represented by the interquartile range. The waters of all lakes are characterized by a low content of mineral phosphorus. Figure 8 shows the phosphate content in Lake Maselgskoe, which ranges from 0.9 to 12 μgP/L in winter and from 0.01 to 12 μgP/L in summer.

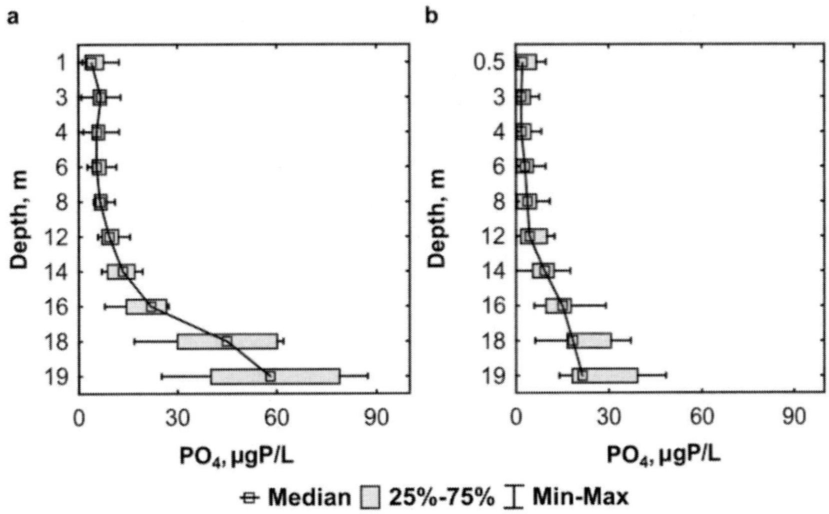

Figure 8. The phosphate concentration variability at the deep station of the Maselgskoe Lake in winter (a) and summer (b).

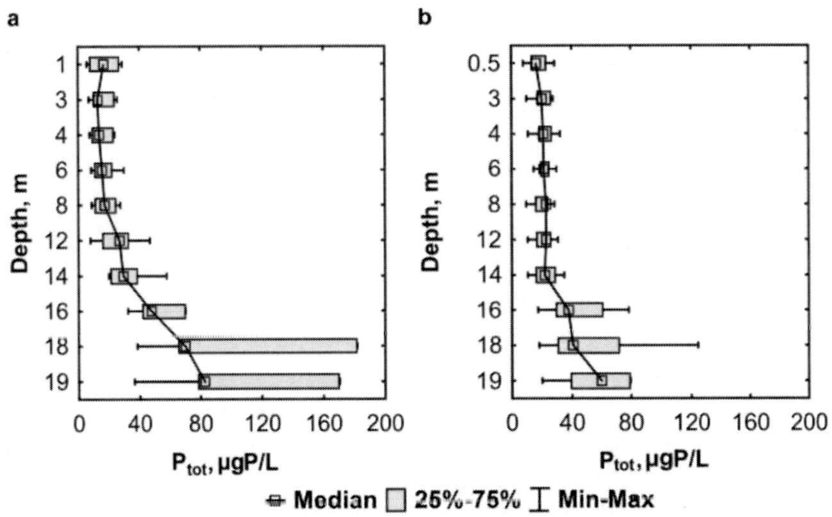

Figure 9. The total phosphorus concentration variability at the deep station of Maselgskoe Lake in winter (a) and summer (b).

From a 16-meter depth in winter and from 14 m in summer, the phosphate content increases to 182 μgP/L in winter and to 220.5 μgP/L in summer in the bottom waters for the entire observation period.

The content of total phosphorus in the photic layer ranges from 10 to 32 μgP/L in summer and from 6 to 30 μgP/L in winter, gradually increasing to 221-277 μgP/L in the bottom layer (Figure 9). The variability of phosphate concentrations is lowest in the surface layers and increases towards the bottom, reaching maximum values in the bottom layers in winter.

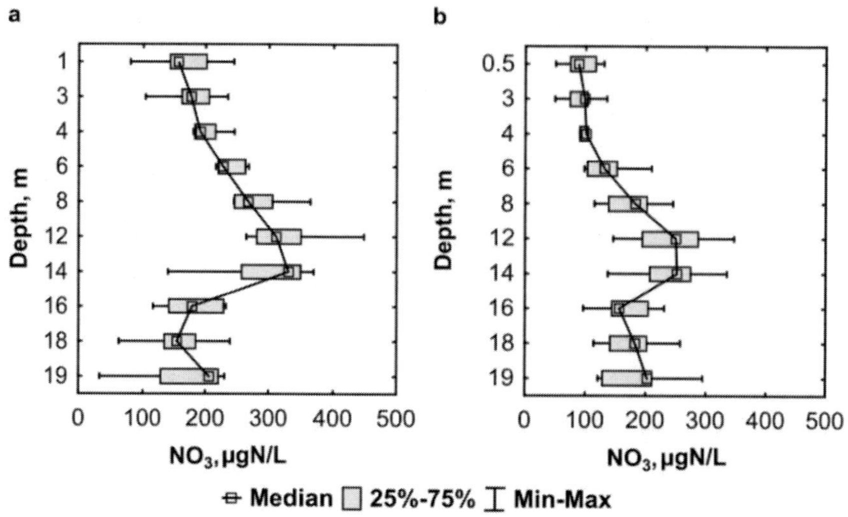

Figure 10. The vertical distribution of dissolved nitrates (2009-2017) in Lake Maselgskoe in winter (a) and summer (b).

The variability of total phosphorus concentration increases in the bottom horizons, which is probably due to variable conditions of spring and autumn mixing.

Among the mineral forms of nitrogen, nitrates account for the largest share in all lakes and all seasons. Figures 10-11 clearly show that the distribution of nitrates and ammonium nitrogen over the water column is controlled by the content of dissolved oxygen (Figure 6). In winter and summer, the concentration of nitrates in the water is almost uniform; down to a depth of 6 meters, the nitrate content in the upper layers did not exceed 160 μgN/L in summer and 260 μgN/L in winter. Below the depth of 6 meters, the nitrate content increases to 367-469 μgN/L.

Starting from a depth of 14 meters (at which the content of dissolved oxygen decreases to almost zero values), the content of dissolved nitrates decreases to values of 47.8-177 μgN/L (Figure 10), and the content of ammonium nitrogen sharply increases to 680-800 μgN/L (Figure 11), presumably due to anaerobic conditions, which were always observed during these seasons over the entire observation period, when nitrogen recovery processes take place.

In the epilimnion, phytophages, filter-feeding zooplankton (Cladocera and Calanoida), filter-feeding Rotifera, nauplii Copepoda and Cladocera, I-IV Copepoda stages are concentrated. Psychrophilic species of Copepoda (representatives of the genus *Cyclops*) remain in the lower layers of the hypolimnion; this distribution is also typical for stenothermal Rotifera. The concentration of zooplankton in the metalimnion zone was also noted. Migrations of zooplankton to the meta- or hypolimnion in deep stratified water bodies are an adaptive form of behavior of aquatic organisms, which contributes to the maintenance of populations in a stable state and reduces competitive relationships in the zoocenosis (Feneva et al., 2016; Lynch, 1979). Depending on the year of observation, Cladocera was mainly concentrated in the epilimnion (0-3 m), but also exhibited some accumulations in the bottom layers (14-18 m). Copepoda are abundant in the metalimnion and hypolimnion (10-14 m). For Rotifera, the maximum abundance was recorded in the hypolimnion (16-18 m). Figure 29 shows the proportion of the main groups of zooplankton in different layers of Lake Maselgskoe in 2013-2015.

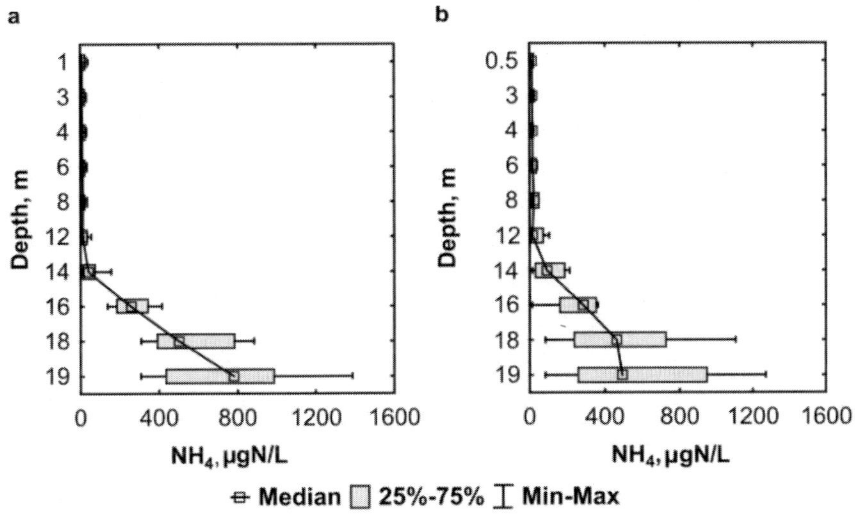

Figure 11. The vertical distribution of dissolved ammonium in Lake Maselgskoe (2009-2017) in winter (a) and summer (b).

At the same time, the concentration of ammonium is quite low up to 14 m depth, with an average value of 17 µgN/L during all seasons. The variability of ammonium concentrations remains very low up to the anaerobic layer (Figure 11). Below14 meters, the variability of the concentrations of ammonium gradually increases, exhibiting the highest variability for the highest values, may be due to rapid changes of conditions in the anaerobic zone.

Overall, the distribution of nitrate and ammonium concentration in Lake Maselgskoe clearly shows the dependence of the distribution of $N\text{-}NO_3^-$ and $N\text{-}NH_4^+$ on the dissolved oxygen content. Starting from a depth of 14 meters (at which a sharp decrease in the content of dissolved oxygen occurs), the content of dissolved $N\text{-}NO_3^-$ decreases by 1.5-2 times, and the content of $N\text{-}NH_4^+$ sharply increases by 1-2 orders of magnitude, i.e., under anaerobic conditions, which were observed over the entire observation period, strong nitrogen reduction occurs. The difference in the $N\text{-}NO_3^-$ concentration between winter and summer indicates their consumption by Hydrobionts in the upper layers in summer season and mineralization and the release from decaying organic matter in winter.

The nitrite content is minimal during all seasons, in the photic layer of Lake Maselgskoe it the $N\text{-}NO_2^-$ ranges from 0.01 to 2.0 µg N/L in winter and from 0.01 to 1.0 µg N/L in summer. In the bottom layers, the nitrite content increases to 14.0 µgN/L during both seasons. This can be explained by the incompleteness of the nitrification process under conditions of low oxygen content in the bottom layers.

There was little variability in the aerobic layers of total nitrogen, with median values of 400 μgN/L in summer and 580 μgN/L in winter (Figure 12). In winter, the variability of values is higher. Below 14 m towards the bottom, the N_{tot} increases by a factor of 2-3 in summer and 2-4 in winter.

The amount of N_{org} in the epilimnion of Lake Maselgskoe in summer averaged 63% of Ntot, decreasing in the hypolimnion to 35%. In winter, the amount of N_{org} in the epilimnion was 53% of N_{tot}, in the hypolimnion - up to 40%.

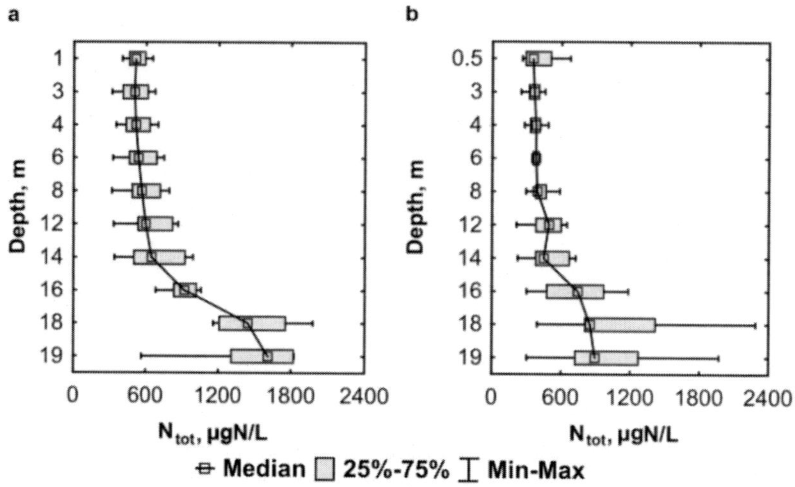

Figure 12. Variability of the total nitrogen concentration in winter (a) and summer (b) in Lake Maselgskoe over the period from 2009 to 2017.

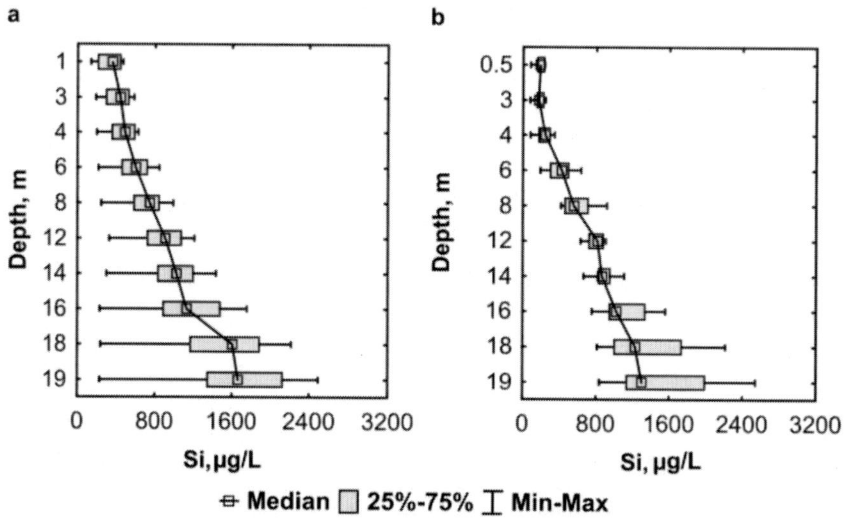

Figure 13. The variability of Si concentration in the Maselgskoe Lake during winter (a) and summer (b).

Overall, seasonal dynamics in the concentration of nitrogen compounds is controlled by accumulation in the end of the winter period and consumption during the summer season.

The content of dissolved silicon compounds in Lake Maselgskoe also depends on the hydrological season. There is an increase in Si concentration in winter and a decrease in summer during its active consumption by living organisms (Figure 13). In summer, the concentration of silicon decreases in the surface layer to 80-90 μg/L, whereas in the bottom layers it is more than an order of magnitude higher and amounts to 1200 μg/L. In winter, the values of silicon concentration vary from 300 μg/L at the surface to 2300 μg/L at the bottom.

The vertical distribution of nutrients is influenced by the conditions of water stratification. In the epilimnion, the content of the main nutrients is almost uniform along the water column; an increase occurs in the bottom layers of deep-sea areas in winter and summer. Due to the sufficient oxygenation of the upper water layers in the lake, mainly oxidized forms of nitrogen prevail. The bottom layers of the deep-water station are characterized by an increase in nutrient concentrations due to the processes of mineralization of organic matter during sedimentation and diagenetic transformation of bottom sediments. The active processes of mineralization in the studied lake are indicated by a decrease in the proportion of the content of organic forms at the expense of inorganic forms of nitrogen with increasing depth. A decrease in the concentration of phosphates, nitrates and silicon in summer indicates the intensive involvement of mineral forms of nutrients in the biological cycle.

Nutrients (N, P, Si) Concentration in Lake Vilno

The water of Lake Vilno is characterized by a low content of mineral phosphorus in all seasons: generally, below 7.5 μgP/L in the water column, increasing towards the bottom horizon to 9.0-13.0 μgP/L (Figure 14).

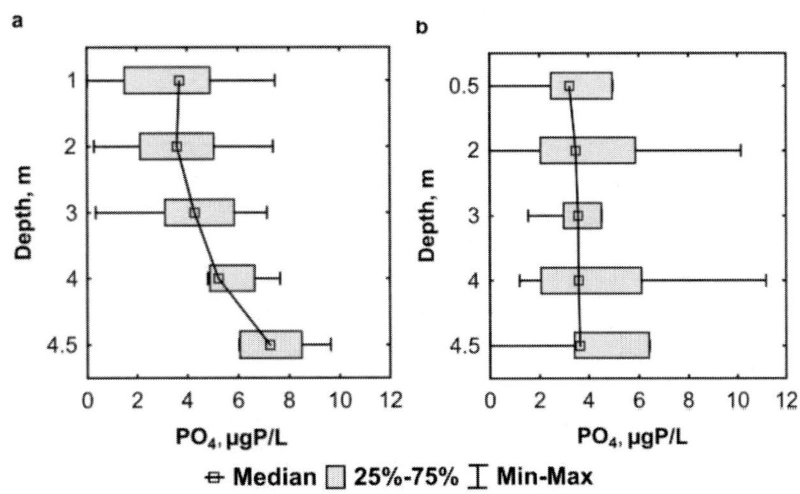

Figure 14. The vertical variability of phosphate concentration at the deep station of Vilno Lake in winter (a) and summer (b).

The content of total phosphorus in the water column is the highest in summer (from 12 to 58 μgP/L) and the lowest in winter (from 7.3 to 28.7 μgP/L), Figure 15.

The variability of phosphates and total phosphorus concentration is associated with good periodic mixing of water under the influence of wind conditions. In contrast, during calm summer weather, a jump in temperature and oxygen often occurs, which affects the concentrations in the bottom layers.

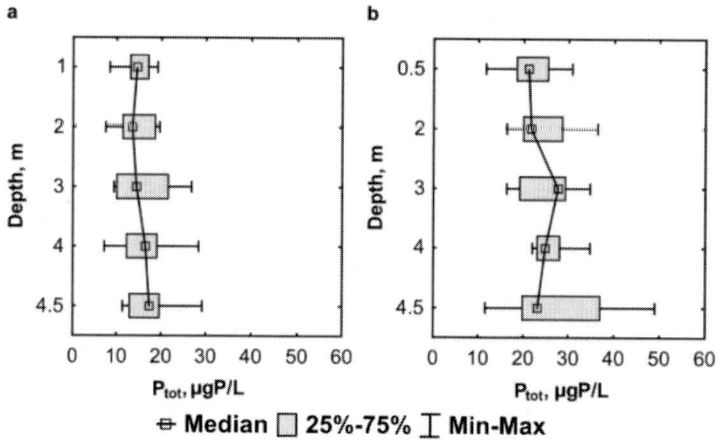

Figure 15. The vertical variability of total phosphorus at the deep station of the Vilno Lake in winter (a) and summer (b).

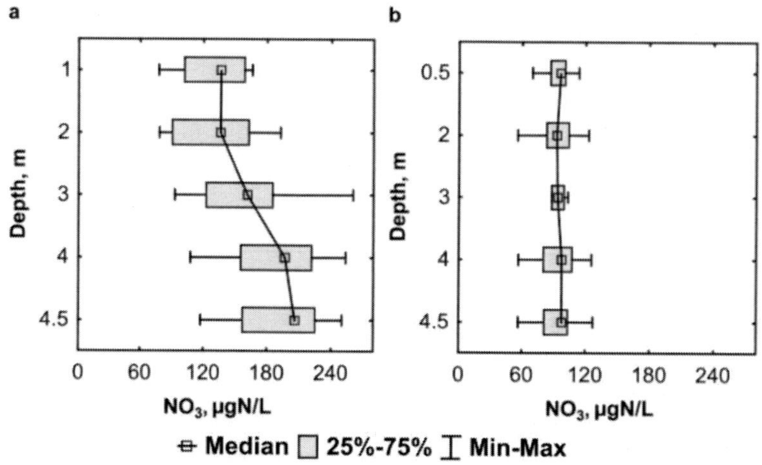

Figure 16. The variability of dissolved nitrates (2009-2017) in Lake Vilno during winter (a) and summer (b).

In summer, the dissolved oxygen, nutrients, pH, and specific conductivity are evenly distributed over the entire water column. The distribution of nitrogen forms is semi-linear (Figures 16-17). The concentration of nitrogen compounds slightly increases in the winter, under ice due to the destruction of OM in the water column and insufficient supply of dissolved oxygen as a result of lack of water mixing. The concentration of nitrates in water varied from 56 to 136 µgN/L in summer and from 78 to 261 µg N/L in winter. The difference in the content of nitrates in the winter and summer seasons is due to N consumption by hydrobionts in the upper layers in the summer season and accumulation from mineralized OM in the winter season (Figure 16).

The concentration of ammonium is quite low in the water column during all seasons: the average values are 22 μgN/L in summer and 36 μgN/L in winter (Figure 17). During the period of winter stratification in the near-bottom anaerobic layer, the concentration of ammonium increases to 122-343 μgN/L.

The concentration of nitrates and ammonium during different seasons depends on the biological activity, as confirmed by their increase in winter, due to the accumulation from mineralized OM, and a decrease in summer, due to their consumption by aquatic organisms (Figures 16, 17).

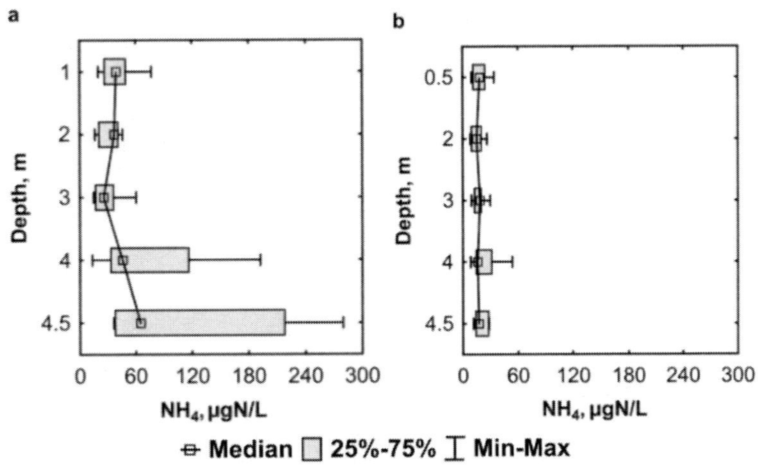

Figure 17. The vertical distribution of dissolved ammonium (2009-2017) in Lake Vilno (a – winter; b – summer).

The total nitrogen content is almost constant during winter and summer and amounted to 391 μgN/L in summer and 512 μgN/L in winter, Figure 18. In the bottom layer, the concentration of total nitrogen increases slightly to 480-528 μgN/L in summer and up to 662-786 μgN/L in winter.

The large variability of total nitrogen concentrations in winter is associated with formation of the ice cover, which prevents gas exchange with the atmosphere. As a result, there is almost no vertical mixing which leads to temperature and oxygen stratification. The greatest variability of N_{tot} is observed in the anoxic zone (Figure 18). The nitrite content is very low in all seasons, sometimes their absence is observed, except for the bottom horizons, where there is a slight increase in the nitrite content, which is explained by the incompleteness of the nitrification process under conditions of low oxygen content. The water column nitrite values did not exceed 1–2 μgN/L, and sometimes were lower than the detection limit.

Overall, seasonal dynamics of nitrogen compounds was characterized by accumulation in the end of the winter period and consumption during the summer growing season.

The content of dissolved silicon also depends on the hydrological season. There was an increase in silicon content in winter, due to mineralization of diatoms and macrophytes, and a decrease in summer during their active consumption by plankton and periphyton (Figure 19).

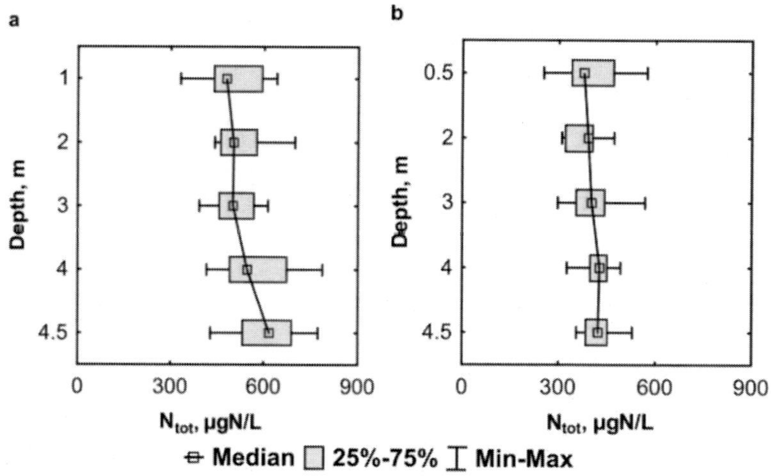

Figure 18. Variability of the total nitrogen concentration in winter (a) and summer (b) in Lake Vilno (2009-2017).

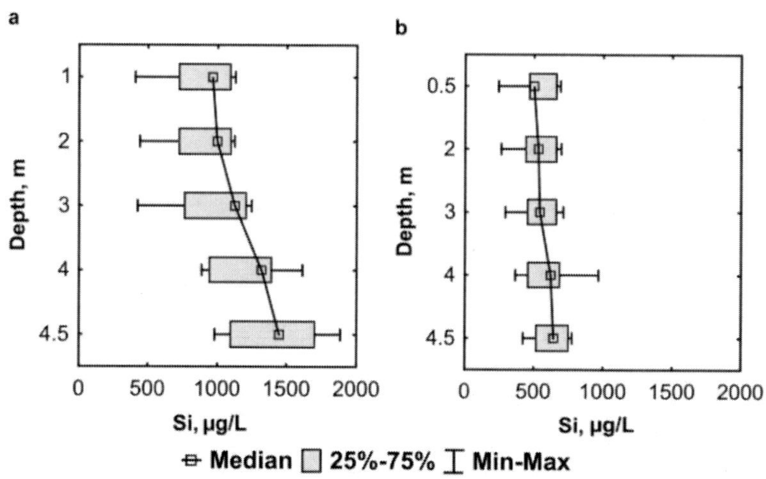

Figure 19. The concentration of Si at the deep station of Lake Vilno in winter (a) and summer (b).

In winter, the values of silicon concentration in Lake Vilno vary from 980 μg/L (median value) in the surface horizon to 2254 μg/L in the bottom layer. In summer, the concentration of silicon decreases in the surface layer to an average of 550 μg/L and does not exceed 828 μg/L in the bottom horizons.

The content of nutrients in water, primarily the content of total nitrogen and total phosphorus, are known to be very effective indicator of the trophic status of water bodies. Based on the total phosphorus content in the water of the studied lakes, the lakes can be classified as mesotrophic (Datsenko, 2007).

The ratio of the main nutrients used by primary aquatic organisms is essential for overall stability of lake ecosystem (Datsenko, 2007; Samarina, 2008; Shaporenko, 2006). We calculated the Ntot/Ptot ratios for lakes Vilno and Maselgskoe, as an average value over a 10-year interval during different hydrological seasons. The Ntot/Ptot ratios for lakes Vilno and Maselgskoe for the period from 2008 to 2017 have the same values and are equal to 33 for each lake, which is much higher than the Redfield ratio (16: 1) (Redfield et al., 1963). Such a high ratio indicates either the limiting role of phosphorus in the development of phytoplankton in these lakes, or the predominance of nitrogen compounds in the lake water. In summer, a very low content of phosphates and a high content of nitrates were often observed, which indicates a phosphorus limitation for aquatic organisms.

Due to the sufficient oxygenation of the upper water layers in lakes, mainly oxidized forms of nitrogen prevail. The bottom layers of the deep-water station are characterized by high concentration of nutrients due to active mineralization of organic matter during sedimentation and upward diffusion of these nutrients from bottom sediments. A decrease in the concentration of phosphates, nitrates and silicon in the summer periods of observations indicates the intensive involvement of mineral forms of nutrients in the biological cycle.

Our research is consistent with the findings of other researchers. The authors of (Oliver S. et al., 2017) quantified the trends in water quality in lakes using observations of nutrients and chlorophyll (Chl) from the Lakes Database of the Northeastern United States (LAGOS-NE). LAGOS-NE was used to quantify changes in lake water quality from 1990 to 2013. The authors concluded that, despite significant changes in the environment, water quality in lakes in the Midwest and Northeast United States did not change.

Spatial and Temporal Dynamics of Iron Distribution in Lake Maselgskoe and Lake Vilno

Numerous studies demonstrated that lakes with a slow water exchange serve, on the one hand, as transit systems in the general circulation of substances in nature, on the other hand, as a kind of systems for the purification and burial of elements in bottom sediments (Halvorson, 1927; Fatchikhina, 1948; Ecology, 1999). The presence of iron and iron-organic complexes in water, on the one hand, increases the retention capacity of waters, and on the other hand, facilitates the migration of most micro- and macroelements from primary watercourses to seas and oceans (Artemev, 1993; Levshina, 2012; Gordeev, 2012)]. Iron in boreal aquatic ecosystems exhibits elevated background concentrations and represents one of the main specificity of the surface waters (Moiseenko et al., 1997). A number of studies reported distribution of iron in lakes in the summer in various regions of the world (Konshin, 1959; Kuznetsov, 1970). Systematic observations of Yoshimura (Yoshimura, 1931, 1936; Ohle, 1934) demonstrated that in the epilimnion of various lakes, with the exception of dystrophic ones, the concentration of iron does not exceed 0.2-0.3 mg/L, and in oligotrophic lakes it is often below the detection limits.

In most oligotrophic lakes, there is a thin layer of water directly above the bottom, in which the iron content is increased due to a decrease in the oxygen content in this layer. In eutrophic and dystrophic lakes, iron stratification is always well pronounced. In acidotrophic lakes, under anaerobic conditions at the bottom, stratification may not be observed (Kuznetsov, 1970). Depending on the oxygen regime of the lake, oxidation of ferrous compounds of iron can occur either in the surface layer of sediments, as is the case in oligotrophic lakes, or in the water column of mesotrophic and eutrophic lakes. The biogeochemical behavior of Fe and its effect on aquatic ecosystems is determined not only by its speciation in the aquatic environment (Varshal et al., 1979; Linnik, 2009).

The lakes considered in this study are intermediate systems in the general transport of organic matter and associated elements from primary streams through streams, rivers, lakes to the seas and oceans, where iron, which has a high affinity for complexation with DOM (dissolved organic matter), plays an important role as natural sorbent and transporter of many trace elements in form of iron organic complexes and colloids. The binding of iron into fulvate complexes occurs mainly in the sediments (Glagolev, 1958; Boyle et al., 1977; Florence et al., 1982; Pokrovsky et al., 2002; Sigee, 2005).

Iron can be present in different oxidation states, entering into various interactions with mineral and organic suspensions, forming complexes with dissolved organic matter, and participating in the metabolism of iron bacteria. Assessment of concentration of iron and its speciation in surface waters of the boreal zone is important for the study of biogeochemical processes in aquatic ecosystems. The section presents seasonal dynamics of iron and its speciation in lakes Maselgskoe and Vilno.

Iron in Lake Maselgskoe

Analysis of literature data (Fedorova, 1964) and our studies indicated an increased iron content in the bottom layer of two deep-water reference stations of Lake Maselgskoe (Figure 20).

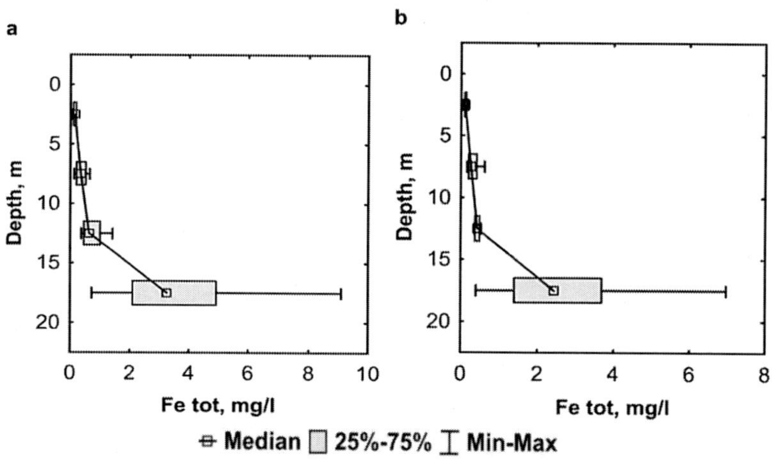

Figure 20. Long-term dynamics of distribution of total iron (Fe tot) in deep-water reference stations of Lake Maselgskoe (a – winter; b – summer; c - Lake Pezhikherye summer).

Figure 21 shows examples of vertical distribution of various forms of iron during winter stagnation, summer stratification and off-season homothermy of two deep-water reference stations of Lake Maselgskoe. There is an increase in the concentration of dissolved and particulate Fe March and July from the surface to the bottom. At the same time, the iron content in the water column is unevenly distributed. The deep reference stations of Lake Maselgskoe have a clearly pronounced bottom anaerobic layer. Iron (II) gradually predominates from 14 m, reaching 90-95% at the bottom. The maximum concentration of ferrous iron in the lake water during the study periods was 6.53-8.3 mg/L, and that of ferric iron was 0.6-0.7 mg/L. In the bottom layers, where the oxygen content is low (percentage oxygen saturation: March - 1.6%, July - 0%), the opposite process occurred - the reduction of iron by microorganisms and their metabolic products (i.e., Kuznetsov, 1952; Fedorova, 1964; Brekhovskikh et al., 2015). In this case, the reduced iron diffuses into the upper layers of the reservoir, but since this process is very slow, the concentration of dissolved iron in the bottom layers is much higher than at the surface. This is confirmed by a close correlation between the concentration of iron and the number of iron bacteria in Lake Glubokoe (Shaporenko et al., 2006).

An important indicator of the medium that affects the distribution of iron is pH. When water is acidified, acidic hydrolysis of iron hydroxides occurs, which transforms Fe_{tot} into a dissolved form. As a result, an increase in the concentration of total dissolved iron (Fe (II + III)) (Figure 21) is observed during the period of summer stratification at the deep-water station of Lake Pezhikherye with a decrease in pH from 7.3 at the surface to 6.1 at the bottom. The percentage of oxygen saturation of the surface layer of water is quite high and amounts to 78% in winter and 104% in summer.

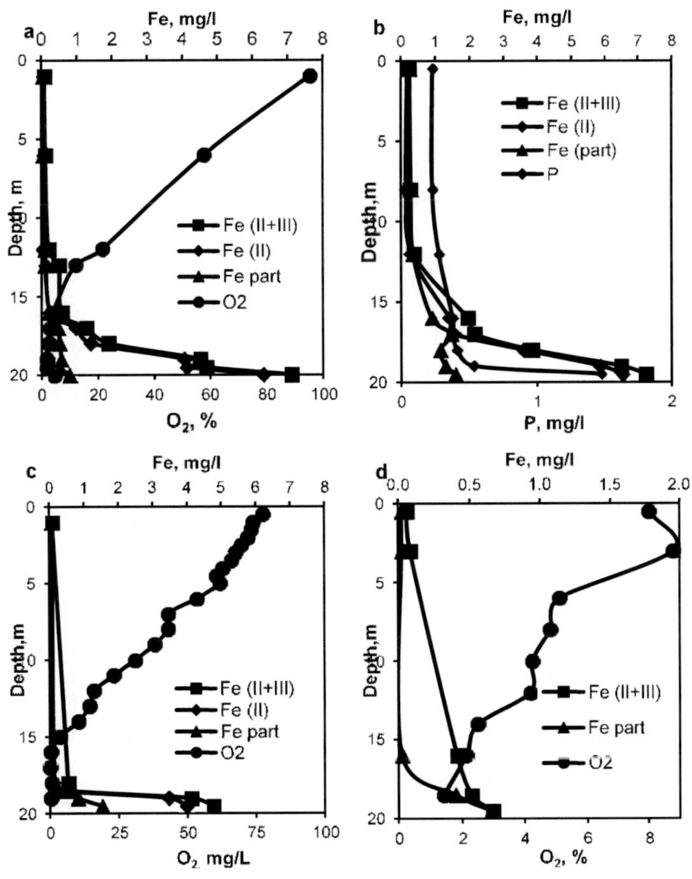

Figure 21. Distribution of iron forms at deep-water reference stations of Lake Maselgskoe (a-winter, b-summer), Lake Pezhikherye (c-winter, d-summer).

During the period of summer stratification, the content of dissolved oxygen starts to decrease from a depth of 4.5-6 meters, reaching only 1-20% at a depth of 17-20 meters. During the period of winter stratification, the oxygen content decreases from a depth of 12-20 m and reaches values of 1-8% at the bottom. The concentration of iron from the surface to the bottom changes accordingly as follows: Fe (II) increases from 0.19 to 6.5 mg/L in the summer; Fe (II + III) increases from 0.23 to 7.2 mg/L and from 0.09 to 7.1 mg/L in the winter. Particulate form (Fe_{part}) increases from 0.66 to 1.58 mg/L in the summer and from 0.04 to 1.17 mg/L in the winter. The total iron concentration (Fe tot) changes from 0.89 to 8.8 mg/L in the summer (July) from the surface to the bottom, and from 0.13 to 7.89 mg/L in the winter (March). An increase in the total iron content towards the bottom (Figure 21) occurs due to an increase in the proportion of dissolved and suspended iron as a result of iron upward diffusion from bottom sediments.

Oxygen deficiency in the bottom layers of the deep-water part of Lake Maselgskoe during the summer and winter stratification favors the influx of ferrous iron from the sediments into the bottom layers.

It is known that increasing the concentration of iron can reduce the availability of phosphorus by precipitating phosphates under oxidizing conditions in fresh waters, i.e., iron actually inhibits the process of eutrophication of water bodies. It was found that under oxidizing conditions of the surface layer, the minimum concentration of Fe corresponds to the minimum concentration of PO_4^-.

The reducing conditions of the bottom horizons of deep-water stations contribute to an increase in all nutrients, including iron (x 10 times), phosphates (x 3.5 times), Figure 21 b. Therefore, bottom sediments act as an important regulator of the nutrient content in the reservoir. Similar results were obtained when studying lakes Glubokoe and Chernoe (Facchina, 1948; Shaporenko et al., 2006).

In these lakes, after the establishment of summer stratification and disappearance of oxygen in the bottom layers, the concentration of iron and nutrients began to increase due to their transfer from bottom sediments to the water column. The same phenomenon was observed in winter, when the concentration of iron, for example, in Lake Glubokoe reached 4-14 mg/L in the bottom layer, while Fe concentration in surface layers ranged from 0.2 to 0.4 mg/L.

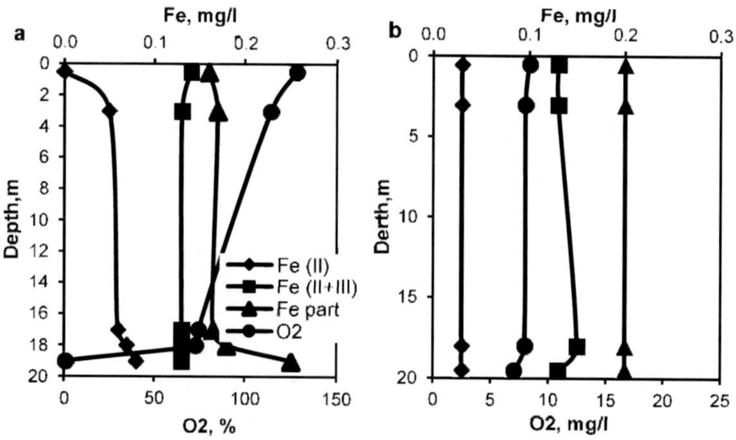

Figure 22. Distribution of iron forms during spring periods of homothermy at deep-water reference stations of Lake Maselgskoe (a) and Lake Pezhikherye (b).

During the period of off-season homothermy, when the main hydrological and hydrochemical indicators (pH, T, mineralization, etc.) become similar across the water column, the distribution of Fe also becomes uniform over the water column and typically does not exceed 0.2 mg/L (Figure 22 b).

In the shallow part of Lake Maselgskoe (4.5 m), the waters are well mixed and weakly stratified. In July, the distribution of pH and water saturation with oxygen is uniform and amounts to 6.75-6.77 and 102% and 52% O_2. During the period of winter stratification, the percentage of oxygen saturation of the bottom layers decreases from 88% to 76%. As can be seen from Figure 23, in the absence of oxygen deficiency, the main form of iron in the period of summer stratification is particulate one. In winter, when anaerobic conditions are formed, the main form of Fe becomes dissolved. There is a general tendency towards an increase in the iron content towards the bottom horizon. This can be explained by involvement of Fe in the C cycle (Kuznetsov, 1952), and an increase in the proportion of dissolved and suspended matter at a shallow water station in July which is associated with production and destruction processes of organic matter (OM).

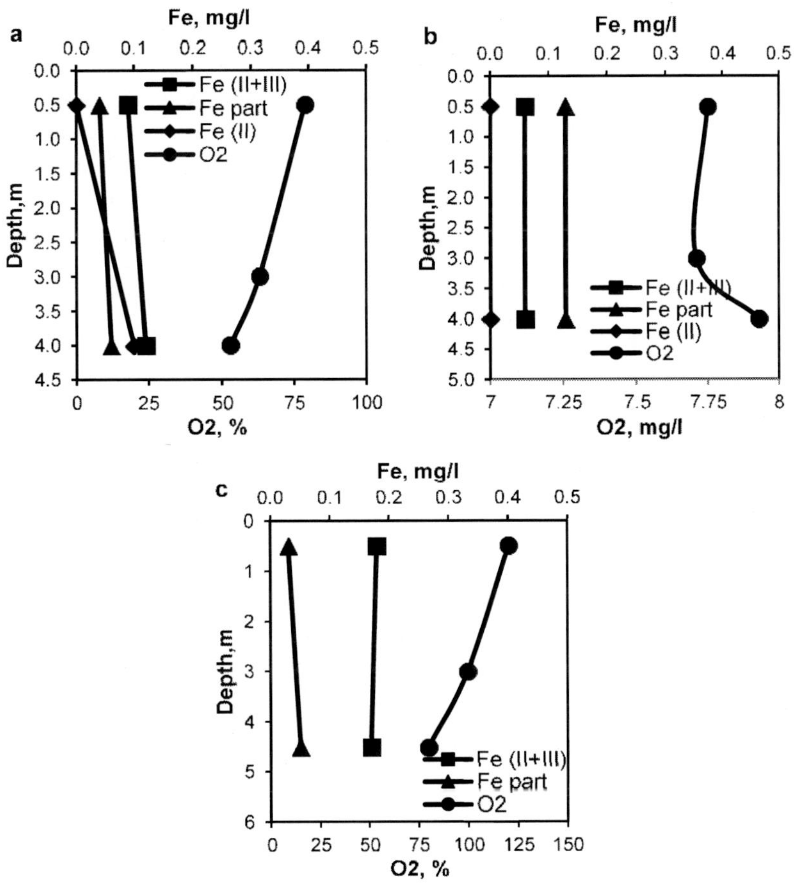

Figure 23. Distribution of iron forms at the shallow-water station Lake Maselgskoe in winter (a), summer (b), and spring (c).

An indirect indicator of the presence of dissolved organic matter is the color index. The color of water reflects to a greater extent the content of allochthonous organic matter, which has a humic nature. The color of water is also strongly influenced by the presence of iron, mostly in the colloidal state. According to the chromaticity indices, Lake Maselgskoe and Lake Pezhikherye are medium-colored lakes, with an average chromaticity index of 69° and 64° in Cr-Co scales, respectively. Note that in the presence of humic acids, the concentration of metal ions existing in the form of complexes is much higher than the concentration of free ions. As such, metal migration in lacustrine environments require accounting for the complexing role of humic acids.

Iron in Lake Vilno

The distribution of iron in the 5-m water column of Lake Vilno is characterized by the following concentrations of total dissolved iron (Fe II + III)) throughout the water column during the periods of summer and winter stratification: up to 0.2 mg/L and up to 1.5 mg/L, respectively, with a predominance of suspended iron (Fe part) in summer, and dissolved iron in winter (Figure 24).

In summer, reference stations V1 and V2 are characterized by sufficient oxygen saturation to the very bottom (from 95% at the surface; up to 84% at the bottom) and a pH range of 7.76 to 7.62. The absence of oxygen deficiency in the summer period contributes to the predominance of suspended forms of iron over dissolved forms; the main share in the total dissolved iron is taken by the Fe (III) form (Figure 24 a). In winter, oxygen concentration has a pronounced stratification from 86.4% at the surface to 0.6% at the bottom, with a pH of 7.3 to 7.23 units. Single measurements of the iron content in the bottom water from the gravitational sampling tube (in the 20-25 cm layer above the water-sediment interface) in winter demonstrated a 10-fold increase in the iron content with a predominance of dissolved iron over suspended iron, and Fe (II) begins to dominate in the total fraction of dissolved iron. In the established reducing environment, bottom sediments, apparently, are an additional source and main regulator of nutrients, including iron (Figure 24 b).

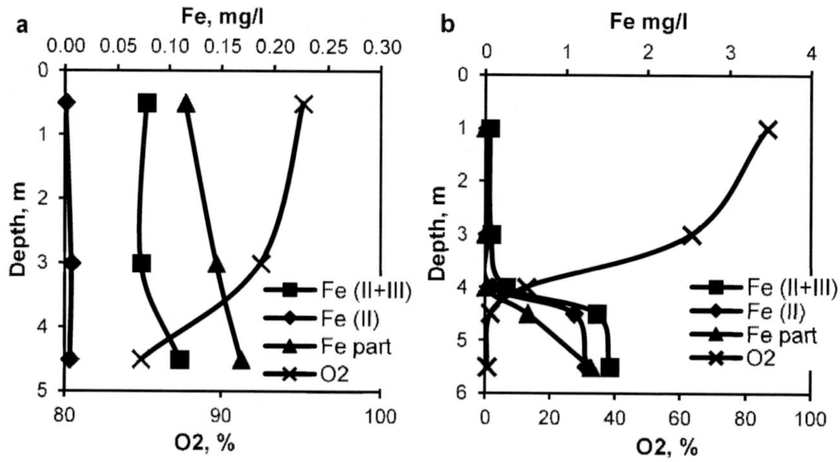

Figure 24. Distribution of iron forms at the V1 station in lake Vilno in 2016 (a - summer; b - winter).

Analysis of the long-term dynamics of total forms of iron in the winter and summer periods at the reference station V1 showed their increase from surface to bottom, i.e., there is a process of accumulation of this element in bottom sediments; the concentration of iron in the winter period is an order of magnitude higher than that in the summer (Figure 25).

During the period of intra-season homothermy, when the water is O_2-saturated till the bottom, the concentration of iron decreases and the nature of the distribution changes to a more uniform from the surface to the bottom.

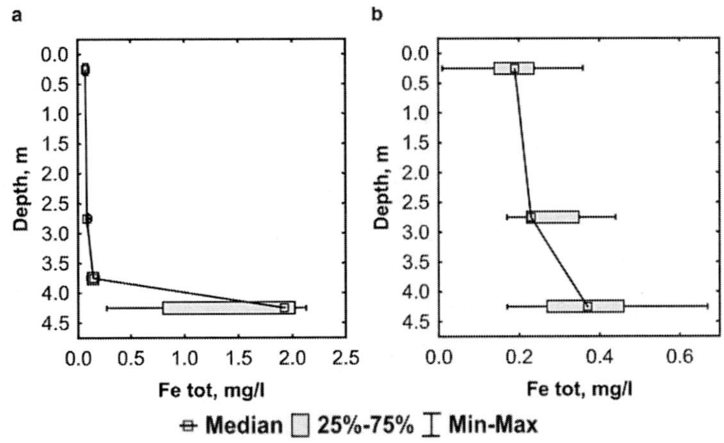

Figure 25. Long-term dynamics of total iron distribution in lake Vilno at the reference station V1 (a – winter; b – summer).

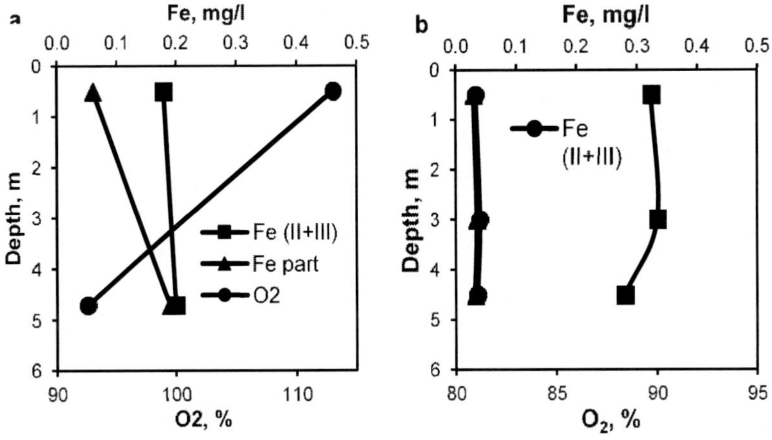

Figure 26. Distribution of iron forms at the reference station V1 of lake Vilno during periods of homothermy (a-spring; b-autumn).

The concentration ranges in the spring from 0.18 mg/L to 0.2 mg/L and in the fall from 0.03 mg/L to 0.035 mg/L (Figure 26). Fe (II) was not detected in water during the period of lake overturn. The hydrological and hydrochemical parameters of the water column during this time showed a low variability of the main indicators at the surface and at the bottom. For example, in the fall of 2012, they were, respectively: O_2 from 89.7% to 88.4%, T from 1.2°C to 1, 3°C. Vilno refers to medium-color waters with an average color of 61° Gr-Co.

To sum up, the spatial and temporal distribution of iron forms in the water of Lake Vilno has been studied in the winter over 8 years' period from 2008 to 2016, in the summer over 9 years' period, from 2009 to 2017. During the periods of spring and autumn overturn, one-time studies were carried out in May 2010 and November 2012.

It is shown that the predominant form of iron is the dissolved form in winter, and the particulate form in summer. There is a process of iron accumulation in bottom sediments, which is reflected in an increase in all forms of iron with depth. Oxygen deficiency in the bottom water layer in winter contributes to an additional supply of iron from bottom sediments via diffusion to the water column. The concentration of iron species in winter is an order of magnitude higher than that in summer.

Dissolved Carbon in Lakes Maselgskoe and Vilno

Aquatic ecosystems serve as one of the most important natural regulators of global geochemical cycles (water, carbon). The study of the evolutionary processes of the chemical composition of land surface waters in response to ongoing environmental changes is today at the center of world environmental science. The study of the biogeochemical carbon cycle is extremely relevant in the context of the ongoing global changes. An increase in the concentration of dissolved organic carbon (DOC) in surface waters due to climate warming and cease of atmospheric acidification is observed throughout the world, including the Scandinavian countries, the British Isles, the northern and eastern regions of the United States (Worrall et al., 2004; Evans et al., 2005; Caron et al., 2006; Sarkkola et al., 2006). The DOC increase leads to modification of boreal surface waters (Reuss et al., 1987; Skjelkvåle et al., 2001; Davies et al., 2005; Köhler et al., 2008). This should inevitably change the nature of DOM and bioavailability (bioaccumulation).

The cycle of organic matter is one of the most important processes occurring in a reservoir. Cyclic processes of organic matter are of great importance, both for the global biosphere, and for its terrestrial and aquatic ecosystems (Neal et al., 2008). Biogeochemical processes are conjugated cycles driven by the carbon cycle. In all ecosystems, stoichiometric aspects governing the recycling of nutrients are fundamental. The conjugated system of cycles with carbon includes the stoichiometrically related cycles of nitrogen and phosphorus; the C: N: P is typically equal to 106: 16: 1 (in weight equivalent - C_{40}: N_7: P_1). The catabolic cycles of sulfur and iron are linked to the carbon cycle. The interaction of biogeochemical cycles in natural waters and the corresponding ratios between nutrients when absorbed by various aquatic organisms largely determine the productivity of freshwater and marine ecosystems (Karim et al., 2001).

Autotrophic organisms control the degree of development and functional activity of bacterioplankton, which, in turn, is one of the main factors in the transformation and accumulation of organic matter created during photosynthesis. In this regard, a study of production and destruction processes in the context of the biogeochemical carbon cycle makes it possible to determine the rate of formation and decomposition of organic matter in the limnic system, as well as to reveal the mechanisms that control the processes of capture (primary production) and release (respiration of plankton, mineralization and decomposition of the allochthonous organic matter in the water column) of carbon dioxide in the water bodies.

Lakes Maselgskoe and Vilno are located at the beginning of the catchments, therefore, intrabasin processes with a minimal input of allochthonous organic matter prevail in them. Thus, one can assume the predominance of autochthonous organic matter in the studied lakes.

To monitor the processes that control the biogeochemical cycle of carbon, reference stations MG and PG with maximum depths and a station in the shallow southern part of the MUG were selected in Lake Maselgskoe (Figure 1). This makes it possible to assess the intrabasin processes occurring in the lake and track their temporal dynamics. The shallow southern part of Lake Maselgskoe is similar to Lake Vilno in morphometric parameters (maximum depths up to 5-6 m, average depth (3.4-3.6 m) and the contribution of bog waters to water recharge (about 11%). As a result, with a water transparency of 2.9-4 m, characteristic of these lakes, the photic layers of the southern part of Lake Maselgskoe and Lake Vilno reach the bottom sediments. In the seasonally stratified parts of Maselgskoe, primary production processes are limited not by the photic layer, but by a sharp decrease in the content of dissolved oxygen (less than 3 mg/L) below 6 - 8 m depth.

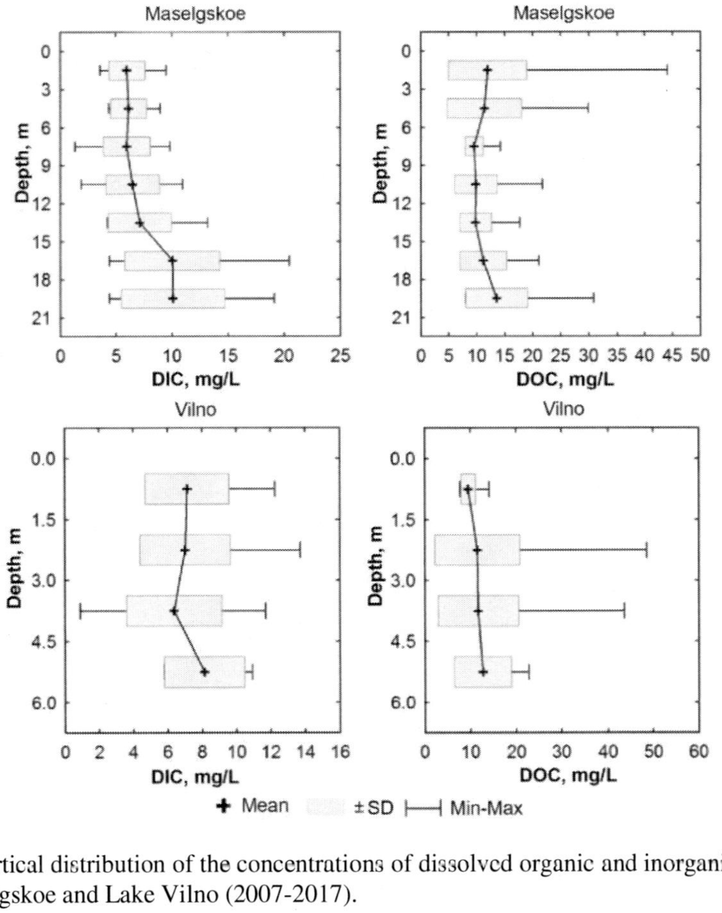

Figure 27. Vertical distribution of the concentrations of dissolved organic and inorganic carbon in Lake Maselgskoe and Lake Vilno (2007-2017).

In Lake Maselgskoe, where the anaerobic layer is quite pronounced, the contribution of aerobic bacterial mineralization to the total OM mineralization varies considerably. The ratio of the primary production of OM to its mineralization (A/R) exceeds 1 on average, which suggests that the lake ecosystem is characterized by the accumulation of organic matter (Shirokova et al., 2008; Kokratskaya et al., 2012; Vorobyeva et al., 2013).

The concentrations of dissolved organic carbon (DOC) and inorganic carbon (DIC) in lakes Maselgskoe and Vilno of Kenozersky National Park varied over a wide range from 8 to 49 mg/L (with average values of 10-14 mg/L) and from 0.9 to 20 mg/L (with average values of 6 - 10 mg/L (Figure 27). In the seasonal dynamics of DOC, some concentration trends can be noted during the study periods (Pokrovsky et al., 2012; Vorobyeva et al., 2017). Seasonally, the vertical distribution of dissolved organic and inorganic carbon concentrations demonstrates heterogeneity in the distribution of dissolved organic carbon with depth in summer and winter.

The main trend is the production of DOC in the summer, which is confirmed by studies of the activity of the phytoplankton community, see below. As a result, in the summer the DOC concentration in Lake Maselgskoe is in a wider range from 8 to 31 mg/L (mean value 11.35 ± 5.1). During the period of winter stratification, the DOC concentration is generally lower than that in the summer period (8-22 mg/L, average value 10 ± 2.5). Stratification of the lake's water column also entails variations in the distribution of a number of physicochemical and biological parameters. The vertical distribution of dissolved carbon forms is subject to more significant variations in the stratified biotope of the reference station (MG, PG) of Lake Maselgskoe in the summer. During the period of winter stagnation of waters, there is a tendency for a uniform increase in dissolved carbon to the bottom horizon. At the same time, the features of the vertical distribution of inorganic carbon in the summer and winter periods, which are manifested in a slight increase in concentrations with depth, indicate the ongoing processes of mineralization of organic matter. Nevertheless, the increase of organic carbon concentration with depth indicates the accumulation of organic matter in the bottom sediments of the lake, which in turn is consistent with the intensity of production and destruction processes. In the shallow part of Lake Maselgskoe (station MUG) and in Lake Vilno (stations V1 and V2), an opposite tendency is observed: in summer, the content of DOC and DIC is distributed almost evenly throughout the water column with a slight increase towards the bottom (8.5-18.5 mg/L (average value 10 ± 2.6), 4.8-8.1 mg/L (average value 5.7 ± 1.1, respectively). During the winter, at a depth of 2 to 4 m, interannual variability of the concentrations of DOC and DIC is within the range of 8 to 49 mg/L (average is 12.5 ± 10.5) in Maselgskoe, and 0.9 to 12 mg/L (average is 7.2 ± 2.5) in Vilno, as illustrated by Figure 27 which shows the vertical distribution of DOC and DIC for the period 2007-2017 in these lakes. In the studied lakes, for the periods of spring and autumn mixing of waters, a fairly uniform distribution of DOC and DOC along the water column is observed: 7.6 to 14.1 mg/L (average is 9.3 ± 1.7) in Maselgskoe; 4.4 to 13.7 mg/L (average is 7.6 ± 2.9) in Vilno.

Analysis of the ratio of seasonally averaged values of organic and inorganic dissolved carbon in lakes is characterized by a slight dominance of the organic component. In winter, there is mainly a dominance and an increase in the DIC with depth, which indicates the predominant processes of mineralization of dissolved and suspended organic matter under the conditions of the under-ice regime. The ratio of organic and inorganic carbon, in turn, confirms the data on the intensity of production and destruction processes in their seasonal aspect and, in general, indicates the heterotrophic status of ecosystems (Shirokova et al., 2008; Kokratskaya et al., 2012; Vorobyeva et al., 2013).

In biogeochemistry, to characterize the sources of organic matter, the Corg/Norg ratio is used; for aquatic phytoplankton, macrophytes, and their humification products it is determined to be ~ 20 (Wolfe et al., 2002; Ilina et al., 2014). The high values of the C_{org}/N_{org} ratio suggest the dominant role of allochthonous sources of organic matter, such as soil, bog and forest waters, in feeding lakes. For the soils of the boreal taiga, this ratio was determined to be about 100 (Ilina et al., 2014); the authors of publications (Dymov et al., 2003; Moiseev et al., 2007) give ratios from 40 to 80.

Table 2. Statistical characteristics of organic nitrogen (Norg) and Corg/Norg ratios in lakes for the period 2007-2017

Depth, m	N	Mean	Min	Max	SD	N	Mean	Min	Max	SD
Maselgskoe (2007-2017) N_{org}, μgN\L						Maselgskoe (2007-2017) C_{org}/N_{org}				
(0;3]	24	342	117	831	142	24	40	13	104	21
(3;6]	7	330	180	437	85	7	44	22	100	27
(6;9]	9	287	114	547	132	9	45	18	101	24
(9;12]	13	279	152	540	115	13	46	19	102	24
(12;15]	11	280	72	614	186	11	55	19	133	37
(15;18]	24	477	44	1350	308	24	30	10	58	14
(18;bottom]	7	534	197	845	241	7	37	13	87	25
Vilno (2007-2017) N_{org}, μgN\L						Vilno (2007-2017) C_{org}/N_{org}				
(0;1,5]	12	368	133	691	170	12	36	17	76	18
(1,5;3,0]	15	300	66	700	151	15	56	15	156	44
(3,0;4,5]	15	295	163	464	77	15	48	23	155	32
(4,5;bottom]	3	316	270	400	74	3	66	25	96	37

The Corg/Norg ratio for Lake Maselgskoe varies in a wide range of 13 to 133 (average value is 43 ± 26), for Lake Vilno the range of ratio is from 15 to 156 (average value is 47 ± 34) (Table 2). During periods of stratification in the bottom layers (from 16 m to the bottom), the Corg/Norg ratios are more stable (mean values 32 ± 17).

The variation in a wide range of C_{org}/N_{org} ratios in the photic layer of the stratified biotope of the reference station of Lake Maselgskoe is explained mainly by the heterogeneity in the distribution of dissolved organic carbon, as mentioned above. For the periods of spring and autumn mixing of waters, it is typical with a fairly uniform distribution of dissolved carbon and the C_{org}/N_{org} ratios range from 12.6 to 41, with an average value of 25.0 ± 8.7. It is possible that an increase in the Corg/Norg ratio, which manifests itself mainly in the summer period, can be associated with active production and destruction processes, which leads to the release of Norg by phytoplankton, periphyton, and bacterioplankton vital activity occurring in both humic and low DOC lakes (Wahenfeldt et al., 2009; Tremblay et al., 2006). On the other hand, carbon compounds can coagulate in the water column and undergo dissolution in deeper layers (Wahenfeldt et al., 2008 a, b)

In lakes Maselgskoe and Vilno, low-molecular forms of DOC (less than 1 kDa or 1 nm) account for 20-30% of the total dissolved concentration in March and May, while the role of suspended forms and large subcolloidal particles (0.025 μm - 20 μm) is insignificant. The relative amount of organic colloids (1 kDa - 0.45 μm) decreases 2-3 times during the summer period (July), while the low molecular weight fraction (<1 kDa) makes up to 80% of the total DOC in the water column. The predominance of potentially "bioavailable" (low molecular weight) and a low proportion of colloidal forms indicates the autochthonous nature of organic carbon. Allochthonous organic carbon is mainly presented in the form of high molecular weight colloids (Wolfe et al., 2002; Pokrovsky, et al., 2002). The ecosystems of lakes Maselgskoe and Vilno are more dynamic compared to that of Lekshmozero in terms of seasonal fluctuations in DOC concentrations. Further comprehensive analysis of variations in the concentrations of forms of dissolved carbon, depending on the intensity of production and destruction processes, concentrations of heterotrophic bacterioplankton, will probably allow a better understanding of the variations and dynamics of the components of the carbon cycle.

Zooplankton Community Composition in Lakes Maselgskoe and Vilno

Zooplankton communities are an important structural unit of aquatic ecosystems. A significant proportion of matter and energy in water bodies passes through trophic chains. Zooplankton is the main consumer of the primary organic matter formed in water bodies during photosynthesis. Organic matter is transformed in the bodies of aquatic organisms and transferred to a new trophic level. The study of the structure, dynamics and functional characteristics of planktonic communities makes it possible to understand the processes occurring in water bodies, as well as the role of aquatic organisms in the biological circulation and transformation of energy.

Hydrobiological studies of water bodies in the southern part of the Kenozersky National Park (Lake Maselgskoe and Vilno) were carried out in winter and summer. Zooplankton communities of the studied lakes are represented by three groups of organisms: Rotifera, Cladocera, and Copepoda. The modern planktonic fauna of Lake Vilno comprises 38 species and forms of invertebrates, including 21 species of Cladocera, 8 species of Copepoda, and 9 species of Rotifera. The zooplankton community of Lake Maselgskoe is represented by 54 species, of which 11 are Copepoda, 30 are Cladocera and 13 are Rotifera.

Copepoda and Rotifera dominate in winter. Zoocenoses are represented by winter species and year-round forms. The species diversity of the zooplankton community was determined for Rotifera - *Kellicottia longispina, Kerattela quadrata, Polyarthra dolichoptera, Keratella cochlearis, Filinia longiseta.* In February, these species inhabited the entire water column, during this period their reproduction begins, and egg-bearing females are found in the samples. Of the Copepoda, *Cyclops strenuus, Cyclops kolensis, and Eudiaptomus gracilis* dominated in the lakes. Copepoda were observed in all horizons during the uner-ice period, but some species peaked at a certain depth. In Lake Maselgskoe, *Eudiaptomus gracilis* are numerous in a layer of 3-12 m; in Vilno, this species is recorded at a depth of 3 m. *Cyclops spp* mainly remained in the lower layers of the lake. In February-March, the zooplankton of the lakes is actively developing; a large number of oviparous females, as well as nauplii, were noted in the populations of Copepoda. The number of nauplii, depending on the year of observation in Lake Maselgskoe, ranged from 11 to 96% of the total number of Copepoda. The nauplii Copepoda were distributed evenly throughout the entire water column, but their greatest number was noted in the 3-7 m layer. In Lake Vilno, the nauplii were confined to the upper 3-meter layer, and their number corresponded to 16-23% of the total Copepoda abundance. Cladocera in the winter plankton of water bodies are represented by the species *Daphnia longispina, Daphnia cristata, Bosmina longirostris.*

In winter, as a result of the formation of an ice cover, the temperature and oxygen regimes of the water bodies change, which affects the structure, distribution and development of plankton communities. The quantitative indicators of the development of zooplankton in winter are significantly lower than those in summer, since most species of zooplankton organisms drop out of the community. However, the number and biomass of rotifers and Copepoda in winter can reach summer values. In the under-ice period, the main contribution to the total abundance and biomass of the zooplankton community is made by Copepoda (up to 98%). The average values of the quantitative indicators of winter zooplankton in the studied lakes changed significantly from year to year. In Lake Maselgskoe, during the study period, the average values of abundance varied from 5430 to 47220 ind./m^3 and biomass from 0.01 to 0.13 g/m^3, respectively. In Vilno, quantitative indicators varied within the range from 300 to 66650 ind./m^3

and from 0.002 to 0.45 g/m³, respectively. The interannual dynamics of quantitative indicators in winter is shown in Figure 28.

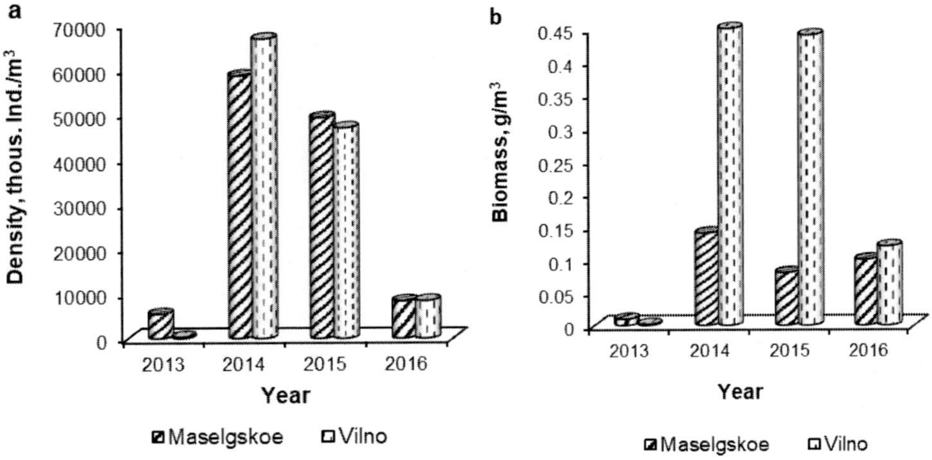

Figure 28. Interannual population dynamics (a) and biomass (b) of zooplankton of lakes in winter (2013-2016).

Zooplankton communities were in a balanced state in winter. This is confirmed by the indicators of the Shannon-Weaver species diversity index. For Lake Vilno, the index values corresponded to 2.25-2.40 bits (in number) and for Lake Maselgskoe they ranged from 2.2 to 2.7 bits (in terms of numbers). In February-March 2016, the species diversity index for Lake Vilno was 1.0 bit (in abundance) and for Lake Maselgskoe it was 1.84 bit (in abundance). The early freeze-up contributed to the preservation of a higher heat content of water masses, which largely determined the temperature conditions in water bodies and influenced the structure and development of zooplankton communities during this period.

The summer is characterized by maximum water warming and the formation of stable thermal stratification. At the beginning of summer, cold-water species migrate from the water column to the bottom layers, their number decreases. The planktonic fauna is enriched by the Cladocera (genus *Daphnia, Bosmina, Chydorus, Polyphemus, Ceriodaphnia, Diaphanosoma*), which determine the species diversity of zoocenoses. In the summer, the dominant assemblage of zooplankton is represented by the following species: *Mesocyclops leuckarti, Eudiaptomus gracilis, Thermocyclops crassus, Diaphanosoma brachyurum, Bosmina coregoni, Bosmina longirostris, Daphnia cucullata, Daphnia galechnata. Ceriodaphnia reticulata, Asplanchna priodonta, Kellicottia longispina, Kerattela cochlearis.*

In deep stratified lakes, the vertical distribution of zooplankton in summer is not uniform. The formation of the vertical structure of zoocenoses in water bodies occurs under the influence of a complex of abiotic and biotic environmental factors. The main factors influencing the formation of biotopic heterogeneity in the distribution of zooplankton within the water body are temperature stratification of the water column and food resources. Thermal differentiation of the water column contributes to the formation of additional ecological niches in water bodies, allowing thermophilic and psychrophilic zooplankton complexes to develop under the most optimal conditions (Vezhnovets et al., 2011; Rechkalov et al., 2011).

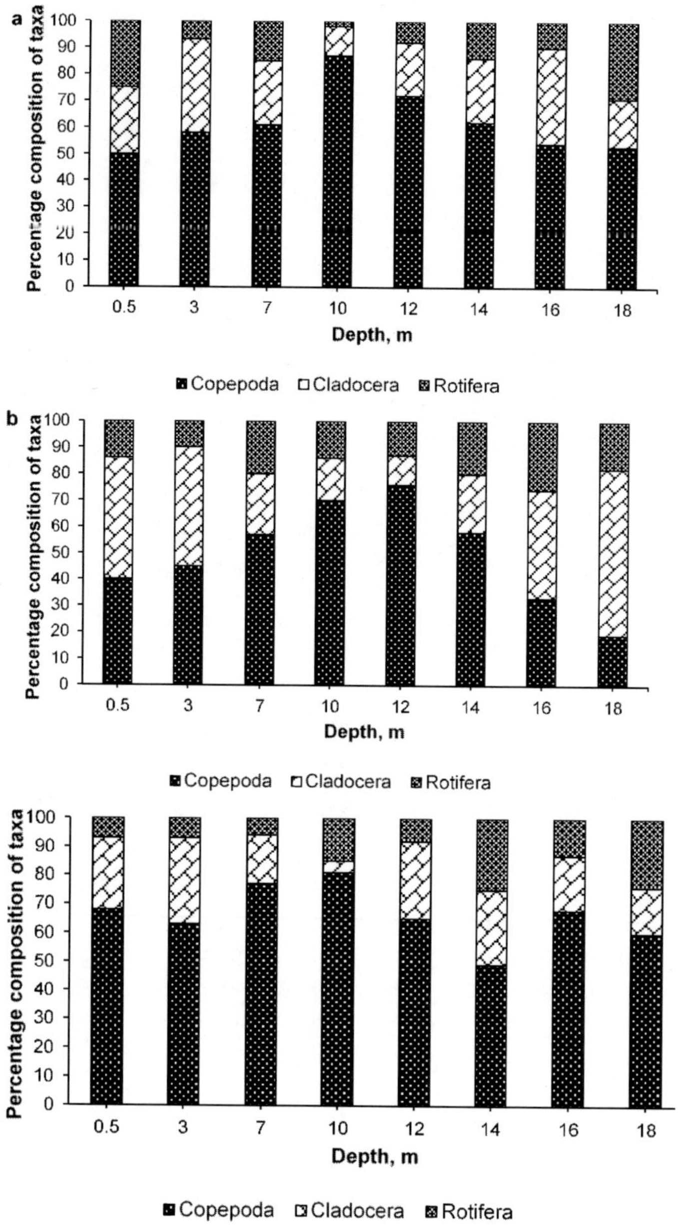

Figure 29. Vertical distribution of various taxa of zooplankton in the lake Maselgskoe: a - July 2013, b - July 2014, c - July 2015 (mean value).

In the epilimnion, phytophages, filter-feeding zooplankton (Cladocera and Calanoida), filter-feeding Rotifera, nauplii Copepoda and Cladocera, I-IV Copepoda stages are concentrated. Psychrophilic species of Copepoda (representatives of the genus *Cyclops*) remain in the lower layers of the hypolimnion; this distribution is also typical for stenothermal Rotifera. The concentration of zooplankton in the metalimnion zone was also noted. Migrations of zooplankton to the meta- or hypolimnion in deep stratified water bodies are an adaptive form of behavior of aquatic organisms, which contributes to the maintenance of populations in a stable state and reduces competitive relationships in the zoocenosis (Feneva et al., 2016; Lynch,

1979). Depending on the year of observation, Cladocera was mainly concentrated in the epilimnion (0-3 m), but also exhibited some accumulations in the bottom layers (14-18 m). Copepoda are abundant in the metalimnion and hypolimnion (10-14 m). For Rotifera, the maximum abundance was recorded in the hypolimnion (16-18 m). Figure 29 shows the proportion of the main groups of zooplankton in different layers of Lake Maselgskoe in 2013-2015.

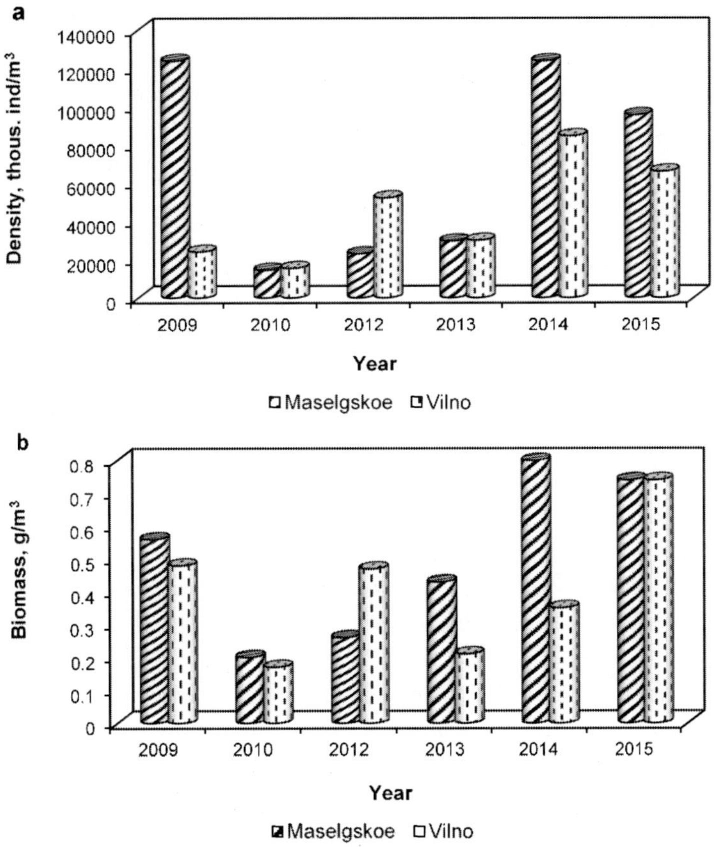

Figure 30. Interannual dynamics of lake zooplankton abundance (a) and biomass (b) in the summer (2009-2015).

The average values of the quantitative indicators of summer zooplankton in the studied water bodies varied significantly depending on the year of observation. In summer, Copepoda and Cladocera dominated in numbers in the lakes. Copepoda provided the main contribution to the biomass. Overall, the development of zooplankton in Lake Maselgskoe in summer is characterized by the following quantitative indicators: the number varied from 14,500 to 125,000 individuals/m^3, and the biomass from 0.2 to 0.8 g/m^3. In Lake Vilno, the abundance of zooplankton varied from 15,400 to 84530 ind./m^3, and the biomass from 0.17 to 0.74 g/m^3. The interannual dynamics of quantitative indicators in the summer period is shown in Figure 30.

To assess the ecological state of zooplanktocenoses in different years of observations, the Shannon - Weaver species diversity index was calculated. The index values of 2.96-4.30 bits

(Lake Vilno) and 2.54-4.23 bits (Lake Maselgskoe) indicate a relatively stable and balanced state of the zooplankton communities of the studied lakes.

To sum up, complex seasonal studies of the ecosystems of lakes Maselgskoe and Vilno showed that the taxonomic composition of zooplankton communities is typical for cold water bodies of the boreal zone. The species structure and quantitative development of zooplankton in both lakes are rather similar. The ecological state of lakes Vilno and Maselgskoe can be assessed as sustainable.

Phytoplankton Community Composition of Maselgskoe and Vilno lakes

The main source of autochthonous organic matter in water bodies is due to the photosynthetic activity of phytoplankton, which provides the basis for all subsequent stages of the production process in the water body. The indicators of the development and functioning of planktonic algocenoses are widely used in hydroecological studies to characterize the development and state of autotrophic communities (Mineeva, 2007). Microalgae form the basis of trophic chains of the planktonic biocenosis and largely determine the ecological situation in the aquatic environment. At the same time, phytoplankton is highly sensitive to pollution of aquatic ecosystems. The development of phytoplankton is significantly influenced by the nutrients and can cause the eutrophication of water bodies. The content of nutrients and the ratio were optimal for the development of phytoplankton in the studied lakes. The ratio N_{tot}/P_{tot} for lakes Maselgskoe and Vilno (> 16) in the summer season indicates the limiting role of phosphorus in the development of phytoplankton communities in lakes.

The water transparency in Lake Maselgskoe in summer was within 2.9-4.0 m (average 3.6 m); in winter it ranged from 2.1 to 5.0 m (average value 3.8 m). In Lake Vilno, the transparency ranged from 1.9 to 3.8 m (average 3.1 m in summer) and from 2.5 to 5.0 m (average 3.42 m in winter). In Lake Vilno and in the littoral part of Lake Maselga, the content of dissolved oxygen throughout the water column in all seasons is optimal for the development of phytoplankton. In the deep-water part of Lake Maselgskoe, during periods of winter and summer stagnation, the development of phytoplankton is not limited by the photic layer, but by a sharp decrease in the content of dissolved oxygen at the 6-8 m deep.

The study of the qualitative and quantitative content of phytoplankton communities in Maselgskoe and Vilno lakes was carried out from 2007 to 2016. The phytoplankton of the studied lakes Maselgskoe and Vilno is characterized by a pronounced seasonality. In each season, one or more groups of algae prevail in the lake. In winter (March), the abundance and diversity of phytoplankton in lakes depends on the thickness and condition of the snow cover. In March 2009, in Lake Maselgskoe, with a thickness of wet translucent snow less than 35 cm, when the level of sunlight becomes sufficient for photosynthesis of algae, even under the ice at a depth of 1-6 m, small flagella are found - representatives of Euglenophyta (genus *Trachelomonas*), Cryptophyta (genus *Cryptomonas*) and Chlorophyta (genus *Chlamydomonas*, etc). Psychrophilic representatives of Bacillariophyta are numerous and diverse (5-8 species) with a predominance of *Aulacoseira islandica* species (up to 54% of the total number and 37.3% of the total biomass) and *Asrerionella formosa*. In terms of abundance and biomass, diatoms (Bacillariophyta) dominated up to 82% of the total phytoplankton content. The abundance and biomass of phytoplankton reached 530 thousand cells/L and 644 μg/L, respectively.

With sufficient snow cover under the ice in Lake Maselgskoe and Lake Vilno, phytoplankton is poor due to lack of light. At the 1 to 3 m depth, the abundance and biomass were in the range of 7.5-154 thousand cells/L and 17-82 μg/L and strongly decreased in the lower horizons. Phytoplankton are mainly represented by Bacillariophyta, Cryptophyta, and Chlorophyta.

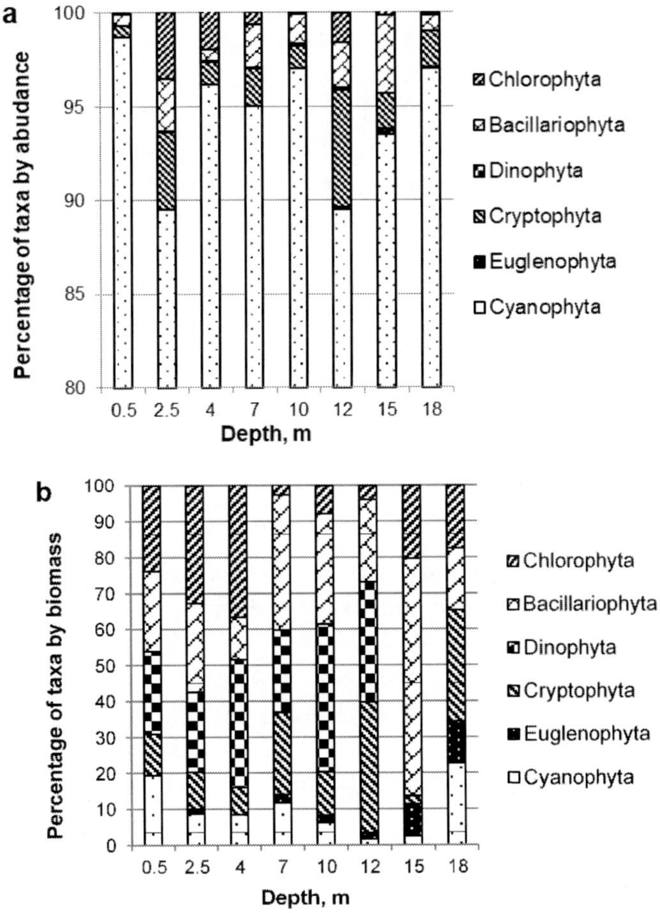

Figure 31. Vertical distribution of various phytoplankton taxa by abundance (a) and biomass (b) in Lake Maselgskoe (July 2015).

During the period of ice breaking up before the establishment of temperature stratification, which usually happens when the upper water layer warms up to 10-12°C, the intensive development of a psychrophilic complex of Bacillariophyta, Dinophyta, Cryptophyta, Chlorophyta begins. Diatoms (Bacillariophyta) dominated in terms of number (up to 96%) and biomass (up to 85%), respectively. The genus *Aulacoseira (A.islandica)* dominates in Lake Maselgskoe, and *Tabellaria fenestrata* dominates in Lake Vilno. Spring phytoplankton is evenly distributed throughout the water column with slight variations in abundance (112.5-265 thousand cells/L) and biomass (522.2-941.9 μg/L).

In the deep part of Lake Maselgskoe, during the period of summer stagnation, the most diverse and abundant phytoplankton is up to a depth of 6-8 m, where the dissolved oxygen content is more than 3-4 mg/L. In the surface horizon, the abundance and biomass of phytoplankton reach 1730 thousand cells/L and 1300 μg/L, respectively. In summer, in Lake Maselgskoe, when the water warms up to 18-22 °C in the water column with the optimal content of dissolved oxygen, the abundance and biomass vary within 433-17314 thousand cells/L and 187.5-1300 μg/L, respectively. In summer phytoplankton, Cyanophyta dominated in numbers - the genus *Anabaena* (*A. lemmermanii*), *Aphanocapsa* (*A.delicatissima and A.inserta*), *Cyanodictyon sp*, reaching 68.4-98% in number. Biomass was dominated by Chlorophyta (*Botryococcus braunii* up to 16.4-53.9%) and the Bacillariophyta complex (*Tabellaria fenestrata, Aulacoseira ambigua, Cyclotella + Stephanodiscus spp.*), Cryptophyta (*Cryptomonas sp*), Dinophyta (*Peridinium cinctum, Ceratium hirundinella*). With a decrease in temperature below 10°C in the water column (at the depths of 8-14 m), among the diatoms (Bacillariophyta), there are psychrophilic species *Aulacoseira islandica*. An example of the vertical distribution of various phytoplankton taxa by abundance (a) and biomass (b) in Lake Maselgskoe is shown in Figure 31, during July 2015.

Due to the shallowness of Lake Vilno, the bottom layer is an active component of the photic layer, where solar energy can freely penetrate and an increase in the concentration of nutrients is observed compared to the surface horizon. As a result, summer phytoplankton is diverse and abundant to the bottom, the abundance and biomass reach values of 766-6027 thousand cells/L and 604-10140 μg/L. Until the bottom horizon, Cyanophyta dominate in abundance (up to 92%), mainly the species *Aphanocapsa delicatissima* or the *Aphanocapsa delicatissima* complex, *Microcystis aeruginosa*; in recent years, the species *Gloeotrichia echinulata* has joined the mass species. In the bottom layer in summer, Raphydophyta (*Gonyostomum semen*) often develops. In terms of biomass, the complex of taxa Cyanophyta, Chlorophyta and Dinophyta) predominates, in the bottom layer the maximum biomass can be up to 96% Raphydophyta (*Gonyostomum semen*).

Raphydophyta, in particular, the species *Gonyostomum semen*, began to occur more and more often in the water bodies of Northwest Russia. Until the end of the last century, they were rare in northern water bodies, but they were quite widespread in water bodies of the temperate zone, mainly in lakes with a pH of 5.5 to 6.8 (Cronberg et al., 1988; Trifonova, 2015).

For the summer phytoplankton in lakes Maselgskoe and Vilno, the species diversity is characteristic of the Northwest of Russia. The presented taxa are predominantly cosmopolitan forms, most of the species are oligo-β and β-mesosaprobes. The biomass of summer phytoplankton makes it possible to determine the trophic status of the studied lakes as mesotrophic.

Long-Term Dynamics of Heterotrophic Bacterioplankton (HB) in Ecosystems of Small Lakes Maselgskoe and Vilno

This section presents the results of studies (2006-2016) of the interannual and seasonal dynamics of the abundance and ecological and trophic structure of heterotrophic bacterioplankton (HB) in two boreal lakes, differing in morphometric, hydrological and hydrochemical characteristics (Lake Vilno (White Sea basin) and Maselgskoe Lake (Baltic Sea basin) of Kenozero National Park.

The study of the quantitative content of various ecological-trophic groups of heterotrophic bacteria showed their low abundance in the lake water, which is due to the low concentration of organic substances and nutrients. The lowest HB concentrations were observed during the under-ice period. Figure 32 shows the interannual dynamics of HB population during the periods of winter stratification. In winter, the abundance of heterotrophic bacterioplankton in the water of Lake Maselgskoe varied over a wide range. For the period 2007-2014 in Lake Maselgskoe, their fluctuations were: EB - 1-1120 CFU/ml (median 48 CFU/ml), FOB - 5-1460 CFU/ml (median 80 CFU/ml) and OB - 5-1310 (median 50 CFU/ml). In Lake Vilno, the ranges were: EB - 2-890 CFU/ml (median 70 CFU/ml), FOB - 10-2000 CFU/ml (median 140 CFU/ml) and OB - 10-2670 (median 60 CFU/ml).

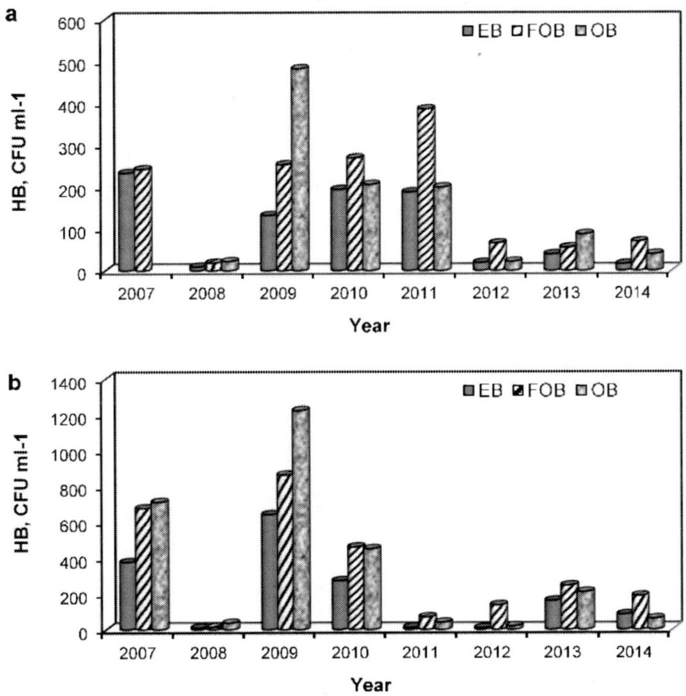

Figure 32. Interannual dynamics of quantitative indicators of various ecological and trophic groups of HB (CFU/ml) in the water of lakes Maselgskoe (a) and Vilno (b) in the winter period 2007-2014 (weighted mean value). EB, eutrophic bacteria; FOB, facultative oligotrophic bacteria, and OB, oligotrophic bacteria.

As can be seen from the data presented, the median values of the abundance of all groups of bacteria were only tens of CFU/ml, which indicates the completion of the transformation of easily oxidized organic matter by the end of the winter stratification period.

The minimum HB concentrations were recorded in March 2008 and 2011. In autumn 2007, the air temperature transition from 4 °C to negative values occurred within three weeks. Winter was the warmest during the study period. Average air temperature for November-March 2007-2008 was -4.8 °C (-5.5 °C in 2006-2007 and -9.1 °C in 2005-2006) (data of meteorological observations at the HMS Kargopol). An earlier transition of air temperature through zero in 2007 and 2011 led to earlier freeze-up and lower heat loss by water bodies, which led to higher values at the bottom and weighted average water temperatures in lakes in March 2008 and 2012. These climatic factors allowed the microbocenosis of lakes by the end of the period stratification to complete basically the mineralization of OM formed in the summer period. In the shallow lake Vilno, low values of HB abundance were also observed in 2011.

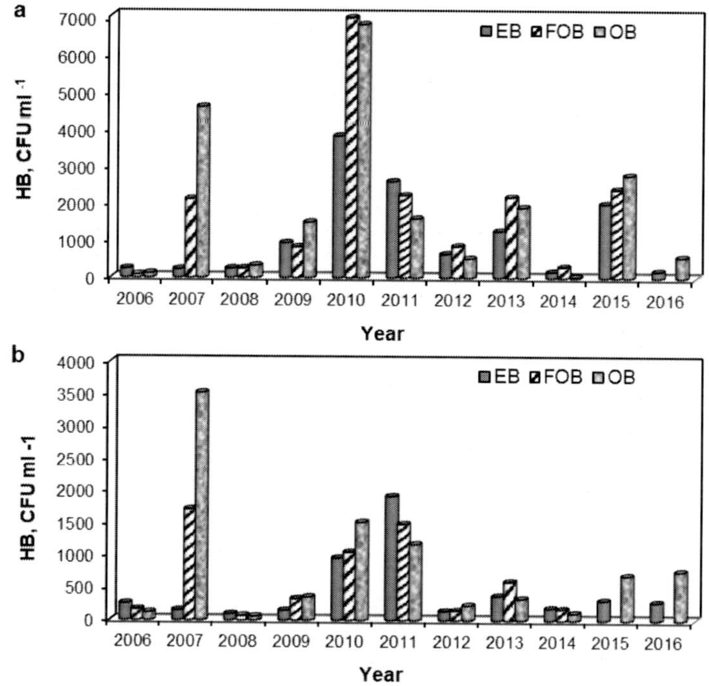

Figure 33. Interannual dynamics of quantitative indicators of various ecological and trophic groups of HB (CFU/ml) in the water of Lake Maselgskoe (a - surface (0.5 m), b - weighted average values over the water column (0-20 m) in the summer over the period from 2006 to 2016.

In summer, the number of all HB groups increases several times, which reflects both a favorable temperature regime and an increase in the intensity of mineralization of the organic matter newly formed during photosynthesis. Concentrations of EB, FOB, OB in the summer periods of 2006-2016 varied within wide limits: in the water of Lake Maselgskoe, respectively, from 40 to 4960 CFU/ml (median 170 CFU/ml), from 30 to 7040 CFU/ml (median 250 CFU/ml), from 20 to 6860 CFU/ml (median 300 CFU/ml), in the water of Lake Vilno, respectively, from 50 to 4930 CFU/ml (median 370 CFU/ml), from 100 to 4740 CFU/ml (median 420 CFU/ml), from 70 to 6800 CFU/ml (median 375 CFU/ml). However, the weighted average values presented in Figures 33 and 34 were not high and remained typical for clean waters. In the ecological-trophic structure of HB, mainly FOB prevailed. In general, the concentrations of various ecological-trophic groups differed insignificantly, maximum 2 times. The data obtained indicate the availability for heterotrophs of organic compounds having various origins and indicate the active self-cleaning activity of bacterioplankton.

The maximal concentrations of heterotrophic bacterioplankton in the surface layer were observed in July 2010. These were associated with the abnormally hot July observed in the European part of Russia. The ecosystem of the lake reacted to the increase in water temperature (25-26°C) by increasing the intensity of production and destruction processes, as evidenced by a sharp increase in the concentrations of all groups of heterotrophic bacterioplankton in the surface layer (Figure 33a).

The bacterioplankton of the water column also reacted to the increase in water temperature (Figure 33b), but not as intensely as the surface one, because the deep-water layers (hypolimnion) of the stratified lake are less subjected to phytoplankton bloom compared to the surface layer. The weighted average values of the EB, FOB and OB content in the full water column were 4, 7 and 5 times, respectively, lower than the corresponding values in the surface layer.

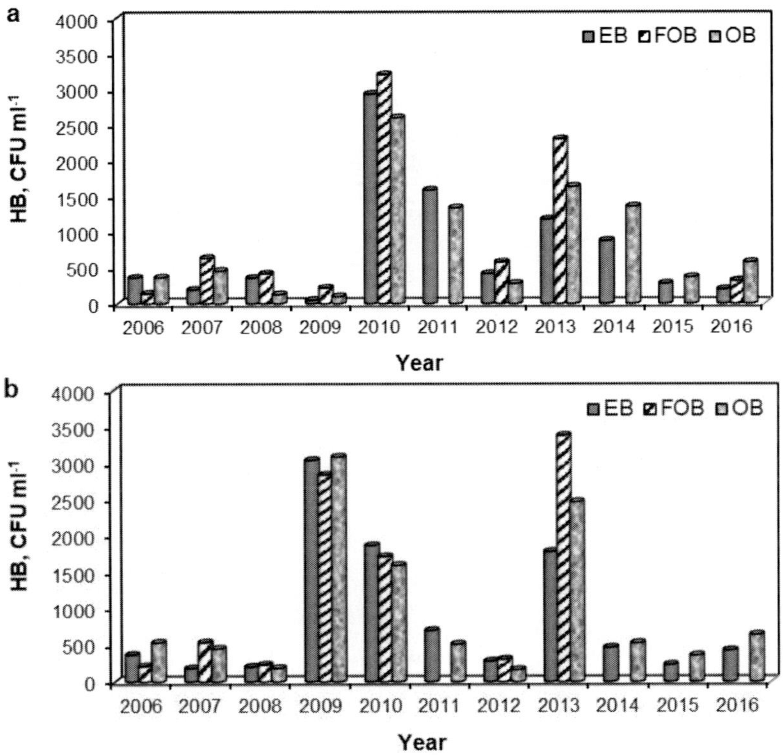

Figure 34. Interannual dynamics of quantitative indicators of various ecological and trophic groups of heterotrophic bacterioplankton (HB, CFU/mL) in the water of Lake Vilno (a - surface; b - weighted average values by the water column) in the summer period 2006-2016.

In the shallow lake Vilno, this increase in HB concentration between surface layer and the entire water column was only 1.5-1.9, which indicates the lower resistance of the unstratified lake to temperature changes.

Shirokova et al. (2013) also noted an increase in the intensity of bacterial mineralization of dissolved organic matter with an increase in the water temperature for a boreal lake in the summer of 2010, which reduced the concentration of DOM in the water column by more than 30% compared to near-normal summer temperatures.

At the same time, there were no significant changes in the concentration of nutrients (Zabelina et al., 2014). Microbiological studies have shown that heterotrophic bacterioplankton maintains the stability of lake ecosystems by rapidly recycling nutrients, reducing their flow into the bottom layers and sediments of the lakes.

Overall, seasonal dynamics of heterotrophic bacterioplankton, studied in 2009, differed for stratified and non-stratified lakes, for both water column weighted averages and the surface layers (Figure 35).

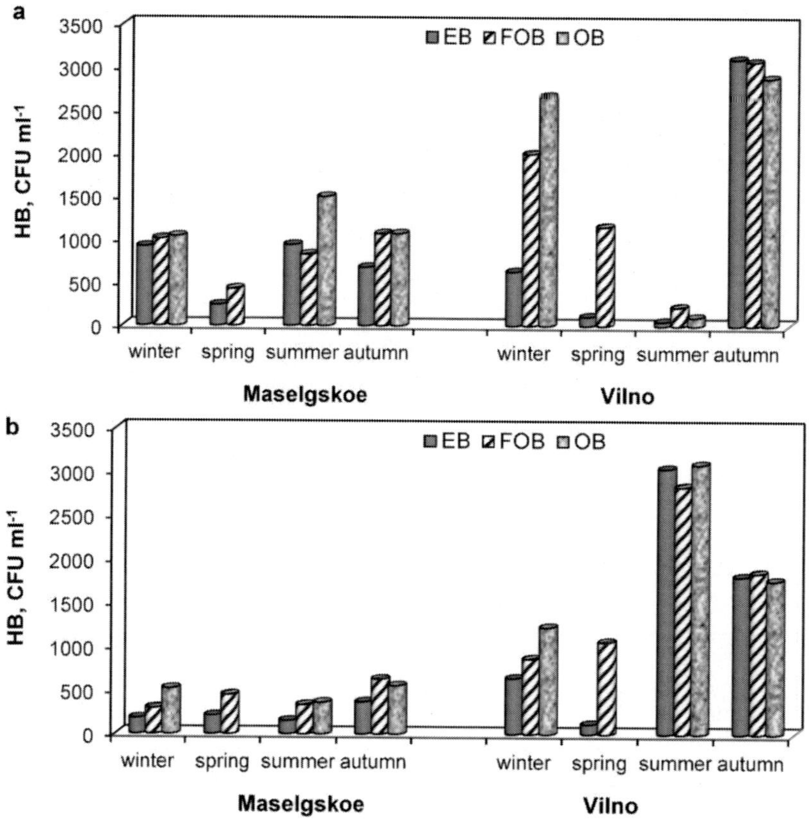

Figure 35. Seasonal dynamics of quantitative indicators of various ecological and trophic groups of HB (CFU/ml) in the surface water (a), and the weighted average values (b) for the water column of lakes Maselgskoe and Vilno.

For the deep stratified lake Maselgskoe, seasonal differences in 2009 are insignificant for both the surface layer and the entire water column. In the shallow, non-stratified lake Vilno, a pronounced seasonal dynamics was observed: the maximum content of all groups of heterotrophic bacterioplankton in the surface layer was found in July. In the entire water column, this maximum was observed in November, when the amount of particulate organic matter in the lake increases due to the death of plankton.

It should be noted a high variability in the content of heterotrophic microorganisms in studied lakes, associated mainly with the morphometry of the lake, seasonal and intrabasin processes. There are no literature data and archival data on microbiological studies for small lakes of the KNP. In general, microbiological indicators in the studied lakes are typical for clean waters.

Primary Production - Organic Matter Mineralization in Lakes Maselgskoe and Vilno

Within a context of carbon cycle, study of production - mineralization (respiration) of OM allows determining rate of OM degradation in limnic system and assess the mechanisms of uptake (primary production) and release of DOM, POM) of CO_2 in lakes. To characterize biogeochemical processes of carbon cycle in lakes, we studied OM degradation and primary productivity in lakes Maselgskoe and Vilno on the deep water (reference) stations in 2009 during several seasons (Table 3). These studies demonstrated that intensive OM production starts after spring flood, when waters are enriched in nutrients and phytoplankton development initiates. PP achieves the maximum in summer (August). In most studies performed over summer period, production was dominated over destruction which is typical for lakes that are capable to accumulate quickly the organic matter and therefore most vulnerable to eutrophication.

In the deep part of the Maselgskoe Lake, which is characterized by well pronounced vertical stratification, the production processes are limited by photic layer depth, given that the rate of OM formation by phytoplankton decreases with depth. Aerobic respiration in lakes varies over the water column. Thus, in the Maselgskoe Lake, the degradation rate of organic matter increased in lower horizons of the photic layer, despite some hypoxia zone. The aerobic respiration in the Maselgskoe Lake plays a subordinary role; Overall, the production - mineralization processes directly depend on the seasons, climate conditions, local wind environments.

The vertical distribution of PP and R values during 2009 demonstrate OM formation in the photic layer, destruction of OM over full depth of the water column and occurrence of both processes during winter, under ice. The PP achieved 6.87 mg C m^{-3} day^{-1} in Maselgskoe and was not detected in Vilno. In spring, with water temperature raise and the beginning of phytoplankton activity, both OM production and its respiration were activated. In May, the PP in Maselgskoe achieved 39.3 mg C m^{-3} day^{-1} whereas the aerobic destruction of OM increased to 280 mg C m^{-3} day^{-1}, compared to 11.3 mg C m^{-3} day^{-1}under ice in winter. In summer, there was further increase in OM production. The respiration of OM was minimal in winter (11.3 mg C m^{-3} day^{-1}), achieving maximum in spring (315 mg C m^{-3} day^{-1}) and decreasing again in summer down to 36.8 mg C m^{-3} day^{-1}. It is important to note that the measurements were often conducted during mass plankton activity so that the PP corresponds to maximal values of the lake ecosystem.

The contribution of bacterioplankton activity to overall OM respiration ranged from 22 to 75% depending on season and given ecosystem. Overall, the bacterioplankton was actively degrading OM in all lakes. In spring, during temperature rise and delivery of allochthonous organic matter, the bacterial degradation of OB also increased. In Maselgskoe, where the anaerobic zone is clearly pronounced, the aerobic respiration plays a subordinary role (22 to 54%). The ratio of Primary Productivity to Respiration also varied depending on season, from 0.1 to 0.9. However, the mean value of this ratio was below 1 in all lakes, which allows to conclude that considered aquatic ecosystems are in stable status.

Table 3. Seasonal dynamics of integral parameters of primary production and respiration of organic matter in lakes measured in 2009

Season	Primary production (ΣPP), mgC/(m^3×day)	Aerobic respiration (ΣR), mgC/(m^3×day)	Bacterioplankton destruction (Db), %	ΣPP/ΣR
Maselgskoe (deep reference station)				
July 2007	260.6	16.42	95	15.9
March 2008	113	36.1	53	3
July 2008	186.8	67.4	84	7.2
March 2009	6.87	11.29	22	0.6
May 2009	39.31	279.79	54	0.1
July 2009	33.91	36.81	38	0.9
Vilno				
March 2009	0	46.8	50	0.1

CONCLUSION

Analysis of long-term data (2006 - 2017) obtained in the course of complex seasonal studies of ecosystems of Lake Maselgskoe and Lake Vilno in the southern part of the Kenozersky National Park, which included the study of the spatial-temporal dynamics of hydrological and hydrochemical indicators, qualitative and quantitative characteristics of biological communities, demonstrated that the current ecological status of these lakes can be assessed as stable.

The investigated lakes Maselgskoe and Vilno are located in the zone of excessive moisture. The predominant role in their nutrition belongs to atmospheric precipitation, which, with a small catchment area, makes them low-mineralized and, therefore, causes a low buffer capacity of their waters with a predisposition to acidification (decrease in pH value). Water belongs to the bicarbonate class of the calcium group. The active reaction (pH) of the environment of aquatic ecosystems during the periods of research was characterized mainly as neutral. According to the vertical temperature distribution in the summer period, the studied lakes are represented by 3 types: epithermal (Vilno), metathermal (Maselgskoe), and hypothermic (Pezhikherye, which is a relatively isolated northern part of Lake Maselgskoe).

During the studied periods, a homothermy is observed for shallow lake Vilno and the shallow part of Maselgskoe. The dissolved oxygen concentration changes slightly from the surface to the bottom layer (Vorobieva et al., 2013; Vorobieva et al., 2017). At deep areas of Lake Maselgskoe, the vertical profile of dissolved oxygen concentration is similar to that of water temperature. Lake Maselgskoe shows constant temperature stratification and an oxygen deficiency during periods of winter and summer stratification (anoxic zone deeper than 12-14 m). Lake Vilno is characterized by a favorable oxygen regime in all seasons. In winter, weak stratification is observed in the bottom layer. The content of dissolved oxygen in the bottom horizons of the lake does not fall below 1.0-8% oxygen saturation.

Seasonal changes in the content of nutrients in the water of lakes Maselgskoe and Vilno reflect a tendency of their accumulation during the under-ice period and a decrease during the open water season. The very low content of phosphates (and sometimes their absence) and the high content of nitrates in the summer period in the Vilno and Maselgskoe lakes indicate a limitation of phosphorus compounds for the development and life of aquatic organisms, while they have an excess of easily assimilated nitrogen in the form of nitrate ion. The ratio N_{tot}/P_{tot} also indicates the limiting role of phosphorus in the studied lakes. The organic form of nitrogen predominates in the water of lakes, constituting up to 60% of N total, while the N-nitrate predominates over other mineral forms. This is typical for water bodies with a predominance of autochthonous organic matter.

The ratio of the seasonally averaged organic and inorganic dissolved carbon concentration is characterized by a slight dominance of the organic component. In winter, an increase in DIC concentration with depth is observed, which in turn indicates the prevailing processes of mineralization of dissolved and suspended organic matter under the conditions of the under-ice regime. The ratio DOC/DIC is consistent with results on the intensity of production and destruction processes in their seasonal aspect and, in general, indicates the heterotrophic status of ecosystems.

In March and May, low-molecular forms of DOC (less than 1 kDa or 1 nm) account for 20-30% of the total dissolved carbon concentration. The relative amount of organic colloids (1 kDa - 0.45 μm) decreases 2-3 times during the summer period (July), while the low molecular weight fraction (<1 kDa) makes up to 80% of the total DOC in the water column. The predominance of potentially "bioavailable" (low molecular weight) and a low proportion of colloidal forms indicates the autochthonous nature of organic carbon, whereas allochthonous organic carbon is presented mainly in the form of colloids with a high molecular weight. Ecosystems of lakes Maselgskoe and Vilno are highly dynamic in terms of seasonal fluctuations in DOC concentrations, given that DOC is controlled mainly by autochthonous processes in the water column, such as primary phytoplankton production, respiration of heterotrophic bacterioplankton, and anaerobic processes.

The C_{org}/N_{org} ratio for Lake Maselgskoe varies in a wide range between 13 and 133 (average value is 43 ± 26), for Lake Vilno the interval of the ratio is 15-156 (average value is 47 ± 34). For the periods of spring and autumn mixing of waters, the C_{org}/N_{org} ratios are more evenly distributed (12.6-41.0 the average value is 25.0 ± 8.7).

The study of spatial and temporal dynamics of the qualitative and quantitative characteristics of phyto- and zooplankton communities showed that the ecological state of the studied lakes can be assessed as stable. The trophic status of the reservoirs of Maselgskoe and Vilno, according to the main hydrobiological indicators, is defined as mesotrophic. To assess the ecological status of zooplanktocenoses in different years of observations, the Shannon-Weaver species diversity index was calculated. The index values (2.54-4.30 bits) indicate a relatively stable and balanced state of the zooplankton communities of both lakes.

However, there is a potential vulnerability of lake ecosystems due to the low buffer capacity of waters. In the deep-water areas of Lake Maselgskoe, where the anaerobic layer is pronounced during periods of stratification, aerobic destruction plays an insignificant role (22-54%). During periods of stagnation, the processes of anaerobic degradation of OM (sulfate reduction and methanogenesis) actively occur in the bottom horizons and bottom sediments of Lake Maselgskoe (Kokratskaya et al., 2012; Zabelina et al., 2012). In addition, in the bottom horizons of the deep-water areas of the Maselgskoe lakes, where oxygen deficiency zones are formed during periods of stratification, favorable conditions are created for the upward inflow of mineral compounds of phosphorus, nitrogen, and reduced forms of iron from bottom sediments into the water column, which can have a significant effect on changing the trophic status of this lake.

We conclude that the status of the ecosystems of lakes Maselgskoe and Vilno is chiefly determined by intrabasin processes. Biological communities are the main factor in the sustainability of ecosystems. At the present stage, hydrobionts of the studied ecosystems are in an active state, responding to changes in biotic and abiotic conditions. This allows maintaining ecosystems in a state of ecological balance under conditions of a changing climate and increasing anthropogenic influence.

ACKNOWLEDGMENTS

This work was supported by the State Task AAAA-A18-118012390200-5. We are grateful to O.A. Pavlova for determining the quantitative and qualitative composition of phytoplankton.

REFERENCES

Artemev, V.E. *Geochemistry of organic matter in the river-sea system,* 1993: 204 p. (In Russian).

Boyle, E.A., Edmond, J.M., Shoikovitz, E.R. 1977. The mechanism of iron removal in estuaries. *Geochimica et Cosmochimica Acta*, 41 (9): 1313-1324.

Cronberg, G., Lindmark, G., Bjork, S. Mass. 1988. Development of the Flagellate Gonyostomum semen (Raphydophyceae) in Swedish Forest Lakes. *Hydrobiologia*, 161: 217-237.

Datsenko, Yu. S. *Eutrophication of reservoirs. Hydrological and hydrochemical aspects,* 2007: 252. (In Russian).

Davies, J., Jenkins, A., Monteith, D., Evans, C.D., Coopera, D.M. 2005. Trends in surface water chemistry of acidified UK Freshwaters, 1988-2002. *Environmental Pollution*, 137: 27-39. doi: 10.1016/j.envpol.2004.12.029.

de Caritat, P., C. Reimann, C., M. Äyräs, M., Niskavaara, H., Chekushin, V.A., Pavlov, V.A., 1996. Composition of stream water from eight catchments on the Kola Peninsula (NW Russia) and in neighbouring areas of Finland and Norway 1: levels and sources. *Aquatic Geochem*, 2: 149–168.

Dymov, A.A., Dzhangurov, E.V., Startsev, V.V. 2003. Soils of the northern part of Polar Ural: morphology, physic-chemical properties, stocks of nitrogen and carbon. *Pochvovedenie (Russ. Soil. Science)*, 5: 507-516.

Ecology of overgrown lake and problems of its restoration, 1999: 222. (In Russian).

Evans, C., Monteith, D., Cooper, D. 2005. Long-term increases in surface water dissolved organic carbon: observations, possible causes and environmental impact. *Environmental Pollution*, 137: 55-71.

Facchina, O.E. 1948. Dynamics of phosphorus and iron in the Black lake. *Hydrochemical materials*, 15: 180-204.

Fedorova, E.I. *Lake iron accumulations within the Baltic crystal shield,* 1964: 164-193. (In Russian).

Feneva, I. Yu. 2016. Vertical distribution of planktonic organisms of crustaceans in meso-eutrophic lakes with different thermal conditions. *Zoological journal*, 95(2): 153-166. (In Russian).

Florence, T.M. 1982. The speciation of trance elements in waters. *Talanta*, 29(5): 345-364.

Glagolev, M.A. 1958. Forms of element migration in river waters. *Reports of the USSR Academy of Sciences*, 121 (6): 1052-1055. (In Russian).

Gordeev, V.V. *Geochemistry of the river-sea system,* 2012: 452. (In Russian).

Halvorson H.O., Starkey R.L. 1927. Starkey Studies on the transformation of iron in nature. Part II. Concerning the importance of microorganisms in the solution and reduction of iron Soilrson. *Science,* 4: 4563-4578.

Ilina, S.M., Drozdova, O.Y., Lapitskiy, S.A., Alekhin., Y.V., Demin V.V., Zavgorodnyaya, Y.A., Shirokova, L.S., Viers, J.,. Pokrovsky, O.S. 2014. Size fractionation and optical properties of dissolved organic matter in the continuum soil solution-bog-river and terminal lake of a boreal watershed. *Organic Geochemistry*, 66: 14-24.

Karim, A., Veizer, J., Barth, J. 2001. Net ecosystem production in the great lakes basin and its implications for the North American missing carbon sink: A hydrologic and stable isotope approach. *Global and Planetary Change,* 61: 15-27.

Köhler, S.J., Kothawala, D., Futter, M.N., Liungman, O. Tranvik, L. 2008. In-lake processes offset increased terrestrial inputs of dissolved organic carbon and color to lake. *PLOS One*, 8(8). https://doi.org/10.1371/journal.pone.0070598.

Konshin, V.D. 1959. Hydrochemistry of iron in surface waters of the land. *Proceedings of the Moscow technical Institute of the fishing industry*, 10: 21-42. (In Russian).

Kuznetsov, S.I. lake Microflora and its geochemical activity, *Leningrad*, 1970: 440. (In Russian).

Levshina, S.I. 2012. Distribution of iron in surface waters of the middle and lower Amur basin. *Water resources,* 39(4): 377-384. (In Russian).

Linnik, P.N. 2009. About some features of migration of metals in surface reservoirs in aerobic and anaerobic conditions. *Ecological chemistry*, 18(3), 129-141: (In Russian).

Mineeva, N.M. 2007. Results of research of primary production of plankton of Volga reservoirs. *Biology of internal waters*, 2: 42-49. (In Russian).

Moiseenko, T.I., Dauvalter, V.A., Kagan, L. Ya. 1997. Mountain lakes as indicators of air pollution. *Water resources management*, 23(5): 669-682. (In Russian).

Moiseev, B.N., Alyabina, I.O. 2007. Estimation and mapping of carbon and nitrogen balance in the main biomes of Russia. *Izvestiya RAN. Series geographical*, 5: 1-12. (In Russian).

Neal, C., Lofts, S., Evans, C.D. Reynolds, B; Tipping, E; Neal, M. 2008. Increasing iron concentrations in UK upland waters. *Aquatic Geochemistry*, 14: 263-288. doi: 10.1007/s10498-008-9036-1.

Ohle, W. 1934. Chemische und physikalische Untersuchungen norddeutschen Seen. *Archiv fur Hydrobiolie,* 26: 1234-1245.

Oliver, S., Collins, S., Soranno, P.A., Wagner, T; Stanley, E.H., Jones, J.R; Stow, C.A; Lottig, N.R. 2017. Unexpected stasis in a changing world: Lake nutrient and chlorophyll trends since 1990. *Global Change Biology*, 23(12): 5455-5467. doi: 10.1111/gcb.13810.

Varshal, G.M., Kosheeva, I., Ya., Sirotkina, I.S. et al. (1979). Study of organic surface waters and their interaction with metal ions. *Geochemistry*, 4: 598-607. (In Russian).

Pokrovsky, O.S., Dupré, B., Schott, J. 2005. Fe-Al-organic colloids control of trace elements in peat soil solutions. *Aquat. Geochem.,* 11: 241-278.

Pokrovsky, O.S., Schott, J. 2002. Iron colloids/organic matter associated transport of major and trace elements in small boreal rivers and their estuaries (NW Russia). *Chemical Geology*, 190:, 141-179.

Pokrovsky, O.S., Shirokova, L.S., Zabelina, S.A., Vorobieva, T.Y, Moreva, O.Y., Klimov, S.I., Chupakov, A.V., Shorina, N.V., Kokryatskaya, N.M., Audry, S., Viers, J., Zoutien, C., Freydier, R. 2012. Size fractionation of trace elements in a seasonally stratified boreal lake: control of organic matter and iron colloids. *Aquatic Geochemistry*, 18: 115–139, doi: 10.1007/s10498-011-9154-z.

Rechkalov, V.V., Golubok, O.V. 2011. Vertical distribution of zooplankton of thermally stratified lakes of the Chelyabinsk region. *Bulletin of Chelyabinsk state University. Ecology. Nature Management*, 5 (220): 110-124. (In Russian).

Redfild, A.C., Ketchum, B.H., Richards, F.A. 1963. *The influence of organisms on the composition of sea water. The sea*. John Wiley and Sons, 2: 26-49.

Reuss, J., Cosby, B., Wright, R. 1987. Chemical processes governing soil and water acidification. *Nature*, 329: 27-32.

Roederer, J.G. 1991. Understanding the Arctic: Research policies and responsibilities. In: Sturges, W.T., Ed.: *Pollution of the Arctic Atmosphere*. N.Y., Elsewhere: 1-11.

Samarina, V.P. 2008. Spatiotemporal variability of biogenic substances in the river water. Oskol. *Water resources*, 35(3): 364-369. (In Russian).

Shaporenko, S.I., Shilkret, G.S. 2006. Stability and variability of hydrochemical parameters of the Glubokoe Lake in the conditions of a natural reserve. *Water resources*, 33(4), 459-474. (In Russian).

Shirokova L.S., Vorobyeva T. Ya., Zabelina S.A., Moreva E.Yu., Klimov S.I. 2008. Characteristics of production and destruction processes of small lakes of the Arkhangelsk region. *Modern problems of science and education*, 5: 11-17. (In Russian).

Shirokova, L.S., Pokrovsky, O.S., Kirpotin, S.N., Dupré, B. 2009. Heterotrophic bacterio-plankton in thawed lakes of northern part of Western Siberia controls the CO_2 flux to the atmosphere. *International Journal of Environmental Studies*, 66 (4): 433-445. doi: 10.1080/00207230902758071.

Shirokova, L.S., Pokrovsky, O.S., Moreva, O.Y., Chupakov, A.V., Zabelina, S.A., Klimov, S.I., Shorina, N.V., Vorobyeva, T.Ya. 2013. Decrease of concentration and colloidal fraction of organic carbon and trace elements in response to the anomalously hot summer 2010 in a humic boreal lake. *Science of the Total Environment*, 463–464: 78-90. http://dx.doi.org/10.1016/j.scitotenv.2013.05.088.

Shorina, N. V. 2016. Study of iron distribution in lakes of the boreal zone of North-West Russia (Arkhangelsk region)//*Lake ecosystems: biological processes, anthropogenic transformation, water quality: V international conference* - Naroch. Minsk: BSU: 351-352. (In Russian).

Shorina, N. V., Chupakov, A.V., Vorobyova, T. Ya. 2015. Distribution of iron in small lakes of the North-West of Russia (Arkhangelsk region): *Sixth international Vereshchagin Baikal conference:* 199-200. (In Russian).

Sigee, D.C. 2005. Front Matter, in *Freshwater Microbiology: Biodiversity and Dynamic Interactions of Microorganisms in the Aquatic Environment*. Chichester: John Wiley & Sons, Ltd, UK. 524.

Skjelkvåle, B.L., Stoddard, J.L., Andersen, T. 2001. Trends in surface water acidification in Europe and North America (1989-1998). *Water Air Soil Pollution*, 130: 787-792.

Smith, L.C., Macdonald, G.M., Velichko, A.A., Beilman, D.W., Borisova, O.K., Frey K.E., Kremenetsky, K.V., Sheng, Y. 2004. Siberian peatlands as a net carbon sink and global methane source since the early Holocene. *Science*, 303: 353-356.

Tremblay, L., Benner, R. 2006. Microbial contributions to N-immobilization and organic matter preservation in decaying plant detritus. *Geochimica et Cosmochimica Acta*, 70: 133-146.

Trifonova I.S. 1998. Phytoplankton composition and biomass structure in relation to trophic gradient in some temperate and subarctic lakes of northwestern Russia and the Prebaltic. *Hydrobiologia*, 369/370: 99-108.

Vezhnovets, V.V., Semenchenko, V.P. 2011. Influence of temperature increase on the vertical distribution of zooplankton in a mesotrophic stratified lake. *Reports of the national Academy of Sciences of Belarus*, 55(5): 72-75.

Von Wahenfeldt E., Bastviken D., Tranvik L.J. 2009. Microbially induced flocculation of allochthonous dissolved organic carbon in lakes. *Limnology and Oceanography*, 54(5): 1811-1818.

Vorobyeva, T. Ya., Klimov, S.I., Shorina, N. In. Moreva, O. Yu. Shirokova, L.S., Zabelina, S.A. 2009. Hydrochemical studies of lakes Maselgskoe and Vilno (Kenozero national Park). *Problems of regional ecology*, 6: 57-61. (In Russian).

Vorobyeva, T. Ya., Shirokova, L.S., Moreva, O. Yu., Klimov, S.I., Shorina, N.V., Zabelina, S.A. 2017. A comprehensive study of the current state of lakes Kargopolsky sector of Kenozersky national Park//*Abstracts of the international conference "wildlife of the Arctic: the conservation of biodiversity, ecosystem assessment"* – Arkhangelsk: 54-56. (In Russian).

Vorobyeva, T. Ya., Moreva O. Yu., Sobko E.I., Shirokova, L.S., Klimov, S.I., Shorina., N.V., Zabelina, S.A., Pokrovsky, O.S., Chupakov, A.V., Ershova, A.A. 2013. Assessment of the ecological status of lakes of Kenozersky National Park (Arkhangelsk oblast). *News of Samara scientific center of RAS*, 825-831. (In Russian).

Wolfe, A.P., Kaushal, S.S., Fulton, J.R., McKnight, D.M. 2002. Spectrofluorescence of sediment humic substances and historical changes of lacustrine organic matter provenance in response to atmospheric nutrient enrichment. *Environmental Science and Technology*, 36: 3217-3223. doi: 10.1021/es011215r.

Worrall F., Harriman R., Evans C.D., Watts, C.D., Adamson, J. Neall, C; Tipping, E., Burt. T., Grieve, I.; Monteith, D., Naden, P.S., Nisbet, T., Reynolds, B., Stevens, P. 2004. Trends in dissolved organic carbon in UK rivers and lakes. *Biogeochemistry*, 70: 369-402.

Yoshimura S. 1931. Contribution to the knowledge of hydrogen iron concentration of the lake waters of Japan. Second report. *Proceedinds of the Jmperial Academy Japan*, 7(5): 2315-2321.

Yoshimura S. 1936. Contribution to the knowledge of iron in lake waters of Japan. *Second report. Jap. J. of Geol.a.Geogr*, 13(1): 1431-1439.

Zabelina S.A., Kokratskaya N.M., Shirokova L.S. Pokrovsky, O.S., Savvichev, A.S., Vorobyeva, T. Ya., Chumakov, A.V., Klimov, S.I. 2012. Anaerobic processes of destruction of organic matter in the lakes of the Arkhangelsk region//*Materials of the all-Russian Symposium with international participation "Organic matter and biogenic elements in inland reservoirs and marine waters"*, Petrozavodsk, Republic of Karelia, Petrozavodsk: Karelian scientific center of the Russian Academy of Sciences: 193-196. (in Russian).

In: Lake Water
Editors: Oleg S. Pokrovsky, Yulia Bespalaya et al.
ISBN: 978-1-53619-275-9

Chapter 7

ALGAE OF LAKES IN THE EUROPEAN NORTH OF RUSSIA

Andrey Sharov[1,*] and Dmitriy Denisov[2]
[1]Papanin Institute for Biology of Inland Waters,
Russian Academy of Sciences, Borok, Russia
[2]Institute of North Industrial Ecology Problems,
Kola Science Center, Russian Academy of Sciences, Apatity, Russia

ABSTRACT

Based on the study of the spatio-temporal aspects of the development of phytoplankton in the lakes of the North and North-West of the ER (large lakes - Imandra, Onega and Chudsko-Pskovskoe and small lakes in the Arctic and Subarctic), the features of phytoplankton structure and dynamics under the influence of natural and anthropogenic factors (eutrophication, pollution with heavy metals, acidification, thermification) are described. The species composition and quantitative characteristics of algae in large lakes in the North of the European Region, small Arctic lakes and lakes in the Arctic regions have been investigated. It has been shown that diatoms predominate in arctic water bodies in terms of species diversity, and green algae and diatoms predominate in boreal ones. In terms of biomass, diatoms dominate in all cold-water lakes, with the exception of small Arctic lakes, where golden algae are in the lead. The features of the reorganization of algocenoses in response to the anthropogenic factors are revealed. It is proved that in northern water bodies the complex action of heavy metals and nutrients does not lead to the inhibition of phytoplankton, and the effect of acidification in combination with heavy metals enhances the toxic effect of the latter. In response to acidification, there is an increase in the variability of the dynamics of phytoplankton biomass. Algal communities are sensitive to all ongoing changes: new adaptive mechanisms are formed, the production characteristics of communities and the ecological preferences of individual species change. This is of decisive importance for bioindication of the quality of the environment and for the hydroecological monitoring.

Keywords: phytoplankton, periphyton, biomass, species composition, dominating species, abiotic factors

* Corresponding Author's Email: sharov@ibiw.ru.

INTRODUCTION

Anthropogenic impact on water bodies is due to global changes in the atmosphere, transboundary transport of pollutants and local impact of industrial and domestic wastewater. Current trends in climate change are manifested in changes in weather characteristics and in an increase in the frequency and intensity of extreme climatic events (climatic variability). Global modern climatic fluctuations and local anthropogenic impacts lead to the restructuring and functioning of aquatic ecosystems. Freshwater ecosystems of Polar Regions are especially sensitive to the influence of climatic and anthropogenic factors. Typical for the polar latitudes are cold-water lakes with an average water temperature in the summer of <15 ° C, which, as a rule, are of the oligotrophic type with high water quality.

The North of the European Territory of Russia (ETR) is characterized by a developed industry based on the extraction and processing of natural resources. The region is rich in water bodies that are used as a source of drinking water supply and a sink for wastewater. Many lakes are of recreational and fishing importance. However, studies of the ecological consequences of anthropogenic influence on the aquatic ecosystems of this region are of a local and irregular nature.

The most important components of aquatic ecosystems include algocenoses, as the main primary producer of organic matter, the basis of all trophic interactions, and an indicator of changes in environmental conditions.

The first data on phytoplankton in northern water bodies of Europe date back to the 19th century, the period of active inventory of flora and fauna (Wahlenberg, 1812; Nylander, Saelan, 1859; Richard, 1889; Kihlman, 1889; etc.). Detailed studies of individual water bodies were carried out at the beginning of the 20th century (Levander, 1905; Arnoldi, Alekseenko, 1915; Roll, 1923, Voronikhin, 1935; Kosinskaya, 1935; etc.). The main attention was paid to the floristic composition of algae. Since the middle of the last century, studies have begun on the structure of the algal flora of water bodies in northern latitudes, namely considering the influence of anthropogenic factors (Komulainen et al., 2006).

To date, numerous studies have been carried out, thanks to which the patterns of anthropogenic transformations of the environment and the response to anthropogenic stress of biological systems, including phytoplankton communities, have become clear (Moriarty, 1983; Yakovlev, 1988; Forbes, Forbes, 1994; Newman, Jagoe, 1996; Alimov, 2000; Sharov, 2004; Moiseenko et al., 2009; and others). Based on the restoration of aquatic ecosystems after pollution reduction (Cairs, 2005; Falk et al., 2006; Moiseenko et al., 2009; Moiseenko, Sharov, 2019), the phenomenon of oligotrophization was noted (Willen, 1987; Jappesen et al., 2002; Özkan et al., 2016, etc.).

A review of studies on freshwater algal communities is provided in the works of the authors (Yakovlev and Kashulin, 2012; Annotated ..., 2010; 2012; 2013). The results of studies of cyanoprokaryotes in the Northern ETR are presented in the works of D.A. Davydov and E.N. Patova (Patova, Davydov, 2006; Davydov, 2008; 2010). At present, the staff of the PABSI KSC RAS have carried out detailed studies of the taxonomic composition of cyanoprokaryotes of various terrestrial habitats on the territory of the Lapland State Nature Biosphere Reserve (Borovichev et al., 2010). S.S. Shalygin compiled an annotated list of identified taxa of cyanoprokaryotes, including 140 species, including 64 new for the Murmansk region, 26 of them are new for Russia; their distribution and ecology were clarified; an illustrated atlas of cyanoprokaryotes was created and posted on the Internet: https://picasaweb.google.com/nqcool in the public domain.

Numerous studies have been devoted to diatoms. The first studies were associated with the search for diatomites, an expedition of the Geomorphological Institute took place, which resulted in a work now considered classical (Poretsky et al., 1934). After the opening of the Kola branch of the USSR Academy of Sciences, large-scale studies of the Quaternary geology of the Kola Peninsula began in the region, in which diatom analysis was one of the main tools for geochronology and typification of deposits of various genesis. As a result of these studies, information was obtained on the taxonomic composition of diatom paleofloras that inhabited different types of water bodies of the Kola Peninsula in the past (Armand et al., 1969). Modern studies of diatoms are devoted to paleoecological reconstructions of the climate and environment in the historical past, as well as bioindication of water quality and the state of ecosystems. A detailed reconstruction of the historical development of different types of small lakes in the Khibiny mountain massif, including the dynamics of changes in pH, salinity, and trophic status of waters under the influence of the apatite industry and aerotechnogenic pollution, was carried out (Denisov et al., 2006; Denisov, 2007). On the example of mountain lakes, the consequences of the fallout of acid-forming compounds of anthropogenic origin with precipitation are shown (Moiseenko et al., 1997). For the first time, the relationship between the change in the integral indicators of the diatom complexes of a mountain lake and the dynamics of solar activity over the past 900 years was revealed (Denisov, 2012). Diatom assemblages of different periods of the Holocene and their transformation in response to changes in the environment and climate were studied; on the basis of diatom assemblages of the Kola Peninsula, various aspects of bioindication were considered, and a new scheme for assessing the overall level of anthropogenic load was proposed using the method of graphic comparison (Razumovsky, 1998).

In recent years, the first special study of Holocene diatoms from bog deposits of the Kola Peninsula, which is of great importance for paleogeographic reconstructions, was carried out (Shilova et al, 2019). It was shown that the successive change of diatom complexes took place in accordance with the general patterns of development of the natural environment in the Kola region during the Holocene; a high sensitivity of lacustrine-bog ecosystems to changes in environmental and climate conditions is noted, which can be successfully used in solving various paleogeographic problems. A number of works are devoted to the study of the diatom periphyton of streams, rivers, bogs, and the upper layers of sediments of lakes (Shilova et al, 2019). Typical fouling species dominate in rivers and streams (80-90%), the proportion of benthic forms is significantly lower. Representatives of the Fragilariophyceae class predominate (according to O.S. Shilova - up to 73%), many of which are typical rheophilic species developing in bulk at high flow rates and good water aeration (in the studied samples up to 56%). Watercourses, the catchment area of which is characterized by a high degree of swampiness, is inhabited by algae characteristic of small acidified lakes.

The most significant event in the field of diatom research can be considered to be the publication of the monograph by L. Ya. Kagan (2012). The work presented for the first time a systematic description of diatom complexes from sediments of ancient and modern water bodies of the Euro-Arctic region, reported the results of reconstruction of changes in the environment and climate in the historical past, and compiled a conclusion of diatom analysis for 100 different objects. This provided a characterization of long-term transformation processes in marine and freshwater ecosystems. Each object corresponds to a set of permanent diatom preparations that are permanently stored at the INEP KSC RAS. The collection contains systematized taxonomic lists of discovered diatoms, as well as original micrographs of typical representatives.

At present, detailed complex ecological studies of algal communities of different types of water bodies of the Kola Peninsula are being continued by the laboratory of aquatic ecosystems of the INEP KSC RAS. Based on algal communities of plankton and periphyton, the lakes were typified according to trophic status, level of photosynthetic pigments, biomass, species diversity, and saprobity index (Denisov, 2010). The seasonal dynamics of the chlorophyll content in small lakes experiencing anthropogenic pressure from the effluents of the apatite industry has been studied, and the main factors regulating the development of algae in mountain water bodies have been determined (Denisov et al., 2006; 2009; 2018). The current trends in the succession of freshwater algal communities in the ETR have been identified, and modern regional aspects of bioindication of the state of ecosystems of water bodies based on algal communities have been analyzed (Kashulin et al., 2017). A number of works are devoted to cyanobacteria (Davydov, 2010; Borovichev et al., 2010; Davydov, 2008; Davydov et al., 2020), and algocenoses of the Green Belt of Fennoscandia (Kozhin et al., 2019). An ecological database of the BEAR DB was developed which included algae of the Euro-Arctic region (Denisov, Kosova, 2017) and the CRIS web service: Cryptogamic Russian Information System (Melekhin et al., 2019).

Based on the results of a survey of 100 water bodies of various landscapes of the Murmansk region, the peculiarities of the species composition and structure of the algal flora were determined, the ranges of quantitative indicators and the consequences of anthropogenic transformation of ecosystems were identified. In eutrophied water bodies, the saprobity index S reaches 3.5, and the content of chlorophyll "a" is 32 mg/ m^3, which is extremely high for the Arctic zone (Denisov, 2010; Denisov, Barinova, 2015; 2016). The main modern trends in algocenoses changes in the Euro-Arctic region under anthropogenic load and climatic changes have been identified (Kashulin et al., 2017).

However, until now, insufficient attention has been paid to the joint study of the transformation of algal communities under the influence of climate change and a complex of direct and indirect anthropogenic effects on aquatic ecosystems located in high latitudes.

Phytoplankton Diversity of Different Water Bodies in the European Northern of Russia

During the study period (1993-2020), 781 species, varieties and forms of algae were identified in phytoplankton of lakes in the north of the European territory of Russia and East Antarctica, among them: Bacillariophyta - 368, Chlorophyta - 175, Cyanobacteria - 127, Chrysophyta - 62, Dinophyta - 19, Euglenophyta - 16, Cryptophyta - 14 (Table 1).

Table 1. The number of algal species in the plankton of the studied water bodies

Divisions	Lage Lakes						Small lakes		
	Ladoga Lake	Onega Lake	Vygozerskoe reservoire	Chudsko-Pskovskoe Lake	Imandra Lake	Umbozero Lake	Arctic-Subarctic Lakes	Boreal zone (Karelia)	Lake Kenozero
Cyanobacteria	38	103	11	43	27	1	14	20	16
Chrysophyta	23	59	25	2	26	5	23	9	5
Bacillariophyta	52	365	122	44	107	37	47	32	23
Xantophyta	6	10	2	1	-	-	3	1	-
Cryptophyta	7	10	2	9	4	1	8	4	4
Dinophyta	9	15	11	2	3	1	8	3	3
Euglenophyta	1	16	5	1	2	-	-	1	-
Chlorophyta	101	111	79	80	55	6	30	35	29
Charophyta	17	25	7	4	10	3	15	6	5
Total	254	714	264	186	234	54	148	111	85

A comparison of the species composition of the dominant complexes of phytoplankton communities in various cold-water lakes shows that the species richness of algae is mainly determined by the size and characteristics of the catchment area (Sharov, 2017). The total number of taxa in the species composition of phytoplankton in the studied lakes ranged from 13 in small lakes in the Arctic to 714 in Lake Onega and, as a rule, increased with an increase in the catchment area and the amount of incoming nutrients.

Data on phytoplankton biomass in arctic and subarctic lakes are scarce. Studies of the lakes of the Kola Peninsula (Letanskaya, 1974; Nikulina, 1975; Sharov, 2004), Bolshezemelskaya tundra (Trifonova, 1990), Canada (Lund, 1962; Moore, 1978), Iceland (Jónsson, 1977), northern Scandinavia (Heinonen, 1980; Holmgren, 1983) showed that the phytoplankton biomass in naturally clean water bodies is low, the average values for the growing season, as a rule, do not exceed 1 mg/ l. According to the available literature data (Letanskaya, 1974; Nikulina, 1975b; Yakovlev, 1991), the biomass of the summer phytoplankton of the tundra and forest-tundra lakes of the Kola Peninsula at the end of the 60s. were low (0.06-0.25 mg/ l), the maximum values were typical for lakes of the northern taiga (up to 0.56 mg/ l for Lake Vyalozero in August 1968, a biomass of 0.61-2.96 mg/ l was noted). The dominant phytoplankton groups are diatoms, cyanobacteria and dinophytes. However, as later studies showed, during eutrophication and in subarctic water bodies, the algal biomass increases tens of times (Trifonova, 1990; Sharov, 2004; Denisov and Kashulin, 2018).

An analysis of the relationship between the characteristics of phytoplankton and the size of lakes showed that the total number of species is related to the area of the lake (Figure 1) and its catchment area. Spearman's rank correlation coefficient ($Rs = 0.66$, $p = 0.0002$) shows that this relationship is statistically significant. For phytoplankton biomass, this relationship is also statistically significant ($Rs = 0.55$, $p = 0.003$, $n = 27$).

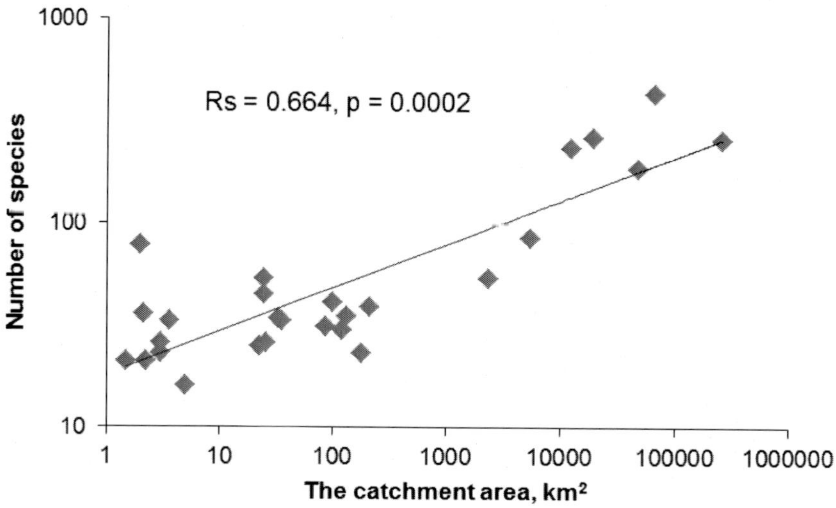

Figure 1. Dependence of the number of phytoplankton species on the catchment area of lakes.

It is known that the species richness of phytoplankton decreases in the latitudinal direction from tropical to polar zones. It increases with increasing water volume and decreasing mineralization (Alimov et al., 2013). The largest number of phytoplankton species is observed in lakes with a capacity (average/maximum depth) of 0.4 and a water salinity of 0.03 g/L (Alimov et al., 2013).

The phytoplankton of large cold-water lakes is dominated by diatoms (Table 2). In small lakes, the role of diatoms decreases, and the dominants are more often golden, cryptophytic and green, sometimes dinophytic algae and cyanobacteria become dominant (Sharov, 2004).

In recent decades, for some aquatic ecosystems, there has been a tendency for the dominance of green algae and cyanobacteria, as well as the periodic mass development of their individual representatives. Compared to the data of the 90s, the share of green algae (*Pseudosphaerocystis lacustris* (Lemm.) Nov., *Acutodesmus acuminatus* (Lag.) Tsar., *Scenedesmus quadricauda* (Turpin) Bréb., *Pandorina morum* (Müll.) Bory de Saint-Vincent, *Kirchneriella obesa* (G.S.West) West & G.S.West, etc.) in the communities increased.

No strong cyanobacterial bloom in the study areas was registered in 1993–2006 (biomass <0.1 mg/L). However, in some shallow bays, the cyanobacterial biomass reached 0.5 mg/L. In 2006-2014 in the lake Imandra, Denisov and Kashulin (2018) observed bloom spots of cyanobacteria *Dolichospermum* (Anabaena) *lemmermannii (*Richter) Wacklin, Hoffmann & J. Komárek. The biomass reached extremely high values for the Subarctic - over 85 mg/L, Figure 2.

It was found that these phenomena are accompanied by calm conditions; they are preceded by relatively high air temperatures. The accumulation of cyanobacteria is confined to bays, where the film forming at the surface is driven by a slight wind. Obviously, these processes are intensified by a favorable combination of meteorological and hydrological-hydrochemical conditions, along with a sufficient amount of nutrients such as phosphorus. Indeed, areas of the water area where cyanoprokaryotes blooming was observed are confined to the zones of distribution of effluents from apatite-nepheline production.

Figure 2. Different Cyano-HABs stages in Imandra Lake caused by *Dolichospermum lemmermannii.*

Phenomena of mass development of cyanobacteria from Lake Imandra are characterized by a number of specific features. First, they are limited to relatively isolated areas of the lake water area, and extend to open water areas only during calm periods. The mass development of cyanobacteria begins when the water warms up to 16 ° C and then continues until freeze-up. Second, the formation of dense clusters of cyanobacteria is partly due to a light wind directed towards the bay, followed by calm conditions. Third, the phenomenon of local "blooming" is not always accompanied by a high proportion of cyanoprokaryotes in the phytoplankton that develops in the water column. During the periods of their mass development, the death of juvenile fish was observed. This phenomenon is currently local in nature. Sections of the water area, where there is a massive development of cyanoprokaryotes, can be considered as a model, illustrating the scenario of the development of Arctic freshwater ecosystems in the future with an increase of nutrient load and further climate warming.

The productivity of phytoplankton in Lake Imandra has increased significantly: the maximum values of the summer biomass of algae in the areas of the water area subject to anthropogenic eutrophication have increased by an order of magnitude since the end of the last century. At the same time, the rates of nutrients turnover also increased. Large-cell diatom plankton, which develops in mass at the beginning of summer, is practically not consumed by small zooplankton. Dying diatoms relatively quickly become a source of nutrients for other groups of algae and cyanoprokaryotes. Along with this, in the coastal zones of mass development of cyanoprokaryotes, an abnormally high biomass is formed in an extremely short period of time, which is also practically inaccessible for higher trophic levels. Rapid lysis of colonies leads to an increase in the concentration of nutrients available to photoautotrophs in the water; an additional source of nutrients is also fish that dyes during this period.

During the research period 1993–2016 the complex of dominant species with the highest frequencies of occurrence and domination consisted of 15 taxa from several divisions of diatoms, golden, peridinia, cryptophytic algae and cyanobacteria (Table 2). According to the codes of the morphofunctional classification (Reynolds et al., 2002) of freshwater phytoplankton and the highest average relative biomass of the dominant species, phytoplankton associations of the studied lakes can be attributed to 9 types: A, B, C, E, N, P, Lo, H1, and Y.

Table 2. The relative average biomass (mg/L) of the dominant phytoplankton species and the corresponding codes for the morphofunctional phytoplankton classification

Species	Code	Lage Lakes						Small lakes		
		Ladoga Lake	Onega Lake	Vygozerskoe reservoire e	Chudsko-Pskovskoe Lake	Imandra Lake	Umbozero Lake	Arctic-Subarctic	Boreal zone (Karelia)	Lake Kenozero
Aulacoseira islandica	B	20	32	16	8	15	2	0.5	1	1
Aulacoseira distans	C	2	2	2	0.1	2	4	1	0	0.1
Aulacoseira italica	B	6	10	1	0	4	8	0.1	0	1
Aulacoseira granulata	P	0.1	1	0	6	0	0.1	0	0.1	12
Asterionella formosa	C	7	9	13	0.1	9	42	2	3	11
Cyclotella ssp.	B	2	1	3	2	5	5	5	1	2
Stephanodiscus ssp.	B	5	0.5	0.1	4	9	0	0.1	0.2	0.3
Rhizosolenia longiseta	A	0.1	0.5		0	6	2	2	22	7
Tabellaria fenestrata	N	9	11	28	38	8	5	8	15	40
Dinobryon ssp.	E	1	4	0	0.2	11	1	30	6	0,5
Ceratium hirundinella	L_O	0.1	0.5	1	3	5	0	0	2	1
Peridinium ssp.	L_O	0.5	3	0.5	0	10	1	10	7	2
Aphanizomenon flos-aquae	H1	23	0.1	0	14	0.1	0.5	0	0	0
Anabaena ssp.	H1	2	2	1	3	2	1	0.3	1	0.1
Cryptomonas ssp.	Y	17	3	1	3	9	0.1	3	1	0.1

During the period 1993–2016, the complex of dominant species consisted of 15 taxa from several divisions of diatoms, golden, peridinia, cryptophytic algae and cyanobacteria (Table 2). According to the codes of the morphofunctional classification (Reynolds et al., 2002) of freshwater phytoplankton and the highest average relative biomass of the dominant species, phytoplankton associations of the studied lakes can be attributed to 9 types: A, B, C, E, N, P, Lo, H1, and Y.

Group A consists of species that live in highly transparent lakes with low nutrient content. Group B is represented by species living in mixed, mesotrophic medium-sized lakes, tolerant to light deficiency, sensitive to an increase in pH, to the onset of stratification, and a decrease in silicon content. Type C is composed of species that prefer mixed, shallow lakes, tolerant to light deficiency, and sensitive to depletion of silicon concentration in water. The N group is the species of mesotrophic epilimnion, sensitive to stratification and tolerant to nutrient deficiency. The group P includes species that prefer the eutrophic epilimnion, tolerant to medium illumination and carbon deficiency, and sensitive to stratification and a decrease in the silicon content in water. The group Y comprises usually representatives of small enriched lakes, phagotrophs, tolerant to low illumination. The group E includes the species that usually live in small oligotrophic lakes are sensitive to oxygen deficiency and tolerant to low nutrient content. The H1 group is represented by diazotrophic species tolerant to low nitrogen and carbon content, sensitive to mixing of the water column, low phosphorus and low illumination. In the Lo group, representatives of different types of lakes are sensitive to mixing.

The total phytoplankton biomass varied widely, from a few µg/L in Arctic lakes to 34 mg/L in the boreal lake Pskov. During eutrophication in the areas where domestic wastewater flows into the subarctic lake (Imandra), the biomass reached high values (20 mg/L). In most lakes, the summer phytoplankton biomass usually did not exceed 3 mg/L (Table 3).

Table 3. Average phytoplankton biomass and concentration of chlorophyll "*a*" in the studied lakes during the growing season

Divisions	Lage lakes						Small lakes		
	Ladoga Lake	Onega Lake	Vygozerskoe reservoire	Chudsko-Pskovskoe Lake	Imandra Lake	Umbozero Lake	Arctic-Subarctic Lakes	Boreal zone (Karelia)	Lake Kenozero
Cyanobacteria	0.5±0.1	0.1±0.1	0.04±0.01	1.3±0.1	0.1±0.02	<0.01	0.1±0.01	0.2±0.01	0.02±0.01
Chrysophyta	<0.01	0.1±0.1	<0.01	0.1±0.1	0.1±0.01	<0.01	0.2±0.01	0.1±0.01	0.02±0.01
Bacillariophyta	0.5±0.1	1.4±0.3	0.8±0.2	1.2±0.3	0.5±0.03	0.89±0.1	0.1±0.03	2.1±0.1	2.54±0.1
Cryptophyta	0.6±0.1	0.2±0.1	<0.01	0.1±0.1	0.4±0.05	<0.01	0.02±0.01	0.1±0.01	<0.01
Dinophyta	<0.01	0.1±0.1	0.14±0.03	0.1±0.1	0.1±0.01	0.1±0.01	0.1±0.01	0.1±0.01	0.1±0.01
Chlorophyta	0.2±0.1	0.1±0.1	0.1±0.01	0.2±0.1	0.3±0.02	<0.01	0.04±0.01	0.5±0.07	0.2±0.01
Total biomass, mg/L	1.9±0.4	2.0±0.5	1.12±0.3	2.9±0.7	1.37±0.3	0.93±0.2	0.8±0.2	3.17±0.7	2.8±0.5
Chl a, µg/L	5.2±1.1	2.7±0.7	4.1±0.5	21.1±3.5	2.6±0.5	1.2±0.3	0.6±0.2	5.4±1.2	4.5±1.2

The phytoplankton study of the largest in Fennoscandia Lake Imandra has been conducted since 1992. For more than half a century, Lake Imandra was used as a source of technical and drinking water supply, in the interests of recreation and tourism, and fishing. In 1930-1940, the development of mineral deposits has started, which led to the emergence of a complex of negative phenomena and overall pollution of the reservoir. Along with the vulnerability due to its geographic position at high latitudes, the ecosystem of Lake Imandra also has significant ecological stability due to long evolution and the large size of the reservoir. Over a long period of wastewater pollution of copper-nickel and apatite-nepheline production, the structural and functional organization of algocenoses in the reservoir has undergone significant changes (Moiseenko, Denisov, 2019).

Long-term changes in the composition and structure of phytoplankton in Lake Imandra caused by anthropogenic changes in the environment are mostly pronounced in the northern reach (Figure 3). Based on the research results in different years, four main periods of the ecosystem state of the most polluted northern reach of the lake - Bolshaya Imandra - were identified: natural conditions, the period of maximum pollution and ecosystem degradation, pollution reduction, the current state of the ecosystem. Indicators of the state of phytoplankton communities during different periods are presented in Table 4.

The natural conditions for the formation of aquatic ecosystems in the Arctic basin include high contribution of atmospheric precipitation, thin soil and vegetation cover, low temperatures and, accordingly, slow processes of chemical weathering and the cycle of elements. They are observed in ultra-fresh and oligotrophic water bodies, a typical example of which was Lake Imandra. This reservoir was characterized by high oxygen saturation (to the bottom) due to the inlet of mountain rivers that do not freeze in winter. The phytoplankton communities were predominantly Arctic cold-water species.

The Period of Maximum Pollution

The male water pollution, which began in the forties of the last century, peaked by the 1970s and 1980s. The influence of wastewater from a metallurgical enterprise led to an increase in the content of heavy metals in the reservoir, especially nickel and copper. The anthropogenic influx of nitrogen and phosphorus with domestic wastewater into the lake has led to an increase in both the total content of these elements and their mineral forms (in comparison with the natural content), especially in the areas that receive industrial and domestic wastewater. In terms of phosphorus content, the lake shifted to the mesotrophic state, and its small bays - to the eutrophic state (Moiseenko, Denisov, 2019). The species composition of phytoplankton has changed: the range of fluctuations in the maximum and minimum values of the abundance and biomass of communities has increased (Table 4).

In response to the decrease in anthropogenic load, which has been observed since the early 1990s, in the 2000s, a tendency towards an improvement in the quality and the state of the ecosystem was observed (Moiseenko et al., 2009; Moiseenko, Sharov, 2019). Recolonization of the lake by some inhabitants of the Arctic waters was observed, which is confirmed by the secondary replacement of the dominant complexes (Table 4). There was an increase in the individual mass of organisms and in the biodiversity index of planktonic communities. The accumulated nutrients were involved in the biological cycle in the ecosystem, as evidenced by the excess of total phosphorus concentrations over mineral P. A decrease in the concentration of bioavailable forms of nutrients including Si is a consequence of their utilization by diatoms, which occupy a dominant position during the recolonization period, but with a significantly higher abundance compared to natural ones.

Table 4. The main indicators of phytoplankton of Lake Imandra (Bolshaya Imandra reach) in different periods

Indicators	Pre-industrial period (up to 1940s)	Intensive pollution and degradation (1980-1990s)	Decrease in pollution (1995-2005)	Stabilization of anthropogenic pressures (2010-2018)
Biomass, g/ m^3	0,1	3,6 (1,6-4,6)	3,4 (0,2-8,4)	3,8 (0,2-19,9)
Number, ind. 10^6/L	0,1	3,8 (0,6-7,2)	3,2 (0,3-10,4)	4,4 (1,5-22,9)
H (Shannon index), bit/ind	3,2	2,5 (1,8-2,7)	3,1 (1,3-2,9)	2,8 (1,2-3,4)

In the modern period (after 2010), a new modification of the ecosystem occurred. Despite a decrease in the concentration of toxic heavy metals, a high concentration of nutrients remains, which led to the formation of mesotrophic phytoplankton communities. Under these conditions, the number of abnormally high quantitative indicators of phytoplankton has increased.

Lake Imandra is characterized by a significant abundance and taxonomic diversity of diatoms, which are the dominant group of plankton algae in terms of abundance and biomass for most of the hydrobiological summer (Figure 4).

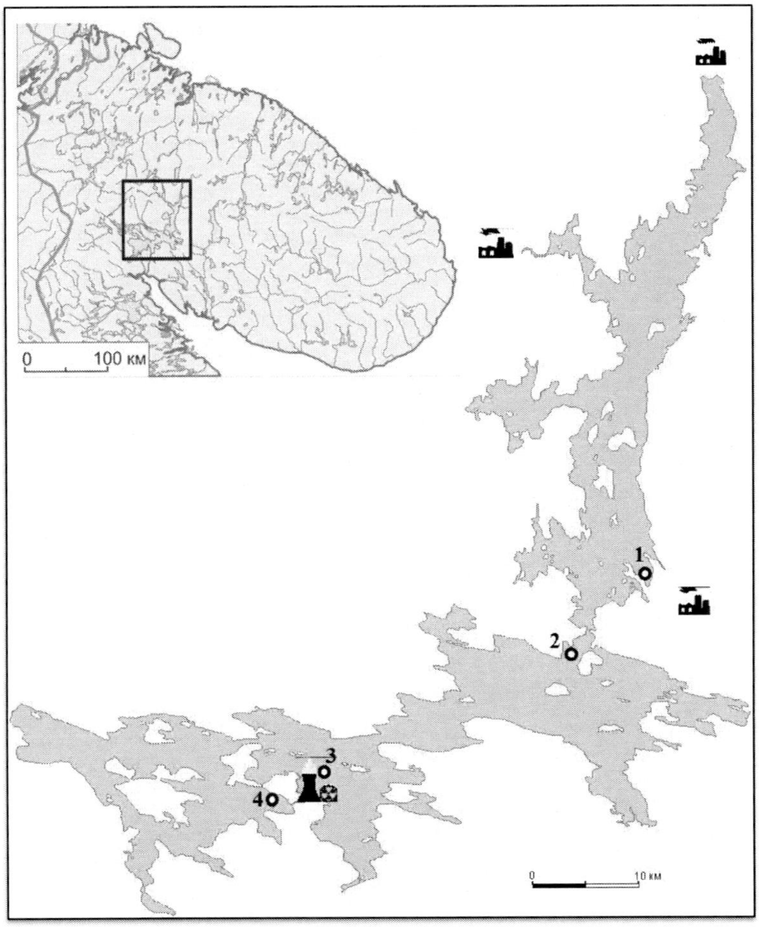

Figure 3. Lake Imandra and location of phytoplankton sampling stations for the study of seasonal dynamics.

The proportion of diatoms in the composition of communities varies from 56 to 100%. The Shannon-Weaver species diversity index varies from 0.21 to 3.2 bit/ ind. The highest species diversity occurs in bays, which are estuarine sections of rivers.

The seasonal dynamics of phytoplankton in the lakes of the Kola Peninsula was studied for the lake Imandra, lake Chuna (Sharov, 2004), tundra lakes Zelenetskoe and Akulkino (Nikulina, 1975), and lakes of northern Karelia Krivoe and Krugloye (Nikulina, 1975, 2016). Fragmentary data were available for Lake Onega (Petrova, 1987; Trifonova, 1990; Chekryzheva, Sharov, 2011), Lake Ladoga (Petrova, 1987; Letanskaya, Protopopova, 2012), Vygozerskoye reservoirs (Chekryzheva, 2011), Chudsko-Pskovskoe (Sharov, Andreeva, 2015), etc.

Seasonal dynamics of common phytoplankton species is determined by different types of adaptive strategies, the amount and ratio of nutrients for phytoplankton is of major importance. One of the mechanisms for regulating the seasonal succession of phytoplankton in northern lakes is associated with the influx of N, P, and Si with melt water. During the flood period, the ratio of N/ P and Si/ P in water increases and at the end of the spring period it decreases. In many water bodies, green algae develop at a N/ P ratio of 20-50, and cyanobacteria develop at

5-10 (Reynolds, 2006). However, in cold water lakes, this pattern is not observed. During the spring outbreak of phytoplankton, the content of N and P naturally decreases, and during the summer minimum it increases.

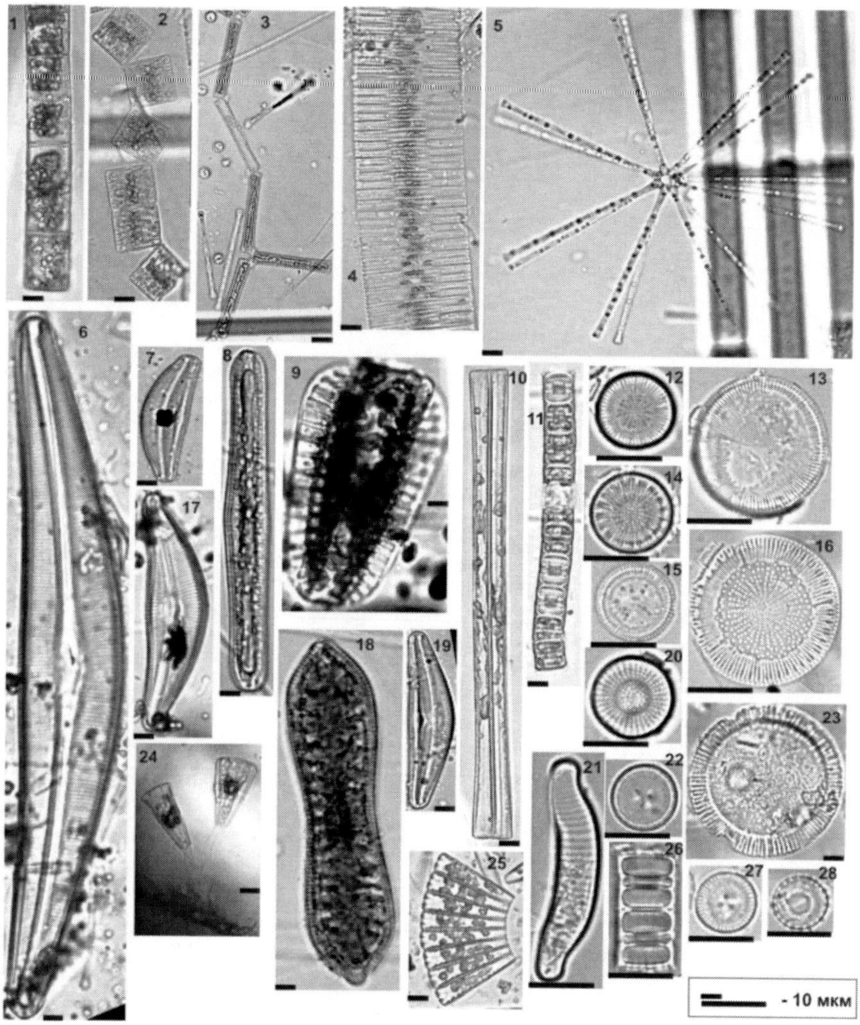

Figure 4. Bacillariophyta from Lake Imandra. 1. – *Aulacoseira islandica* (Müll.) Simons.; 2 – *Tabellaria flocculosa* (Roth) Kütz.; 3 – *Diatoma tenuis* Ag.; 4 – *Fragilaria crotonensis* Kitton; 5 – *Asterionella formosa* Hass.; 6 – *Cymbella aspera* (Ehrb.) Cleve; 7 – *C. tumida* (Bréb.) Van Heurck; 8 – *Epithemia parallela* (Grun.) Ruck & Nakov; 9 – *Surirella splendida* (Ehrb.) Kütz.; 10 – *Ulnaria ulna* (Nitzsch) Compère; 11, 26, 28 – *Aulacoseira alpigena* (Grun.) Kramm.; 12 – *Cyclotella radiosa* (Grun.) Lemm.; 13 – *Pantocsekiella schumannii* (Grun.) Kiss & Ács.; 14 – *Cyclotella bodanica var. lemanica* (Müll. ex Schroter) Bach.; 15 – *Cyclotella kuetzingiana var. planetophora* Fricke; 16 – *Lindavia bodanica* (Eulenst. ex Grun.) T.Nakov, Guillory, Julius, Theriot & Alverson; 17 – *Cymbella cymbiformis* Ag.; 18 – *Cymatopleura solea* (Bréb.) Smith; 19 – *Cymbella helvetica* Kütz.; 20 – *Stephanodiscus astraea* (Kütz.) Grun.; 21 – *Eunotia arcus* Ehrb.; 22, 27 – *Pantocsekiella ocellata* (Pant.) Kiss & Ács; 23 – *Ellerbeckia arenaria* (Moore ex Ralfs) Crawf.; 24 – *Gomphonema acuminatum* Ehrb.; 25 – *Meridion circulare* (Greville) Ag.

The high Si content decreases in spring during the development of diatoms. Silicium becomes limiting for algae when its concentration is below 0.10 mg/L (Reynolds, 2006).

Seasonal dynamics of phytoplankton biomass in the studied water bodies is characterized by one or several maxima. Spring development of phytoplankton begins under ice. One spring biomass peak was observed in lakes Onega and Ladoga, associated with the development of the diatom *Aulacoseira islandica* (O. Müll.) Simonsen. In small lakes in northern Karelia and the mountain tundra of the Kola Peninsula, the spring peak is associated with the development of dinophytic algae (Sharov, 2004; Nikulina, 2016).

Seasonal changes in phytoplankton in Lake Imandra are different in different parts of the water area (Figure 5). Thus, the western part of the Yokostrovskaya Imandra reach and the Babinskaya Imandra reach are characterized by a "classical" type of seasonal changes with a pronounced maximum of biomass during July (Figure 5, c, d). At the same time, in the southern part of the Bolshaya Imandra Reach and the northern part of the Yokostrovskaya Imandra Reach, the seasonal dynamics of phytoplankton biomass is drastically different (Figure 5, a, b). Two main scenarios of seasonal changes were identified, which are determined by the trophic status of waters.

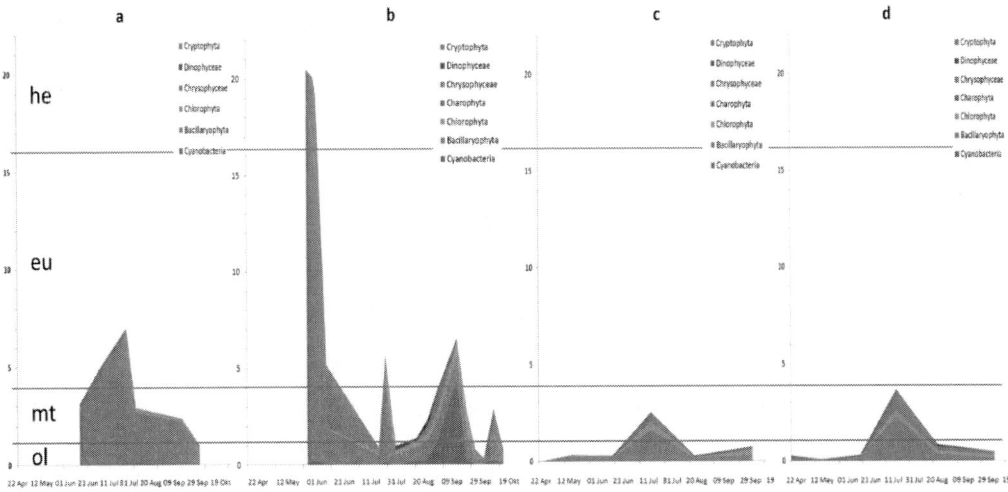

Figure 5. Seasonal dynamics of phytoplankton biomass (g/m^3) and trophic status of waters in 4 stations of the lake Imandra. a - St. 1; b - St. 2; c - St. 3; d - St. 4; ol - oligotrophic, mt - mesotrophic, eu - eutrophic, he - hypereutrophic.

Characteristics of seasonal dynamics of mesotrophic waters are described below (Trifonova, 1990). At station 2, at the beginning of the hydrobiological summer, in late May - early June, the rapid development of diatom plankton begins, dominated by *Aulacoseira islandica*, sometimes along with chrysophyte algae of the genus Dinobryon. At this time, the maximum biomass levels for the season are noted (Figure 5, b). In late June - early July, the biomass of diatoms decreases, and the proportion of other groups of algae, mainly Chlorophyta, increases. During the same period, an increase in the number of cyanoprokaryotes is possible. Their maximum proportion in the plankton composition occurs from July to September, when their biomass can reach 4.3 g/m^3. Thus, the "classical" July maximum of phytoplankton biomass in the Yokostrovskaya Imandra reach is poorly expressed; an increase in phytoplankton biomass in autumn is also characteristic. The main reason for such differences is the eutrophication of the waters of the northern part of the reach with nutrients entering through the strait. At station 1, which receives wastewater from apatite production, the June development of diatom plankton is inhibited by a significant amount of suspended matter, which is supplied by the Belaya River. Diatoms retain their dominant positions until freeze-up; the maximum biomass occurs in July. The share of golden and other groups of algae in the composition of communities is minimal. The green algae are not typically planktonic, but, as a rule, are scraps of Ulothrix zonata filaments that enter the water column from the littoral zone of the estuarine section of the Belaya River (Figure 5, a).

Seasonal Dynamics of Oligotrophic Waters

In this case, the phytoplankton dynamics characteristic of large Arctic lakes remains, with a pronounced July maximum biomass. The higher biomass values at station 4 as compared to station 3 are explained by the influence of the heated waters of the Kola NPP.

In the lakes of the North, the spring period begins in mid-April and lasts until the establishment of thermal stratification at the end of June. The spring peak of phytoplankton biomass is observed under ice, when cold-loving peridineas dominate; *Aulacoseira islandica*, sometimes cyanobacteria and small flagellate algae (cryptomonads) are subdominants. After the opening of the reservoir, a decrease in the total biomass of phytoplankton is observed, without change of dominant species.

The summer period begins with the establishment of stratification and ends in August with the onset of intense wind mixing. At the beginning of summer, a decrease in the total biomass of phytoplankton is noted, which is probably associated with the active development of peaceful zooplankton and sedimentation of diatoms. In late July - early August, a slight increase in biomass is noted. Diatoms and green algae dominate.

The autumn period lasts from the end of August until freezing and is characterized by an increase in the biomass of diatoms. Sometimes there is a massive development of cyanobacteria.

The beginning of the winter period coincides with freeze-up, lasts from November to April and is characterized by a weak development of phytoplankton. However, until the end of December, one can observe the presence of diatoms at all horizons, in particular, species of the genus Cyclotella, whose biomass reaches 1 mg/ L. The total biomass and number of algal species are usually higher in the bottom layer of water than in the surface one.

After the opening of mountain lakes, the first maximum of phytoplankton biomass is observed due to the development of diatoms and golden algae. In July, there is a significant decrease in the number and biomass of algae. The second maximum is usually observed in August-September. During this period, in the composition of algocenoses, a significant proportion, along with diatoms, were chara and green algae.

In mountain water bodies, a single summer maximum can be expressed in July. The second maximum can be short-term, or completely absent, which is largely determined by external conditions. Small lakes are characterized by the presence of fouling and benthic forms in plankton. The most abundant species are diatoms - *Aulacoseira alpigena* (Grun.) Kramm.; *A. islandica* (Müll.) Simons.; *A. subarctica* (Müll.) Haworth; *A. distans* (Ehrb.) Simons; *Asterionella formosa* Hass., *Planothidium lanceolatum* (Bréb.ex Kütz.) Lange-Bert., *Fragilaria capucina* Desm.; almost all lakes contain: *Tabellaria flocculosa* (Roth) Kütz. and *T. fenestrata* (Lyngb.) Kütz .; golden - *Dinobryon divergens* Imhof, *D. bavaricum* Imhof,

Mallomonas punctifera Korsh.; peridiniaceae - *Ceratium hirundinella* (Müll.) Bergh, *Peridinium umbonatum var. goslaviense* (Wolszynska) J. Popovsky & L. Pfeister; desmidian genera *Cosmarium* and *Staurastrum* are abundant among charots, green ones are represented by the following species: *Pandorina morum* (Müll.) Bory, *P. charkowiensis* Korsch., *Eudorina elegans* Ehrb. Less abundant are cyanobacteria, cryptophyte, and euglena algae. Some species found in plankton are shown in Figure 6.

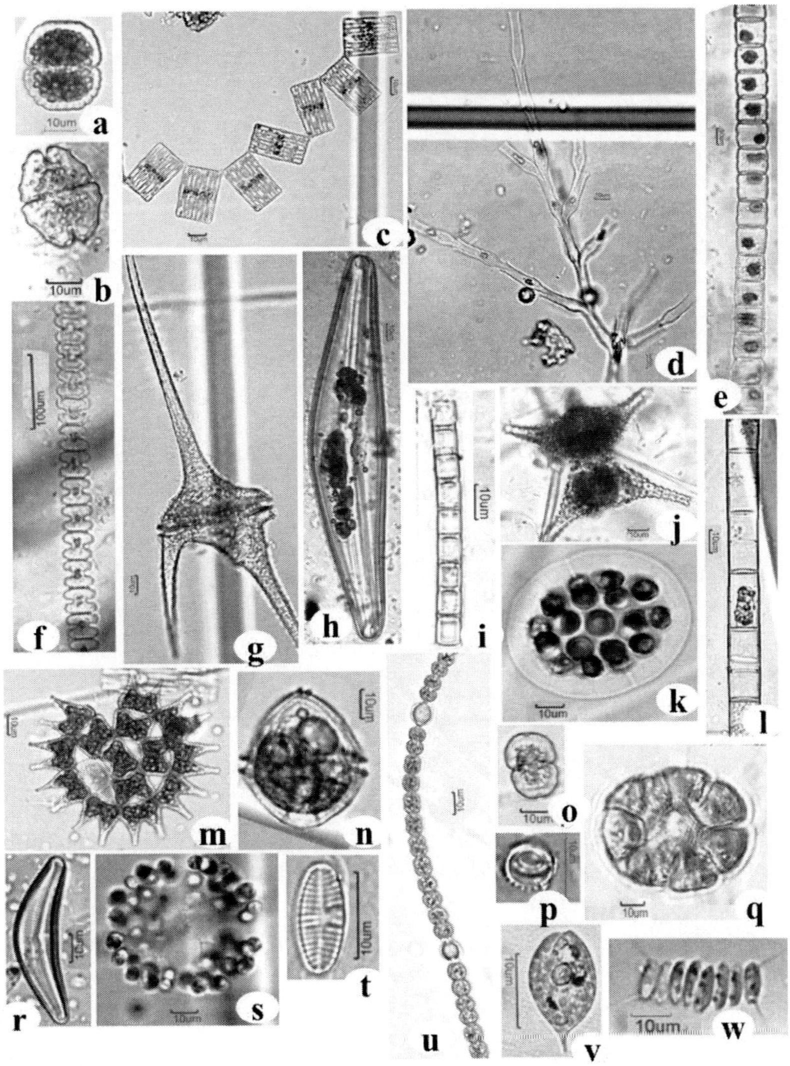

Figure 6. Some representatives of phytoplankton of small water bodies of the North of the European Russia: a – *Cosmarium subcostatum*; b – *Euastrum elegans*; c – *Tabellaria flocculosa*; d – *Dinobryon divergens*; e – *Hyalotheca dissiliens*; f – *Spondylosium planum*; g – *Ceratium hirundinella*; h *Cymbella aspera*; i – *Aulacoseira subarctica*; j – *Staurastrum anatinum*; k – *Pandorina charkowiensis*; l – *Aulacoseira islandica*; m – *Pediastrum duplex*; n – *Peridinium umbonatum var. goslaviense*; o – *Cosmarium laeve*; p – *Aulacoseira alpigena*; q – *Pandorina morum*; r – *Cymbella arctica*; s – *Snowella litoralis*; t – *Planothidium lanceolatum*; u – *Dolichospermum planctonicum*; v – *Phacus caudatus var. tenuis*; w – *Desmodesmus quadricaudatus*.

Phytoperiphyton in oligotrophic flowing water bodies also develops abundantly starting from April. On the runoff of mountain oligotrophic lakes, as well as on ice-free areas of rivers and streams, the golden algae *Hydrurus foetidus* (Villars) Trevisan develops in bulk, and the substrate coverage can reach 60-80% (Figure 7).

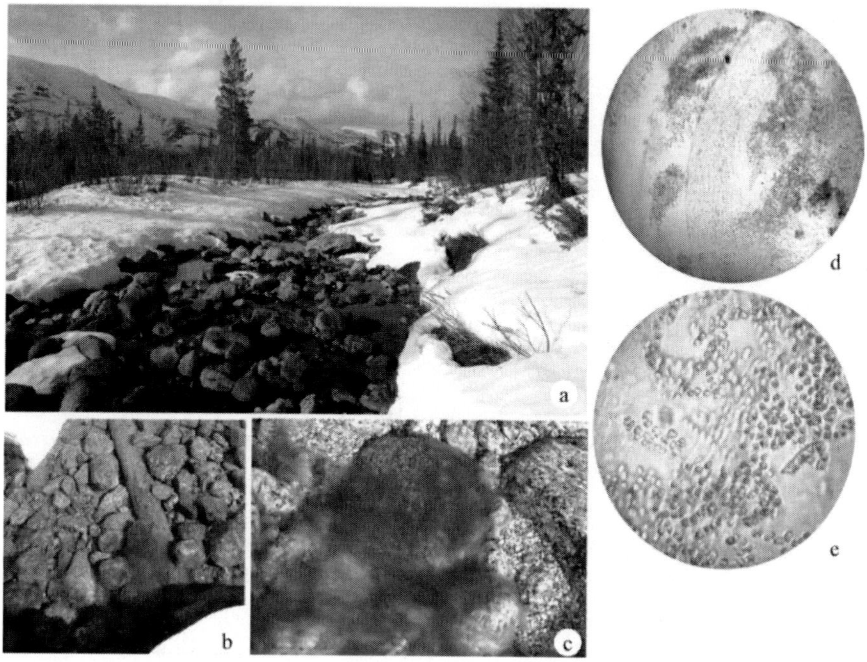

Figure 7. Spring phytoperiphyton of oligotrophic water bodies of the Khibiny mountain range: a - typical habitat; b, c - appearance of fouling of *Hydrurus foetidus* d, e - colonies of *H. foetidus* under a microscope.

After opening the lakes from ice, algae completely disappear from the littoral, and begin to reappear closer to the end of summer - in August and September. The high transparency of the water allows the development of rich flora on fluffy and stony substrate at depths of 10 - 15 m, represented mainly by diatoms. Their growth at these depths, with varying degrees of photosynthetic activity, apparently occurs all year round. The appearance of typical fouling of the littoral of oligotrophic lakes is shown in Figure 8. Depending on the mobility of the water edge and the type of substrate, algae can form both a continuous layer of fouling, which is typical, first of all, for diatoms (Figure 8, a), and separate tufts, on which green and chara algae can grow (Figure 8, c). Numerous representatives of benthic diatoms are often associated with the upper layer of silt deposits and detritus particles: diatoms of the genus Surirella, desmidia of the genera Cosmarium, Closterium. Typical rheophilic fouling plants, demanding a high oxygen content, are formed in flowing lakes in the drainage area. At the end of summer (August-September), as a rule, filamentous charoses of the genus Zygnema dominate, forming dense mats on the stony substrate of the rapids of rivers and the drains of flowing oligotrophic lakes (Figure 8, c).

A significant anthropogenic factor stimulating the development of periphyton algae is the presence of self-flowing hydrogeological wells, the waters of which have elevated pH values (up to 8.0 and higher). In the area of such wells, filamentous green algae (*Ulothrix zonata*) develop abundantly even in the cold season, when they can use ice as a substrate (Figure 8, f).

Algae and cyanoprokaryotic communities in freshwater lakes of arch Spitsbergen are poorly studied due to limited inaccessibility of the region (Davydov et al., 2013; Thomasson, 1958, 1961; Willen, 1980; Skulberg, 1996). Three comparatively large lakes of West Spitsbergen Island, in the western part of Nordenskjold Land on the western coast of Grenfjord Bay and the southern coast of Isfjord Bay were investigated. In the period from 13 to 15 July 2011, samples of plankton and periphyton were collected in lakes Linne, Congress and Stemme. The investigated lakes are of karst-glacial origin. The lakes are fed by glacier melt water, and to a lesser extent atmospheric precipitation - snow and rainwater, as well as groundwater.

Figure 8. Phytoperiphyton of water bodies of various habitats: a - oligotrophic lake of the northern taiga zone (Kalevaevskoye) - mainly diatom periphyton; b, d - eutrophied lake (Bol. Vudyavr) - green and diatoms; c - mountain river (Kuna) - charovye; e - mountain lake of the lichen tundra zone (Heart-shaped) - *Oedogonium sp.*; f - fouling in the area of a self-flowing well (valley of Lake Mal.Vudyavr) - *Ulothrix zonata.*

It was shown that the planktonic communities of the lakes are very poor in composition and abundance of algae - the number does not exceed 50-150 thousand cells/ l. This is due to the harsh environmental conditions of water bodies, primarily low temperatures and low content of the main nutrients in the water. The main limiting factor is low temperature (2-3°C), which prevents the intensive development of algae in plankton. The poorest in composition and abundance of cyanoprokaryotes and algae was the plankton of Lake Linne, whose waters are the coldest with high flow rates. In the lake Stemme, plankton numbers are slightly higher; this lake is located lower in relief and has a more favorable temperature regime compared to other surveyed lakes. Diatoms are found in plankton more often than other algae, followed by the golden algue, whereas representatives of cyanoprokaryotes in the samples were noted only singly. Diatoms in plankton are represented by 101 taxa, comprising the genera Nitzschia, Cyclotella, Pinnularia. Cyanoprokaryotes are represented by 20 species, leading in terms of species diversity in the genera Pseudanabaena, Leptolyngbya, Oscillatoria. Golden algae are represented by species of the genus Dinobryon.

Most of the species recorded in plankton are not typically planktonic species and most likely they enter the water column as a result of intense wind mixing from the periphyton and benthos; almost all of them are encountered in both plankton and non-planktonic communities. In the lake Linne, there are no dominants in plankton. *Pinnularia viridis* (Nitzsch) Ehrb is assigned to subdominants (up to 18.8%), followed by *Navicula medioconvexa* Hust. (15%); *Pinnularia subcapitata* Greg. (9.4%). Golden algue are represented by *Dinobryon sp.*; cyanoprokaryotes are represented by single filaments *Leptolyngbya cf. angustissima* (W. West et G.S. West) Anagn et Komarek. Epilithon samples collected on rocks in the coastal zone are brownish leathery crusts, the base of which is formed by cyanoprokaryotes. Here, the dominant species are *Leptolyngbya cf. angustissima*, *Oscillatoria anguina* Bory ex Gom., *Calothrix cf. parietina* (Nag.) Thuret ex Born. et Flah., *Aphanothece castagnei* (Breb.) Rabenh., *Aphanothece stagnina* (Spreng.) A. Braun, *Chroococcus sp.*, *Synechocystis cf. aquatilis* Sauv. The diatoms are represented by *Nitzschia perminuta* (Grun.) M. Perag (22.8%); *Achnanthidium minutissimum* (Kutz.) Czarnecki (15.2%); *Denticula subtilis* Grun. (10.8%); *Epithemia adnata* (Kutz.) Breb. (11.9%).

In the lake Congress, algae of plankton are more diverse, but their abundance is also very low; in a high-quality net sample, the diatoms *Pinnularia subcapitata* Greg., *Aulacoseira alpigena* (Grun.) Kramm. were noted. Among golden algae, *Dinobryon sp.* dominated; the cyanobacteria were represented by filaments *Pseudanabaena catenata* Lauterb., *Jaaginema subtilissimum* (G. Schmid) Anagn. et Komarek. The latter species is found in hydrogen sulfide springs, which indicates that the lake is replenished with hydrogen sulfide waters located in the catchment area of this lake. In periphyton and benthos, *Epithemia sorex* Kutz has relatively high abundance; *Hannaea arcus* (Ehrb.) R.M. Patrick; *Encyonema minutum* (Hilse) Mann; *Pinnularia subcapitata* Greg., As well as all the above-mentioned species of cyanoprokaryotes, as well as *Leptolyngbya cf. angustissima*.

In the lake Stemme, the plankton is dominated by *Planothidium frequentissimum* (Lange-Bert.) Round & Bukht. Subdominants: *Encyonema minutum* (Hilse) Mann. Of the cyanoprokaryotes, *Leptolyngbya cf. angustissima* and *Oscillatoria tenuis* C. Ag. ex Gom. In samples of epilithon and benthos collected in small warmed bays, a massive development of cyanoprokaryotes was noted, forming biofilms on the bottom and stones, the basis of which is formed by *Oscillatoria tenuis, Oscillatoria anguina, Leptolyngbya cf. angustissima, Phormidium chlorinum* (Kutz.ex Gom.) Anagn., *Leptolyngbya sp.* Diatoms in the periphyton include *Planothidium frequentissimum* (Lange-Bert.) Round & Bukht., *Aulacoseira alpigena* (Grun.) Kramm.; *Achnanthidium minutissimum* (Kutz.) Czarnecki; *Meridion circulare* (Greville) C. Ag., *Tabellaria flocculosa* (Roth) Kutz.; *Hannaea arcus var. arcus* (Ehrb.) Patrick.

A study of the phytoplankton of glacial mountain lakes showed low diversity and very low abundance of algae in planktonic communities. The development of phytoplankton is limited by low water temperatures and water poverty in nutrients, which is generally typical for water bodies in the Arctic and the Alpine zone. In ~~the of~~ the structure of phytoplankton, the leading role belongs mainly to diatoms, whereas the cyanoprokaryotes are represented by single cells. Communities of benthos and periphyton are more diverse in species composition and abundance of algae; they are formed by cyanoprokaryotes and diatoms.

In connection with the need to develop an international network of protected areas in the North of the ETR, studies of the biodiversity of the so-called Green Belt of Fennoscandia (GBF) are of particular relevance. In total, 26 lakes and 7 watercourses were studied within the GBF. 472 taxa of algae and cyanoprokaryotes with a rank below the genus were identified. Among large taxonomic categories, Bacillariophyta (311) and Chlorophyta (69) are characterized by the highest species richness. The less rich in Cyanoprokaryota (37), Chrysophyta (22), and Charophyta (15) species, the smallest number of taxa are recorded for Cryptophyta (9), Dinophyta (7) and Xantophyta (2).

The phytoplankton of the large lake-river system of the river Groove in the period from 2011-2012 omprised 124 taxa of algae and cyanoprokaryotes with a rank below the genus. The dominant taxa were *Asterionella formosa* Hass.; *Fragilaria capucina* Desm.; *Aulacoseira alpigena* (Grun.) Kram.; *Urosolenia eriensis* (Smith) Round & Craw.; *Dinobryon divergens* Imh., *Paulschulzia tenera* (Korsh.) Lund.; *Pseudopediastrum boryanum* (Turp.) Hegew.; *Pandorina morum*; *Eudorina elegans*; *Achroonema proteiforme* Skuja; *Dolichospermum planctonicum* (Brunn.) Wack., Hoff. & Kom. The species composition varies greatly in different parts of the lake-river system. Thus, an increase in the proportion of Chlorophyta in the composition of communities from south to north was noted. The most abundant algae in all studied areas were Bacillariophyta, Cyanoprokaryota, and Chrysophyta; In Kuetsjärvi, a significant proportion (up to 51%) was Chlorophyta. Dinophyta (5%) and Charophyta (<1%) were few in all areas.

We also studied algocenoses of different types of lakes on the territory of the ZPF belonging to the Murmansk region. 201 taxa of algae and cyanoprokaryotes with a rank below the genus were discovered during the period 2012-2015. Typical representatives were: *Aulacoseira alpigena, Asterionella formosa, Dinobryon elegans f. glabra* Korsh., *D. Divergens, Crucigenia tetrapedia* (Kirch.) Kuntze, *Oocystis rhomboidea* Fott, *Botryococcus braunii* Kütz., *Chroococcus disperses* (Keiss.) Lemm., *Microcystis sp.* The diatom periphyton turned out to be the most diverse and comprised 229 taxa below the genus. The common species were *Tabellaria fenestrata* (Lyngb.) Kütz., *T. flocculosa, Brachysira brebissonii* Ross, *B. vitrea* (Grun.) Ross, *B. zellensis* (Grun.) Round & Mann, *B. styriaca* (Grun.) Ross, *Denticula tenuis* Kütz., *Frustulia rhomboides* (Ehrb.) De Toni, *Achnanthidium minutissimum* (Kütz.) Czarnecki. Diatom communities of the studied lakes on the territory of the GBF are represented by species characteristic of low mineralized, oligotrophic water bodies, with a neutral or slightly acidic reaction.

ALGAE CENOSES RESPONSE TO A NATURAL AND ANTHROPOGENIC FACTORS

Climate Change

The ongoing climate warming enhances the eutrophication of waters and causes the massive development of phytoplankton algae and cyanobacteria in summer even in northern water bodies (Denisov et al., 2017), which was not observed prior 1990s (Sharov, 2004). For example, the lake Onega, the characteristics of phytoplankton were studied over a 20-year period in connection with climatic variability (Sharov et al., 2014). Correlation analysis revealed significant ($p < 0.05$) positive relationships between the content of chlorophyll "*a*" and water temperature ($Rs = 0.66$; $p = 0.03$) and negative ones between the content of Chl *a* and the duration of the ice-free period ($Rs = -0.53$; $p = 0.05$).

It was revealed that the total number of phytoplankton depends on the length of the ice-free period ($Rs = -0.89$; $p = 0.006$). The largest abundance of phytoplankton occurred when the ice cover period of the lake was the longest (Sharov et al., 2014). At the same time, no reliable correlations were found between climatic variables and the total phytoplankton biomass in Onega and Peipsi-Pskov lakes due to strong seasonal fluctuations and the possible influence of other factors (Sharov et al., 2014; Sharov and Andreeva, 2015).

In the lake Chudsko-Pskovskiy revealed that the biomass of diatoms in August negatively correlated with water temperature ($Rs = -0.80$; $p = 0.01$). The average annual nitrogen content correlated with the water level ($Rs = 0.74$; $p = 0.02$). The water level in Lake Peipsi was linked ($Rs = 0.54$; $p = 0.003$) with the mean monthly index of the North Atlantic Oscillation (NAO) in August (Sharov, Andreeva, 2015). It is obvious that in low-water years was fewer nutrients enter the lake from the catchment area than in high-water years, reducing the degree of phytoplankton development.

The number of planktonic cyanobacteria in Lake Chudsko-Pskovsky increased significantly ($Rs = 0.89$; $p = 0.006$) in years with a high NAO index. Close positive relationships ($Rs = 0.83$; $p = 0.02$) were found between the P concentration in water in August and the summer NAO index, and negative relationships with NAO in March ($Rs = -0.82$; $p = 0.001$).

The content of chlorophyll "*a*" ($Rs = -0.78$; $p = 0.01$) and the intensity of primary phytoplankton production were inversely related to the amount of precipitation in August (Sharov et al., 2014), which in turn wass related to the duration of sunshine ($Rs = -0.77$; $p = 0.01$).

Multiple regression analysis confirms a close relationship between the NAO and regional climatic variables (water temperature, precipitation, ice-free period) at $p < 0.01$, and also between the Arctic Oscillation Index (AO) and these climatic variables at $p < 0.02$. In addition, this analysis (at $p < 0.05$) shows that the concentration of chlorophyll "a" in water is determined mainly by global indicators that depend on water temperature and is associated with the duration of the ice-free period (Sharov and Andreeva, 2015).

Algocenoses of small water bodies, devoid of a pronounced surface runoff are highly dependent on meteorological factors and the composition of atmospheric precipitation. Sharp changes in temperature conditions and hydrological regime, along with positive temperature anomalies, can cause massive development of algae. There was a massive development of filamentous chara (*Zygnema sp.*) algae in the water column, with the simultaneous development of zooplankton in September 2010, in small oligotrophic lakes of the Khibiny mountain range. The development of phytoplankton in lakes apparently begins simultaneously with the spring snowmelt in April. High transparency of ice allows algae to start vegetating even under meter-thick ice. In samples taken in late March - early April from mountain oligotrophic lakes (Maly Vudyavr, Travyanoe), living diatoms were found, the cells of which were rich in fatty inclusions.

In August 2012, the staff of the Pasvik Nature Reserve discovered non-typical massive fouling of a stony substrate in shallow waters in the area of the Yaniskoski reservoir runoff of the Pasvik lake-river system (Patsojoki). It was found that they were formed by colonies of the diatom *Didymosphenia geminata* (Lyngb.) Schmidt, the cells of which in the colony grow on long polysaccharide stems, which gives the fouling a beige-light brown color. For r. Paz was the first to record a phenomenon known as "Brown plague: Didymo," which has been a significant problem for cold water flowing waters around the world in recent years and is associated with modern climate change (International ..., 2013). In 2019, a similar phenomenon was recorded in the south of the Murmansk region on the Kutsajoki River (Yovskoye reservoir basin) (Figure 9).

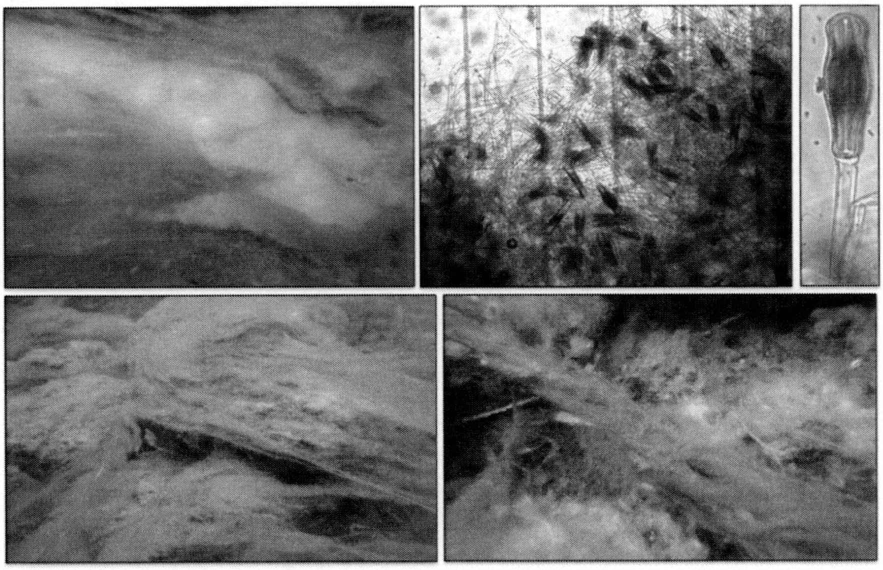

Figure 9. Mass development of *Didymosphenia geminata* - "Didymo" in the Kutsajoki River.

The Effect of Anthropogenic Factors

The structure of ecosystems that are not subject to anthropogenic influences changes over time depending on changes in external and internal factors. These are evolutionary and related successional processes. Similarly, structural and functional changes occur under anthropogenic influences, for example, eutrophication, pollution, and acidification. In these cases, the processes of structural and functional changes proceed at higher rates than in natural conditions. A characteristic feature of the anthropogenic influence on large lakes (Ladoga, Onega, Imandra) is mixed pollution by household and industrial wastewater containing both nutrients and toxic substances. During the period of maximum anthropogenic loads, changes in water chemistry occurred in all three lakes according to the same type: the transparency decreased, the pH, the content of sulfates, chlorides, and nutrients increased, and contaminants with toxic properties appeared (Moiseenko and Sharov, 2019).

The results of the conducted studies show that in the water bodies of the North, during the long-term effect of technogenic pollution, qualitative and quantitative changes occurred in all trophic links of ecosystems (Moiseenko et al., 1999; Kashulin et al., 2018). The transformation of phytoplankton communities is determined by the intensity of the influence of various anthropogenic factors: acidification, eutrophication, toxic and thermal pollution, as well as their complex impact. In most cases, under anthropogenic influence, the variability of phytoplankton biomass dynamics increased. It is known that the smallest variability in biomass dynamics is characteristic of Arctic and tropical water bodies (Alimov et al., 2013).

The North of the ETR is distinguished by a high level of industrial and urban development, aquatic ecosystems are in one or another stage of transformation caused by the direct input of various pollutants, the development of eutrophication processes, disturbance of the landscapes of catchment areas, and aerial industrial pollution. Such changes could not but affect the state of algal communities, which are the most sensitive and dynamic element of water bodies. Structural and functional indicators of algal communities in many lakes are currently significantly different from the natural, characteristic of these natural and climatic conditions. Small lakes are the most vulnerable to anthropogenic impact. Many of them are located in the immediate vicinity of industrial enterprises, in urban areas and experience powerful multifactorial effects, including the direct release of toxic and biogenic components into the aquatic environment. In each of these reservoirs, specific algocenoses are formed. Species composition, community structure and quantitative characteristics of these algocenoses are radically different from those developed before industrial impact on the region, and the direction of succession is determined by the type and intensity of impact.

Eutrophication

Anthropogenic load determines the heterogeneity of phytoplankton distribution over the water area of large lakes. Small lakes tend to experience indirect anthropogenic impacts through air pollution and changes in the nature of activities in the catchment area. With the combined impact of anthropogenic factors, the greatest impact on phytoplankton is provided by domestic wastewater. As a result of this input, the total biomass increases and the species diversity of phytoplankton decreases, accompanied by a restructuring of the dominant complexes: an increase in the role of cryptophytes, dinophytes, and green (volvox) algae. An intensive development of cyanobacteria was observed in the Ladoga and Onega lakes. Structural changes in the phytoplankton community occurred in the direction of the development of species characteristic of eutrophic water bodies: cyanobacteria - *Oscillatoria tenuis* C. Agardh ex Gomont, *Oscillatoria planctonica* Woloszynska, *Anabaena lemmermanii, Anabaena hassalii* (Kütz.) Wittz., *Aphanizomenon flos-aquae, Microcystis aeruginosa*; зеленые - *Ankistrodesmus pseudomirabilis* Korschik, *Dictyosphaerium pulchellum* Wood, *Coelastrum microporum* Näg., *Pandorina morum* (Müll.) Bory, *Eudorina elegans* (Brebisson) Kützing; cryptophytic - *Cryptomonas sp.* Golden algae and dinoflagellates widespread in northern water bodies were rare. Among the diatoms, the common species from the genera Asterionella and Tabellaria were practically absent. Representatives of the genera Fragilaria, Diatoma, Aulacoseira, Stephanodiscus were widespread.

Changes in the structure of phytoplankton and its quantitative indicators associated with anthropogenic eutrophication are most noticeable in the Petrozavodsk and Kondopoga bays of Lake Onega, the Volkhov Bay of Lake Ladoga and the Bolshaya Imandra reach.

As a result of anthropogenic eutrophication, the phytoplankton biomass is, on average, an order of magnitude higher than the background values, and in short periods of mass development it can reach extremely high values (up to 50 g/m^3). Thus, in lakes affected by the effluents of the apatite industry, elevated concentrations of nitrates and phosphates supplied with effluents from the apatite production lead to the massive development of diatoms: *Stephanodiscus hantzschii* Grun. in Cl. et Grun., *Fragilaria capucina var. rumpens* (Kütz.) Lange-Bert., *F. capucina* Desm., *Staurosira venter* (Ehrb.) Kobayasi, etc. A characteristic feature of eutrophication processes is the intensive development of fouling and benthic species in the water column. The structure of algal communities includes species that prefer highly mineralized waters and pH> 7.0 - *Surirella brebissonii* Kramm. et Lange-Bert., *Melosira varians* Ag. In eutrophied water bodies, the development of algae occurs practically in all seasons, and continues under ice even during the polar night. Thus, at the minimum values of phytoplankton biomass in January (<0.01 g/ m^3), diatoms were found in the samples: their cells were rich in fatty acid inclusions (Figure 10). The presence of algae in the water column in winter allows zooplankton to develop.

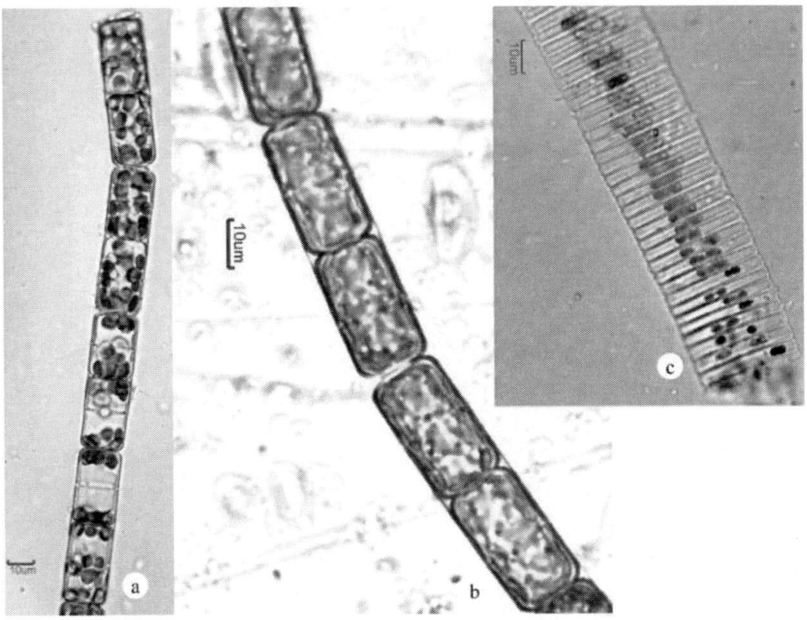

Figure 10. Representatives of winter plankton of eutrophied lakes: a, b – *Melosira varians*; c – *Fragilaria capucina*.

Phytoperiphyton also develops abundantly in eutrophied flowing water bodies all year round. At a time when the lake is covered with ice, fouling develops abundantly in the drainage area, in ice-free areas. Already starting from March - April on the drain of the lake Bolshoi Vudyavr (Khibiny mountain range, Murmansk region) green filamentous algae - *Ulothrix zonata* (Weber et Mohr) Kütz., were developed, forming long colonies together with diatoms (*Melosira varians, Fragilaria capucina, Ulnaria ulna* (Nitzsch) Compaère, (Ehrb.) Patrick

(Figure 11). After the opening of the lake from ice in June, the littoral is completely covered with abundant periphyton in the beginning of July, which forms phytomass that is extreme for the subarctic. *U. zonata* forms the basis of these overgrowths; various diatoms are less abundant. In addition, in the periphyton communities, *Gomphonema acuminatum* Ehrb., + *var. coronatum* (Ehrb.) Ehrb., *Encyonema minutum* (Hilse) D. G. Mann and various representatives of the order Fragilariales occurred. By the end of July, the fouling disappears from the littoral, and its development resumes by the end of summer, in August and September. It was during this period that the most stable phytoperiphyton communities were formed. The proportion of diatoms, both in abundance and in biomass, in the autumn periphyton is much higher than in the spring and summer. The development of algae in eutrophied water bodies continues until late autumn; some representatives of fouling species retain their vital functions under ice for a long time.

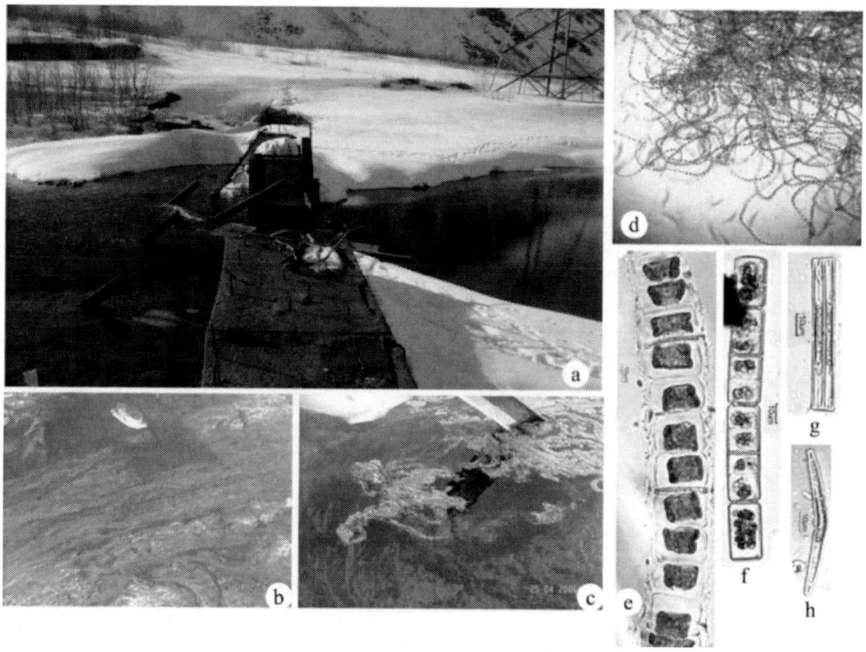

Figure 11. Spring phytoperiphyton of a small eutrophied lake: a - lake runoff; b, c - appearance of fouling; d,e – *Ulothrix zonata*; f – *Melosira varians*; g – *Ulnaria ulna*; h – *Hannaea arcus*.

Toxic Pollution

Pollution with heavy metals (an increase in the concentration of Ni by 90 times and Cu by 6 times in water relative to background ones) does not have a pronounced toxic effect on the level of phytoplankton biomass in the studied lakes (Moiseenko and Sharov, 2019). The high content of nutrients in water supplied with municipal wastewater probably reduces toxicity and has a stimulating effect on phytoplankton. In the structure of phytoplankton, a decrease in diversity occurs, and the proportion of green algae, mainly of the genera Scenedesmus and Pandorina, increases (Sharov, 2005). The threshold concentrations of metals in relation to algae differ by orders of magnitude in the case of the presence of organic substances and complex compounds in the water. In the presence of high concentrations of heavy metal compounds in the phytoplankton community, the suppression or disappearance of some species is observed

(Petrova et al., 2002). The experience of field observations even in extremely polluted water bodies shows that, although the dominant phytoplankton species are changing, the community does not degrade completely (Sharov, 2005). Despite the depletion of the species composition and the general suppression of the community under the influence of heavy metals, its complete degradation is noted only occasionally, even at extremely high concentrations of heavy metals (Petrova, 1990).

High values of the total phytoplankton biomass for northern water bodies (17 mg/L) were recorded in Lake Kuetsjärvi, while the concentration of Ni in water was 50-80 µg/L (Sharov, 2005). Experimental studies (Polyak et al., 2011) showed that a statistically significant effect of the action of metals on cyanobacteria was observed at Cu concentrations > 6–7 µg/L, Ni > 50–85 µg/L. Such high concentrations are rarely found in natural water bodies.

Lake Kuetsjärvi is one of the most polluted water bodies in Northern Europe. Long-term observation of the state of its ecosystem allows us to distinguish three large periods of development of phytoplankton communities, differing in the composition of the dominant taxa and quantitative indicators: I - 1994 - 1998; II - 2007 - 2011; III - 2012 - 2015. The average values of phytoplankton biomass and chlorophyll "*a*" content for the entire study period corresponded to the β-mesotrophic trophic status (Table 5).

In the period I, diatoms dominated in the southern part, and green algae (*Pandorina morum* (Müll.) Bory, 1827) dominated in the northern part, while golden algae (*Dinobryon sociale* (Ehrb.) Ehrb., 1834) were numerous throughout the whole lake. In period II, there were changes in the structure of community - the proportion of cyanoprokaryotes (*Microcystis pulverea f. delicatissima* W. & GSWest) Elenk., 1938) and representatives of the family Fragilariaceae as well as the content of chlorophyll "*a*" increased. As in the previous period, the share of green algae was high (up to 39%) in the northern part of the lake. Further changes in phytoplankton communities (period III) were associated with an increase in the abundance of cyanoprokaryotes (*Pseudanabaena sp.*), especially towards the end of the growing season, and green algae (up to 51%), whose proportion increased not only in the northern, but also in the southern part of the lake. The proportion of gold and diatom algae decreased, along with an increase in biomass and chlorophyll "*a*" content. The trend towards an increase in the biomass of phytoplankton in Lake Kuetsjärvi from 1994 to 2015 testifies the intensification of productivity in the reservoir: the average biomass value has doubled (Table 5). At the same time, the maximum values of phytoplankton biomass for the season also increased (up to 10.68 g/m^3).

The dynamics of phytoplankton during the hydrobiological summer (in 2012-2013) was characterized by a high level of biomass (up to 7 g/m^3) starting from June, which is typical for mesotrophic water bodies, with the largest share (over 40%) of green algae. In July, the biomass remained at the same level, the proportion of green algae increased (over 53%). The most abundant families were representatives of Volvox and Chlamydomonas; the proportion of diatoms decreased, whereas dinophytes and golden algae practically disappeared. In July, cyanoprokaryotes (*Pseudanabaena sp.*) also developed. By August, the biomass of algae decreased, mainly due to a slowdown in the vegetation processes of green algae (up to 1.4 g/m^3), golden algae completely disappeared from the composition of the communities, the proportion of cyanoprokaryotes decreased whereas the abundance of diatoms remained at the same level.

Table 5. Dynamics of some indicators of phytoplankton communities in Lake Kuetsjärvi during different periods (B - biomass, g/ m^3; Chl "a" - chlorophyll "*a*" content, mg/ m^3; S - saprobity index; K - quality class; T - trophic status of waters)

Indicators	Research period		
	1996-1998	2007-2011	2012-2015
Dominant taxa	*Melosira varians* *Pandorina morum* *Asterionella formosa* *Diatoma tenuis* *Dinobryon sociale* *Eudorina sp.*	*Asterionella formosa* *Fragilaria tenera var. nanana* *Staurosira construens* *Diatoma tenuis* *Microcystis pulverea f. delicatissima* *Dinobryon bavaricum*	*Asterionella formosa* *Pseudosphaerocystis lacustris* *Mucidosphaerium pulchellum* *Fragilaria capucina* *Pseudanabaena sp.*
B, g/m^3	1.33 (1.06-2.05)	1.93 (1.33-2.65)	2.75 (1.23-10.68)
Chl «*a*», mg/m^3	4.26 (2.71-5.95)	4.76 (0.35-4.04)	5.35 (0.46-28.34)
S	1.92 (1.02-1.44)	1.27 (1.17-1.52)	1.42 (1.30-1.88)
K	III	II	II
T	β-Mesotrophic		

The combined effect of climate change and environmental pollution is complex, causing profound restructuring in Arctic freshwater ecosystems that reduce their stability. The most significant changes are observed in phytoplankton communities: over the past 20 years, the proportion of green algae and cyanoprokaryotes has increased along with a decrease in the proportion of diatoms and aureus typical for the Arctic zone, while the average phytoplankton biomass became several times higher than the background values. The values of the saprobity index indicate a change in the water quality class from III to II. This does not indicate an intensification of the self-purification processes of the reservoir, but, on the contrary, illustrates an increase in the toxic load, at which species resistant to pollution are actively developing. This is confirmed by long-term trends towards an increase in the content of heavy metals in water. A powerful modulator of the observed changes is also the warming of the Arctic climate, which can increase the consequences of water eutrophication. A decrease in the saprobity index illustrates changes in the hydrochemical conditions of water formation, in particular, the ratio of toxic and trophic load on the lake against the background of climate warming. Summer dynamics of phytoplankton in Lake Kuetsjärvi also has a course not typical for subarctic lakes, due to the intensive development of green algae, which begin to actively vegetate from June.

Using an example of diatom complexes of the lakes sediments in the zone of influence of non-ferrous metallurgy enterprises, the dependence of phytoplankton taxonomic diversity on the load of toxic metals was shown. Thus, at present, the taxonomic diversity of diatom complexes of the studied lakes sediments is largely determined by the intensity of anthropogenic load, which confirms the dependence ($r = -0.77$) of the Shannon-Weaver diversity index (H ') on the coefficient of copper pollution, a dminant pollutant in the study area. At the same time, “preindustrial” diatom complexes do not exhibit such a dependence ($r = 0.09$). In lakes with high pollution coefficients of Cu, Ni, Cd: there was a sharp change in the dominant species of diatoms against the background and a general decrease in species richness and taxonomic diversity.

The presence of toxic metals in the aquatic environment at concentrations that are many times higher than the "background" ones leads to inhibition of photosynthetic activity. As a result, algae that develop in water bodies become capable of heterotrophic (mixotrophic) nutrition (Barinova et al., 2006). If nutrients are present in excess, the biomass of algae can reach extremely high values for subarctic lakes, exceeding the background values by two orders of magnitude. For example, there was a massive development of *Ceratium hirundinella* (Müll.) Bergh in the lake Verkhnee Staroe, experiencing pollution by industrial enterprises of ferrous metallurgy, at the end of summer. The phytoplankton biomass reached 60 g/m^3, while the concentration of photosynthetic pigments was low, at the level of 1.5 g/m^3. In lakes located in the immediate vicinity of metallurgical plants, a massive development of golden algae of the genus Dinobryon, also capable of mixotrophy, was observed; cells lacking chloroplasts were abundant in the samples.

Phytoplankton communities, unique for the Arctic zone, develop in the lakes of urbanized areas. A survey was carried out of the lakes located in the largest city of the North of the ETR - Murmansk in the period from 2018 to 2020. In total, 132 taxa of algae and cyanoprokaryotes with a rank below the genus were identified in the plankton: diatoms (Bacillariophyta) - 63; golden (Chrysophyceae) - 6; green (Chlorophyta) - 24; dinoflagellates (Dinoflagellata) - 8; charovy (Charophyta) - 12; cryptophyta (Cryptophyta) - 5; Euglenophyta (Euglenophyta) - 1; and cyanobacteria (Cyanobacteria) - 13. The proportion of main taxonomic categories and the level of biomass are shown in Figure 12, whereas some typical representatives are illustrated in Figure 13.

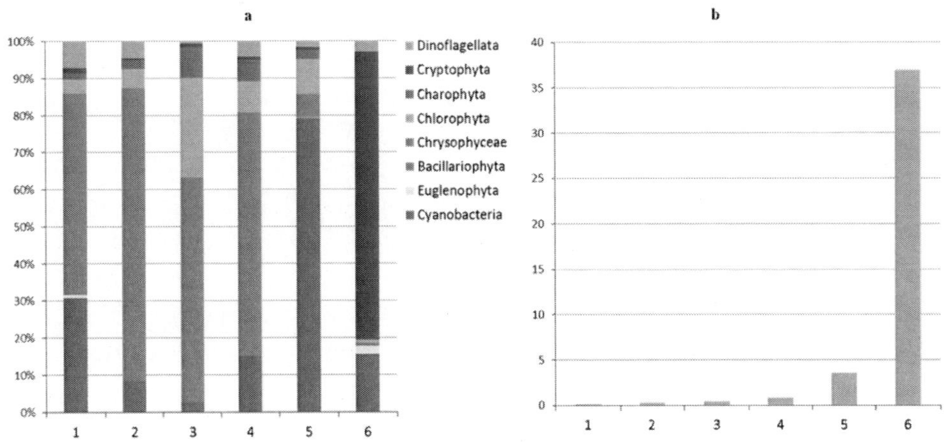

Figure 12. Communities of phytoplankton of lakes in Murmansk: a - ratio of the number of large taxonomic categories; b - biomass (g/m^3). Lakes: 1 - Srednee, 2 - Yuzhnoe, 3 - Okunevoe, 4 - Severnoye, 5 - Semenovskoye, 6 - Ledovoe.

The structure of phytoplankton communities in the studied lakes is characterized by the presence of a significant proportion of cyanoprokaryotes, including species that can cause water bloom and are potentially toxic to aquatic organisms and humans, including: *Aphanizomenon gracile; Dolichospermum solitarium; D. affine* (Algae ..., 2006; Kokociński et al., 2013; Skulberg et al., 1994). Colonies of collar protozoa (Choanoflagellata) were recorded together with colonies of *Asterionella formosa*, which is an indicator of eutrophication (Zolotarev, 2013; Kopylov and Kosolapov, 2011).

The ecosystems of lakes Semenovskoe and Ledovoe are most transformed under anthropogenic impact. The first lake is dominated by cyanoprokaryotes: *Aphanizomenon gracile, Dolichospermum spp.*, and the second lake is dominated by the cryptomonads *Cryptomonas marssonii* Skuja (Figure 12, a). Such a structure of phytoplankton communities, which is not typical for the Arctic regions, has been recorded for the first time for water bodies of the Murmansk region. The algoflora of plankton in other lakes as a whole retains the features of natural waters.

The revealed features of the species composition and structure of communities make it possible to conclude on high bioindication potential of phytoplankton in the studied water bodies. This is due, first of all, to the active development of cyanoprokaryotes, which prefer eutrophic water bodies and elevated water salinity. This indicates a significant transformation of the ecosystems of the lakes of Murmansk in comparison with the conditionally background lakes of the tundra and forest-tundra zones of the region (Denisov and Barinova, 2015).

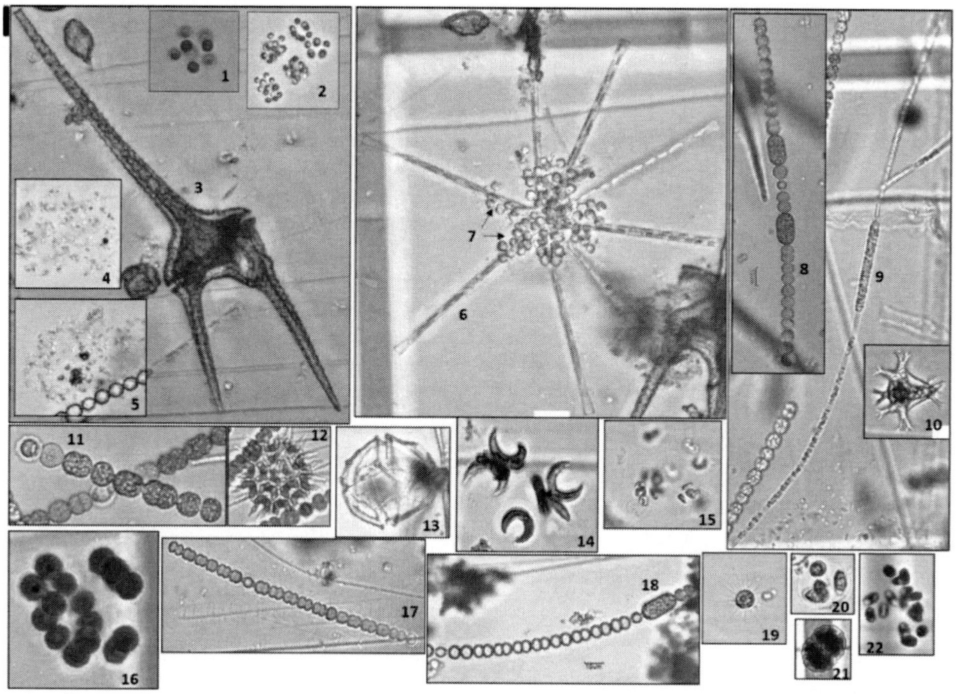

Figure 13. Some typical representatives of algae and cyanoprokaryotes in the lakes of Murmansk: 1,2 - *Sphaerocystis planctonica* (Korsh.) Bourrelly; 3 - *Ceratium hirundinella* (O. Müll.) Dujardin; 4.5 - *Anathece clathrata* (West & G.S. West) Komár., Kastovsky & Jezberová; 6 - colony of *Asterionella formosa* Hass. and Choanoflagellata attached to it (7); 8 - *Dolichospermum affine* (Lemm.) Wacklin, L.Hoffmann & Komár.; 9 – *Aphanizomenon gracile* Lemm; 10 – *Pseudostaurastrum lobulatum* (Nägeli) Bourrelly; 11 – *Dolichospermum solitarium* (Klebahn) Wacklin, L.Hoffmann & Komárek; 12 – *Pediastrum boryanum* (Turpin) Meneghini; 13 – *Peridinium willei* Huitfeldt-Kaas; 14 – *Selenastrum bibraianum* Reinsch; 15 – *Messastrum gracile* (Reinsch) T.S.Garcia; 16 – *Planctococcus sphaerocystiformis* Korsh.; 17 – *Dolichospermum viguieri* (Denis & Frémy) Wacklin, Hoffm. & Komár.; 18 – *Anabaena echinospora* Skuja; 19 – *Golenkinia radiata* Chodat; 20 – *Tetrachlorella alternans* (G.M.Smith) Korsh.; 21 – *Cosmarium subprotumidum* Nordstedt; 22 – *Dictyosphaerium ehrenbergianum* Näg.

Acidification

When studying the combined effect of toxic metals and low pH on the biomass and species composition of phytoplankton in a number of lakes (Sharov, 2005), despite high concentrations of copper and nickel, no clear correlation between biomass and metal content was found. Species diversity (Figure 14) serves as a better indicator of acidification than biomass, which has not been significantly correlated with pH.

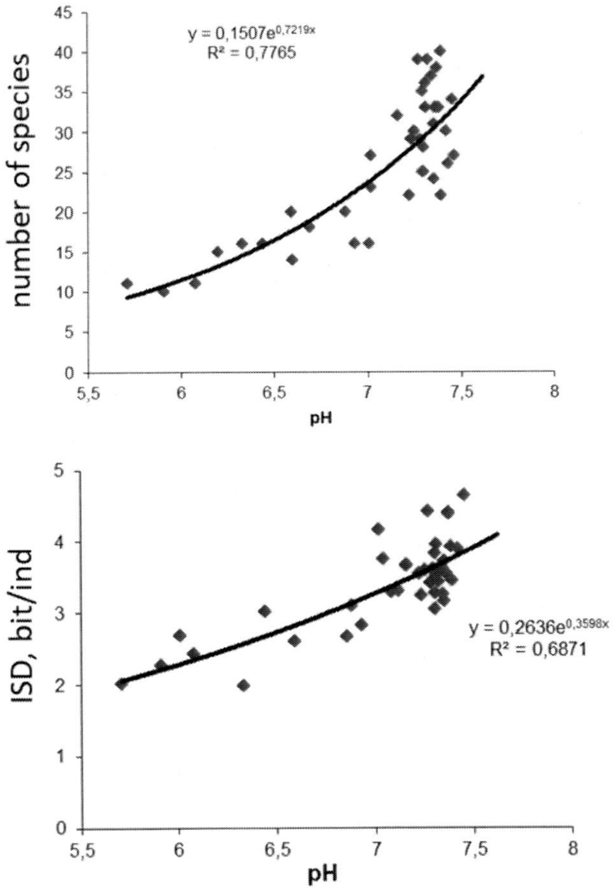

Figure 14. Dependence of the number of species (in the sample) of phytoplankton and the index of species diversity (ISD) on the pH of lake water in the North of the ETR ($p < 0.02$) (Sharov, 2005).

At low water pH values (<5), the phytoplankton biomass can sometimes reach unusually high values (Findlay and Kasian, 1986; Sharov, 2005), which is probably associated with a decrease in the pressure of predatory zooplankton on algae. Since the acidification of water bodies leads to a decrease in the species richness of zooplankton, monospecific communities of resistant species (*Eudiaptomus gracilis* (Sars), *Bosmina longirostris* (O.F. Müller)) remain. Accordingly, the variability of phytoplankton biomass dynamics increases. This was noted during the study of small anthropogenically acidified lakes in the northwest of the Kola Peninsula (Sharov, 2005).

Thermal Pollution

In the heating zone as a result of discharge into the lake Imandra warm waters of the Kola NPP, sizable changes affected all parts of the ecosystem. But the most striking manifestations were noted in the phytoplankton community, where both quantitative and structural changes occurred. The maximum heating of the surface water layer in summer reaches 26°C, while in the unheated zone the temperature rarely reaches 18°C. The area of influence extends over a radius of about 5 km. The temperature difference reaches 10ºC in spring and 6ºC in summer (Sharov, 2005). Thermal pollution in subarctic conditions leads to an increase in the abundance and structural changes of phytoplankton, similar to those that occur during eutrophication. Along with a general increase in the number of planktonic algae, the formation of thermophilic complexes is noted, in which *Ceratium hirundinella* predominates (Sharov, 2005). High values of phytoplankton biomass were recorded in the warmest part of the lake - at the mouth of the spillway. Diatoms are dominated by *Rhizosolenia longiseta, Asterionella formosa* Hass., *Diatoma mesodon* (Ehr.) Kütz. *Tabellaria fenestrata* (Lyngbye) Kütz. *Fragilaria capucina* Desm is quite common, and *F. crotonensis* Kitt., whose colonies can consist of several hundred cells. It is known that representatives of the genus Fragilaria are thermophilic; they generally develop in heated water layers under stratification conditions (Trifonova, 1990; Razumovsky, 1998).

Overall, the species composition of phytoplankton in the zone of influence of heated waters corresponds to the typical subarctic type with a predominance of diatoms and golden algae in abundance and biomass. The phytoplankton biomass in areas of the lake with undisturbed temperature regime was not much lower than that in the zone of influence of heated waters. The distribution of phytoplankton over the water area in July is largely determined by the total nitrogen content. Compared with the results of studies of the 90s of the last century, there is a significant decrease in the average biomass of phytoplankton, which indicates the stabilization of the ecosystem and confirms some trends towards a decrease in the nutrient load on the reservoir (Denisov et al., 2017).

A significant factor determining the development of plankton in the zone of influence of heated waters is the intensive artificial circulation of water masses between the reaches of Lake Imandra during the operation of the cooling system of the Kola NPP. Despite the constant supply of nutrients as a result of the operation of fish cages, intensification of eutrophication processes is not observed due to the active hydrodynamic redistribution of water masses in the heating zone due to a constant flow of water from the outlet channel. Probably, the active utilization of nutrients by algae is facilitated by the increased water temperature, the early beginning of the growing season, and the abundant development of higher aquatic vegetation and periphyton.

The zone of influence of the heated waters of the Kola NPP can probably be considered as a model water area for studying the response of aquatic ecosystems to possible climate warming. Due to thermofication in the heating zone as a result of discharge into the lake Imandra warm waters of the Kola NPP, changes affected all parts of the ecosystem. However, the most striking manifestations were noted in the phytoplankton community, where both quantitative and structural changes took place. The maximum heating of the surface water layer in summer reaches 26°C, while in the unheated zone the temperature rarely reaches 18°C. The area of influence covers the water area with a radius of about 5 km. The temperature difference reaches 10ºC in spring and 6ºC in summer (Sharov, 2005). Thermal pollution in subarctic conditions leads to an increase in the abundance and structural changes of phytoplankton, similar to those that occur during eutrophication. Along with a general increase in the number of planktonic algae, the formation of thermophilic complexes is noted, in which *Ceratium hirundinella* predominates (Sharov, 2005). High values of phytoplankton biomass were recorded in the warmest part of the lake - at the mouth of the spillway. Diatoms are dominated by *Rhizosolenia longiseta, Asterionella formosa* Hass., *Diatoma mesodon* (Ehr.) Kütz., *Tabellaria fenestrata* (Lyngbye) Kütz., *Fragilaria capucina;* Desm is quite common and *F. crotonensis* Kitt., whose colonies can consist of several hundred cells. It is known that representatives of the genus Fragilaria are thermophilic; they generally develop in heated water layers under stratification conditions (Trifonova, 1990; Razumovsky, 1998).

Over the last decade, the species composition of phytoplankton in the zone of influence of heated waters corresponded to the typical subarctic type with a predominance of diatoms and golden algae in abundance and biomass. The level of phytoplankton biomass in areas of the lake with a natural temperature regime during the period of intensive vegetation was not much lower than in the zone of influence of heated waters. The distribution of phytoplankton over the water area in July is largely determined by the total nitrogen content. Compared with the results of studies of the 90s of the last century, there is a significant decrease in the average biomass of phytoplankton, which indicates the stabilization of the ecosystem and confirms some trends towards a decrease in the nutrient load on the reservoir (Denisov et al., 2017).

A significant factor determining the development of plankton in the zone of influence of heated waters is the intensive artificial circulation of water masses between the reaches of Lake Imandra during the operation of the cooling system of the Kola NPP. Despite the constant supply of nutrients as a result of the operation of fish cages, intensification of eutrophication processes is not observed due to the active hydrodynamic redistribution of water masses in the heating zone via a constant flow of water from the outlet channel. Probably, the active utilization of nutrients by algae is facilitated by the increased water temperature, the early beginning of the growing season, and the abundant development of higher aquatic vegetation and periphyton.

The zone of influence of the heated waters of the Kola NPP can probably be considered as a model water area for studying the response of aquatic ecosystems to possible climate warming.

Reduced Anthropogenic Load

Water bodies are capable of restoration, transition to a new equilibrium state of the system after a decrease in anthropogenic load. Based on the generalization of long-term studies of large lakes in the North of the European Region (Ladoga, Onega and Imandra), a retrospective analysis of their condition is made and the main modifications of aquatic ecosystems under anthropogenic loads and their decrease were revealed (Moiseenko, Sharov, 2019). During the period of decreasing anthropogenic load on the lakes, starting from the 1990s, tendencies towards an improvement in water quality have emerged. The concentration of toxic substances in the water of the studied lake bays decreased. For example, in the Imandra Lake, the concentration of nickel decreased from 150 to 10 μg/L; in the Kondopoga Bay of Lake Onega and Lake Ladoga, the concentration of lignosulfates and phenols (markers of pollution by wastewater from the pulp and paper industry) have also decreased (Moiseenko, Sharov, 2019).

The content of common forms of phosphorus in the Ladoga and Onega lakes decreased: in the lake Imandra, it remained at the same level; total nitrogen concentrations have only slightly decreased in all three lakes. The concentrations of bioavailable phosphates and nitrates decreased significantly (by 6-20 times), which indicates their more active utilization in the trophic structure of the ecosystem. The dynamics of silicon is even more interesting: Si concentration continues to systematically decrease from year to year, which is associated with its active absorption by diatoms.

The abundance of phytoplankton decreased in the Volkhov Bay of Lake Ladoga and the Kondopoga Bay of Lake Onega; in Lake Imandra it remains at the same level. The average biomass in the lakes varied from 1.7 to 3.4 mg/ L, and the concentration of chlorophyll "a" ranges from 3.6 to 7.9 µg/L. There is still a high abundance of Cryptomonas sp., Stephanodiscus sp. and *Aulacoseira islandica*. The structure of the biomass differs from the natural state by a larger proportion of cryptophyte and green algae. The high biomass of phytoplankton in Lake Imandra, recorded over recent decades, is associated with the massive development of diatoms and cyanobacteria (Denisov et al., 2017).

After a decrease in anthropogenic impact, the ecosystems do not return to their original natural state (Moiseenko, Sharov, 2019). A new stage with characteristic features is being formed. A decrease in the supply of total phosphorus from the drainage basin does not lead to a rapid drop in the concentration of chlorophyll "*a*" and the abundance of phytoplankton. At the same time, the relative amount of cryptophyte algae and cyanobacteria increases.

Conclusion

The species richness of planktonic algae in cold water bodies is mainly determined by the size and nature of the catchment area. In the studied lakes of the North of the European Region, 781 phytoplankton taxons with a rank below the genus were found. In the phytoplankton of large lakes, diatoms predominate in terms of the number of species and biomass. In small lakes, their diversity and quantitative characteristics decrease. At the same time, the proportion in the total biomass of golden, cryptophytic, green, and sometimes dinophytic algae and cyanobacteria increases.

The phytoplankton biomass during the study period varied within wide limits - from several mg/L in Arctic lakes to 34 mg/L in Lake Pskovskoe. During eutrophication, even in a subarctic lake, it could reach 20 mg/L in the areas where domestic waste water input. Thermal pollution in subarctic conditions leads to an increase in the abundance and structural changes of phytoplankton, similar to those that occur during eutrophication. The average growing biomass of phytoplankton in naturally clean cold-water lakes, as a rule, does not exceed 1 mg/L.

Long-term comprehensive studies of algocenoses indicate that they are characterized by significant taxonomic diversity, which largely determines the general gene pool of the Euro-Arctic region. The formation and development of algal communities occurs under the influence of anthropogenic load, associated mainly with the activities of industrial enterprises. Another significant factor is global and regional climatic changes, which are most pronounced at high latitudes. The study of long-term changes in ecosystems has become most relevant in recent decades, in the era of intensive anthropogenic transformation of the biosphere. As such, special algological studies are becoming increasingly important.

The long-term eutrophication of the waters of the Euro-Arctic region, as well as the intensification of these processes in recent years acquire a global character. There are radical restructuring of the structural and functional characteristics of algocenoses, and a change in the direction and rate of successions. These result from interdependent action of a complex of regulatory factors. Among these factors, the most significant is anthropogenic, currently enhanced by anomalous dynamics of the regional climate system.

Algal communities are sensitive to all ongoing changes: new adaptive mechanisms are formed, the productivity of communities and the ecological preferences of individual species also change. This becomes of decisive importance for bioindication of the quality of the environment and the tasks of hydroecological monitoring.

Climatic changes can lead to the transformation of the structural and quantitative characteristics of aquatic communities. In the water bodies of the North of Russia, climate change affects primarily the ice regime. This in turn can affect the length of the growing season and the seasonal cycles of aquatic organisms. At the same time, no significant correlations were found between climatic characteristics (temperature, precipitation, ice-free period, NAO, etc.) and phytoplankton biomass. Climate change was reflected in the increase in seasonal and interannual dynamics of phytoplankton biomass.

Possible climate warming in the northern regions may slow down the processes of lakes recovery, since the input and circulation of nutrients will intensify, more thermophilic species characteristic of eutrophic waters will predominate, and typical Arctic species will not recover their numbers. During the period of decreasing anthropogenic load on the lakes, starting from the 1990s, tendencies towards an improvement in the quality of water in terms of hydrochemical indicators and restoration of ecosystems have emerged. However, with a decrease in the concentration of toxic and biogenic substances in the water of the studied lake bays, there was no significant drop in the concentration of chlorophyll "*a*" in the water and the restoration of the phytoplankton structure to background values. The structure of phytoplankton biomass changes towards the prevalence of cyanobacteria, green and cryptophytic algae.

Acknowledgments

This research was funded by research project 0226-2019-0045 (results generalization) and partially supported by the RFBR grants Arctic 18-05-60125 (large lakes algae study), 18-05-60142 (climatic changes impact analysis), 19-04-01000 (algae under the ice) and RSF grant 19-77-10007 (hydrochemical analysis and fieldwork)."

References

Algae causing "blooming" of water bodies in the North-West of Russia. 2006. Russian Acad. Sciences, Botanical Institute im. V. L. Komarova. Moscow: Association of scientific. ed. KMK, (Moscow: Nauka).

Alimov A. F. 2000. *Elements of the theory of the functioning of aquatic ecosystems*. SPb, Nauka, 147 p.

Alimov A. F., Bogatov V. V., Golubkov S. M. 2013. *Production hydrobiology.* Nauka, Saint-Petersburg.

Annotated ecological catalog of lakes in the Murmansk region: central and southwestern regions of the Murmansk region (*basins of the Barents and White seas and the Gulf of Bothnia of the Baltic Sea*). 2013. Kashulin N. A., Sandimirov S. S., Dauvalter V. A., Kudryavtseva L. P., Terent'ev P. M., Denisov D. B., Vandysh O. I., Koroleva I. M., Valkova S. A., Kashulina T. G. Apatity, Publishing House of the Kola Scientific Center of the Russian Academy of Sciences.

Annotated ecological catalog of lakes in the Murmansk region: southeastern part (*White Sea basin*). 2012. Kashulin N. A., Sandimirov S. S., Dauvalter V. A., Kudryavtseva L. P., Terentyev P. M., Denisov D. B., Vandysh O. I., Valkova S. A. Apatity, Publishing House of the Kola Scientific Center of the Russian Academy of Sciences.

Annotated ecological catalog of the lakes of the Murmansk region: the eastern part (*the Barents Sea basin*). 2010. Kashulin N. A., Sandimirov S. S., Dauvalter V. A., Kudryavtseva L. P., Terentyev P. M., Denisov D. B., Vandysh O. I., Valkova S. A. Apatity, Publishing house of the Kola Scientific Center of the Russian Academy of Sciences.

Armand A. D., Armand N. N., Grave M. K. et al. 1969. Consolidated stratigraphic scheme of Quaternary (anthropogenic) deposits of the Kola Peninsula in the light of the latest data. *Basic problems of geomorphology and stratigraphy of the anthropogen of the Kola Peninsula*. L: Nauka: 7-24.

Arnoldi V., Alekseenko M. 1915. Materials for the flora of algae in Russia. III. *Lakes of Lapland. Tr. Islands of nature testers at Kharkiv unte*. Kharkov, 43 (2): 43.

Borovichev E. A., Shalygin S. S., Davydov D. A. 2010. Supplement to the flora of cyanoprokaryotes and liverworts of the Lapland reserve (Murmansk region). Scientific notes of Petrozavodsk State University. Series: *Natural and technical sciences*, 8 (113): 7-10.

Cairns J. Jr. 2005. Restoration Ecology and Ecotoxicology, *Handbook of Ecotoxicology* (eds. Hoffman, D. J., Rattner, B. A., Burton, G. A. & Cairnce, J. Jr.). N. Y, Lewis Publishers: 1015–1029.

Chekryzheva T. A. 2011. Phytoplankton of the Vygozersk reservoir, Republic of Karelia (Russia). *Gidrobiol. Journal. Kiev,* 47(5): 13–21.

Chekryzheva T. A., Sharov A. N. 2011. Phytoplankton. *Atlas of Lake Onega* (edited by N. N. Filatov). Petrozavodsk: KarRC RAS: 111–112.

Davydov D. A. 2010. *Cyanoprokaryotes and their role in the process of nitrogen fixation in terrestrial ecosystems of the Murmansk region*. Moscow: GEOS.

Davydov D. A. 2008. *Terrestrial cyanoprokaryotes on the territory of the Polar-Alpine Botanical Garden-Institute* (Khibiny, Kola Peninsula). Byull. MOIP. Biological department, 113(1): 72–75.

Davydov D. A., Patova E. N., Shalygin S. S., Vilnet A. A., Novakovskaya I. V. 2020. The problem of the hidden diversity of cyanoprokaryotes in the Arctic territories. *Theoretical and Applied Ecology,* 1: 110–117. https://doi.org/10.25750/1995-4301-2020-1-110-116.

Denisov D. B. 2007. Changes in the hydrochemical composition and diatom flora of bottom sediments in the zone of influence of mining production (Kola Peninsula). *Water Resources,* 34(6): 719–730. https://doi.org/10.1134/S0097807807060097.

Denisov D. B. 2010. Ecological features of algal communities of different types of subarctic water bodies. *Bulletin of the Kola Scientific Center of the Russian Academy of Sciences*, 1: 48–55.

Denisov D. B. 2012. Reconstruction of the development of the ecosystem of a small mountain subarctic water body over the past 900 years (for example, Lake Akadmicheskoe, Khibiny, Kola Peninsula). Proceedings of the KSC RAS "*Applied Ecology of the North*." Issue 1. Apatity: 126–147.

Denisov D. B., Barinova S. S. 2015. Diversity of algae and environmental indicators in the Arctic lakes of the Kola Peninsula, Russian North. *Questions of modern algology*, 2(9). URL: http://algology.ru/742.

Denisov D. B., Barinova S. S. 2016. Phytoperiphyton and environmental indicators in the rivers of the Kola Peninsula, Russian Arctic North. *Questions of modern algology*, 1(11). URL: http://algology.ru/ 957.

Denisov D. B., Dauvalter V. A., Kashulin N. A., Kagan L. Ya. 2006. Long-term changes in the state of subarctic water bodies under anthropogenic load (according to diatom analysis). *Biology of internal waters,* 1: 24–30.

Denisov D. B., Kashulin N. A. 2018. Algal communities of freshwater ecosystems of the Kola North in the context of global environmental transformations. *Bulletin of the Kola Scientific Center of the Russian Academy of Sciences*, 1: 23–35.

Denisov D. B., Kosova A. L. 2017. Development of a database on diatoms in the Euro-Arctic region. *Proceedings of the Kola Scientific Center of the Russian Academy of Sciences*, 6-5(8): 45–52.

Denisov D. B., Valkova S. A., Kashulin N. A. 2018. Algal communities and macrozoobenthos of aquatic ecosystems of the Khibiny mountain range (Kola Peninsula). *Bulletin of the Kola Scientific Center of the Russian Academy of Sciences*, 1(10): 23–35.

Denisov D. B., Valkova S. A., Kashulin N. A. 2009. *Modern trends in the change in the biota of freshwater ecosystems in the Murmansk region*. Vestnik MGTU, 12(3): 525–538.

Denisov D. B., Valkova S. A., Terentyev P. M., Sandimirov S. S., Vandysh O. I. 2017. The current state of the ecosystem of the lake. Imandra in the zone of influence of the Kola NPP (Murmansk region). *Water: chemistry and ecology,* 6: 41–51.

Falk D. A., Palmer M., Zedler J. 2006. *Foundation of Restoration Ecology: the Science and Practice of Ecological restoration*. Island Press.

Findlay D. L., Kasian S. E. M. 1986. Phytoplankton community responses to acidification of lake 223, Experimental Lakes Area, northwestern Ontario. *Air and Soil Pollut,* 30(3-4): 719–726.

Forbes V. E., Forbes, T. L. 1994. *Ecotoxicology in Theory and Practice*. London, Chapman & Hall Ltd.

Heinonen P. 1980. *Quantity and composition of phytoplankton in Finnish inland waters*. Vesien-tutkimuslaitoksen julkaisuja: 1-91.

Holmgren S. K. 1983. Phytoplankton biomass and algal composition in natural, fertilized and polluted subarctic lakes. *Acta Univ. Upsal. Uppsala Diss. Fac. Sci.* 674: 1-16.

International Didymo Conference: new horizons in science and management 2013 (March, 12-13, 2013, Providence, Rhode, Island). Hosted by the Invasive Species Action Network and the Northeast Aquatic Nuisance Species Panel Fiscal management provided by the Northeast Aquatic Nuisance Species Council: 44.

Jeppesen E., Jensen J. P., Søndergaard M. (2002). Response of Phytoplankton, Zooplankton, and Fish to Re-Oligotrophication. An 11 Year Study of 23 Danish Lakes. *Aquatic Ecosystem Health & Management* 5(1): 31–43. https://doi.org/10.1080/14634980260199945.

Jónsson G. St. 1977. *Phytoplankton in Lake Thingvallavatn 19754-75.* M.Sc. thesis. Univ. of Iceland, Reykjavik.

Kagan L. Ya. 2012. *Diatoms of the Euro-Arctic region Annotated collection (ancient and modern marine and freshwater).* Apatity: Publishing House of the Kola Scientific Center of the Russian Academy of Sciences.

Kashulin N. A., Dauvalter V. A., Denisov D. B., Valkova S., Vandysh O., Terentjev P., Kashulin A. 2017. Selected aspects of the current state of freshwater resources in the Murmansk region, Russia. *Journal of Environmental Science and Health,* Part A: 1-9. http://dx.doi.org/10.1080/10934529.2017.1318633

Kashulin N. A., Dauvalter V. A., Denisov D. B., Valkova S. A., Vandysh O. I., Terent'ev P. M., Zubova E. M., Koroleva I. M., Kosova A. L., Cherepanov A. A. 2018. Complex studies of freshwater ecosystems in Fennoscandia. *Proceedings of the Kola Scientific Center of the Russian Academy of Sciences,* 9-6(9): 34–86.

Kihlman A. O. 1889. *Berichteiner naturwissenschaftlichen Reise durch Russland Lapland im Jahre 1889* [*Report of a scientific trip through Russia Lapland in the year 1889*]. Fennia, 3(6): 1–40.

Kokociński, M., Mankiewicz-Boczek, J., Jurczak, T., Spoof L., Meriluoto J., Rejmonczyk E., Hautala H., Vehniäinen M., Pawełczyk J., Soininen J. 2013. *Aphanizomenon gracile* (Nostocales), a cylindrospermopsin-producing cyanobacterium in Polish lakes. *Environ Sci. Pollut. Res.,* 20: 5243–5264.

Komulainen S. F., Antipina G. A., Vislyanskaya I. G., Ieshko T. A., Lak G. Ts., Chekryzheva T. A., Sharov A. N., Shelekhova T. S. 2006. *Bibliography of works on algae in the European North of Russia (Republic of Karelia, Murmansk region).* Petrozavodsk: Karelian Scientific Center of the Russian Academy of Sciences.

Kopylov A. I., Kosolapov D. B. 2011. *Microbial "loop" in planktonic communities of marine and freshwater ecosystems.* IBVV Izhevsk: Knigograd.

Kosinskaya E. K. 1935. *Materials for the flora of algae on the Kola Peninsula.* - Tr. BIN of the USSR Academy of Sciences.

Kozhin M. N., Borichev E. A., Belkina B., Davydov E. A., Denisov D. B., Isaeva L. G., Konstantinova N. A., Melekhin A. V., Popova K. B., Urbanavichyus G. P., Khimich Yu. R. 2019. *History and main results of the study of cryptogamic organisms of the Green Belt of Fennoscandia within the Murmansk region.* Transactions of the Karelian Scientific Center of the Russian Academy of Sciences, 4: 64-88. https://doi.org/10.17076/them1024.

Letanskaya G. I. 1974. Phytoplankton and primary production of lakes on the Kola Peninsula. *Lakes of various landscapes of the Kola Peninsula.* Leningrad, Part 2: 78–119.

Letanskaya G. I., Protopopova E. V. 2012. The current state of phytoplankton of Lake Ladoga (2005-2009). *Inland Water Biology,* 4: 17–24. https://doi.org/10.1134/S1995082912040104.

Levander K. 1905. Zur Kenntniss des Planktons einiger Binnenseen in Rusisch –Lapland [To know the plankton of some inland lakes in Russian Lapland]. *Festschrift f. Palmrn. Bd* 1, 11: 1–40.

Lund J. W. G. 1962. Phytoplankton from some lakes in northern Saskatchewan and from Great Slave Lake. *Canad. J. Bot.,* 40: 1500–1514.

Melekhin A. V., Davydov D. A., Borovichev E. A., Shalygin S. S., Konstantinova N. A. 2019. CRIS – service for input, storage and analysis of the biodiversity data of the cryptogams. *Folia Cryptog. Estonica, Fasc.* 56: 99–108. https://doi.org/10.12697/fce.2019.56.10.

Moiseenko T. I., Sharov A. N., Vandish O. I., Kudryavtseva L. P., Gashkina N. A., Rose C. 2009. Long-term modification of arctic lake ecosystem: reference condition, degradation and recovery. *Limnologica,* 39(1): 1–13. https://doi.org/10.1016/j.limno.2008.03.003.

Moiseenko T. I., Dauvalter V. A., Kagan L. Ya. 1997. Mountain lakes as indicators of air pollution. *Water Resources,* 23(5): 669–682.

Moiseenko T. I., Denisov D. B. 2019. Is it possible to restore the Arctic lake ecosystem after long-term pollution? *Arctic: ecology and economics,* 4(36): 16–25. https://doi.org/10.25283/2223-4594-2019-4-16-25.

Moiseenko T. I., Vandysh O. I., Yakovlev V. A., Lukin A. A., Sharov A. N. 1999. Changes in the biodiversity of surface waters in the North under conditions of acidification, eutrophication and toxic pollution. *Water Resources,* 4: 492–501.

Moiseenko T., Sharov A. 2019. Large Russian Lakes Ladoga, Onega, and Imandra under Strong Pollution and in the Period of Revitalization: A Review. *Geosciences* 9, 492. https://doi.org/10.3390/geosciences9120492.

Moore J. W. 1978. Distribution and abundance of phytoplankton in 153 lakes, rivers and pools in the Norhwest Territories. *Can. J. Bot.*, 58(15): P. 1766–1773.

Moriarty F. 1989. *Ecotoxicology: The Study of Pollutants in Ecosystems*, London; New-York: Academic Press Inc.

Newman M. C., Jagoe Ch. H. 1996. *Ecotoxicology: a Hierarchical Trieatment.* N.Y.: Levis publishers Ltd.

Nikulina V. N. 1975. Phytoplankton. *Biological productivity of northern lakes. Part 2. Lakes Zelenetskoye and Akulkino*. Leningrad: 37–52.

Nylander W., Saelan Th. 1859. *Hebarium musci Pennici*. Helsingfors.

Özkan K., Jeppesen E., Davidson T. A., Bjerring R., Johansson L. S., Søndergaard M., Lauridsen T. L., Svenning J. C. 2016. Long-Term Trends and Temporal Synchrony in Plankton Richness, Diversity and Biomass Driven by Re-Oligotrophication and Climate across 17 Danish Lakes. *Water*, 8, 427.

Patova E. N., Davydov D. A. 2006. Cyanoprokaryota diversity and ecology of the European sector of the Russian Arctic. *Algological research: current state and future prospects.* Materials of the I All-Russian scientific-practical conference (November 16-18, 2006, Ufa). - Ufa: Publishing house BSPU: 89–91.

Petrova N. A., Antonov S. E., Protopopova E. V. 1992. Structural and functional characteristics of phytoplankton. *Lake Ladoga - criteria for the state of the ecosystem.* SPb.: 119–145.

Petrova N. A., 1987. The phytoplankton of Ladoga and Onega lakes and its recent successional changes. *Arch. Hydrobiol. Beih. Ergebn. Limnol.*, 25: 11–18.

Polyak Y., Zaytseva T., Petrova V., Medvedeva N. 2011. Development of Mass Cyanobacteria Species under Heavy Metals Pollution. *Hydrobiological Journal* 47(3):75.

Poretskij V. S., Zhuze A. P. Sheshukova V. S. (1934). *Diatoms of Kola peninsula in collection with microscopic structure Kola Diatomite*. Works Geomorphological institute AN USSR. Leningrad. 8: 96–210.

Razumovsky L. V. 1998. Assessment of the general level of anthropogenic load by the method of graphic comparison on the example of diatom assemblages from the lakes of the Kola

Peninsula. *Problems of ecological monitoring and modeling of ecosystems*. T. XVII. SPb: Gidrometeoizdat: 188–206.

Reynolds C. S. 2006. *The ecology of phytoplankton*. Cambridge University Press, Cambridge.

Reynolds C. S., V. Huszar, C. Kruk, L. Naselli-Flores, S. Melo. 2002. Towards a functional classification of the freshwater phytoplankton. *Journal of Plankton Research,* 24(5): 417–428.

Richard I. 1889. Notes sur les peches effectuees par M. Ch. Rabor dansles lacs Enara, Imandra et dans le Kolozero [Notes on the fisheries carried out by Mr. Ch. Rabor in lakes Enara, Imandra and in Kolozero]. *Bull. Sos. Zool. France*. 14: 104.

Roll Ya. V. 1923. *Desmidium algae found in water bodies of Lapland, Olonets province*. Vologda: 5–63.

Sharov A. N. 2005. The effect of anthropogenic pollution on phytoplankton of the subarctic waterbodies (Kola Peninsula). *Inland water biology* 2, 61–68.

Sharov A., Andreeva I. 2015. Phytoplankton responses to climate change in the large lakes of the Baltic Sea basin. *Acta Biologica Universitatis Daugavpiliensis* 15 (2): 349–358.

Sharov A. N. 2017. Assessment of the water quality of different types of cold-water lakes by phytoplankton indicators. *Regional ecology,* 2(48): 40–47.

Sharov A. N. 2004. *Phytoplankton of water bodies of the Kola Peninsula*. Petrozavodsk: Karelian Scientific Center of the Russian Academy of Sciences.

Sharov A. N., Berezina N. A., Nazarova L. E., Poliakova T. N., Chekryzheva T. A. 2014. Links between biota and climate-related variables in the Baltic Region using Lake Onega as an example. *Oceanologia.* 56(2): 291–306. https://doi.org/10.5697/oc.56-2.291.

Shilova O. S., Romanenko F. A., Kolka V. V., Denisov D. B. 2019. Changes in the natural environment in the Northern Khibiny (Kola Peninsula) in the Holocene according to the diatom analysis of lacustrine sediments. *Geomorphology,* (3): 91–101. https://doi.org/10.31857/S0435-42812019391-101

Skulberg O. M., Underdal B., Utkilen H. 1994. Toxic waterblooms with cyanophytes in Norway — current knowledge. *Algological Studies,* 75. Stuttgart: 279–289.

Skulberg O. M. 1996. Terrestrial and limnic algae and cyanobacteria Part 9 /A. Elvebakk and P.Prestrud (eds.). *A catalogue of Svalbard plants, fungi, algae and Cyanobacteria*. Skrifter, 198: 383–395.

Thomasson K. 1958. Zur planktonkunde Spitzbergens [To plankton science in Svalbard], 1. *Hydrobiologia,* 12(2-3): 226–236.

Thomasson K. 1961. Zur planktonkunde Spitzbergens [To plankton science in Svalbard], 2. *Hydrobiologia,* 18(3): 192–198.

Trifonova I. S. 1990. *Ecology and succession of lacustrine phytoplankton*. Leningrad.

Voronikhin N. N. 1935. *Algae and their groups in lakes Imandra and Notozero* (Kola Peninsula). Tr. Bot. inst. Acad. Sciences of the USSR, Spore plants, ser. II, no. 2.M.-L.: 107–150.

Wahlenberg C. 1812. *Flora Lapponica*. Berlin.

Willen T. 1980. Phytoplankton from Lakes and Ponds on Vestspitsbergen//*Acta phytogeographica suecica,* 68: 173–188.

Willen E. 1987. Phytoplankton and reversed eutrophication in Lake Malaren, Central Sweden, 1965-1983. *Br. Phycol. J.,* 22: 193–208.

Yakovlev V. A. 1988. *Assessment of the quality of surface waters in the Kola North by hydrobiological indicators and biotesting data* (*practical recommendations*). - Apatity.

Yakovlev V. A. 1991. *Hydrobiological studies of inland waters of the Kola North*. Apatity: Kolsk Publishing House. AN SSSR.

Yakovlev V. A., Kashulin N. A. 2012. On the history of the initial stages of limnological research in the Kola North. *Bulletin of the Kola Scientific Center of the Russian Academy of Sciences*, 4: 116–138.

Zolotarev V. A. 2013. Ecological functions of collar flagellates in connection with the hydrochemical parameters of reservoirs. *Water: chemistry and ecology,* 3: 40–44.

In: Lake Water
Editors: Oleg S. Pokrovsky, Yulia Bespalaya et al.
ISBN: 978-1-53619-275-9

Chapter 8

BIOLOGY AND GEOGRAPHICAL DISTRIBUTION OF REPRESENTATIVES OF THE FAMILIES POLYPHEMIDAE AND CERCOPAGIDAE (ONYCHOPODA, CLADOCERA) IN THE NORTHERN AND ARCTIC REGIONS OF THE EUROPEAN PART OF RUSSIA

Larisa Litvinchuk*
Zoological Institute of the Russian Academy of Sciences,
Saint Petersburg, Russian Federation

ABSTRACT

This chapter describes geographical distribution and biology of representatives of the families Polyphemidae and Cercopagidae (Onychopoda, Cladocera) in lakes of the European North and the Arctic. The Karelo-Kola and Baltic limnobiological regions are the most studied areas. For all species of the genus *Bythotrephes*, significant geographic variability in body length and relative length of the caudal appendage was noted. *B. crassicaudus* has the greatest body length, number and diameter of resting eggs. The development time of the more northern population of *Polyphemus pediculus* from the White Sea basin is longer, the density is higher, and, at the same time, the body length and fecundity of parthenogenetic and gamogenetic females are lower than in the population from the eastern part of the Gulf of Finland.

Keywords: Cladocera, Onychopoda, geographical distribution, population biology and reproduction

* Corresponding Author's Email: larisalitvinchuk@yandex.ru.

Introduction

The northern and arctic regions of the European part of Russia are characterized by a huge number of fresh water bodies of various sizes and origins (Vincent, 2008; Wrona et al., 2005). S. V. Gerd in 1956 proposed a scheme for biolimnological zoning of the North of Russia. In the European part, the Baltic, Karelo-Kola, Upper-Volga, North-Dvina and Pechora regions were distinguished.

At present, the study of the northern water bodies is mainly represented by data on the distribution of aquatic animal faunas. Species biology data are fragmented. Planktonic crustaceans, in particular the predatory Cladocera (Onychopoda), are one of the important groups of organisms in water bodies of various types, constituting a significant part of zooplankton. Understanding the biological characteristics of populations of species of this group is important for expanding information on the nature of distribution of species and the history of their dispersal and origin.

Representatives of the genera *Bythotrephes* Leydig, 1860 and *Polyphemus* (O. F. Müller, 1785) are included of the Families Cercopagidae Mordukhai-Boltovskoi, 1968, and Polyphemidae Baird, 1854 respectively, Order Onychopoda Sars, 1865 (= Superfamily Polyphemoidea Brooks, 1959).

Species of the genus *Bythotrephes* (*B. brevimanus* Lilljeborg, 1901 [= *B. balticus*], *B. cederstroemii* Schoedler, 1863, *B. crassicaudus* Lilljeborg, 1890 [= *B. arcticus*] and *B. longimanus* Leydig, 1860), which were previously considered as separate species, were later assigned to one polymorphic species with a wide range (northern part of the Palaearctic) - *Bythotrephes longimanus* (Mordukhai-Boltovskoy, Rivier, 1987; Rivier, 1998). However, this point of view was discussed (Mäemets, 1966; Pejler, 1975; Nilsson, 1979). It should be noted that no study has studied material from the entire area inhabited by representatives of this genus.

During the analysis of allozyme variability, it was found that the genus *Bythotrephes* consists of five species (Litvinchuk, Litvinchuk, 2016). In the Northern and Arctic regions of the European part of Russia, this genus is represented by three species, *B. brevimanus*, *B. cederstroemii* and *B. crassicaudus*.

B. cederstroemii (Leksozero) and *B. crassicaudus* (Dalnie Zelentsy) form one cluster, and *B. brevimanus* (Vyshtinets) forms another one. These data are in very good agreement with clustering by morphological characters, where the same type of branching is preserved (Litvinchuk, Litvinchuk, 2016).

The rarest and most northern species of the genus *Bythotrephes* is *B. crassicaudus*. It is distinguished by its larger body size and inhabits small tundra water bodies. It can be assumed that the formation of the *B. crassicaudus* species is associated with the emergence of thermokarst (fishless) shallow water bodies in the Arctic region, where the ancestors of *B. crassicaudus* were able to evolve towards a sharp increase in their size.

Two other species, *B. brevimanus* and *B. cederstroemii*, are widespread in the North of the European part of Russia (Figure 1). Here, in a number of water bodies, they contact each other and form hybrid populations (Litvinchuk, Litvinchuk, 2016).

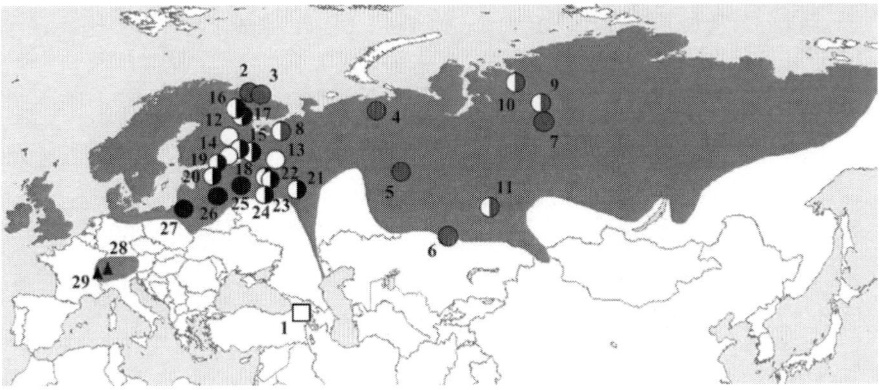

Figure 1. The range of *Bythotrephes* (gray area) with geographic location of studied lakes. *B. transcaucasicus* (white square) – locality 1; *B. crassicaudus* (grey circles) – localities 2-7; samples from the contact zone of *B. crassicaudus* and *B. cederstroemii* (semi-grey circles) – 8-11; *B. cederstroemii* (white circles) – 12-14; samples from the contact zone of *B. cederstroemii* and *B. brevimanus* (semi-black circles) – 15-24; *B. brevimanus* (black circles) – 25-27; and *B. longimanus* (black triangles) – 28-29 (modified from Litvinchuk and Litvinchuk, 2016).

Polyphemus pediculus is a Holarctic species, and recent work by molecular researchers has shown the complex genetic structure of this species (Xu et al., 2009). As a rule, *P. pediculus* inhabits the littoral zone of water bodies of various types. The biological features of the littoral populations of *P. pediculus* from the Rybinsk Reservoir (Upper Volga) have been well studied by L. G. Butorina (1971). At the same time, pelagic populations of *P. pediculus* were recorded in some water bodies - in Lake Sandal (Olonetskaya ..., 1923), in oligotrophic and dystrophic water bodies of the Solovetsky Islands (Linko, 1900; Zakhvatkin, 1927), in Lake Michigan, North America (Butorina, 1978). According to our data, *P. pediculus* also inhabits the pelagic part of the Neva Bay and the eastern part of the Gulf of Finland (Baltic Sea), and Lake Krivoe, located on the coast of the White Sea. There are no literature data on the biological characteristics of the pelagic populations of *P. pediculus* from the eastern part of the Gulf of Finland and Lake Krivoe.

It is important to verify the features of the geographical distribution of species of genus *Bythotrephes* and *P. pediculus* in the Northern and Arctic regions of the European part of Russia, and to study the features of their biology, such as density, population structure, fecundity, and others.

MATERIAL AND METHODS

Zooplankton samples were collected in May-November 1998 to 2020 on the water bodies of Baltic, Karelo-Kola, Upper-Volga regions (Figure 1, except №№ 1, 4-11, 28 and 29):

- *Murmansk region:* reservoirs in the vicinity of Murmansk, Kildinskoe Lake, Pityevoe Lake (Dalnie Zelentsy);
- *Republic of Karelia:* Cherlivoe Bolshoe Lake, Krivoe Lake, Krugloe Lake, Ladoga Lake (Valaam Island), Leksozero, Onega Lake (Petrozavodsk), Starushechie Verkhnee Lake, Starushechie Lake, Suojarvi Lake;

- *Leningrad region:* Filkino Lake, the Gruzinka River flood, the eastern part of the Gulf of Finland, Khimozi Lake, the Neva Bay, Sestroretskiy Razliv Lake, Vuoksa River (Priozersk);
- *Kaliningrad region:* Vyshtinetskoe Lake;
- *Pskov region:* Chudskoye Lake;
- *Novgorod region:* Valdai Lake, Edrovo Lake, Piros Lake;
- *Tver region:* Bologoye Lake;
- *Vologda region:* Khotavets Lake (Darwin Reserve), Lake Vozhe, Svyatoe Lake (village Tavenga Vozhegsky district).

The material for research was also collected during two scientific expeditions (August and October 2000) along the system of the Upper Volga reservoirs (Ivankovskoe, Uglichskoe, Rybinskoe and Gorkovskoe).

Table 1. Abiotic factors changes in the eastern part of the Gulf of Finland (Primorsk) in the beginning of May – mid October 2009-2019 years. T – water temperature, S – water salinity, Tr – transparency of water

	Abiotic factors in sampling dates											
	05.5	15.5	31.5	15.6	30.6	15.7	31.7	15.8	31.8	15.9	30.9	15.10
T, °C	14.50 ± 3.54	11.86 ± 3.24	14.15 ± 2.58	15.77 ± 1.75	17.22 ± 2.06	19.01 ± 2.88	20.38 ± 3.61	19.67 ± 1.85	17.58 ± 1.76	15.40 ± 2.57	12.25 ± 1.77	10.93 ± 0.81
S, ‰	2.20	0.74	0.66	1.80 ± 1.64	2.66	2.74	2.82 ± 0.57	2.27 ± 0.66	2.65 ± 0.05	2.05 ± 0.13	3.01	1.27 ± 1.46
Tr, м	1.50	1.55 ± 0.33	1.36 ± 0.21	1.13 ± 0.13	1.23 ± 0.36	1.46 ± 0.42	1.34 ± 0.24	1.09 ± 0.18	1.10 ± 0.28	1.26 ± 0.51	2.70	1.97 ± 0.23

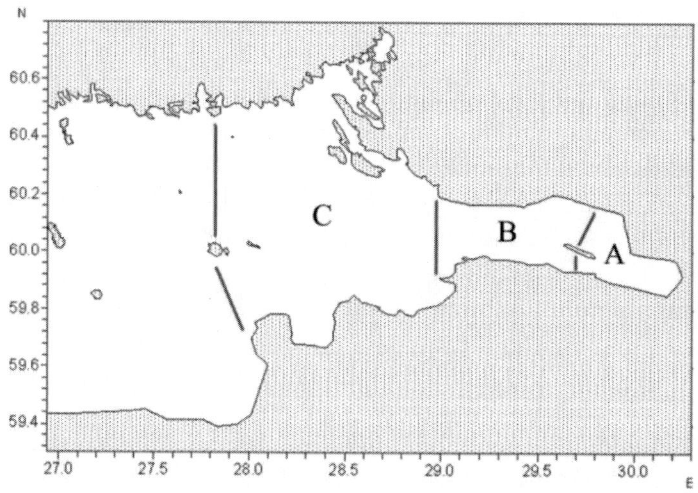

Figure 2. Map of zones of the eastern part of the Gulf of Finland (the Baltic Sea): A – Neva Bay, B – the Inner Estuary, C – the Outer Estuary.

Some materials were provided by my colleagues: Anninskoe Lake (L. A. Stepanova), Seliger Lake (E. S. Makartseva), Severnaya Dvina (I. K. Rivier).

Zooplankton samples were taken by vertical hauls with the Juday quantitative plankton net (diameter 0.2 m, mesh size 125 μm) from bottom to surface. The samples were collected during the daytime, preserved in 4% formaldehyde solution and examined in the laboratory with a light microscope.

Table 2. Water temperature in horizon 0-5 m (total station depth is 30 m) in Krivoe Lake (the White Lake basin) in mid-May – the end of October of 2002-2019 years

Year/Days and	Water temperature in sampling dates (°C)											
months	15.05	31.05	15.06	30.06	15.07	31.07	15.08	31.08	15.09	30.09	15.10	31.10
2002	1.40	7.00	13.94	14.00	14.70	15.30	14.67	15.38	8.80	6.90	4.00	2.50
2003	4.00	7.50	9.93	16.32	17.55	16.80	15.48	14.53	12.92	5.30	4.50	3.70
2004	5.10	5.40	8.33	12.60	15.90	19.00	15.00	13.10	10.40	8.00	6.00	3.50
2005	3.80	5.70	11.00	12.50	18.30	18.00	15.80	13.10	11.50	6.30	4.80	3.50
2006	5.10	7.55	12.17	13.30	12.20	14.22	15.55	12.40	11.20	9.10	6.80	4.10
2007	3.50	9.00	12.67	15.90	13.30	17.03	18.00	13.68	7.80	9.80	6.50	5.50
2008	2.00	7.00	14.90	13.23	14.40	14.40	9.50	9.00	8.00	5.60	5.70	3.40
2009	8.00	8.60	8.00	12.67	13.00	14.12	15.50	13.55	13.65	5.61	3.00	4.70
2010	7.95	9.58	9.50	11.19	14.50	16.58	14.10	11.50	12.00	8.60	5.60	4.50
2011	4.50	9.80	13.50	14.80	16.60	19.50	14.60	15.17	13.10	10.80	7.80	5.60
2012	4.40	7.90	14.00	11.60	14.50	18.20	–	13.70	11.60	10.10	7.45	4.75
2013	6.40	9.50	13.00	–	–	–	–	15.50	–	–	–	3.63
2014	–	6.50	–	–	–	18.08	–	14.30	12.65	–	–	2.90
2015		9.70	–	–	–	13.80	–	13.75	13.10	–	–	5.70
2016	9.20	12.20	–	16.87	–	16.10	–	12.35	10.70	–	–	5.50
2017	–	3.80	8.00	10.07	15.95	12.00	–	13.00	9.60	–	–	4.15
2018	–	10.00	13.07	–	20.63	24.70	–	16.00	14.60	–	–	5.00
2019	–	7.40	10.20	–	15.09	14.42	–	14.97		–	–	3.49
Mean water temperature for all seasons, °C (X ± SD)	5.03 ± 2.32	8.01 ± 2.01	11.48 ± 2.35	13.47 ± 2.05	15.47 ± 2.25	16.60 ± 2.95	14.82 ± 2.15	13.61 ± 1.68	11.35 ± 2.02	7.83 ± 1.99	5.65 ± 1.48	4.23 ± 0.97

Dash – measuring of temperature not carried out.

In the Baltic Sea Onychopoda were collected on the Neva Bay (freshwater, 0.06‰), eastern part of the Gulf of Finland (the inner estuary and outer estuary; slightly brackish-water, from 0.1 to 2.00‰) of the Baltic Sea (Figure 2; abiotic data are in Table 1). Long-term and seasonal observations were carried out on the Katerlahti bay on the shore of which the city Primorsk is located (eastern part of the outer estuary, 60° 36 ′ N, 28° 61 ′ E, Vyborgsky district).

Research Collections of the Zoological Institute of Russian Academy of Sciences, St. Petersburg, Russia (ZISP is the Zoological Institute RAS) were studied.

In the Karelian region long-term and seasonal observations were carried out on the Krivoe Lake (66° 21´ N 33° 38´ E, water temperature data are in Table 2).

RESULTS AND DISCUSSION

Geographical Distribution and Biology of Species of Genus *Bythotrephes* in the Northern and Arctic Regions of the European Part of Russia

The study of species of the genus *Bythotrephes* has been going on for more than a hundred years. *Bythotrephes crassicaudus* was first found and described from the territory of Russia (the Kola coast of the Barents Sea, Sars, 1890; Lilljeborg, 1901). Therefore, it is not surprising that most of the sites of this species (five out of seven) in northwest Russia are located on the Arctic coast of the Murmansk region (Figure 3). However, almost all of these finds (Korabelny settlement, Murmansk, Heretik Island, and Kildin Island) date from the late 19th and early 20th centuries and were not subsequently confirmed. Moreover, our attempts to find this species in tundra reservoirs on the coast of the Kola Bay (near the city of Murmansk) did not give a positive result. Nevertheless, in the area of the Murmansk Biological Station (Dalnie Zelentsy), this species is common and has been constantly recorded since the 40s.

In addition to the tundra coast of the Murmansk region, *B. crassicaudus* was also recorded in the taiga zone of Karelia. This find was made in the summer of 1894 by the Finnish explorer K. E. Stenroos (1897). It could have been considered dubious if the author had not indicated the body size of the individuals found here (4-6 mm without the length of the cauda needle). Only one species, *B. crassicaudus*, has such dimensions. This circumstance allowed V. M. Rylov (1916) referred this find to "*Bythotrephes longimanus* var. *arctica*." Lake Urgani, or Unusozero, where this find was made, is located in the Rugozero region northwest of Ondozero. Its area is 7.6 km^2, the maximum length is 6.5 km, and the width is 1.7 km (Grigoriev & Gritsevskaya, 1959). It has a shallow depth (up to 3 m), and its entire surface is covered with aquatic vegetation (Stenroos, 1897). The bottom is heavily silted, the shores are waterlogged. It contains fish (pike, roach, crucian carp and others). There is no doubt that this find point requires additional study.

The only find of *Bythotrephes* in the west of the Arkhangelsk region was made in July-August 1998 in the near-delta water bodies of the Northern Dvina River (IK Riviere, pers. communication). Numerous individuals of *B. crassicaudus* were captured here. However, among them there was one female, which outwardly practically did not differ from the rest of the individuals, but had a weak bend on the caudal appendage, covered with teeth. Similar individuals ("*B. cederströmii* var. *robustus*") were previously recorded in northern Sweden and Norway (Lilljeborg, 1901), but have never been described from the territory of Russia. Undoubtedly, this population belongs to *B. crassicaudus*, and the appearance in it of individuals with a bend on the tail is associated with the hybridization of this species with *B. cederstroemii*, a species apparently carried out by the current of the Northern Dvina from higher taiga water bodies, as is observed, for example, in the Pechora river basin (Vekhov, 1987).

In the course of analyzing the map and the list of points of *B. cederstroemii* finds in the Murmansk region (Figure 3), we found that this species is common here and occurs throughout its territory, with the exception of small tundra water bodies of the Arctic coast of the Barents Sea inhabited by *B. crassicaudus*. However, despite the wide distribution of this species, not all suitable water bodies are inhabited by it. In the summer of 1998, we investigated a rather large lake Kildinskoye, located somewhat south of the city of Murmansk, but *Bythotrephes* was not found here.

B. cederstroemii was first discovered in the Murmansk region by V. M. Rylov (1916). In a sample from Murdozero, one severely damaged specimen with a pronounced bend in the caudal appendage was found. Subsequently, this species was found in the Murmansk region by other authors (Petrovskaya, 1966; Makartseva, 1974; Volkova et al., 1975; Drabkova et al., 1975; Lomaeva & Svirskaya, 1977; Reshetnikov, 1980).

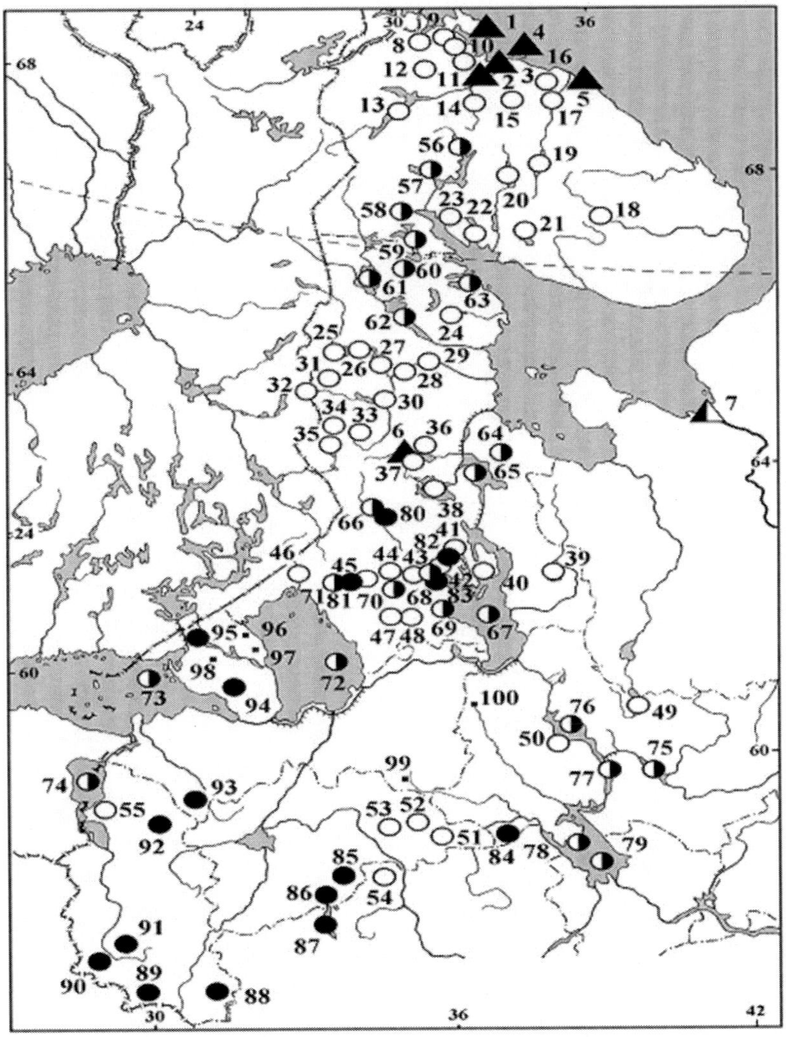

Figure 3. Geographical distribution of species of genus *Bythotrephes* in the Northwest of Russia. *B. crassicaudus* (black triangles) – localities 1-6; samples from the contact zone of *B. crassicaudus* and *B. cederstroemii* (semi- black triangles) – 7; *B. cederstroemii* (white circles) – 8-56; samples from the contact zone of *B. cederstroemii* and *B. brevimanus* (semi-black circles) – 57-79; *B. brevimanus* (black circles) – 80-98.

B. cederstroemii is apparently very rare in the Leningrad and Pskov regions. Together with *B. brevimanus* and hybrids, it is found only in large bodies of water, such as the Gulf of Finland, Pskov-Chudskoe, Onega and Ladoga lakes, as well as in the upper reaches of the Neva River.

In the west of the Vologda Oblast, *B. cederstroemii* (together with B. brevimanus and hybrids) was found in water bodies adjacent to the Sheksna River and the North Dvinsky Canal. In addition, we met this species in Lake Vozhe (basin of the Onega River).

Table 3. Geographical distribution of species of genus *Bythotrephes* and *Polyphemus pediculus* in the Northern and Arctic regions of the European part of Russia

Taxon	Water body	Literature source
Order Onychopoda Sars, 1865 (= Superfamily Polyphemoidea Brooks, 1959)		
Family Polyphemidae Baird, 1854		
Genus *Polyphemus* (O. F. Müller, 1785)		
P. pediculus (Linnaeus, 1761)	The White Sea basin	
	Lakes	
	Aittajarvi	Kulikova, 2010
	Alanjarvi	Kulikova, 2010
	Alojarvi	Kulikova, 2010
	Bolshoe	Zakhvatkin, 1927
	Byalozero	Kulikova, 2010
	Bygozero	Kulikova, 2010
	Cherlivoe Bolshoe	Our data
	Chernivozero	Kulikova, 2010
	Devichya lamba	Kulikova, 2010
	Distrophnoe 1	Zakhvatkin, 1927
	Distrophnoe 3	Zakhvatkin, 1927
	Dolgonkoe	Zakhvatkin, 1927
	Elmozero	Kulikova, 2010
	Engozero	Kulikova, 2010
	Evzhozero-2	Kulikova, 2010
	Gormozero	Kulikova, 2010
	Idel	Kulikova, 2010
	Ikshozero Bolshoe	Kulikova, 2010
	Indel	Kulikova, 2010
	Kalozero	Kulikova, 2010
	Kamennoe	Kulikova, 2010
	Kamennoe	Zakhvatkin, 1927
	Kanozero	Kulikova, 2010
	Kento	Kulikova, 2010
	Keret	Kulikova, 2010
	Khapijarvi	Kulikova, 2010
	Khapkhijarvi	Kulikova, 2010
	Kimasozero	Kulikova, 2010
	Klyukvennoe	Zakhvatkin, 1927
	Kojvas	Kulikova, 2010
	Kontokki	Kulikova, 2010
	Korpangi	Kulikova, 2010
	Korpijarvi	Kulikova, 2010
	Kovdozero	Kulikova, 2010
	Kostomukshskoe	Kulikova, 2010
	Krivoe	Our data; Ivanova, 1975; Kulikova, 2010
	Krugloe	Our data; Ivanova, 1975; Kulikova, 2010
	Kuito Nizhnee	Kulikova, 2010
	Kuito Srednee	Kulikova, 2010
	Kuito Verkhnee	Kulikova, 2010
	Kurijarvi (Kem River basin)	Kulikova, 2010
	Kurojarvi (Kenti River basin)	Kulikova, 2010
	Kyargozero	Kulikova, 2010
	Ladvo Srednee	Kulikova, 2010
	Ladvo Verkhnee	Kulikova, 2010

Taxon	Water body	Literature source
	Lapushistoe	Zakhvatkin, 1927
	Lazarevskoe	Kulikova, 2010
	Lomcheozero	Kulikova, 2010
	Lomozero	Kulikova, 2010
	Luvozero	Kulikova, 2010
	Malata Lominoga	Zakhvatkin, 1927
	Maloye	Zakhvatkin, 1927
	Marya-Shereka	Kulikova, 2010
	Maslozero	Kulikova, 2010
	Matkozero	Kulikova, 2010
	Mergubskoe	Kulikova, 2010
	Mezh-Liminoga	Zakhvatkin, 1927
	Mochazhina	Zakhvatkin, 1927
	Neris Nizhnee	Kulikova, 2010
	Neris Verkhnee	Kulikova, 2010
	Nyuk	Kulikova, 2010
	Okunevoe	Kulikova, 2010
	Ondozero	Kulikova, 2010
	Ondskoe Resevoire	Kulikova, 2010
	Paanajarvi	Kulikova, 2010
	Panojarvi	Kulikova, 2010
	Pasmozero	Kulikova, 2010
	Peralampi	Kulikova, 2010
	Pinezhsky region swamps	Zaytseva, Galanina, Philippov, 2017
	Ploskoe	Zakhvatkin, 1927
	Poppalijarvi	Kulikova, 2010
	Prigubnoe	Zakhvatkin, 1927
	Pulozero	Kulikova, 2010
	Pushtosjarvi	Kulikova, 2010
	Pyaozero	Kulikova, 2010
	Rokzhozero	Kulikova, 2010
	Roppomo	Kulikova, 2010
	Rugozero	Kulikova, 2010
	Segozero	Kulikova, 2010
	Semchozero	Kulikova, 2010
	Seniozero	Kulikova, 2010
	Seniozero Verkhnee	Kulikova, 2010
	Shavan	Kulikova, 2010
	Shuchya lamba	Kulikova, 2010
	Shuezero	Kulikova, 2010
	Sobachie	Zakhvatkin, 1927
	Sonozero	Kulikova, 2010
	Stanovoe	Zakhvatkin, 1927
	Starushechie Verkhnee	Our data
	Starushechie	Our data
	Sudno	Kulikova, 2010
	Sukozero	Kulikova, 2010
	Sumozero	Kulikova, 2010
	Suojarvi	Our data
	Syargozero (Segezha River basin)	Kulikova, 2010
	Syarki-Yarvi	Kulikova, 2010
	Tiksheozero	Kulikova, 2010
	Topozero	Kulikova, 2010
	Torphyanoe	Zakhvatkin, 1927
	Tukhkozero	Kulikova, 2010

Table 3. (Continued)

Taxon	Water body	Literature source
	Umbozero	Kulikova, 2010
	Vanchozero	Kulikova, 2010
	Velikoe (Varzuga River basin)	Kulikova, 2010
	Voitskoe	Kulikova, 2010
	Vyagozero	Kulikova, 2010
	Yagodnoe	Zakhvatkin, 1927
	Yangozero	Kulikova, 2010
	Yulijarvi (Kenti river basin)	Kulikova, 2010
	Yulijarvi (Kem River basin)	Kulikova, 2010
	Yurikkojarvi	Kulikova, 2010
	Yushkojarvi	Kulikova, 2010
	Zelenoe	Zakhvatkin, 1927
	Zelenoe 1	Zakhvatkin, 1927
	Zelenoe 2	Zakhvatkin, 1927
	Rivers	
	Byg Nizhniy	Kulikova, 2010
	Chirko-Kem	Kulikova, 2010
	Elma	Kulikova, 2010
	Kem	Kulikova, 2010
	Kenti	Kulikova, 2010
	Keret	Kulikova, 2010
	Kolvitsa	Kulikova, 2010
	Kontokki	Kulikova, 2010
	Kuzema	Kulikova, 2010
	Kyamenka	Kulikova, 2010
	Kyatka	Kulikova, 2010
	Lakhna	Kulikova, 2010
	Luva (Vonga)	Kulikova, 2010
	Luvenga	Kulikova, 2010
	Nogeuksa	Kulikova, 2010
	Nizhema	Kulikova, 2010
	Olanga	Kulikova, 2010
	Onega	Zmetnaya, Plakueva, 2018
	Orchezhoya	Kulikova, 2010
	Pila	Kulikova, 2010
	Pista	Kulikova, 2010
	Polga	Kulikova, 2010
	Pulonga (Karelia)	Kulikova, 2010
	Segezha	Kulikova, 2010
	Severnaya Dvina	Zmetnaya, Plakueva, 2018
	Shigerendga	Kulikova, 2010
	Shuya Belomorskaya	Kulikova, 2010
	Sopa	Kulikova, 2010
	Suma	Kulikova, 2010
	Tyanuksa	Kulikova, 2010
	Umba	Kulikova, 2010
	Unezhma	Kulikova, 2010
	Vyala	Kulikova, 2010
	Vyanya	Kulikova, 2010
	Vyg Verkhniy	Kulikova, 2010
	Yuriy Stream	Kulikova, 2010
	Ladoga Lake basin	
	Lakes	
	Akhvenjarvi	Kulikova, 2012

Taxon	Water body	Literature source
	Ala-Tolvajarvi	Kulikova, 2012
	Antonievskoe	Kulikova, 2012
	Chernoye	Kulikova, 2012
	Germanovskoe	Kulikova, 2012
	Hjalmozero	Kulikova, 2012
	Iso-Ijjarvi	Kulikova, 2012; Ryabinkina, Kulikova, Ryzhkov, 2012
	Jccbtdj	Kulikova, 2012
	Kargiozero	Kulikova, 2012
	Kaskozero	Kulikova, 2012
	Khelmijarvi	Kulikova, 2012
	Khirvasjarvi	Kulikova, 2012
	Kjuljajarvi	Kulikova, 2012
	Kortelanjarvi	Kulikova, 2012
	Kotajarvi	Ryabinkina, Kulikova, Ryzhkov, 2012
	Kuikkajarvi	Ryabinkina, Kulikova, Ryzhkov, 2012
	Ladoga	Our data
	Latvajarvi	Kulikova, 2012
	Leksozero	Kulikova, 2012
	Muonionjarvi	Kulikova, 2012
	Paikjarvi	Ryabinkina, Kulikova, Ryzhkov, 2012
	Pairijarvi	Kulikova, 2012
	Parikanjarvi	Kulikova, 2012; Ryabinkina, Kulikova, Ryzhkov, 2012
	Ruokojarvi (Tokhmajoki River basin)	Kulikova, 2012; Ryabinkina, Kulikova, Ryzhkov, 2012
	Ruokolampi	Kulikova, 2012; Ryabinkina, Kulikova, Ryzhkov, 2012
	Saiunejarvi	Kulikova, 2012
	Sarijarvi	Kulikova, 2012
	Sarijarvi (Janisjoki River basin)	Kulikova, 2012
	Sarijarvi (Tolvajoki River basin)	Kulikova, 2012
	Sasajarvi	Kulikova, 2012
	Simniakovskoe	Kulikova, 2012
	Sirkoinjarvi	Kulikova, 2012; Ryabinkina, Kulikova, Ryzhkov, 2012
	Sisyajarvi	Kulikova, 2012
	Suistamonjarvi	Kulikova, 2012; Ryabinkina, Kulikova, Ryzhkov, 2012
	Suuri-Kuokhajarvi	Kulikova, 2012
	Taivajarvi	Kulikova, 2012
	Tolvajarvi	Kulikova, 2012
	Tulos	Kulikova, 2012
	Turianjarvi	Kulikova, 2012; Ryabinkina, Kulikova, Ryzhkov, 2012
	Uksunjarvi (Uksujarvi)	Kulikova, 2012; Ryabinkina, Kulikova, Ryzhkov, 2012
	Valkijarvi (Uksujoki River basin)	Kulikova, 2012; Ryabinkina, Kulikova, Ryzhkov, 2012
	Varpajarvi	Kulikova, 2012; Ryabinkina, Kulikova, Ryzhkov, 2012
	Vedlozero	Kulikova, 2012
	Vozhjarvi	Kulikova, 2012
	Yanisjarvi	Kulikova, 2012; Ryabinkina, Kulikova, Ryzhkov, 2012

Table 3. (Continued)

Taxon	Water body	Literature source
	Yimiajarvi	Kulikova, 2012 Ryabinkina, Kulikova, Ryzhkov, 2012
	Yurikajarvi	Kulikova, 2012
	Rivers	
	Mijnalanjoki	Kulikova, 2012; Ryabinkina, Kulikova, Ryzhkov, 2012
	Nyalma	Kulikova, 2012
	Pensanjoki	Kulikova, 2012; Ryabinkina, Kulikova, Ryzhkov, 2012
	Sarkajoki	Kulikova, 2012; Ryabinkina, Kulikova, Ryzhkov, 2012
	Svir (the source of the river)	Kulikova, 2012
	Soanjoki	Kulikova, 2012; Ryabinkina, Kulikova, Ryzhkov, 2012
	Sulla	Kulikova, 2012
	Tokhmajoki	Kulikova, 2012; Ryabinkina, Kulikova, Ryzhkov, 2012
	Uksunjoki	Kulikova, 2012; Ryabinkina, Kulikova, Ryzhkov, 2012
	Ulmosenjoki	Kulikova, 2012
	Ulismaysenijoki	Ryabinkina, Kulikova, Ryzhkov, 2012
	Veljakanjoki	Kulikova, 2012; Ryabinkina, Kulikova, Ryzhkov, 2012
	Villitsa	Kulikova, 2012
	Vuoksa	Kulikova, 2012
	Yanisjoki	Kulikova, 2012; Ryabinkina, Kulikova, Ryzhkov, 2012
	Onega Lake basin	
	Lakes	
	Aglimozero	Kulikova, 2007
	Bachanskoe	Kulikova, 2007
	Belozero	Kulikova, 2007
	Chelmozero	Kulikova, 2007
	Chuzhmozero	Kulikova, 2007
	Chukozero	Kulikova, 2007
	Chumbaozero (Chumbas)	Kulikova, 2007
	Gakhkozero	Kulikova, 2007
	Ivinskiy Razliv	Kulikova, 2007
	Kedrozero	Kulikova, 2007
	Keldolampi	Kulikova, 2007
	Khizhozero	Kulikova, 2007
	Kodozero	Kulikova, 2007
	Kolodozero	Kulikova, 2007
	Koppolozero	Kulikova, 2007
	Kosmozero	Kulikova, 2007
	Krestovaya lamba	Kulikova, 2007
	Kumchozero Nizhnee (Kumchozero)	Kulikova, 2007
	Ladmozero	Kulikova, 2007
	Livozero	Kulikova, 2007
	Lizhmozero	Kulikova, 2007
	Lososinskoye (Lososinnoye)	Kulikova, 2007
	Matchozero (Buche-landa)	Kulikova, 2007
	Melnichnoe 2	Kulikova, 2007

Taxon	Water body	Literature source
	Monastyrskoe	Kulikova, 2007
	Myagrozero	Kulikova, 2007
	Nemozero Maloye	Kulikova, 2007
	Nigozero	Kulikova, 2007
	Onega	Kulikova, 2007
	Ostrechie	Kulikova, 2007
	Padmozero	Kulikova, 2007
	Pichozero	Kulikova, 2007
	Pigmozero Nizhnee	Kulikova, 2007
	Pigmozero Verkhnee	Kulikova, 2007
	Punozero	Kulikova, 2007
	Putkozero	Kulikova, 2007
	Ravdozero	Kulikova, 2007
	Rapsudozero	Kulikova, 2007
	Rindozero	Kulikova, 2007
	Ryndozero	Kulikova, 2007
	Sandal	Kulikova, 2007
	Sarozero Srednee	Kulikova, 2007
	Sosnovoe	Kulikova, 2007
	Stolpozero	Kulikova, 2007
	Syuzikozero	Kulikova, 2007
	Tarismozero (Tarasmozero)	Kulikova, 2007
	Tungozero	Kulikova, 2007
	Tyagozero	Kulikova, 2007
	Uros	Kulikova, 2007
	Valgomozero	Kulikova, 2007
	Vangozero	Kulikova, 2007
	Vashozero	Kulikova, 2007
	Vatchelskoe	Kulikova, 2007
	Vekkhozero	Kulikova, 2007
	Velikozero	Kulikova, 2007
	Vendurskoe	Kulikova, 2007
	Volozero Maloye	Kulikova, 2007
	Volozero Nizhnee (Volozero)	Kulikova, 2007
	Vuontelenjarvi	Kulikova, 2007
	Yandomozero	Kulikova, 2007
	Rivers	Kulikova, 2007
	Andoma	Kulikova, 2007
	Ileksa	Kulikova, 2007
	Isselga	Kulikova, 2007
	Izhmuksa Yuzhnaya	Kulikova, 2007
	Kodacha	Kulikova, 2007
	Kumsa	Kulikova, 2007
	Lizhma	Kulikova, 2007
	Lososinka	Kulikova, 2007
	Megra	Kulikova, 2007
	Neglinka	Kulikova, 2007
	Neleksa	Kulikova, 2007
	Nemina	Kulikova, 2007
	Philippa	Kulikova, 2007
	Povenchanka	Kulikova, 2007
	Putkozerka	Kulikova, 2007
	Pyalma	Kulikova, 2007
	Sapenitsa	Kulikova, 2007
	Shuya	Kulikova, 2007
	Svir (the source of the river)	Kulikova, 2007

Table 3. (Continued)

Taxon	Water body	Literature source
	Sunna	Kulikova, 2007
	Tambitsa	Kulikova, 2007
	Tuba	Kulikova, 2007
	Unitsa	Kulikova, 2007
	Vilga	Kulikova, 2007
	Vodla	Kulikova, 2007
	The Baltic Sea basin	
	Neva Bay	Our data
	Eastern part of the Guf of Finland	Our data
	Filkino Lake	Our data
	Khimozi Lake	Our data
	Lake Sestroretskiy Razliv	Our data
	The Barents Sea basin	
	Zelenetskoe	Korobtsova, 1975
	Akulkino	Korobtsova, 1975
	Bobrovoe 1	Korobtsova, 1975
	Bobrovoe 2	Korobtsova, 1975
	Bobrovoe 3	Korobtsova, 1975
	Bobrovoe 4	Korobtsova, 1975
	Yarnyshnoe 1	Korobtsova, 1975
	Yarnyshnoe 3	Korobtsova, 1975
	Yarnyshnoe 4	Korobtsova, 1975
	Besstochnoe	Korobtsova, 1975
	Lake with Island	Korobtsova, 1975
Family Cercopagidae Mordukhai-Boltovskoi, 1968		
Genus *Bythotrephes* Leydig, 1860		
B. brevimanus Lilljeborg, 1901	Ladoga Lake basin	
	Lakes	
	Alalampi	Kulikova, 2012; Ryabinkina, Kulikova, Ryzhkov, 2012
	Khiyajarvi	Kulikova T.P., 2012; Ryabinkina, Kulikova, Ryzhkov, 2012
	Kotajarvi	Kulikova, 2012; Ryabinkina, Kulikova, Ryzhkov, 2012
	Leschevoe	Kulikova, 2012
	Loimolanjarvi	Kulikova, 2012
	Matrijarvi	Kulikova, 2012; Ryabinkina, Kulikova, Ryzhkov, 2012
	Nialmozero	Kulikova, 2012
	Piukhiajarvi (Vuoksa River basin)	Kulikova, 2012
	Salmenjarvi	Kulikova, 2012
	Sisyajarvi	Kulikova, 2012; Ryabinkina, Kulikova, Ryzhkov, 2012
	Vakhvajarvi	Kulikova, 2012; Ryabinkina, Kulikova, Ryzhkov, 2012
	Valkiajarvi (Uksuijoki River basin)	Kulikova, 2012; Ryabinkina, Kulikova, Ryzhkov, 2012
	Vedlozero	Kulikova, 2012
	Rivers	
	Svir (the source of the river)	Kulikova, 2012
	Vuoksa	Kulikova, 2012

Taxon	Water body	Literature source
	Volga River basin	
	Lakes	
	Seliger	Our data
	The Baltic Sea basin	
	Lakes	
	Anninskoe	Our data
	Valday	Our data
	Vyshtinets	Our data
B. cederstroemii Schoedler, 1877	Ladoga Lake basin	
	Lakes	
	Lexozero	Our data; Kulikova, 2012
	The White Sea basin	
	Lakes	
	Vozhe	Our data
Hybrids *B. brevimanus* x *B. cederstroemii*	Ladoga Lake basin	
	Lakes	
	Laimolanjarvi	Kulikova, 2012
	Yanisjarvi	Kulikova, 2012; Ryabinkina, Kulikova, Ryzhkov, 2012
	Rivers	
	Svir (the source of the river)	Kulikova, 2012
	The White Sea basin	
	Lakes	
	Bolshoe Cherlivoe	Our data
	Suojarvi	Our data
B. crassicaudus Lilljeborg, 1890	The Barents Sea basin	
	Lakes	
	Akulkino	Korobtsova, 1975
	Kildin	ZISP.6850, 6852
	Pitievoe	Our data
	Zelenetskoe	Korobtsova, 1975
	Sabanty-Kul	ZISP.6848
	Severnaya Dvina	Our data
	Berezovo	ZISP.6849
	Turukhansk	ZISP.6851
	Malozemelskaya tundra Lakes	Vekhov, 1987

Thus, in the Northern and Arctic regions of the European part of Russia, *B. cederstroemii* is the most common and widespread in the north (in the Murmansk region and Karelia). Apparently, it is not uncommon in the west of the Vologda and Arkhangelsk regions, which, unfortunately, have not been sufficiently studied. On the territory of the Leningrad and Pskov regions, this species is rare, and usually occurs in hybrid populations with *B. brevimanus*.

Bythotrephes brevimanus has always been referred to in Russian literature as "*B. longimanus*". In the Murmansk region, like the previous species, it was first discovered by V. M. Rylov (1916), who found it in a significant amount in samples from Lake Imandra. Subsequently, it was found that here it lives only in this lake and adjacent water bodies, where it occurs together with B. *cederstroemii* and hybrids (Pirozhnikov, 1972; Vekhov, 1987; own data).

In Karelia, a survey on the distribution of *B. brevimanus* was made by Z. I. Filimonova (1965). According to her data (with the previously indicated corrections), the distribution of this species (often together with *B. cederstroemii* and hybrids) is limited to the territory lying between the Onega and Ladoga lakes, as well as water bodies adjacent to the White Sea-Baltic Canal and the White Sea coast.

B. brevimanus is common in the Leningrad and Pskov regions, in the south of the Novgorod region and in the west of the Vologda region. Here it is distributed sporadically, inhabiting clean, small and deep lakes, mainly located on geological elevations.

There is a lot of data in the literature on the geographical distribution of representatives of the genus *Bythotrephes* in the North-West of Russia, in the Baltic and Karelian Kola regions (Figure 3, Table 3). There is much less data on the North-Dvina and Pechora limnobiological regions. In the eastern part of the Gulf of Finland *B. brevimanus* was found in all studied regions (Figure 2). *B. cederstroemii* was observed in Neva Bay and Inner Estuary (Table 4).

Bythotrephes brevimanus (Figure 4) inhabits mainly the Baltic region and the southern part of the Karelo-Kola region. *B. cederstroemii* (Figure 5) occurs mainly in the Karelo-Kola region. *B. crassicaudus* (Figure 6) lives in the northern part of the Karelo-Kola region and in the northeast of the North-Dvina and Pechora region.

Table 4. Density (ind/m^3) of species of genus *Bythotrephes* and *Polyphemus pediculus* in Neva Bay and eastern part of the Gulf of Finland in July of 2005

Taxon	Neva Bay	Inner Estuary	Outer estuary	
			Eastern part	Western part
Bythotrephes cederstroemii	13.61	8.75	0.00	0.00
Bythotrephes brevimanus	13.27	53.53	10.42	0.87
Polyphemus pediculus	4.07	9.93	0.00	0.00

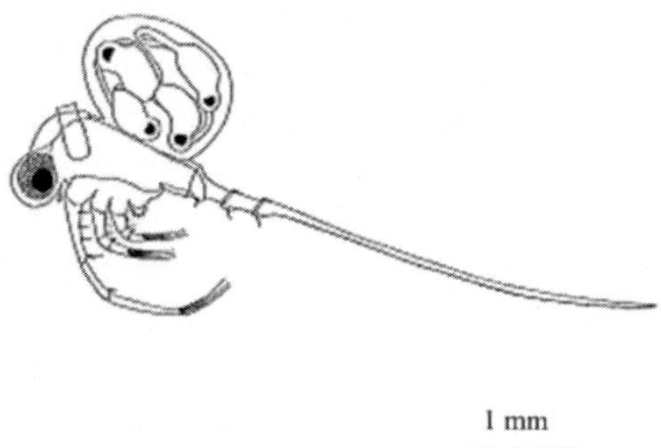

Figure 4. *Bythotrephes brevimanus* Lilljeborg, 1901 (Orig.).

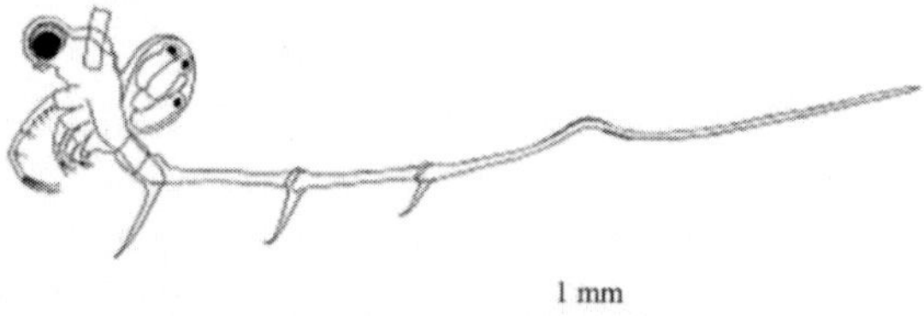

Figure 5. *Bythotrephes cederstroemii* Schoedler, 1863 (Orig.).

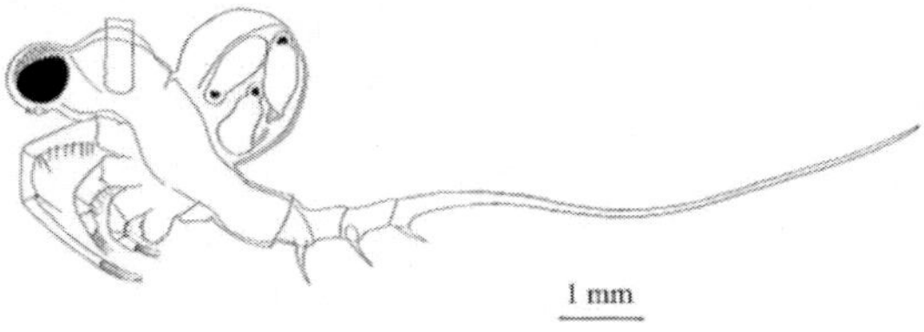

Figure 6. *Bythotrephes crassicaudus* Lilljeborg, 1890 (Orig.)

Seasonal Population Dynamics and Population Structure

The duration of the growing season in representatives of the genus *Bythotrephes* mainly depends on the geographic location of the water body.

In Karelia and southern Siberia, *B. cederstroemii* usually appears in June, and its populations reach their maximum development in late July – August (Filimonova, 1965; Pomerantseva, 1973; Konovalyuk, 1978; Krokhalevskaya, Krokhalevsky, 1980). During this period, the maximum abundance of this species in Syamozero can reach 200 ind./m3 (Filimonova, 1965). The bisexual generation appears in August, and the species disappears from the plankton in late autumn (Pomerantseva, 1973; Konovalyuk, 1978; own data).

According to our data, *Bythotrephes cederstroemii* were found in the eastern part of the Gulf of Finland from mid-August to mid-September in small density (up to 2.91 ind./m3).

In Austria, northern Germany, and Belarus, the first individuals of *B. brevimanus* usually appear in June (Ischreyt, 1939; Cheremisova, 1960; Pechen, 1970; Flossner, 1972; Herzig, 1985). In Belarus, this species reaches the highest abundance (Lake Obsterno, up to 200 ind./m3) in July-August (Cheremisova, 1960; Pechen, 1970; Semenchenko, 1996). In the Baltic States and Belarus, the bisexual generation appears in July-August (Ischreyt, 1930; Cheremisova, 1960; Pechen, 1970), and in Austria only in October (Ischreyt, 1939; Herzig, 1985). In Belarus, the last individuals of this species disappear in September-October (Cheremisova, 1960; Pechen, 1970), in northern Germany in October-November (Flossner, 1972), and in Austria (Lake Mond) in mid-January (Herzig, 1985).

B. brevimanus was recorded in the eastern part of the Gulf of Finland from late May to mid-September (our data). Moreover, at the end of May, both parthenogenetical individuals of the first generation (0.97 ind./m^3), born from resting eggs, and parthenogenetical individuals of the second and subsequent generations (5.83 ind./ m^3), born by parthenogenetic females, were found. The density of *B. brevimanus* was low and reached 18.46 ind./m^3.

On the Kola Peninsula, *B. crassicaudus* appears in June, its populations reach their maximum development in July – August, and the bisexual generation appears in July (Ivanova, 1975; own data). During the period of maximum development, its abundance in lakes of the European subarctic is on average 65.2 ind./m^3, and the maximum does not exceed 125 ind./m^3 (Vekhov, 1987). In northern Sweden, this species disappears from plankton at the end of August (Ekman, 1904). Thus, the duration of the stay of individuals of this species in the plankton of Arctic water bodies usually does not exceed 2-2.5 months (Vekhov, 1987).

It should be noted that in the hybrid populations, the highest abundance was noted among representatives of the genus *Bythotrephes*. In the Rybinsk reservoir (*B. cederstroemii* × *B. brevimanus*), its maximum indices were 700 ind./m3 (Riv'er, Grigorovich, 1999), and in the Vilyui reservoir (*B. crassicaudus* × *B. cederstroemii*) - 2800 ind./m^3 (Sokolova, 1979).

Features of the Development of Latent Eggs

As can be seen from the previous section, the cycle length for different species of the genus *Bythotrephes* can vary greatly depending on the latitude of habitat. As a consequence, it can be assumed that the reproductive strategy in different species may differ. This can be expressed in a different ratio of the phases of bisexual and parthenogenetic reproduction, as well as in different periods of wintering and the tactics of emergence of individuals of the first generation from resting eggs. Most authors (Zozulya, 1977; Vekhov, 1981; Yan and Pawson, 1998) indicate that individuals of the first generation can be found in plankton only at the very beginning of the growing season. However, special studies carried out in Lake Mond, Austria, showed that the release of juveniles of *B. brevimanus* from resting eggs in this population occurs almost throughout the season, from early June to mid-August (Herzig, 1985). It is possible that different periods of development of representatives of the first generation from resting eggs are a mechanism for protecting the population from unfavorable factors.

We have carried out studies to study the developmental features of latent eggs in various species of the genus *Bythotrephes*. For this, latent eggs of bitotrefes (*B. brevimanus* × *B. cederstroemii*) were collected from Lake Peipsi (August 8, 2000) and the Upper Volga reservoir system (August 10-11 and September 20 - October 1, 2000), as well as *B. crassicaudus* from Lake Pit'evoe (July 15, 2000). The eggs were removed from brood chambers of gamogenetic females and placed in a refrigerator at a constant temperature of 8°C. This is the optimum temperature for the development of latent eggs in representatives of the genus *Bythotrephes* (Zozulya, 1977).

The first individuals of *B. crassicaudus* from Lake Pit'evoe hatched from eggs in early April and continued to hatch until mid-July. The borning of all individuals from Lake Peipsi took place throughout May. Resting eggs from the Upper Volga reservoirs, collected at different times, had different periods of development. Individuals from resting eggs collected in August hatched from April 10 to May 13, and in September from June 13 to November 20, 2001. Moreover, in the experiment, young crustaceans borned from all resting eggs (100%) collected in August in Lake Peipsi and the Upper Volga reservoirs. Most (72.2%) individuals from Lake Pit'evoe also successfully developed from resting eggs. Of the eggs collected in September in the Upper Volga reservoirs, only 35% of the individuals were borned. It should be noted that living embryos at different stages of development remained in resting eggs in a dormant state (Lake Pitievoe and the Upper Volga reservoirs) until at least March 2002.

These observations may indicate that in the harsh Arctic conditions, latent eggs do not develop simultaneously, but throughout the entire growing season, and some of the resting eggs remain dormant until the next season. In the Upper Volga, individuals of the first generation were borned from all resting eggs formed in July-August only at the beginning of the next growing season. Eggs laid at the end of the growing season seem to form "safety pools." It should be noted that the experiment continues at the present time (October 2020).

Average Body Length

The largest number of samples is represented by *B. crassicaudus*. According to these data, individuals of this species from Kildin Island in the Barents Sea reach the greatest body length (up to 5.0 mm; Table 5).

The same sizes of individuals of this species were given by V. Lilleborg and S. Ekman (Lilljeborg, 1901; Ekman, 1904) for northern Sweden and N.V. Vekhov (1981) for the north-east of the European part of Russia. However, according to K.E. Stenros and J.O. Sars (Stenroos, 1897; Sars, 1903), this species (from Karelia and northern Norway) can reach a length of 6.0 mm.

The average body length in this species can vary within wide limits (3.2-4.4 mm). Apparently, there is a relationship between the body length and the geographical location of the population, since the largest individuals are common in the north of the range (Kildin and Severnaya Dvina).

Body length among the three studied samples of *B. brevimanus*, individuals from Lake Anninskoe (1.2 mm on average) sharply differed from individuals from lakes Vyshtinetskoe (1.8 mm) and Seliger (1.6 mm) in their very small sizes. However, these values did not go beyond the variability of this species. The variability in the relative length of the tail in this species (as in the two previous ones) was minimal.

The body length noted in *B. cederstroemii* specimens from Leksozero and Vozhe lakes did not exceed 1.3-2.0 mm (Table 5). Such values of this trait bring these populations closer to those from the Engozero and Senezhskoe lakes and sharply distinguish them from the Swedish representatives of this species, which are characterized by a significantly greater body length (1.9-4.0 mm). The relative length of the tail shows great geographic variability.

To summarize, it should be noted that, for all studied species of the genus *Bythotrephes*, significant geographic variability in body length and in the majority and relative length of the caudal appendage was noted.

Table 5. Variability of morphometric characters of *Bythotrephes crassicaudus, B. brevimanus* и *B. cederstroemii* (adult females)

N	Population	n	Character [X ± σ (min-max)]	
			Body length (mm)	Cauda length/body length (mm)
B. crassicaudus				
1	Kildin	5	4.4 ± 0.5 (3.7-5.0)	1.9 ± 0.7 (1.7-2.0)
2	Dalnie Zelentsy	18	3.4 ± 0.2 (2.8-3.7)	2.0 ± 0.2 (1.7-2.4)
3	Severnaya Dvina	10	4.4 ± 0.2 (4.1-4.6)	1.8 ± 0.2 (1.5-1.9)
4	Berezovo	1	3.2	1.9
5	Turukhansk	4	3.8 ± 0.5 (3.3-4.5)	1.9 ± 0.0 (1.9-1.9)
B. brevimanus				
6	Vyshtynetskoe	30	1.8 ± 0.1 (1.5-2.0)	2.6 ± 0.2 (2.2-3.3)
7	Anninskoe	6	1.2 ± 0.1 (1.1-1.4)	2.6 ± 0.3 (2.2-3.0)
8	Seliger	2	1.6 (1.5-1.6)	2.6 (2.5-2.7)
B. cederstroemii				
9	Vozhe	1	1.8	3.9
10	Lexozero	33	1.6 ± 0.1 (1.3-2.0)	4.4 ± 0.5 (3.4-5.4)

Fecundity of Parthenogenetic Females

Sven Ekman (1904) provided the first information about the fecundity of parthenogenetic females in *B. crassicaudus* (northern Sweden – 7-9), *B. brevimanus* (southern Sweden and Denmark - 4) and *B. longimanus* (Switzerland - 4). However, long before that, Konrad Beck (Beck, 1883) indicated that *B. brevimanus* from England has a fecundity of 10-12 embryones. The German researcher Wagler (Wagler, 1927) noted no more than 5 embryos in *B. brevimanus* from Germany, and V. M. Rylov (1915) for this species from Lake Ivan (Belarus) - up to 6 embryones.

The geographical variability of this trait in *B. brevimanus* was described in detail by G. Ischreyt (1930). This author noted the smallest number of embryones in individuals from the North German Lake Höft (average 3.0, span of 1-5 embryones). The highest fecundity was noted in the Latvian lake Poussin (on average 6.1, span of 3-10 embryones). E. D. Mordukhai-Boltovskaya (1957) noted for representatives of this genus from the Rybinsk reservoir (*B. cederstroemii* × *B. brevimanus*) from 1 to 10, more often 5-6 embryones.

The highest fecundity was noted in the Latvian lake Poussin (on average 6, span of 3-10 embryos). E. D. Mordukhai-Boltovskaya (1957) noted for representatives of this genus from the Rybinsk reservoir (*B. cederstroemii* × *B. brevimanus*) from 1 to 10, more often 5-6 embryones.

According to M. K. Cheremisova (1960), *B. brevimanus* from Belarus can have from 4 to 14 embryones. *B. longimanus* from Lake Maggiore, Italy, had 4-6 embryos (Zaffagnini, 1964), *B. brevimanus* from Holland had an average of 3.5-3.6 embryos (Ketelaars et al., 1995), and *B. cederstroemii* × *B. brevimanus* from the Great American lakes - an average of 4.9 embryones (Berg, Garton, 1988). For *B. crassicaudus* from Malo- and Bolshezemelskaya tundra N. K. Vekhov (1987) indicates an average of 7 ± 3 embryones, and for *B. crassicaudus* × *B. cederstroemii* from the lower reaches of the Pechora River, an average of 14-19 embryones.

Table 6. The number of embryos in adult parthenogenetic *Bythotrephes* females

Taxon	Population	Date	n	Number of embryos		
				X ± σ	min.	max
crassicaudus	Dalnie Zelentsy	15.07.2000	8	6.8 ± 2.4	3	10
	Severnaya Dvina	18.08.1998	2	4.0	3	5
	Sabanty-Kul	6.06.1899	2	4.0	-	-
cederstroemii	Lexozero	11.08.1999	30	3.6 ± 1.3	1	7
	Vozhe	26.07.1999	1	3	-	-
brevimanus	Vyshtinets	15.07.1999	85	4.4 ± 1.4	2	8
	Anninskoe	-	7	5.7 ± 1.5	4	8
	Seliger	2000	3	7.7 ± 1.5	6	9
cederstroemii × *brevimanus*	Bolshoe Cherlivoe	1995-1996	3	4.0 ± 0	-	-
	Gulf of Finland	2.07.1999	23	5.2 ± 2.2	1	11
	Chudskoe	10.08.2000	142	4.4 ± 1.8	1	10
	Khotavets	11.08.2000	36	5.3 ± 1.3	2	9
	Uglichskoe	19.08.2000	1	8	-	-
	Gorkovskoe	30.09.2000	3	8.7 ± 1.2	8	10
	Rybinskoe (Shumorovskyi)	17.07.1999	14	3.8 ± 1.0	2	5
	Rybinskoe (Vsekhsvyatskoe)	27.06.2000	14	10.4 ± 1.9	8	13
	Rybinskoe (GES)	27.06.2000	6	12.2 ± 2.9	8	15
	Rybinskoe (Babii Gory)	10.08.2000	6	5.8 ± 1.7	3	8

S. S. Zozulya (1977, 1979) was the first to study the seasonal variability of this trait. This author found that the fertility of the first generation *Bythotrephes* from the Rybinsk Reservoir is more than two times higher (12-24) than that of its subsequent generations (1-10). In addition, he noted that in the summer, females form 2-4 embryones, and in the fall - 6-8 embryones. I. K. Riviere (1993), studying the same population, noted that in July 1976 the average fertility was 10, in July 1975 – 5, and in July 1990 - only 3 embryones. Thus, this trait cannot be considered stable, even if taken into account in the same season and in the same water body.

Overall, our data on the fertility of parthenogenetic females from 18 water bodies (Table 6) are within the limits indicated by other researchers.

According to our data (Table 6), fecundity of parthenogenetic female *B. crassicaudus* was 4.0-6.8 embryones, *B. cederstroemii* – 3.0-3.6 embryones, *B. brevimanus* – 4.4-7.7 embryones. Fecundity of parthenogenetic females of hybrides of *B. cederstroemii* × *B. brevimanus* was highter then their parents species.

Fecundity of Gamogenetic Females

Sven Ekman (1904) was the first to review the fecundity of gamogenetic females in representatives of the genus *Bythotrephes*. For *B. crassicaudus*, he indicated 4-6, and in some cases up to 9 resting eggs, for *B. brevimanus* from the south of Sweden and from Denmark – 2 eggs (Muller, 1867; Lilljeborg, 1901), for *B. longimanus* from Switzerland – 2-3, rarely 1 egg (Weismann, 1879). V. Lilljeborg (Lilljeborg, 1901) gave images of *B. crassicaudus* × *B. cederstroemii* (Koresuando, Kola Peninsula) with five resting eggs.

In *B. brevimanus* from Belarus and *B. cederstroemii* × *B. crassicaudus* from the Vilyui reservoir, the number of resting eggs, as a rule, is equal to two (Cheremisova, 1960; Sokolova, 1979), and in representatives of the genus *Bythotrephes* from the Rybinsk reservoir (*B. cederstroemii* × *B. brevimanus*) - two or three, sometimes four (Rivier, 1969, 1993). For *B. crassicaudus* from the vicinity of Vorkuta, N. V. Vekhov (1981) indicated 4-8 wintering eggs. In crustaceans in the Great American Lakes (*B. cederstroemii* × *B. brevimanus*), an average of 3.4-4.0 resting eggs were recorded (Garton and Berg, 1990; Garton et al., 1990), and in some individuals up to 7 eggs (Berg, Garton, 1988). In the *B. brevimanus* population from Holland, H. A. M. Ketelaars et al., (1995) noted females with 1-4 resting eggs.

In this study, fecundity was studied in *Bythotrephes* from 13 samples. In all four studied populations of *B. crassicaudus*, the number of latent eggs in gamogenetic females was significantly higher than in all other populations of other species. In females of this species, an average of 10-11 eggs were observed (Table 7). Only in one sample (with selected large females) the number of resting eggs was slightly lower (7.2). Some *B. crassicaudus* specimens contained up to 14 eggs. This is the first time that such a number of hibernating eggs have been reported in this genus. In the remaining samples, the average number of wintering eggs per female ranged from 1.0 in *B. cederstroemii* from Leksozero to 5.8 in individuals of *B. cederstroemii* × *B. brevimanus* from the Gulf of Finland.

Table 7. Diameter of resting eggs (mm), fertility (embryo / ♀), and body length (mm) in adult gamogenetic females in *Bythotrephes*. The range of variability is indicated in parentheses

			Fecundity		Egg diameter		Body length	
Taxon	Population	Date	n	X ± σ	n	X ± σ	n	X ± σ
crassicaudus	Kildin	1894	5	11.0 ± 1.0 (10-12)	8	0.581 ± 0.018 (0.55-0.60)	5	4.19 ± 0.40 (3.7-4.7)
	Dalnie Zelentsy	07.2000	6	7.2 ± 1.2 (6-9)	71	0.562 ± 0.034 (0.48-0.63)	4	3.21 ± 0.30 (2.8-3.5)
	Severnaya Dvina	09.1998	4	10.0 ± 2.2 (7-12)	28	0.523 ± 0.046 (0.43-0.60)	3	4.35 ± 0.09 (4.3-4.5)
	Turukhansk	-	2	11.0 (8-14)	22	0.523 ± 0.017 (0.50-0.55)	2	3.47 (3.3-4.5)
cederstroemii	Lexozero	08.1999	2	1.0	2	0.413 (0.40-0.43)	2	1.80
crassicaudus × *cederstroemii*	Severnaya Dvina	09.1998	1	8	8	0.434 ± 0.013 (0.43-0.45)	1	3.9
	M. Sovetskoe	08.1998	1	4	4	0.438 ± 0.043 (0.38-0.48)	1	2.7
cederstroemii × *brevimanus*	Ladoga	08.1996	4	5.0 ± 1.2 (4-6)	20	0.456 ± 0.036 (0.40-0.53)	4	2.61 ± 0.31 (2.3-3.0)
	Gulf of Finland	1998-1999	5	5.8 ± 1.5 (4-8)	17	0.450 ± 0.023 (0.40-0.50)	5	2.74 ± 0.52 (2.2-3.4)
	Chudskoe	08.2000	2	4.0	4	0.450 ± 0	-	1.91*
	Khotavets	08.2000	-	-	4	0.450 ± 0	-	1.82*
	Uglicheskoe	08.2000	1	4	4	0.425 ± 0.020 (0.40-0.45)	1	2.2
	Rybinskoe	09.2000	33	2.6 ± 0.9 (1-5)	45	0.440 ± 0.022 (0.38-0.50)	11	2.04 ± 0.30 (1.7-2.7)
	Gorkovskoe	09.2000	57	4.1 ± 1.3 (1-6)	170	0.475 ± 0.026 (0.38-0.53)	26	2.30 ± 0.33 (1.7-2.8)

* - mean for the sample.

Thus, the number of latent eggs can be considered as a diagnostic feature that makes it possible to distinguish gamogenetic females of *B. crassicaudus* from other species of this genus. It can also be noted that the number of resting eggs in species of the genus *Bythotrephes* correlates well with the diameter of resting eggs and body length (Table 7), which also distinguish *B. crassicaudus* from other species of this genus.

Geographical Distribution and Biology of *Polyphemus pediculus* in the Northern and Arctic Regions of the European Part of Russia

According to our and literature data, *P. pediculus* widely distributed in the Northern regions of the European part of Russia (Table 3). At the same time, the Arctic regions are poorly studied and there are few finds. Researchers did not find this species on Vaygach Island and the Novaya Zemlya Archipelago (Vekhov, 1997; Bespalaya et al., in press).

In the eastern part of the Gulf of Finland *P. pediculus* was observed in pelagic zone of the Neva Bay and Inner Estuary (Figure 2). In the period 2009-2019, *P. pediculus* (Figure 7) was found in plankton in the eastern part of the Gulf of Finland from mid-May to late August (Table 8). High densities were recorded from mid-June to late July. The maximum number was registered at the end of June 2017. At this time, the highest values of temperature of water were noted for the seasons (Table 1). The water salinity and transparency varied considerably during all seasons.

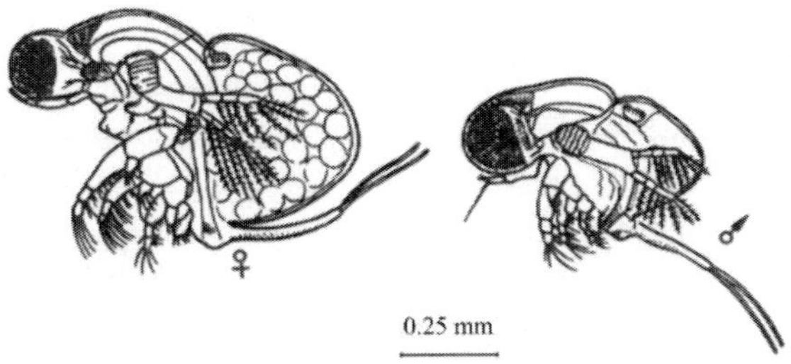

Figure 7. *Polyphemus pediculus* (Linnaeus, 1761), modified from Lilljeborg (1900).

Table 8. Mean density of *P. pediculus* (ind./ m^3) in the eastern part of the Gulf of Finland (Primorsk) in the beginning of May – mid October 2009-2019 years

Year/	Mean density in sampling dates											
Days and months	05.05	15.05	31.05	15.06	30.06	15.07	31.07	15.08	31.08	15.09	30.09	15.10
2009	–	–	0.00	–	0.00	94.07	62.79	–	20.7	–	–	–
2010	–	–	0.00	–	48.30	2.07	5.80	0.00	–	–	–	–
2011	–	–	0.00	0.00	2.72	2.90	12.24	0.00	0.00	0.00	–	0.00
2012	0.00	0.00	0.00	0.00	291.43	58.29	0.00	0.00	0.00	0.00	–	–
2013	–	0.00	28.56	131.14	19.89	119.31	70.96	0.00	0.00	0.00	–	0.00
2014	–	13.54	32.64	2.27	0.97	13.60	233.14	0.00	0.00	0.00	–	0.00
2015	–	2.91	0.00	0.00	38.86	0.97	155.43	0.00	0.00	0.00	–	0.00
2016	0.00	–	0.00	13.60	48.57	9.71	29.14	0.00	0.00	0.00	–	0.00
2017	–	0.00	0.00	1.94	971.43	5.83	10.69	0.00	0.00	0.00	–	0.00
2018	–	1.70	0.00	388.57	3.89	7.77	49.54	0.00	0.00	0.00	–	–
2019	–	0.00	0.00	0.00	97.14	9.71	0.97	0.00	0.00	0.00	0.00	–

Dash - sampling was not carried out.

Table 9. Mean body length (mm) of adult parthenogenetic females of *P. pediculus* in the eastern part of the Gulf of Finland (Primorsk) in the beginning of May – mid October 2009-2019 years

Year/	Mean body length in sampling dates											
Days and months	05.05	15.05	31.05	15.06	30.06	15.07	31.07	15.08	31.08	15.09	30.09	15.10
2009	–	–	0.00	–	0.55	0.00	0.6	–	0.65	–	–	–
2010	–	–	0.00	–	0.7	0.00	0.53	0.00	–	–	–	–
2011	–	–	0.00	0.00	0.68	0.7	0.57	0.00	0.00	0.00	–	0.00
2012	0.00	0.00	0.00	0.00	0.63	0.6	0.00	0.00	0.00	0.00	–	–
2013	–	0.00	0.55	0.58	0.55	0.55	0.62	0.00	0.00	0.00	–	0.00
2014	–	0.6	0.55	0.5	0.5	0.62	0.58	0.00	0.00	–	–	0.00
2015	–	0.00	0.00	0.00	0.6	0.6	0.5	0.00	0.00	0.00	–	0.00
2016	0.00		0.00	0.00	0.6	0.65	0.6	0.00	0.00	0.00	–	0.00
2017	–	0.00	0.00	0.45	0.45	0.65	0.6	0.00	0.00	0.00	–	0.00
2018	–	0.59	0.00	0.55	0.6	0.6	0.68	0.00	0.00	–	–	–
2019	–	0.00	0.00	0.6	0.6	0.65	0.7	0.00	0.00	0.00	0.00	–
Mean body length for all seasons, mm (X ± SD)		0.60 ± 0.01	0.55	0.54 ± 0.06	0.59 ± 0.07	0.62 ± 0.04	0.60 ± 0.06		0.65			

Dash - sampling was not carried out.

The average body length of adult parthenogenetic females of *P. pediculus* during all seasons was 0.59-0.65 mm (Table 9); the decrease in body length in late May – mid June is most likely associated with the large number of juveniles in the populations at the beginning of the growing seasons. Parthenogenetic females with embryos were found during the entire period of population development (Table 10). The maximum number of parthenogenetic embryos was observed in mid-May at the beginning of the development of the population.

Table 10. Mean fecundity of parthenogenetic females (number of embryos per female) of *P. pediculus* in the eastern part of the Gulf of Finland (Primorsk) in the beginning of May – mid October 2009-2019 years

Year/	Mean fecundity of parthenogenetic females in sampling dates											
Days and months	05.5	15.5	31.5	15.6	30.6	15.7	31.7	15.8	31.8	15.9	30.9	15.10
2009	–	–	0.00	–	0.00	15.00	–	–	12.00	–	–	–
2010	–	–	0.00	–	9.00	6.50	0.00	0.00	–	–	–	–
2011	–	–	0.00	0.00	15.00	0.00	11.70	0.00	0.00	0.00	–	0.00
2012	0.00	0.00	0.00	0.00	17.00	14.50	0.00	0.00	–	0.00	–	–
2013	–	0.00	13.30	7.33	4.50	9.50	10.50	0.00	0.00	0.00	–	0.00
2014	–	21.00	19.50	0.00	8.00	10.40	7.67	0.00	0.00	–	–	0.00
2015	–	0.00	0.00	0.00	8.50	12.00	0.00	0.00	0.00	0.00	–	0.00
2016	0.00	–	0.00	11.50	5.75	11.30	9.00	0.00	0.00	0.00	–	0.00
2017	–	0.00	0.00	18.00	0.00	9.50	8.00	0.00	0.00	0.00	–	0.00
2018	–	30.00	0.00	14.00	7.50	9.50	7.75	0.00	0.00	–	–	–
2019	–	0.00	0.00	19.00	9.67	14.50	15.00	0.00	0.00	0.00	0.00	–
Mean fecundity for all seasons (X ± SD)		25.50 ± 6.36	16.42 ± 4.36	13.97 ± 4.79	9.44 ± 4.08	11.27 ± 2.75	9.94 ± 2.69		12.00			

Dash - sampling was not carried out.

Gamogenetic females with resting eggs (in average 2.25-3.50 eggs per female) were found from mid-June to late July (Table 11), which coincided with the maximum abundance of *P. pediculus* during the seasons. In brood pouches of females, we observed between 1 and 6 eggs.

Table 11. Mean fecundity of gamogenetic females (number of resting eggs per female) of *P. pediculus* in the eastern part of the Gulf of Finland (Primorsk) in the beginning of May – mid October 2009-2019 years

Year/	Mean fecundity gamogenetic females in sampling dates											
Days and months	05.5	15.5	31.5	15.6	30.6	15.7	31.7	15.8	31.8	15.9	30.9	15.10
2009	–	–	0.00	–	0.00	0.00	4.00	–	0.00	–	–	–
2010	–	–	0.00	–	4.00	0.00	0.00	0.00	–	–	–	–
2011	–	–	0.00	0.00	0.00	2.00	5.00	0.00	0.00	0.00	–	0.00
2012	0.00	0.00	0.00	0.00	3.70	0.00	3.00	0.00	0.00	0.00	–	–
2013	–	0.00	0.00	3.00	0.00	0.00	0.00	0.00	0.00	0.00	–	0.00
2014	–	0.00	0.00	0.00	0.00	3.00	0.00	0.00	0.00	–	–	0.00
2015	–	0.00	0.00	0.00	0.00	0.00	0.00	0.00	0.00	0.00	–	0.00
2016	0.00	0.00	0.00	1.50	1.00	0.00	2.00	0.00	0.00	0.00	–	0.00
2017	–	0.00	0.00	0.00	2.50	0.00	0.00	0.00	0.00	0.00	–	0.00
2018	–	0.00	0.00	0.00	0.00	0.00	0.00	0.00	0.00	–	–	–
2019	–	0.00	0.00	0.00	5.50	0.00	0.00	0.00	0.00	0.00	0.00	–
Mean fecundity for all seasons (X ± SD)				2.25 ± 1.06	3.33 ± 1.69	2.50 ± 0.71	3.50 ± 1.29					

Dash - sampling was not carried out.

In Lake Krivoe, located on the coast of the White Sea, in pelagic zone on the horizon 0-5 m (deep water part of the lakes) *P. pediculus* was found from mid-May to late September (Table 12).

Table 12. Mean density of *P. pediculus* (ind./ m^3) in horizon 0-5 m (total station depth is 30 m) in Krivoe Lake (the White Lake basin) in mid of May – the end of October of 2002-2019 years

Year/ Days and months	Mean density in sampling dates											
	15.05	31.05	15.06	30.06	15.07	31.07	15.08	31.08	15.09	30.09	15.10	31.10
2002	–	0.00	19.57	18.35	209.19	0.00	484.44	244.67	22.02	–	–	–
2003	–	0.00	–	26.43	44.05	330.30	–	44.04	170.81	0,00	–	0.00
2004	8.81	2.94	0.00	0.00	–	293.60	–	–	51.38	–	–	–
2005	–	0.00	–	–	–	1585.80	–	–	49.59	–	–	–
2006	–	0.00	–	–	–	528.60	–	–	105.72	–	–	–
2007	–	8.81	–	–	–	660.60	–	–	–	22.02	–	–
2008	–	0.00	0.00	–	–	440.00	1468.00	14.68	0.00	–	0.00	–
2009	–	0.00	0.00	–	–	440.00	1468.00	14.68	0.00	–	0.00	–
2010	–	0.00	–	–	–	502.17	–	44.05	–	–	–	0.00
2011	–	0.00	11.01	–	–	493.36	–	0.00	–	–	–	0.00
2012	0.00	0.00	–	–	264.30	5286.00	–	–	70.48	–	–	–
2013	–	0.00	–	–	–	261.00	–	34.80	–	–	–	–
2014	–	0.00	–	–	–	174.00	–	17.40	–	–	–	–
2015	–	0.00	–	0,00	–	34.80	–	43.50	–	–	–	–
2016	8,7	0,00	–	0,00	–	0.00	–	8.70	–	–	0.00	–
2017	–	0,00	–	0,00	–	1566.00	–	52.20	–	–	–	–
2018	–	0.00	–	0.00	–	522.00	–	17.40	–	–	–	–
2019	–	0.00	0.00	–	696.00	–	17.40	69.60	–	–	–	0.00

Dash - sampling was not carried out.

The densities of this species at the beginning of the growing season were low, then, by the end of July, the abundance increased and reached its maximum values (up to 5000 ind./m^3 in 2012). The water temperatures in the lake at the end of July also reached its maximum values for all seasons (Table 2).

At the 5-15 m horizon, the time of *P. pediculus* development during the season was shorter than in the upper worm and well-illuminated layer of 0-5 m. The average water temperature at the 5-15 m horizon at the end of July was 9.14 ± 3.19°C compared to 16.60 ± 2.95°C at the 0-5 m horizon.

According to Zakhvatkin (1927, Solovetskiye Lakes) and Olonetskaya Scientific Expedition data (1923, Sandal Lake), this species was found only in oligotrophic and dystrophic water bodies, not only for the coastal, but also for the pelagic zone. In eutrophic lakes, *P. pediculus* has so far been found only in the coastal zone. In our study *P. pediculus* inhabits the pelagic zone of both oligotrophic and eutrophic water bodies.

The development of the population at the horizon was shorter by a month, from late May to mid-September (Table 13). The peak of its density was in mid-August (294 ind./ m^3 in 2008). However, in mid-September, high values of the density of *P. pediculus* were also observed (308 ind./m^3 in 2012). In the deepest water layers of Lake Krivoe (15-30 m) from late July to mid-September, which is characterized by an average water temperature of 6.60 ± 2.38°C, *P. pediculus* was recorded from late July to mid-September with low densities values (Table 14).

In the shallow part of Lake Krivoe, *P. pediculus* was recorded from late May to late September (Table 15). This species reached its maximum abundance at the end of July (2898 ind./m^3 in 2018).

Table 13. Mean density of *P. pediculus* (ind./ m^3) in horizont 5-15 m (total station depth is 30 m) in Krivoe Lake (the White Lake basin) in mid of May – the end of October of 2002-2019 years

Year/	Mean density in sampling dates											
Days and months	15.05	31.05	15.06	30.06	15.07	31.07	15.08	31.08	15.09	30.09	15.10	31.10
2002	–	0.00	0.00	0.00	0.00	11.30	13.56	7.34	1.13	–	–	–
2003	–	0.00	–	0.00	0.00	22.02	–	55.05	4.41	0.00	–	–
2004	0.00	0.00	0.00	0.00	–	19.56	–	–	34.23	–	–	–
2005	–	0.00	–	–	–	98.00	–	–	0.00	–	–	–
2006	–	0.00	–	–	–	4.41	–	–	0.00	–	–	–
2007	–	0.00	–	–	–	0.00	–	–	–	0.00	–	–
2008	–	0.00	0.00	–	–	147.00	294.00	0.00	0.00	–	0.00	–
2009	–	0.00	–	–	–	0.00	0.00	0.00	4.41	–	0.00	–
2010	–	0.00	–	–	–	0.00	–	0.00	–	–	–	0.00
2011	–	120.30	0.00	–	–	0.00	–	0.00	–	–	–	–
2012	0.00	0.00	–	–	0.00	8.81	–	–	308.35	–	–	–
2013	–	0.00	–	–	–	69.60	–	0.00	–	–	–	–
2014	–	0.00	–	–	–	0.00	–	4.35	–	–	–	–
2015	–	0.00	–	–	–	47.85	–	2.90	–	–	–	–
2016	0.00	0.00	–	–	–	82.65	–	0.00	–	–	–	–
2017	–	4.35	–	0.00	–	4.35	–	0.00	–	–	–	–
2018	–	0.00	–	0.00	–	43.50	–	0.00	–	–	–	–
2019	–	0.00	0.00	–	0.00	–	0.00	0.00	–	–	–	–

Dash - sampling was not carried out.

Table 14. Mean density of *P. pediculus* (ind./m^3) in horizont 15-30 m (total station depth is 30 m) in Krivoe Lake (the White Lake basin) in mid of May – the end of October of 2002-2019 years

Year/	Mean density in sampling dates											
Days and months	15.05	31.05	15.06	30.06	15.07	31.07	15.08	31.08	15.09	30.09	15.10	31.10
2002	–	0.00	0.00	0.00	0.98	0.98	2.94	0.00	1.13	–	–	–
2003	–	0.00	–	0.00	0.00	10.00	–	0.00	3.89	0.00	–	0.00
2004	0.00	0.00	0.00	0.00	–	0.00	–	–	0.00	–	–	–
2005	–	0.00	–	–	–	3.15	–	–	0.00	–	–	–
2006	–	0.00	–	–	–	0.00	–	–	0.00	–	–	–
2007	–	0.00	–	–	–	94.50	–	–	–	0.00	–	–
2008	–	0.00	0.00	–	–	88.20	0.00	0.00	0.00	–	0.00	–
2009	–	0.00	–	–	–	0.00	0.00	0.00	0.00	–	0.00	–
2010	–	0.00	–	–	–	44.00	–	0.00	–	–	–	0.00
2011	–	0.00	0.00	–	–	94.41	–	0.00	–	–	–	0.00
2012	0.00	0.00	–	–	0.00	0.00	–	–	31.47	–	–	–
2013	–	0.00	–	–	–	10.87	–	0.00	–	–	–	–
2014	–	0.00	–	–	–	0.	–	0.00	–	–	–	–
2015	–	0.00	–	–	–	5.80	–	0.00	–	–	0.00	–
2016	0.00	0.00	–	–	–	0.00	–	0.00	–	–	0.00	–
2017	–	0.00	–	0.00	–	0.00	–	0.00	–	–	–	–
2018	–	0.00	–	0.00	–	0.00	–	0.00	–	–	–	–
2019	–	0.00	0.00	–	0.00	–	0.00	0.00	–	–	–	0.00

Dash - sampling was not carried out.

Table 15. Mean density of *P. pediculus* (ind./m^3) in horizont 0-3 m (total station depth is 3.5 m) in Krivoe Lake (the White Lake basin) in mid of May – the end of October of 2003 and 2010-2019 years

Year/ Days and months	Mean density in sampling dates											
	15.05	31.05	15.06	30.06	15.07	31.07	15.08	31.08	15.09	30.09	15.10	31.10
2003	–	12.90	–	11.00	161.48	0.00	–	165.15	484.44	234.88	–	0.00
2010	–	9.79	–	–	–	0.00	–	9.79	–	–	–	0.00
2011	–	0.00	0.00	–	–	831.87	–	14.68	–	–	–	–
2012	–	0.00	–	–	2202.00	1419.07	–	–	39.15	–	–	–
2013	–	7.25	–	–	–	28.98	–	19.32	–	–	–	–
2014	–	0.00	–	–	–	53.13	–	9.66	–	–	–	–
2015	–	0.00	–	–	–	140.07	–	115.92	–	–	–	–
2016	0.00	14.49	–	–	–	173.88	–	0.00	–	–	0.00	–
2017	–	7.25	–	0.00	–	268.07	–	43.47	–	–	–	–
2018	–	0.00	–	0.00	–	2898.00	–	9.66	–	–	–	–
2019	–	0.00	4.83	–	966.00	0.00	483.00	96.60	–	–	–	0.00

Dash - sampling was not carried out.

In the coastal zone of Lake Krivoe (depth 0.5-0.7 m), *P. pediculus* was recorded from mid-May to late September and in the mid of June (2012) reached the highest density values for this species in Lake Krivoe, 8570 ind./m^3 (Table 16).

The average body length of adult parthenogenetic females of *P. pediculus* in Krivoye Lake varied from 0.55 to 0.57 mm (Table 17). Relatively high averaged values of body length were noted in late July, mid-August and mid-September.

Table 16. Mean density of *P. pediculus* (ind./m^3) in horizont 0-0.5 m (total station depth is 0.5-0.7 m) in Krivoe Lake (the White Lake basin) in mid of May – to the end of October of 2011-2012 years

Year/ Days and months	Mean density in sampling dates											
	15.05	31.05	15.06	30.06	15.07	31.07	15.08	31.08	15.09	30.09	15.10	31.10
2011	–	88.10	–	–	2379.24	–	1000.00	–	–	–	–	–
2012	25.00	1000.00	–	–	8571.00	0.00	–	–	50.00	–	–	–

Dash - sampling was not carried out.

When analyzing the age structure of *P. pediculus* (example of the year 2003) considered on the example of 2003, it was found that at the beginning of the growing season, young individuals (up to 80%) with a size of 0.30-0.35 mm prevailed in the population (Table 18). At the end of the season, the population was dominated by adult parthenogenetic females 0.50-0.57 mm in size (up to 67%). The study of the sexual structure of *P. pediculus* in Lake Krivoe showed that at the beginning of the population development, representatives of the parthenogenetic generation predominate. Later, gamogenetic females and males appear in the population. At the end of the growing season, only representatives of the gamogenetic generation were observed in the lake (Table 19).

Table 17. Mean body length (mm) of adult parthenogenetic females of *P. pediculus* in Krivoe Lake (the White Lake basin) in mid of May – the end of October of 2002-2019 years

Year/ Days and months	Mean body length in sampling dates											
	05.05	15.05	31.05	15.06	30.06	15.07	31.07	15.08	31.08	15.09	30.09	15.10
2002	–	0.00	0.00	0.00	0.55	0.55	0.58	0.57	0.00	0.55	–	–
2003	–	0.00	–	0.00	0.00	0.58	0.59	–	0.55	0.6	–	–
2004	0.00	0.00	0.00	0.00	–	–	0.00	–	0.55	0.55	–	–
2005	–	0.00	–	–	–	–	0.00	–	–	0.00	–	–
2006	–	0.00	–	–	–	–	0.00	–	–	0.00	–	–
2007	–	0.00	–	–	–	–	0.55	–	–	–	–	–
2008	–	0.00	0.00	–	–	–	0.5	0.00	0.00	0.55	0.00	–
2009	–	0.00	–	–	–	–	0.00	0.00	–	0.00	–	–
2010	–	0.00	–	–	–	–	0.00	–	0.55	–	–	0.00
2011	–	0.00	0.55	–	–	–	0.55	–	0.00	–	–	–
2012	0.00	0.00	–	–	0.00	0.53	0.00	–	–	0.55	–	–
2013	–	0.00	–	–	–	–	0.55	–	0.00	0.63	–	–
2014	–	0.00	–	–	–	–	0.65	–	0.00	0.55	–	–
2015	–	0.00	–	–	–	–	0.55	–	0.55	–	–	–
2016	0.00	0.00	–	–	–	–	0.00	–	0.00	–	–	–
2017	–	0.00	–	0.00	–	–	0.58	–	0.50	–	–	–
2018	–	0.00	–	0.00	–	–	0.55	–	0.5	–	–	–
2019	–	0.00	0.00	–	0.00	–	0.53	0.00	0.5	0.55	–	–
Mean body length for all seasons, mm (X ± SD)			0.55		0.55	0.55 ± 0.03	0.56 ± 0.04	0.57	0.53 ± 0.03	0.57 ± 0.03		

Dash - sampling was not carried out.

Table 18. Age structure of parthenogenetic females (in %) of *P. pediculus* in Krivoe Lake (the White Lake basin) in the June beginning – the end of October of 2003 year

Date	Body length, mm, of parthenogenetic females			Total number of specimens
	0.30-0.35	0.40-0.45	0.50-0.57	
01.06.2003	0.00	0.00	0.00	0.00
01.07.2003	66.67	33.33	0.00	30.00
15.07.2003	80.00	0.00	20.00	50.00
31.07.2003	29.41	29.41	41.18	170.00
31.08.2003	50.00	25.00	25.00	40.00
15.09.2003	0.00	33.33	66.67	30.00
30.09.2003	28.57	57.14	14.29	70.00
31.10.2003	0.00	0.00	0.00	0.00

Parthenogenetic females with embryos were found in the Krivoe Lake from late May to mid-September. Fecundity of parthenogenetic females of *P. pediculus* at the beginning of the season was maximum (14 embryos per female). In the following months, average fecundity decreased and varied widely (Table 20).

Table 19. Sex structure (in %) of *P. pediculus* in Krivoe Lake (the White Lake basin) in the June beginning – the end of October of 2003 year

Date	Body length, mm				Total number if specimens
	Part. females	Gamogenetic females without resting eggs	Gamogenetic females with resting eggs	Males	
	0.50-0.57	0.50-0.55	0.50-0.55	0.50-0.55	
01.06.2003	0.00	0.00	0.00	0.00	0.00
01.07.2003	0.00	0.00	0.00	0.00	0.00
15.07.2003	100.00	0.00	0.00	0.00	10.00
31.07.2003	35.00	25.00	10.00	30.00	200.00
31.08.2003	100.00	0.00	0.00	0.00	10.00
15.09.2003	15.38	23.08	30.77	30.77	130.00
30.09.2003	3.85	57.69	11.54	26.92	260.00
31.10.2003	0.00	0.00	0.00	0.00	0.00

Table 20. Fecundity of parthenogenetic females of *P. pediculus* in Krivoe Lake (the White Lake basin) in mid of May – the end of October of 2002-2019 years

Year/ Days and months	Mean body length in sampling dates										
	15.05	31.05	15.06	30.06	15.07	31.07	15.08	31.08	15.09	30.09	15.10
2002	–	0.00	0.00	0.00	0.00	–	0.00	0.00	0.00	–	–
2003	–	0.00	–	0.00	0.00	–	–	4.00	0.00	–	–
2004	0.00	0.00	0.00	0.00	0.00	–	–	–	0.00	0.00	–
2005	–	0.00	–	–	0.00	5.80	5.00	–	0.00	–	–
2006	–	0.00	–	–	0.00	–	–	–	0.00	–	–
2007	–	0.00	–	–	0.00	4.80	–	–	–	0.00	–
2008	–	0.00	0.00	–	0.00	–	0.00	5.00	0.00	0.00	0.00
2009	–	0.00	–	–	0.00	–	0.00	–	0.00	–	–
2010	–	0.00	–	–	0.00	4.0	–	9.00	–	–	–
2011	–	14.00	0.00	–	0.00	2.83	–	0.00	–	–	–
2012	0.00	0.00	–	–	3.25	5.0	–	–	6.00	–	–
2013	–	0.00	–	–	0.00	4.64	–	0.00	13.50	–	–
2014	–	0.00	–	–	0.00	4.25	–	0.00	7.00	–	–
2015	–	0.00	–	–	0.00	5.50	–	6.00	–	–	–
2016	0.00	0.00	–	6.67	0.00	4.00	–	0.00	–	–	–
2017	–	0.00	–	0.00	0.00	7.00	–	3.75	–	–	–
2018	–	0.00	–	0.00	0.00	4.50	–	7.00	–	–	–
2019	–	0.00	7.00	–	–	2.75	0.00	8.00	6.50	–	–
Mean fecundity for all seasons (X ± SD)		14.00	7.00	6.67	3.25	4.59 ± 1.19	5.00 ±	6.11 ± 2.00	8.25 ± 3.52		

Dash - sampling was not carried out.

The study of the fecundity of gamogenetic females of *P. pediculus* showed that females with resting eggs are found in the population from mid-July to late September. Per females, 1-3 eggs were observed. The average fertility was low (Table 21). The diameter of resting eggs varied in the range of 0.13–0.21 mm.

Table 21. Fecundity of gamogenetic females of *P. pediculus* in Krivoe Lake (the White Lake basin) in mid of May – the end of October of 2002-2019 years

Year/	Mean body length in sampling dates										
Days and months	15.05	31.05	15.06	30.06	15.07	31.07	15.08	31.08	15.09	30.09	15.10
2002	–	0.00	0.00	0.00	0.00	1.00	1.00	0.00	1.00	–	–
2003	–	0.00	–	0.00	0.00	1.00	0.00	1.00	2.00	–	–
2004	0.00	0.00	0.00	0.00	0.00	–	–	0.00	2.10	–	–
2005	–	0.00	–	–	0.00	–	–	0.00	–	–	–
2006	–	0.00	–	–	0.00	–	–	0.00	–	–	–
2007	–	0.00	–	–	0.00	–	–	–	0.00	2.33	–
2008	–	0.00	0.00	–	0.00	0.00	0.00	0.00	–	0.00	–
2009	–	0.00	–	–	0.00	0.00	–	0.00	–	–	–
2010	–	0.00	–	–	0.00	–	0.00	2.00	–	–	0.00
2011	–	0.00	0.00	–	0.00	1.15	0.00	2.00	–	–	
2012	0.00	0.00	–	–	1.67	2.00	–	3.00	–	–	
2013	–	0.00	–	–	0.00	–	0.00	–	2.00	–	
2014	–	0.00	–	–	0.00	1.00	0.00	–	2.00	–	
2015	–	0.00	–	–	0.00	–	0.00	1.00	–	–	
2016	0.00	0.00	–	–	0.00	1.00	0.00	–	–	–	
2017	–	0.00	–	0.00	0.00	–	0.00	2.00	–	–	
2018	–	0.00	–	0.00	0.00	–	0.00	–	–	–	
2019	–	0.00	0.00	–	–	0.00	0.00	2.00	2.00	–	
Mean fecundity for all seasons (X ± SD)					1.67	1.19 ± 0.40	1.00	1.86 ± 0.69	1.85 ± 0.42	2.33	

Dash - sampling was not carried out.

CONCLUSION

The study of the geographical distribution and functions of the biology of freshwater Onychopoda in the Northern and Arctic regions of the European part of Russia has been carried out. The geographic distribution of *Bythotrephes* and *P. pediculus* species was characterized. A number of differences have been established between populations from different biolimnological areas.

The geographical distribution data analysis has showed that the main part of localities of freshwater Onychopoda in the Northern and Arctic regions of the European part of Russia showed were situated on the Karelo-Kola and Baltic limnobiological regions. In the Severnaya Dvina and Pechora regions, the data on the distribution of Onychopoda are quite scarce. On Vaygach Island and Novaya Zemlya Archipelago Onychopoda were not found.

The presence of freshwater species was noted both in freshwater (0.1-0.6‰) and slightly brackish water (up to 2‰) in the eastern part of the Gulf of Finland of the Baltic Sea.

When studying the body length and length of the caudal appendage of the genus *Bythotrephes*, it was found that for all studied species, significant geographic variability in body length and in the majority and relative length of the caudal appendage was noted. The maximum body length was found in the northernmost species *B. crassicaudus*. This species is characterized by maximum fertility and diameter of resting eggs.

The development time of the more northern population of *P. pediculus* from the White Sea basin is longer, whereas the density is higher. Moreover, the body length and fecundity of parthenogenetic and gamogenetic females are less than in the population from the eastern part of the Gulf of Finland. At the beginning of the growing season, young individuals predominated in the studied populations of *P. pediculus*. Gamogenetic individuals appeared after the population reached maximum numbers.

ACKNOWLEDGMENTS

The author is very grateful to S. N. Litvinchuk for help, support and fruitful discussions during this work. The author is appreciated to A. A. Maksimov, T. R. Eremina, S. M. Golubkov, M. S. Golubkov, M. I. Orlova for help in collecting of zooplankton samples. I express special gratitude to the staff of VOSVOD (Russian Society for Rescue on Waters), Primorsk Rescue Station Of the Leningrad region, to the station chief A.A. Gromov and V.V. Lagow for providing the opportunity to collect samples of zooplankton in the Katerlahti bay (on the shore of which is situated the city Primorsk, Leningrad region). I am very grateful to my colleagues from the Belomorsk Biological Station of ZIN RAS and its chief A.A. Sukhotin (Cape Kartesh, Republic of Karelia); to my colleagues from the State St. Petersburg University N.V. Maksimovich, I.A. Stogov, N.V. Polyakova, T.S. Ivanova, M.V. Ivanov, E.A. Movchan; to my colleagues from the Valdai branch of the FGBU State Hydrological Institute F.Yu. Reshetnikov. A.A. Marunich for help in collecting of zooplankton samples.

The study was supported by the Ministry of Science and Higher Education of the Russian Federation (project no. AAAA-A19-119020690091-0), RFBR 20-04-00918, RFBR 19-04-01000 and State Assignment "Research Collections of the Zoological Institute RAS".

REFERENCES

Beck C. 1883. On some new Cladocera of the England lakes, in *J. Royal Microscop.*, Soc. Ser. 2. 3: 777-785.

Berg D. J., Garton D. W. 1988. Seasonal abundance of the exotic predatory cladoceran, *Bythotrephes cederstroemi*, in western Lake Erie, *J. Great Lakes Res,* 14(4): 479-488.

Bespalaya Y., Aksenova O., Berezina N., Gofarov M., Przhiboro A., Kondakov A., Kurashov E., Litvinchuk L., Sokolova S., Spitsyn V., Shevchenko A., Tsiplenkina I., Travina O., Tomilova A. Preliminary study of the selected benthic fauna in the Arctic lakes of the Novaya Zemlya Archipelago and Vaigach Island (the Russian Arctic). *Polar Biology* (in press).

Butorina L. G. 1971. Biology and life cycle of *Polyphemus pediculis* (L.), in *Biologia i productivnost presnovodnykh organismov*, Leningrad. P. 94-105. [In Russian].

Butorina L. G. 1978. On *Polyphemus pediculis* (L). from lake Michigan, Informatsionnyi Bulleten, *Biologia Vnutrennikh Vod*, 8: 48-52. [In Russian].

Cheremisova K. A. 1960. Observations on the biology of *Bythotrephes longimanus* Leydig and *Leptodora kindti* (Focke), in Tr. Belarusian. *Research Institute of Fisheries*. Minsk. 3: 125-130. [In Russian].

Ekman S. 1904. *Die Phyllopoden, Cladoceren und freilebenden Copepoden der nordschwedischen Hochgebirge, Zool* [*The phyllopods, cladoceras and free-living copepods of the northern Swedish high mountains, Zool*]. Jahrbuchern, Jena, 12(1): 1-170.

Filimonova Z. I. 1965. Lower crustaceans of plankton of the lakes of Karelia, in *Fauna of the lakes of Karelia.* Invertebrates. Moscow, Leningrad: Nauka, P. 111–146. [In Russian].

Flössner D. 1972. Uber familie Polyphemoidea (Onychopoda) [Uber familie Polyphemoidea (Onychopoda)], in *Kiemen- und Blattfusser, Brachiopoda Fischlause*, Branchiura, Jena, P. 386–407.

Garton D. W., Berg D. J. 1990. Occurence of *Bythotrephes cederstroemi* (Schoedler, 1877) in Lake Superior, with evidence of demographic variation within the Great Lakes, *J. Great Lakes Res.*, 16(1): 148–152.

Garton D. W., Berg D. J., Fletcher R. J. 1990. Thermal tolerances of the predatory cladocerans *Bythotrephes cederstroemi* and *Leptodora kindti*: relationship to seasonal abundance in Western Lake Erie, *Can. J. Fish. Aquat. Sci.*, 47: 731–738.

Gerd S. V. 1956. *Experience of biolimnological zoning of lakes of Karelia*. Petrozavodsk: Karelia. Research Center, Russian Academy of Sciences, P. 47–75. [In Russian].

Grigoriev S. V., Gritsevskaya G. L. 1959. *Katalog ozer Karelii* (*Catalog of lakes in Karelia*), Moscow, Leningrad: Nauka, 239 p. [In Russian].

Herzig A. 1985. Resting eggs – a significant stage in the life cycle of crustaceans *Leptodora kindti* and *Bythotrephes longimanus*, Verh. Internat. Verein. Limnol., 22: 3088–3098.

Ivanova M. B. 1975. Planktonic Crustacean. Biological productivity of North Lakes. Part. I. *Lakes Krugloe and Krivoe*, Proc. of Zoological Institute USSR, Ed. Vinberg V. V., Leningrad: Nauka, 57: 76–90. [In Russian].

Ischreyt G. 1930. Uber Korperbau und Lebensweise des *Bythotrephes longimanus* Leydig [On the body structure and way of life of Bythotrephes longimanus Leydig], *Arch. Hydrobiol.*, 21: 241–323.

Ischreyt G. 1939. Uber den *Bythotrephes* subalpiner Seen [Over the Bythotrephes subalpine lakes], *Arch. Hydrobiol.*, 34:. 105–129.

Ketelaars H. A. M., Wagenvoort A. J., Herbst R. F. 1995. Life history characteristics and distribution of *Bythotrephes longimanus* Leydig (Crustacea, Onychopoda) in the Biesbosch reservoirs, *Hydrobiologia*, 307: 239–251.

Konovalyuk E. F. 1978. Features of zooplankton in reservoirs of the middle Ob (Kargasok region), *Vopr. Biology*, Tomsk, P. 70–74. [In Russian].

Korobtsova E. V. 1975. Zooplankton. Biological productivity of the north lakes. Lakes Zelenetskoe and Akulkino. *Proc. of Zoological Institute USSR*, Ed. Vinberg V. V., Leningrad: Nauka, 57: 77–97. [In Russian].

Krokhalevskaya N. G., Krokhalevsky V. R. 1980. Pelyad feeding peculiarities in floodplain water bodies of the lower Ob. *Collection of scientific works of GosNIORKh, Fish nutrition and use of the food supply by them in different types of water bodies*. Leningrad, P. 79–87. [In Russian].

Kulikova T. P. 2012. *Zooplankton in Waters of the Northern Lake Ladoga Drainage Basin*. Ed. A. A. Lukin. Petrozavodsk: Karelia. Research Center, Russian Academy of Sciences, 192 p. [In Russian].

Kulikova T. P. 2007. *Zooplankton in water-bodies of the Lake Onego catchment*, Ed. A. A. Lukin. Petrozavodsk: Karelian Research Centre, Russian Academy of Science, 223 p. [In Russian].

Kulikova T. P. 2010. *Zooplankton in waters of the White Sea drainage basin*, Ed. A. A. Lukin. Petrozavodsk: Karelian Research Centre, Russian Academy of Science, 325 p. [In Russian].

Linko A. K. 1900. O Cladocera solovetchkogo ostrova i Belogo morja (Sur les Cladocères de la mer blanche et des iles Solowetzky) (Cladocera of Solovetskyi Icland and of the White Sea), Trav. Soc. imp. Nat. St. Petersbourg, *Sect. Zool. Physiol.*, 30(11), 19 p. [In Russian].

Lilljeborg W. 1900 (1901). *Cladocera Sueciae, Nova Acta r. Soc. scient. Upsal., Ser. 3,* (reprinted 1982 in three volumes by Almqvist & Wiksell, Stockholm, edited by W. Rodhe & D. G. Frey)., 19: 701 p.

Litvinchuk L. F., Litvinchuk S. N. 2016. Morphological diversity and widespread hybridization in the genus *Bythotrephes* Leydig, 1860 (Branchiopoda, Onychopoda, Cercopagidae). *Arch. Biol. Sci.*, 68: 67–79. https://doi.org/10.2298/ABS150505010L.

Lomaeva T. A., Svirskaya N. L. 1977. Zooplankon of the Serebryansky reservoir, *Fish of the lakes of the Kola Peninsula*, Petrozavodsk, P. 31–44. [In Russian].

Makartseva E. S. 1974. Zooplankton of lakes of various landscapes of the Kola Peninsula, *Lakes of various landscapes of the Kola Peninsula*, Part 2, North-western part of the Kola Peninsula, P. 143–79. [In Russian].

Mordukhai-Boltovskaya E. D. 1957. *Parthenogenetic reproduction of Leptodora kindtii (Focke) and Bythotrephes Leydig*, DAN SSSR, 112(6): 1133–1135. [In Russian].

Mordukhai-Boltovskoy F. D., Rivier I. K. 1987. *Predatory cladocerans (Podonidae, Polyphemidae, Cercopagidae and Leptodoridae) of the world's fauna*. Leningrad: Nauka, 182 p. [In Russian].

Mäemets A. Kh. 1966. About summer zooplankton of the Pskov-Peipsi lake, Hydrobiological research, Tallinn, (*Hydrobiology and fisheries of the Pskov-Peipsi lake*), 4: 80-96. [In Russian].

Müller P. E. 1867. Denmark Cladocera, *Schiodtes Naturhist*, Tidsskrift, Kjobenhavn, 3(5): 53–240.

Nilsson N.-A. 1979. Food and habitat of the fish community of the offshore region of Lake Vanern, Sweden, *Inst. Freshwater Res. Drottningholm Rep.*, 58: 26–139.

Olonetskaya Nauchnaya Expeditsiya: Preliminary Report about *researches 1921*, Verashagin G. Y. et al.,; *Ros. Hydrolog*. Institute, Petrograd: B. i., 1923, 73 p. [In Russian].

Palmer A., Stich H. B., Maier G. 1999. Leptodora kindtii and *Bythotrephes longimanus* in Lake Constance: dynamics, distribution and mortalities, *Vth International Symposium on Cladocera, Programme and book of abstracts*, Plön, P. 37.

Pechen G. A., Kostin V. A., Bregman Yu. E. 1970. Zooplankton production of Lake Drivyaty, *Biological productivity of the eutrophic lake*, M.: Science, P. 89–105. [In Russian].

Pejler B. 1975. On the long-term stability of zooplankton composition, *Inst. Freshwater Res.* Drottningholm Rep., 54: 107–117.

Pellose J. 1934. *Etude sur la faunedes Cladoceres et des Copepodes de la region moyenne des Alpes francaises* [*Study on the Cladoceres and Copepods fauna of the middle region of the French Alps*], Lion, P. 1–256.

Petrovskaya M. V. 1966. *Characteristics of zooplankton in lakes in the Murmansk region, Fish of the Murmansk region*, Murmansk, P. 84–91. [In Russian].

Pirozhnikov P. L. 1972. Biological resources of reservoirs (history of study, state, fishery value), Izv. GosNIORKH (*Biological and commercial resources of reservoirs*), P. 5–78. [In Russian].

Pomerantseva D. P. 1973. Seasonal dynamics of zooplankton, Nauch. tr. Novosib. ped. inst-ta, (*Fish industry of the Ubinsky lake and ways of its development*), 90: 27–34. [In Russian].

Reshetnikov Yu. S. 1980. *Ecology and taxonomy of coregonids*, Moscow: Science, 302 p. [In Russian].

Rivier I. K. 1969. Reproduction of cercopagids (Cladocera, Polyphemidae) of the Caspian Sea, Tr. IBiVV, (*Physiology of aquatic organisms and their role in the cycle of organic matter*), 19(22): 119–127. [In Russian].

Rivier I. K. 1993. The current state of zooplankton in the Rybinsk reservoir, *Tr. IBiVV*, (*Current state of the ecosystem of the Rybinsk reservoir*). 67(70): 205-233. [In Russian].

Rivier I. K. 1998. The predatory Cladocera (Onychopoda: Podonidae, Polyphemidae, Cercopagidae) and Leptodorida of the world, In: H., J., F. Dumont (ed.): *Guides to the identification of the Macroinvertebrates of the continental waters of the world*, Bachuys Publ., 13, 213 p.

Rivier I. K., Grigorovich I.A. 1999. Biology of *Bythotrephes* Leydig (Crustacea, Cladocera, Onychopoda): research results, *Hydrobiol. j.*, 35(5): 13–40.

Ryabinkina M. G., Kulikova T. P., Ryzhkov L. P. 2012. *Zooplankton in the northern Lake Ladoga catchment,* Petrozavodsk: Karelian Research Centre, Russian Academy of Science, 1: 113–125. [In Russian].

Rylov V. M. 1915. To plankton of some lakes in Vitebsk province. Fauna of Copepoda and Cladocera, *Tr. Petrogradsky Island Natural.*, 44(4): 111–162. [In Russian].

Rylov V. M., *To the Cladocera fauna of Russian Lapland*, Tr. Petrogradsky Island Natural., 1916, vol. 45, no. 4. pp. 106-135. [In Russian].

Sars G. O. 1890Oversigt af Norges Crustaceer. Forhandl, *Vidensks. Selsk.*, Christiania, 1: 1–80.

Sars G. O. 1903. On the Crustacean Fauna of Central Asia, *Annual Rep. Zool. Museum Imperator Academy of the Sciences,* 8: 157–187.

Semenchenko V. P. 1996. Analysis of population dynamics and stability of Cladocera populations in a eutrophic lake, *Ecology*, 1: 44–49. [In Russian].

Sokolova V. A. 1979. Zooplankton, *Biology of the Vilyui Reservoir*, Novosibirsk, P. 90–135. [In Russian].

Stenroos K. E. 1897. Zur Kenntniss der Crustaceen-fauna von Russisch-Karelien [To the knowledge of the crustacean fauna of Russian Karelia]. Cladocera, Calanidae, *Acta soc. fauna flora Fennica*, 15(2): 7–71.

Vekhov N. V. 1981. Some features of the biology of *Bythotrephes longimanus* of tundra lakes, *Bulletin of Zoology*, 5: 72–75. [In Russian].

Vekhov N. V. 1987. Distribution features, biology and morphological variability of *Bythotrephes longimanus* (Leydig) s. lat. in the European subarctic, *Biol. Sciences*, 2: 27–35. [In Russian].

Vekhov N. V. 1997. Lower crustaceans (Crustacea, Entomostraca) of water bodies of polar deserts and arctic tundras on the islands of the eastern part of the Barents Sea, *Vestnik zoologii*, 31(1-2): 25–32. [In Russian].

Vincent W. F. 2008. Introduction to the limnology of high-latitude lake and river ecosystems, W. F. Vincent, J. E. Hobbie, J. Laybourn-Parry, W. F. Vincent, J. Laybourn-Parry (eds.), *Polar Lakes and Rivers*, Oxford: University Press, 23 p.

Volkova L. A., Drabkova V. G., Letanskaya G. I., Makartseva E. S., Stalmanova G. A. 1975. *Limnological sketch of the Vyalozero*, Big lakes of the Kola Peninsula, Leningrad: Nauka. P. 87–99. [In Russian].

Wagler E. 1927. Der Blaufelchen des Bodensees. (*Coregonus wartmanni* Bloch.), *Internat. Rev. Hydrobiol. Hydrogr.*, 18(3/4,): 150–190.

Weismann A. 1879. *Das Thierleben im Bodensee* [*Animal life in Lake Constance*], 31 p.

Wrona, F. J., Prowse, T. D., and Reist, J. D. 2005. Freshwater ecosystems and fisheries, in *Arctic Climate Impact Assessment*, Cambridge University Press, P. 353–452.

Xu S., Hebert P. D. N., Kotov A. A., Cristescu M. E. 2009. The non-cosmopolitanism paradigm of freshwater zooplankton: insights from the global phylogeography of the predatory cladoceran *Polyphemus pediculus* (Crustacea, Onychopoda), *Mol. Ecol.*, 18: 5161–5179.

Yan N. D., Pawson T. W. 1998. Seasonal variation in the size and abundance of the invading *Bythotrephes* in Harp Lake, Ontario, Canada, *Hydrobiologia*, 361: 157–168.

Zaffagnini F. Il ciclo reproduttivo parttenogenetico di Bythotrephes longimanus Leydig [The parttenogenetic reproductive cycle of Bythotrephes longimanus Leydig], *Rev. Idrobiol.*, 1964, vol. 3, no. 2-3, pp. 97-109.

Zakhvatkin A. A. 1927. Solovetsky Islands, Brief hydrobiological Issue. Solovetsky Society of Local Investigations. *Materials*, Edition of the USLON Press Bureau, 9: 142 p. [In Russian].

Zaytseva V. L., Galanina O. V., Philippov D. A. 2017. *On the zooplankton of some mire water bodies of Pinega district*, Arkhangelsk region, Russia, Trudy of IBIW RAS, 79(82): 76–81. [In Russian].

Zmetnaya M. I., Plakueva M. V. 2018. Species diversity and structure of the zooplankton community of water currents of the subarctic on the example of the Arkhangelsk region, *Arctic: ecology and economy*, 1(29): 68–83. [In Russian].

Zozulya S. S. 1977. Features of the first generation of *Bythotrephes* that developed from latent eggs, *Biol. Int. Waters, Inf. Bul.*, 33: 3—38. [In Russian].

Zozulya S. S. 1979. *Functional morphology and behavior of Bythotrephes longimanus Leydig (Crustacea, Cladocera)*, Dis. ... k. b. n., Moscow: Moscow State University, 20 p. [In Russian].

In: Lake Water
Editors: Oleg S. Pokrovsky, Yulia Bespalaya et al.
ISBN: 978-1-53619-275-9

Chapter 9

DISTRIBUTION AND FUNCTIONAL ECOLOGY OF MALACOSTRACAN CRUSTACEANS IN RUSSIAN NORTHERN AND ARCTIC LAKES

Nadezhda Berezina[1], Natalya Kalinkina[2] and Alexey Maximov[1]

[1]Zoological Institute of the Russian Academy of Sciences, Saint Petersburg, Russian Federation
[2]Northern Water Problems Institute, Karelian Research Centre, the Russian Academy of Sciences, Petrozavodsk, Russian Federation

ABSTRACT

This chapter describes the distribution and biology of the northern lakes dominant species of amphipods and narrow-clawed crayfish, which have recently spread to the north, the features of their life cycles, metabolism and nutrition in cold habitats. The work also reveals patterns of energy metabolism, feeding of amphipods from different ecological and biographical groups under changing climate in lakes of northern latitudes. The differences in the level of oxygen consumption were significant among representatives of different biogeographic and ecological groups. Eurytherms that are evolutionary adapted to temperature variations may acclimate faster to climate changes than stenotherms and juveniles. Significance of amphipods in foodweb of nortern lakes is very high, because, due to elevated (up to 40%) lipid level they are the main food item in diet of various fish.

Keywords: amphipods, crayfish, species range, life cycle traits, temperature adaptations, metabolism, feeding, trophic webs interactions

INTRODUCTION

Amphipods, by being omnivorous animals, make up a significant proportion of the freshwater benthic biomass and often act as trophic links from primary producers to higher-order consumers in various lakes. They are known as a highly important item in the diet of many fish. Climate change has a high potential to impact the distribution and ecophysiological traits of amphipods in the northern and Arctic lakes, for which small changes in termal regime and nutrient content can lead to a noticeable responses.

An increase in the ice-free period in freezing lakes (Blenckner et al., 2007) and an increase in minimum surface water temperatures (closely related to winter air temperatures) in lakes that usually do not freeze in winter (Woolway et al., 2019) may strongly affect the length of the growing season with a subsequent increase in the productivity of plankton (Kernan et al., 2010) and benthos (Maximov et al., 2020). For example, in the last 20 years, there has been a noticeable transformation of the ecosystem of large Lake Onega (Sharov et al., 2014; Kalinkina et al., 2016) and small lakes (example: Lake Krivoe, Maximov et al., 2021), associated with the action of climatic factors. Climate change in Karelia (including its nortern part) is evidenced by data on air temperature and the amount of atmospheric precipitation. For the period from 1980s to the end 2010s, the average annual air temperature has increased by 1–2°C compared to the previous 40 years. As a result of these changes the annual amounts of precipitation and the duration of the ice-free period on Lake Onega significantly increased (Efremova et al., 2013).

Climate change can lead to a decrease in biomass or even cause the loss of cold-loving (stenothermal) species such as glacial relic crustaceans that breed in cold winter temperatures (Penk et al., 2015). The interspecific differences in life-history variables (brood development time, egg survival, reproductive success, reproductive resting stage, intermoult intervals, seasonal fecundity, reproductive effort, egg size, and growth rates) as an adaptation to variable environments (including temperature) are well-known among gammaridean amphipods (Sutcliffe, 1992; Pöckl, 2007). These variables depend on female size and the length of the intermoult period when egg formation take place. Water temperature and food conditions are considered the most important factors, but any factors directly or indirectly influencing this period may be determining the number and size of the eggs and neonates and reproductive success of populations and life cycle variables (Wilhelm & Schindler, 2000; Berezina, 2011).

In cold waters at all stages of life, a decrease in metabolism and a decrease in energy for growth lead to delays and a decrease in other functions, the onset of late maturity, a decrease in fertility and offspring, as well as an increase in development and life expectancy due to extended periods of growth at a slow pace of development (Portner, 2006). A number of unclear questions remain for aquatic animals (including amphipods) of high latitudes, living at low temperatures, such as interspecies difference, ontogenetic features, the influence of food quality and hunting on the level of metabolism, since all these characteristics are adaptation to life in certain conditions.

Aim of this review to describe the features of the distribution and biology of the northern dominant species of amphipods and narrow-clawed crayfish, which have recently spread to the north, the features of their life cycles, metabolism and nutrition in cold habitats. The work also revealed patterns of energy metabolism, feeding in northern latitudes amphipods, belonging to different ecological and biographical groups, under chanding climate

Material and Methods

The biological traits of crustaceans in small lakes of the Karelia, Murmansk and Archangelsk region, the Novaya Zemlya Archipelago and Vaigach Island were studied. Also, we based on examples from two large lakes such as Onego and Ladoga. The study lakes of the Russian European North and partly of the North-West (without the Kaliningrad region) are located within the territory of the Lapland, Dvina and Baltic provinces of the Palearctic.

The Lapland province (subarctic zone) is a sparsely populated region, characterized by harsh climatic conditions and low productivity of lakes, which are mainly oligotrophic and dystrophic. Several model lakes are located in North Karelia, near the border of the Arctic Circle. In the east, Karelia borders on the Arkhangelsk region (Dvina province), in the south it borders the Leningrad region (Baltic province). The Baltic province covers a large territory of large lakes, such as Onega, Ladoga, Vygoozero. The estuaries of the Baltic Sea rivers (Neva, Narva) and the brackish-water Gulf of Finland were not considered in details and taken here only as a modern center of species introduction and formation. Finally, several model water bodies were considered on the Vaigach and Novaya Zemlya Islands. Studied lakes differ in their genesis, nutrient type, and chemical composition. Most inland lakes are relict and thermokarst, lakes situated along the seacoast are lagoons and glacial lakes.

The experimental studies of the metabolic activity, feeding rates and species interactions of amphipods were carried out taking into account the modern methodology of setting up experiments. The method of stable carbon and nitrogen isotopes was applied to study of the food web interactions.

Results

Distribution and Life Cycle of Species

There are several groups of malocostracan crustaceans which colonized the fresh waters of Arctic and sub-Arctic regions. First of them, this is a group of the Arctic immigrants present species of Arctic origin (*Monoporeia affinis* (Lindström, 1855), *Gammaracanthus loricatus* Sabine, 1821 and *Gammaracanthus lacustris* Sars 1867, *Saduria entomon* (Linnaeus, 1758), *Mysis relicta* Lovén, 1862 species complex), relics of late Cenozoic marine transgressions that migrated from Arctic Ocean through marine or brackish-water refugees and adapted to freshwater habitats (Takhteev et al., 2015). The group of postglacial migrants from Baikal Lake region and inland waters of Eurasia include *Gammarus lacustris* Sars, 1863 and other species of this genus, *Eulimnogammarus spp., Pallasiopsis quadrispinosa* (Sars, 1867).

During the last Ice Age species, the Arctic immigrants, together with *P. quadrispinosa* moved outside of the Arctic zone using system of ice-marginal lakes and colonized the Baltic Sea and continental lakes of Northern Europe, where they are known as "glacial relicts." Since ocean disturbances on land did not occur during glacial periods, but during interglacial periods due to increased water inflow to the ocean and thermal expansion of water, this term is usually taken in quotation marks.

The alien malacostracan species (the amphipod *Gmelinoides fasciatus* (Stebbing, 1899) and the decapods, Chinese mitten crab *Eriocheir sinensis* H. Milne Edwards, 1853 and narrow-clawed crayfish *Astacus leptodactylus* Eschscholtz, 1823) or *Pontastacus leptodactylus* (according to Crandall & De Grave, 2017) have arrived in the northern areas in recent decades through the human activity using direct and indirect vectors for range extension (Berezina et al., 2011). In last cases, the climate warming facilitated the adaptation process of these species.

Gammarus lacustris belongs to the Holarctic group of species and is widely distributed in the Northern Hemisphere on both continents (Vainio & Vainola, 2003; Takhteev et al., 2015). This species inhabits a variety of cold water habitats, including lakes, tundra, ponds, streams, swamps, and others. *G. lacustris* was found in seven of lakes on the Vaigach Island and the Novaya Zemlya Archipelago (Bespalaya et al., in press).

Several studies support that a more prolonged life cycle has been observed in high-elevation and cold-water living *G. lacustris* populations (Bjerknes, 1974; Østbye et., 2018). Besides, the length distribution of *G. lacustris* individuals is associated with the length of the life-cycle where larger animals are found in populations with prolonged life-cycles. For example, *G. lacustris* from lakes on Vaigach Island were found as very big individuals (body length achieved 27 mm). The maximal body length described for *G. lacustris* from mountain Lake Abant (Turkey) was 25 mm (Karaman & Pinkster, 1977). Fecundity (clutch size) and reproductive success may differ in diverse environments due to combined influence of temperature, food and even fish predation (Berezina, 2011). In a study along an elevated gradient, Wilhelm & Schindler (2000) found that *G. lacustris* females at higher altitudes produced larger and smaller eggs, the larger eggs had longer incubation time, and that young from larger eggs were larger.

The life span of *G. lacustris* in small subarctic lakes is about one year. Reproduction begins when the ice melts, but the number of generations and the time of embryogenesis and maturation depend on environmental factors. During June-October, the bulk of its population is concentrated in shallow-water areas 0.2–1.5 m of depth, where there are favorable conditions for feeding, growth, and reproduction. The majority of *G. lacustris* females released juveniles in early July-August, and reproduction stops completely. By August, the "old" individuals die off, and the basis of the population (~ 70% of its total number) was made up of individuals of the new generation with small body length. From late autumn and all winter when the ice covered the coastal area (up to 70 cm thick), the amphipod population located in deeper areas of the lake (1.5–5 m) with a water temperature of 2–4°C. By the end of ice-off the next year, amphipods of the new generation reach definitive sizes (8–14 mm).

Monoporeia affinis belongs to a circumpolar species that inhabits numerous freshwater lakes in the Arctic and sub-Arctic zones. This pontoporeid amphipod was recorded in Svyatoe Lake and Krugloe Lakes (the Novaya Zemlya Archipelago). It also occurs in number of deep-water boreal lakes in the Eastern and Northern Europe and in the Baltic Sea and where is considered as "glacial relict" (Segerstrale, 1957; Spikkeland et al., 2016). The size range of the adult *M. affinis* (7-10 mm) which was recorded in the northern and arctic lakes is typical for freshwater populations (Berezina et al., 2018). The maximal length of *M. affinis* known from marine habitats (near the Sakhalin Island) was 15 mm (Demchenko, 2010).

The deepest small lakes in Karelia are located in its northern part, forming a clearly defined zone of northwestern strike. It has been established that the axis along which the deepest lakes are located coincides with the Central Karelian fault zone. Representatives of the relict fauna (including *M. affinis* and *Gammaracanthus lacustris*) inhabit the bottom of the deep-water lakes, which penetrated into lakes of Karelia from the glacial lakes about 10,000–12,000 years ago. The origin of relict arctic crustaceans determined the optima of their habitat: low temperatures and not very low concentration of salts in the water. For example, abundances of *M. affinis* in different lakes vary noticeably (Table 1). The highest numbers and biomass were observed in the lakes Putkozero, Pertozero and Munozero (1–5 thousand ind./m), where the water salinity reaches 74.4–97.8 mg/L. In other low-salt lakes (19.1–46.7 mg/L), the abundance of *M. affinis* was sizably lower (22–703 ind./m^2).

In contrast to nortern (subarctic) lakes, in Arctic lakes up to 90% of species richness and 70% of benthic biomass is concentrated in the littoral zone, from the lower boundary to the 11 m, with maximum at 3-5 m (Zadelenov et al., 2017). The benthos becomes poorer with depth, and only a complex of eurythermal forms with a biomass aroung 1 g/m^2 is preserved in the lower profundal. The fact is that a too low temperature of fresh waters in the Arctic usually suppresses the development of aquatic flora and fauna, and winter freezing has the most severe effect on lake inhabitants (Greze, 1947).

Table 1. Maximum depth, mean abundance and biomass of the relict amphipod *M. affinis* in the northern and Arctic lakes

Lake (Location)	Depth max, m	Abundance ind./m^2	Biomass, g/ m^2
Pyaozero (Karelia)	49	703	0.93
Putkozerov (Karelia)	42	5240	7.87
Maslozero (Karelia)	75	559	0.83
Pyhäjärvi (Karelia)	26	33	0.09
Segozero (Karelia)	103	22	0.07
Ladmozero (Karelia)	52	118	0.22
Pertozero (Karelia)	40	1082	-
Munozero (Karelia)	50	1085	3.02
Krivoe (Karelia)	32	415	1.2
Onego (Karelia)	120	830 (max 2000)	1.1 (max 7.0)
Ladoga* (2017, depths 6-125 м)	230	20-880	0.03-1.94
Taymyr (Taymyr)**	26	-	3.72
Lower Yenisei River ***	14	3800	5.0
Lama (Taymyr)	254	680	0.44
Melkoe (Taymyr)	25	256	0.38
Pyasino (Taymyr)	24	1293	2.6
Svyatoe (Novaya Zemlya)	5	80	0.06
Krugloe (Novaya Zemlya)	4	20	0.03

According to Greze, 1957*; Kurashov et al., 2018**, Andrianova et al., 2018*** and own data.

Monoporeia affinis differs in many ways from many other species by winter breeding, which proves the arctic origin of this species. In the small northern lakes and Baltic Sea, this species is monocyclic, reaching sexual maturity by autumn in the second year of life. In Arctic lakes, its life span can last from 1 to 4 years. The population of *M. affinis* released 1.5-mm juveniles in late March or early April. By autumn, this generation reaches 5 mm and practically does not grow in winter, resuming growth in the spring of next year. By the end of autumn, sexually mature adult crustaceans (6–8 mm) begin to reproduce. Males in the lake were recorded already from the end of October to the end of December; after mating with females, they die. In December, the females lay eggs and carry them throughout the winter in the marsupium (brood pouch), die off in the spring after the juveniles hatching.

Gammaracanthus loricatus (which was described as subspecies "aestuariorum" by Lomakina in Gurjanova (1951) is a representative of arctic relict amphipods; it is found in river mouths and coastal lakes of the subarctic region. The original marine form *G. loricatus* is considered a high-arctic species and is found off the coast of Spitsbergen, Novaya Zemlya and Siberia. The differences between the forms are in the size and structure of the rostrum and epimers of the abdominal segments. Adults of *G. loricatus* are 45–58 mm of body length, and in nortern lakes they are much smaller (maximum as 40 mm). This species inhabits deepest part of lakes with cold waters and high oxygen saturation.

In all known cases, information on the number of lake populations of the species is absent, or the rarity of its finds is indicated, which is most likely due to the difficulty of catching them using mobile active fishing gear (bottom grabs). For example, in the Ob Bay, representatives of the genus *Gammaracanthus* were found at a depth of 4.5 to 13 m, but, as a rule, they are scarce in bottom grab samples (13-40 ind./m^2), and over several decades they were found only in 7 samples of zoobenthos from three regions of the bay (Stepanova et al., 2011).

In fact, lacustrine representatives of the genus *Gammaracanthus* are not small in number, they are well caught in traps, suspended at the bottom, and trawls. In the subarctic lakes of Karelia, the catch per trap 50 cm in diameter was 10-60 ind. per day at depths from 20 to 30 m. Estimated by Petersen mark method, the abundance of the amphipod *G. loricatus* (22-40 mm body length and 0.187-0.878 g (wet weight) of body) in summer varied in the ranges of 10-50 ind./ m^2. For example, number of returned marked amphipods was 3 of 169 ind. of caught (at area 400 m^2) after release of 200 marked ones, therefore, abundance of *G. loricatus* was evaluated approximately 26 ind./m^2. In October, almost 100% of females with length at least 25 mm (i.e., females at least 1 - year - old) had a marsupium with 20–200 eggs or embryos.

Pallasiopsis quadrispinosa is a common species in nortern and arctic lakes. This amphipod is a glacial freshwater relict in lakes of the European part of Eurasia (Hill et al., 1990). The penetration of *P. quadrispinosa* to the south of the Baltic Sea, the amphipod dwells only in lakes that are located along areas of the last Valdai glaciation (Segestrale, 1957). In northern small lakes and large Lakes Onego *P. quadrispinosa* inhabits deepwater areas (Kalinkina, 2015). In 2017, Lake Ladoga its abundance was 20-120 ind./m^2 at stations with depths 15-38 m (Kurashov et al., 2018). The adult amphipods prefer temperature less than 10°C, living deeper in the summer. Their life cycle has been reported from 1 to 2 years. Reproduction starts when the temperature decreases to 8°C (in October), lasting to next summer (Hill, 1988).

Other the past decades the *P. quadrispinosa* range has notably decreased, in particular, that was documented in lakes of Poland, Lithuania and Germany (Arbačiauskas & Kalytytė, 2010). The disappearance of this and other relict species in Europe may be stimulated by climate warming and lake eutrophication. The other possible reason led to the amphipod *P. quadrispinosa* to significantly reduce its populations in many northern lakes is anthropogenic pollution together with fluctuation of hydrological regime. Example is from the Vygozero Reservoir (created during the construction of the White Sea-Baltic Canal in 1933 by damming up the Nizhny Vyg River and Lake Vygozera). In the 1950s-1960s, there were common three species of amphipods: the *G. lacustris* (*Rivulogammarus scandinavicus* as synonym) and “glacial relics” *P. quadrispinosa, M. affinis* and *Mysis relicta*, and by the 2000s they practically disappeared (Berezina et al., 2013). These crustaceans belong to the psychrophilic oxyphilic species that need high oxygen saturation of water and a low temperature of 6-8°C for successful reproducing. Anthropogenic pollution by Pulp mill wastewater in the northern regions and winter oxygen deficiency (<40%) are the most probable reasons for disappearance of all three species.

Earlier, a representative of *Pallasea* from lakes of the Novaya Zemlya Archipelago was recorded as new species *Pallasea laevis* (Ekman, 1923). According to WoRMS database (http://www.marinespecies.org/) *P. laevis* is an unaccepted name and a synonym of *Pallaseopsis quadrispinosa*. No any representatives of family Pallaseidae were found during recent field survey at Lakes Svvaytoe and Krugloe on the Novaya Zemlya Archipelago, so we have no data to clarify the present status of this species at Arctic lakes. However, recent collections of pallaseid by L. N. Stepanov (2019) in lakes Teunto and Yunto of Yamalo-Nenets region (close to Novaya Zemlya) showed a presense of rich populations of another pallaseid of Baikalian origin such as *Pentagonurus viridis* (Garjajeff, 1901). It may result from a range extension of new element to the Arctic and need detailed studies.

Gammarus oceanicus Segerstråle, 1947 is one of the most abundant species on the coast of the northern seas and it tolerates significant desalination. The species had Palaearctic distribution in marine areas of Arctic seas and estuaries of Baltic Sea. Along the south and west coast of the White and Barents Seas, it is common in litoral areas at the algal belt of Fucaceae. This is probably the most numerous *Gammarus* species on the Arctic coast. It has recently been found further north, in Svalbard waters, probably as a result of climate warming (Węsławski et al., 2018). Sexually mature individuals are rather large, reaching 35 mm in length. *G. oceanicus* population living in a freshwater-influenced shore contained smaller specimens, had a lower proportion of the female population in breeding condition and probably produced fewer broods than a population living on a fully marine shore (Skadsheim, 1984). The breeding period of *G. oceanicus* occurs from December to May.

The three species of *Gammarus zaddachi* group (*G. zaddachi* Sexton, 1912, *G. salinus* Spooner, 1947 and *G. wilkitzkii* Birula, 1897) are basically brackish water species, closely related morphologically, zoogeographically and ecologically. The high-arctic *G. wilkitzkii* seems quite different from the other two, that are confined to a quite restricted area in northwestern Europe (estuarine areas of the Baltic Sea). This is a high-arctic species, occurring along the coasts of Siberia and northern Canada, as well as in the pack ice of the Polar Sea (Węsławski et al., 2018). Freshwater biotope was discovered by Økland et al. (2011) in the freshwater lake Eikeren in inland southern Norway, where a population of *G. zaddachi* lives at 150 m depth.

Intentional introductions of higher crustaceans (amphipods, mysids and decapods) in continental water bodies were carried out in the 1960s-1980s with the aim of enriching the fish food base (Berezina, 2007a). The process of dispersal of higher crustaceans lead to an increase in their importance in the ecosystems of inland water bodies of North-West Russia. At the same time, due to various reasons, a decrease in the number or even extinction of native crustacean species was noted. Thus, until the 1960s, amphipods *Gammarus pulex* (Linnaeus, 1758), *Gammarus lacustris* were common and abundant in the estuary of the river Neva and lakes Onego and Ladoga, and by now they have noticeably narrowed their areas. One of the reasons for the narrowing of the native species ranges is considered to be a competition for food resources, predation from the introduced species as well as the introduction of parasites and diseases for the local fauna.

Gmelinoides fasciatus is a Baikal invasive species in the Baltic Sea basin and northern lakes. Until the early 1960s, the range of this species was limited to the basins of the Siberian rivers (Angara, Barguzin, Irtysh, Lena, Pyasina, Tunguska, Selenga, Yenisei). As a result of intentional introduction to improve the food supply of fish in the 1960s and 1970s, two species of Baikal amphipods (*G. fasciatus* and *Micruropus possolskii* Sowinsky, 1915) appeared in the

Baltic Sea basin, i.e., lakes of the Karelian Isthmus (Berezina, 2007a). Several decades later, *G. fasciatus* spread widely in Karelian lakes, including Onega and Ladoga, and became an abundant component of benthic communities, representing 20 to 90% of the total abundance and biomass (Berezina et al., 2009). Another Baikal amphipod species *M. possolskii*, which was introduced together with *G. fasciatus* in 44 water bodies of the former USSR, by the end of the 2000s, was naturalized in the Upper Ob basin, in the Iriklinsky reservoir and then in Lake Ladoga (Barbashova et al., 2013).

In nortern lakes *G. fasciatus* reproduction lasted from April to September continuously with some periods of the intensive reproduction. Precopula pairs appeared already at 4–5 °C in April. The first generation of *G. fasciatus* juveniles was collected at the beginning of May. The time taken by juveniles of first generation to maturation was 50–60 days at temperatures 18–20°C. Developmental time from hatched juveniles of *G. fasciatus* to the period for initiating oogenesis related on temperature and takes about 1000–1200 degree-days (Panov & Berezina, 2002). In late October (at 3−4°C) no ovigerous females and precopula pairs were observed, the reproduction ceased.

The 19th century was the golden era of crayfish (Astacida) eating in Europe, when all cookery books recommended that everyone should consume at least 20 crayfish for health well-being. This was a primarily reason that stock of crayfish was plentiful in northern European countries (Sweden, Finland, Poland, Lithuania) and also in north-western regions of Russia. From this time several species of crayfish have been introduced in Europe to supplement the native noble crayfish *Astacus astacus* Linnaeus, 1758. The indigenous European crayfish species, the noble crayfish among them, have drastically decreased their geographical range and abundance, and this undesirable process is ongoing all over Europe (Souty-Grosset et al., 2006). As a result of these active stocks and natural spreading, at least three crayfish species (narrow-clawed crayfish *Astacus leptodactilus*, American signal crayfish *Pacifastacus leniusculus* (Dana, 1852), American spiny-cheek *Orconectes limosus* (Rafinesque, 1817)) were appeared in northern regions of Europe.

Astacus leptodactilus is alien species for a majority of European territory and, particularly, for the northern part of Russia (Holdich et al., 2002). Origin area of *A. leptodactylus* is waterbodies of south-western Europe and Asia (Harlioglu, 2004) including basins of the Caspian, Black and Azov seas. After opening of canals connected southern seas (Ponto-Azov-Caspian), the Volga River and waterbodies of the White and Baltic seas basins, this species moved to northern and eastern Europe. Also, from the second half of 20 century *A. leptodactylus* has been introduced as a valuable commercial object to waterbodies more than 14 countries of Europe, and expanding also naturally within each (Skurdal & Taugbol, 2001; Kouba et al., 2014).

In 1950-70s it has also been introduced actively within northern-western Russia for commercial purposes, particularly in small lakes of Karelia and Leningrad province (Gordeev, 1965). Because *A. leptodactilus* is highly fecund and abundant species, can grow to a very large size, up to 500g wet weight (Cukerzis, 1988; Souty-Grosset et al. 2006), it extented its range to the north replacing the noble crayfish *A. astacus* in all waterbodies if they were coexisting. Since the middle of the last century, *A. leptodactylus* has spread widely in small lakes in Karelia: the basin of the Shuya, Padas, Suna, and Lizhma rivers. Also, they were noted in lakes Onega and Ladoga (Gordeev, 1965).

The field survey of Karelian and Archangelsk region lakes in 2009-2016 confirmed the presence of *A. leptodactilys* in the benthic fauna of some lakes: Gomsel'skoe, Gabozero (River

Shuya basin), Kenozero, Lacha (River Onega basin) and lake Bereznik (Severnaya Dvina River basin). In the 19th century, it was already found in the Severnaya Dvina (Baer, 1837). In the 1930s *A. leptodactilys* was noted near Arkhangelsk, i.e., in its river mouth (Birshtejn & Vinogradov, 1934). Recently, narrow-clawed crayfish have been reported from the lakes of Bolshoi Solovetsky Island in the White Sea, there they adapted (Borovikova et al., 2016). In 2020, it was found in Lake Krugloe (Figure1), just 30 km below of the Polar cycle. Thus, the narrow-clawed crayfish turned out to be cold tolerant species. Also, this species extent its range to the east (Siberia) and was recently found in man-made lakes, in the upper reaches of the Ob' River (Popov, 2020). The ongoing climate warming may facilitate the process of intensive dispersal of this crayfish.

Figure 1. Female of crayfish *A. leptodactilys* from coastal lake Lake Krugloe (Nortern Karelia, White Sea), it was recorded in September 2020.

Metabolism

In the light of current climate changes, to which the amphipods of the Polar regions are most vulnerable, the study of their adaptive capabilities becomes an urgent task. In this regard, there is a relevant question whether high latitudes animals, adapted to live at low temperatures, will be able to maintain an optimal level of energy exchange during extreme fluctuation of factors.

The metabolic theory of ecology considered mainly basal metabolism, or metabolic rates of rest, and it was shown that the intensity of oxygen consumption for crustaceans, from tiny cyclops to huge lobsters, is proportional to their body mass to the power of 0.759 (Suschenya et al., 1986). Halcrow & Boyd (1967) showed that the oxygen consumption of *Gammarus oceanicus* during normal spontaneous activity differed from the basal rate by almost 2 times in the temperature range 5-15°C (Q_{10} 1.3 and 2.4, respectively). Many authors studying certain types of amphipods during one or another activities (moving, eating) obtained deviations of the natural values from this calculated average value of the equation.

Since the metabolism of organisms engaged in routine activities (active metabolic rates) is not only about maintaining tissue activity, its change depending on the body weight can differ significantly from changes for maintenance metabolism, i.e., basal metabolic rate, or metabolic indicators of rest (Glazier, 2005; Hudson et al., 2013; Reuman, 2013). Futher, it was shown that the rate of respiration and the metabolism intensity of poikilothermic animals of each species depends on various natural factors in the habitat (primarily on the water temperature and trophic conditions), and it is determined by the specific behavior, life stage and the way of feeding (Glazier, 2014).

Since low temperatures contribute to a slowdown in the rate of the main physiological processes, in species living in southern and temperate latitudes, and transplanted into low-temperature conditions, aerobic processes generally proceed more slowly than in polar organisms. For example, I.V. Ivleva (1981) measured the levels of energy metabolism of amphipods from different populations, and found a lower level of metabolism in thermophilic forms compared to those of polar species. Such metabolic traits were confirmed by comparing of field sub-Arctic and boreal populations of the cold-water amphipod *M. affinis* (Suschenya et al., 1986; Arakelova, 2006). This amphipod from sub-arctic lake was characterized by higher energy metabolic rate at low water temperatures than representatives of southernmore populations.

The specificity of the vital activity of Arctic amphipods lies in the fact that they live at extreme values of the temperature range (at low and even negative temperatures). Are they able to maintain an optimal metabolic rate when the temperature of the external environment changes, that is, to show regulatory plateaus in the zone of the lowest temperatures ($Q_{10} \rightarrow 1$)? The hypothesis of a compensatory increase in the rate of aerobic metabolism at low temperatures suggests that physiological processes in the body proceed at the same rate both in species from polar (Arctic and Antarctic) and species from boreal water bodies. The beginnings of this concept can be found in the works of the Danish physiologist August Krogh, who showed that the dependence of the intensity of standard metabolism on temperature obeys a general quantitative pattern, graphically expressed in the form of a smoothly increasing curve, which was called the normal Krogh curve (Krogh, 1914). In the region of low temperatures of this curve, the acceleration of exchange was several times higher than at medium and high temperatures, which meant the presence of a certain form of compensation, i.e., increasing the level of energy metabolism in response to low temperatures.

The concept of the ability of ectothermic animals to increase the metabolic rate when adapting to low temperatures was established in the 1950s and 1960s under the influence of the works of Scholander et al. (1953) and Wohlschlag (1964) and for a long time prevailed in science. It assumed higher metabolic rates in Arctic and Antarctic poikilothermic animals than would be expected from the Van't Hoff and Arrhenius law. These works became a tool in establishing the concept of increased oxygen consumption in polar marine poikilothermic organisms, commonly referred to as "metabolic adaptation to cold."

The concept of "metabolic adaptation" for polar ectotherms (invertebrates and fish) has long remained controversial (Holeton, 1974; Clarke, 1998). Since the 1980s, many researchers (Ivleva, 1981; Ikeda, 1985) have criticized the hypothesis of a compensatory increase in the metabolic rate of poikilothermic animals at low temperatures. As a result, the hypothesis of economical energy consumption in constant cold was put forward (Portner, 2006). For example, it was shown that the main direction of adaptation of fish at high latitudes to life at very low temperatures (Antarctica) is not an increase in the intensity of energy metabolism, but in a more efficient use of assimilated energy for production processes, as compared to inhabitants of lower latitudes (Karamushko, 2016).

Minimization of the base energy costs for growth in cold conditions is now considered as a single all-encompassing principle for the formation of the evolution and functioning of Antarctic marine ectotherms. One of the modern hypotheses supports this idea, suggesting that the effective regulation of metabolism in polar stenotherms at low metabolic rates occurs due to the high density of mitochondria and, possibly, an increase in the level of Arrhenius activation energy (Portner, 2006). This contrasts with the more energy-consuming model of metabolic regulation by increasing metabolism in cold-adapted ectotherms. At the biochemical level, the regulation of metabolism at low temperatures is supported by cold compensation of enzyme kinetic parameters, changes in enzymatic activities and physical properties of membranes, in particular, restructuring of membrane phospholipids (Guderley, 2004).

Crustaceans (including amphipods and decapods) are known for their high demand for oxygen, they have active respiration, which requires constant oxygen renewal. For example, the respiration rate of a mussel (0.01-0.005 mL O_2/g/h) weighing 1-5 g is an order of magnitude lower than that of shrimp of the same mass (0.12-0.13 mL O_2/g/h). Due to this feature, amphipods are not at all adapted to the deterioration of oxygen conditions and die under poor oxygen conditions. Under short-term experimental decrease in oxygen below 4 to 3 mg/L (hypoxia), some species of amphipods were capable of activating anaerobic glycolysis and other anaerobic processes (including lipolysis), which are necessary to maintain the body's energy production (Hervant et al., 1999; Timofeev et al., 2006). Howether, their adaptive biochemical mechanisms did not work at 0.5-1.2 mg O_2/L (close to anoxia), leading to rapid mass death (Timofeev et al., 2006).

The ranges of thermal tolerance of animals are associated with the characteristics of metabolism. For polar stenothermal amphipods, narrow ranges of selected temperatures have been identified, since the need for increased metabolism is greater than the ability of this organism to receive enough oxygen for energy production while the system is adapted to lower environmental loads (Pörtner et al., 2006). This requires understanding the aerobic capacity, that is, the difference between the basic and maximum oxygen consumption, which represents the energy at which the body is able to carry out vital functions: growth, movement and reproduction (Pörtner & Knust, 2007; Pörtner, 2010).

The difference in the level of standard exchange between different amphipod species is associated with adaptation to habitat conditions. It was differed significantlty in representatives of various biogeographic group (Arctic *Gammaracanthus loricatus*, Palaearctic *Gammarus oceanicus*, Holarctic *G. lacustris*) and ecological (littoral, deep-water) groups.

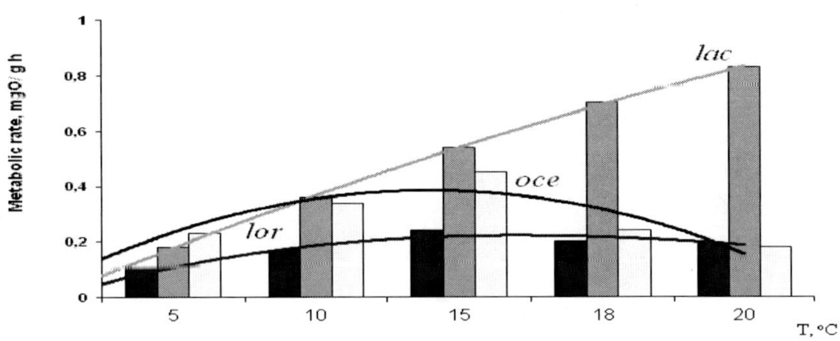

Figure 2. Metabolic rate (standard level) of amphipods: *Gammarus lacustris (lac), G.oceanicus (oce)* and *Gammaracanthus loricatus* (*lor*) from northern high latitudes at various water temperatures.

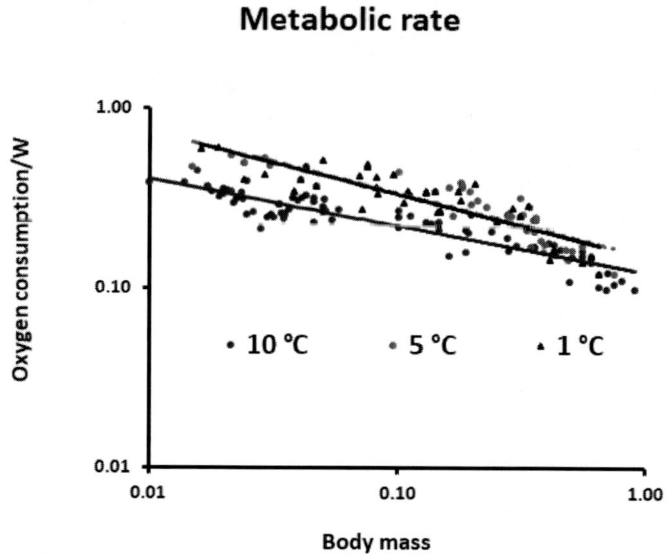

Figure 3. Intensity of energy metabolizm (Oxygen consumption (mkgO/g wet mass/ h) of amphipods (*Gammarus lacustris, G. oceanicus* and *Gammaracanthus loricatus*) from northern high latitudes at various water temperatures.

The temperature optima and maximum of energy metabolism for these species were recorded at different water temperature (Figure 2). In the temperature range of 5-15°C, the metabolic intensity depends on water temperature. Van Hoff's Q_{10} approximation is one of the most widely adopted methods for comparing the effect of temperature on the velocity of a biological process. When Q is solved from the Arrhenius equation, which is thermodynamically based on the 10 frequency of molecular collisions, Q_{10} shows to be a function of temperature and decreases with increasing temperature (Schmidt-Nielsen, 1995). Q_{10} (range of 5-15°C) for *G. loricatus* and *G. oceanicus* were 1.8-2.3 and 1.5-2.2, respectively. The energy exchange of deepwater *G. loricatus* was lower than that of actively swimming intertidal species *G. lacustris*. G. lacustris was more sensetive of them to decrese in temperatutes, its Q_{10} in the temperature range 5-20°C changed from 2.4 to 3.6, that is very similar with the Krogh's curve dynamics (Krogh, 1914). In the zone of low temperatures of 1°C, the metabolic activity of all studied amphipods remained at a level similar to that at 5°C, demonstrating a clear independence on temperature, i.e., $Q10 \rightarrow 1$. Thus, low temperatures do not lead to a change in the level of aerobic metabolism, indicating a developed adaptation of northern amphipods to low temperatures.

Ontogenetic changes in energy metabolism were pronounced when the young individuals reacted sharply to temperature changes, changing the rate of oxygen consumption. The effect of food (predatory feeding) was expressed in a 1.5–3-fold increase in the amount of oxygen consumed by them from the standard level.

Feeding and Role in Food Webs

There is a lot of research on the feeding of various amphipod species. Until recently, most amphipods of the family Gammaridae were classified as detritivorous animals (Moore, 1977), but numerous studies have shown that many members of this family belong to omnivores: predation, along with saprophagy (consumption of dead organisms) and phytophagy, also belongs to the main food strategies (MacNeil et al., 1997). Gammarids are capable to attack benthic and planktonic invertebrates, including their congeners (Dick et al., 2013).

Analysis of the results showed that the dietary spectrum changed with an increase in the size of amphipods, in particular, the frequency of occurrence of food residues of animal origin, most often represented by fragments of crustaceans (Amphipoda and Cladocera), polychaetes and oligochaetes (Annelida), and larvae of dipterans (Chironomidae), increased. Analysis of the food spectra of amphipods of various species from northern lakes showed that these amphipods, being organisms with a mixed diet, use three main food strategies: predation, saprophagy and detritophagy (Berezina, 2007b, Berezina & Maximov, 2016). An exception is *M. affinis* which feed almost exclusively on the surface layer (0 to 0.5 cm) of sediment using organic detritus as food. As a rule, the intensive growth of *M. affinis* coincides with sedimentation of high-quality fresh detritus after spring phytoplankton bloom (Uitto & Sarvala, 1991).

The stable $\delta^{13}C$ and $\delta^{15}N$ isotope ratios and fatty acid biomarkers of the food web, together with visual observations, indicate that *G. lacustris* is a carnivore that feeds mainly on zooplankton and other relict macrocrustaceans (Salonen et al., 2019). The diet of the deep-water amphipod *Gammaracanthus* spp. in study lakes also based on animal food (chironomids, ostracods) and detritus of zoogenic origin (exuvia, the remains of dipterans), while *G. lacustris* living in the coastal lakes has a wider range of food, which includes such components as detritus of various origins, algae and various small invertebrates (protozoa, nematodes, oligochaetes, chironomids, crustaceans). Trophic differentiation depends on size of amphipods and differs between juveniles and adult males and females (Berezina et al., 2017). In the diet of young of all species, it has detritus of various origins. There were no any preferences revealed in the diet of *G. lacustris* and *G. fasciatus*; apparently, these omnivorous crustaceans have very low selectivity, and the range of their food is determined by the availability of resources in the habitat.

When two species of amphipods coexist, large amphipods primarily consume juveniles or small individuals of the prey. Experiments have shown that the predatory pressure of *G. oceanicus* increased with an increase in the number of prey (*G. duebeni*). With an increase in the number of prey from 3 to 15 ind./dm^2, the ecological rations of the predator doubled, amounting to 0.02-0.05 ind./h, or 0.48-1.20 ind./day (Figure 4).

The diets of another amphipod (*G. fasciatus*), preyed upon small isopods (*Asellus aquaticus*), averaged from 0.28 to 1.07 ind./day, depending on prey number (from 5 to 20 ind./dm^2). At the maximum amount of prey, losses from predation were 40% of initial number in two weeks (Pankova & Berezina, 2007).

Losses of prey from predation usually significantly exceed the physiological needs of predator; as a result, the predator can cause significant damage to the prey population. This difference between ecological (number of killed prey) and physiological ration (mass of assimilated food) diet also increases with an increase in the number of prey (Figure 4 and Berezina, 2007b). For example, the diets of *G. loricatus* hunting on *G. lacustris* strictly correlated with the number of last one. The predation index (the amount of food killed but not eaten) increased proportionally to the increase in the number of prey, reaching a maximum (50%) at the number of prey of 25-30 ind./dm^2 (Berezina, 2007b).

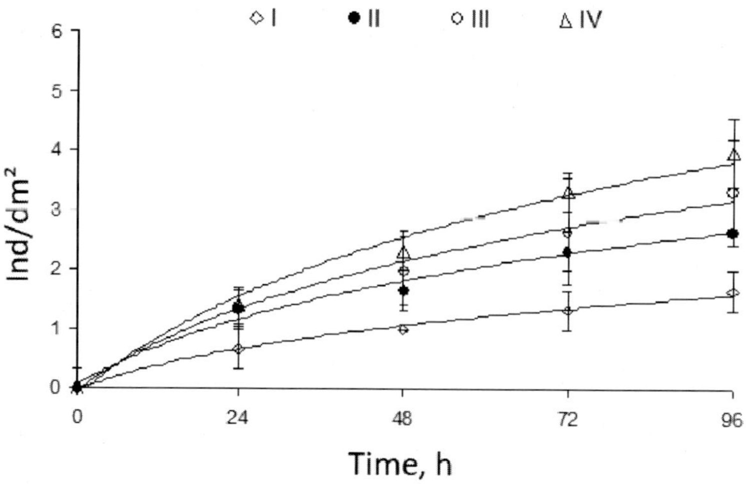

Figure 4. The ecological diet of the predator *Gammarus oceanicus* with different numbers of prey (*G. duebeni*) at exposure of 24, 48, 72 and 96 hours. The abscissa is the exposure time of the experiment; the ordinate is the average number of preys eaten. I - 3, II - 6, III - 10, IV - 15 ind./dm^2.

For many species, a decrease in the intensity of feeding at low temperatures has been shown. However, the rates of biological processes do not always increase with increasing temperature. As predicted by the metabolic theory of ecology, biological systems can actively show such reactions, at least to some extent, because they increase the rate of growth, reproduction, energy consumption and avoidance of predators (Angilletta et al., 2009). Consequently, the effect of temperature on the pace of life may not always and not be completely physically inevitable, but rather represents the result of biological adaptation.

In northern lakes, benthic invertebrates are the most important food resources in the diet of fish, including perch and vendace (Berezina et al., 2018). Amphipods (*G. lacustris, M. affinis*) are the most significant in the diet of perch in summer (> 90%) and autumn (> 40%). In the diet of vendace, these amphipods make up> 50% in June, and in autumn other objects dominate in its food: mollusks (up to 50%) and cladocerans (up to 18%). Amphipods, especially relict species (*M. affinis, Gammaracanthus spp.)* and others present valuable food items because of their high lipid level for many fish and invertebrates (Salonen et al., 2019). During autumn-winter time *Gammaracanthus spp.* are included in diet by fish. *Gammaracanthus* were the main food item for arctic char in Norway. The stomachs of *Salpinus* sampled in Lake Jensenvatn, Svalbard, were found to be filled with *Gammaracanthus* (*Hammar*, 2000), in an amount of 24-392 ind. per 1 fish, which also indirectly indicates a fairly high abundance of these amphipods in the profundal of the lakes. In Arctic communities this species is consumed by white fish, vendace and perch (Stepanova et al., 2011). In low reaches of Yenisei, *M. affinis* constitutes about 40% of food supply of commercial fish (Greze, 1957).

To sum up, at present period amphipods are very important representatives of biota in the biocenoses of northern and Arctic lakes in Russia. The leading role of relicts (*M. affinis, Gammaracanthus spp.*) and Holarctic *G. lacustris* in food webs was revealed. All species of amphipods being mainly omnivory consumers represent useful ecosystem members providing "cleansing" function in the lakes. They process a lot of detritus and enrich bottom substrates with fine organics and nutrients available to producers and other consumers. At the same time, the introduction of some new species of amphipods (and even decapods) to the north of Russia is alarming due to the possible impact on local species and food webs since the alien species are capable to prey upon and compete for food resources with native species.

ACKNOWLEDGMENTS

This study has been partially supported by the by the Ministry of Science and Higher Education of the Russian Federation (AAAA-A19-119020690091-0, **AAAA-A18-118032290035-2**) and the Russian Foundation for Basic Research (grant 19-04-01000).

REFERENCES

Andrianova A. V., Yakubaylik O. E., Shan'ko Y. V. 2018. Modern Data on the Spatial Distribution of the Baikal Amphipods in the Yenisei River and Their Visualization in the Geoinformational Web System. *Russian Journal of Biological Invasions,* 9: 299–312. https://doi.org/10.1134/S2075111718040021.

Angilletta M. J. Jr, Huey R. B., Frazier M. R. 2010 Thermodynamic effects on organismal performance: is hotter better? *Physiological and Biochemical Zoology,* 83 (2): 197−206. https://doi.org/10.1086/648567. PMID: 20001251.

Arakelova E. S. 2006. Energy metabolism and growth of *Monoporeia affinis* Lindstr. in the northern Lake Krivoe (Karelia). *Russian Journal of Ecology,* 37: 167-171.

Arbačiauskas K., Kalytytė D. 2010. Occurrence and Interannual Abundance Variation of Glacial Relict Calanoids *Limnocalanus macrurus* and *Eurytemora lacustris* in Lithuanian Lakes. *Acta Zoologica Lituanica,* 20 (1) : 61−67. https://doi.org/10.2478/v10043-010-0009-4.

Baer K. 1837. Expédition à Novaïa Zemlia et en Laponie [Expedition to Novaïa Zemlia and Lapland]. *Bulletin scientifique publié par l'Académie Impériale des Sciences de Saint-Pétersbourg,* 2: 315−319.

Barbashova M. A., Malyavin S. A., Kurashov E. A. 2013. Finding the Baikal amphipod *Micruropus possolskii* Sowinsky, 1915 (Amphipoda, Crustacea) in Lake Ladoga. *Russian Journal of Biological Invasions,* 3: 16–23. https://doi.org/10.1134/S2075111713040036.

Berezina N. A. 2007a. Invasions of alien amphipods (Crustacea: Amphipoda) in aquatic ecosystems of North-Western Russia: pathways and consequences. *Hydrobiologia,* 590: 15−29. https://doi.org/10.1007/s10750-007-0753-z.

Berezina N. A. 2007b. Food spectra and consumption rates of four amphipod species from the North-West of Russia. *Fundamental and Applied Limnology/Archiv fur Hydrobiologie,* 168 (4): 317−326. https://doi.org/10.1127/1863-9135/2007/0168-0317.

Berezina N. A. 2011. Perch-mediated shifts in reproductive variables of *Gammarus lacustris* (Amphipoda, Gammaridae) in lakes of northern Russia. *Crustaceana,* 84 (5-6): 523-542.

Berezina N., Razinkovas-Baziukas A., Tiunov A. 2017. Non indigenous amphipods and mysids in coastal food webs of eastern Baltic Sea estuaries. *Journal of the Marine Biological Association of the United Kingdom,* 97 (3): 581−590. https://doi.org/10.1017/S0025315416000643.

Berezina N. A., Gubelit Yu. I., Zhakova L. V., Timakova T. M., Petukhov V. A., Sharov A. N. 2013. Structural characteristics of biotic communities in the stony littoral of the Vygozero Reservoir. Proceedings of the Kar SC RAS. Ser. *Environmental studies,* 6. Petrozavodsk: KarRC RAS: 52–62.

Berezina N. A., Petryashev V. V., Razinkovas A., Lesutiene J. 2011. Alien malacostraca in the eastern Baltic Sea: pathways and consequences. In: Galil B. S., Clark P. F., Carlton J. T. (eds.). *the Wrong Place - Alien Marine Crustaceans: Distribution, Biology and Impacts Invading Nature - Springer Series in Invasion Ecology*, 2011, Volume 6, Part 3, 301–322.

Berezina N. A., Strelnikova A. P., Maximov A. A. 2018. The benthos as the basis of vendace, *Coregonus albula*, and perch, *Perca fluviatilis*, diets in an oligotrophic sub-arctic lake. *Polar Biology,* 41 (9): 1789–1799. https://doi.org/ 10.1007/s00300-018-2319-0.

Berezina N. A., Maximov A. A. 2016. Abundance and Food Preferences of Amphipods (Crustacea: Amphipoda) in the Eastern Gulf of Finland, Baltic Sea. *Journal of Siberian Federal University. Biology* 4 (9): 409–426. https://doi.org/10.17516/1997-1389-2016-9-4-409-426 [In Russian].

Berezina N. A., Zhakova L. V., Zaporozhets N. V., Panov V. E. 2009. Key role of the amphipod *Gmelinoides fasciatus* in reed beds of Lake Ladoga. *Boreal Environmental Research,* 14 (3): 404–414.

Bespalaya Yu., Aksenova O., Berezina N. et al. in press. *Biodiversity of freshwater fauna in the arctic lakes of the Novaya Zemlya Archipelago and Vaigach Island* (Russian Arctic). Polar Biology.

Birshtejn J. A., Vinogradov L. G. 1934. Freshwater decapods of USSR and their geographic distribution. *Zoologicheskii Zhurnal,* 13: 39–70 [in Russian].

Bjerknes V. 1974. Life cycle and reproduction of *Gammarus lacustris* G. O. Sars (Amphipoda) in a lake at Hardangervidda, western Norway. *Norwegian Journal of Zoology,* 22: 39–43.

Borovikova E. A., Alekseeva J. A., Bagirov N. E., Makhrov A. A., Popov I. Y. 2016. Genetic identification of a crayfish (*Astacus*) species at the northern edge of their distribution area (Solovetsky Islands, White Sea). *Biochemical Systematics and Ecology,* 65: 205–208.

Clarke A. 1998. Temperature and energetics: an introduction to cold ocean physiology. In: Pörtner H. O., Playle R. C. (eds.) *Cold ocean physiology.* Cambridge, Cambridge University Press: 3–30.

Cukerzis J. 1988. *Astacus astacus* in Europe. In: Holdich D. M., Lowery R. S. (eds.) *Freshwater crayfish: Biology, Management and Exploitation:* 309–340.

Demchenko N. L. 2010. Ecological aspects of the dominant amphipod *Monoporeia affinis* (Amphipoda: Pontoporeiidae) in the infralittoral zone on the northeastern coast of the Sakhalin Island (Sea of Okhotsk). *Zoologica baetica* 21: 143–149.

Dick J. T. A., Gallagher K., Avlijas S. et al. 2013. Ecological impacts of an invasive predator explained and predicted by comparative functional responses. *Biological Invasions,* 15: 837–846. https://doi.org/10.1007/s10530-012-0332-8.

Efremova T., Palshin N., Zdorovennov R. 2013. Long-term characteristics of ice phenology in Karelian lakes. *Estonian Journal of Earth Sciences,* 62 (1): 33–41. https://doi.org/10.3176/earth.2013.04.

Ekman S. 1923. *Süsswasserkrustazeen aus Nowaja Semlja* [*Reports of the Scientific Results of the Norwegian Expedition to Novaya Zemlya*], 10: 1–16.

Glazier D. S. 2005. Beyond the "3/4-power law": Variation in the intra- and interspecific scaling of metabolic rate in animals. *Biological Review* 80: 611–662. https://doi.org/10.1017/S1464793105006834.

Glazier D. S. 2014. Is metabolic rate a universal "pacemaker" for biological processes? *Biological Review,* 90 (2). https://doi.org/10.1111/brv.12115.

Goedkoop W., Johnson R. K. 2001. Factors affecting population fluctuations of the Glacial relict amphipod *Monoporeia affinis* (Lindström) in Sweden's Largest Lakes. *Ambio*, 30: 552–558. https://doi.org/10.1579/0044-7447-30.8.552.

Gordeev O. N. 1965. Malacostracans of Karelian lakes. In: *[Fauna of Karelian lakes. Invertebrates].* Nauka, Moscow, Leningrad, pp. 153-171 [In Russian].

Greze V. N. 1947. The anabiosis of the zoobenthos of the Taimyr Lake and its productivity. *Zoologicheskiy Zhurnal,* 26 (1): 3–8 [in Russian].

Greze V. N. 1957. *The main features of the hydrobiology of Lake Taimyr.* Tr. Of the All-Union Hydrobiological Society, 8: 183–218 [in Russian].

Guderley H. 2004. Metabolic responses to low temperature in fish muscle. *Biological Reviews,* 79: 409–427. https://doi.org/10.1017/S1464793103006328.

Gurjanova E. F. 1951. *Gammaridea of the Seas of the U.S.S.R. and Adjacent Waters.* Isdatelstvo Akademii Nauk SSSR, Moscow, Leningradv (in Russian).

Halcrow K., Boyd C. M. 1967. The oxygen consumption and swimming activity of the amphipod *Gammarus oceanicus* at different temperatures *Comparative Biochemistry and Physiology* 23 (1): 233–242. https://doi.org/10.1016/0010-406x(67)90491-4.

Hammar J. 2000. *Cannibals and parasites: conflicting regulators of bimodality in high latitude Arctic char, Salvelinus alpinus.* Oikos, 88: 33–47. https://doi.org/10.1034/j.1600-0706.2000.880105.x.

Harlioglu M. M. 2004. The present situation of freshwater crayfish, *Astacus leptodactylus* (Eschscholtz, 1823) in Turkey. *Aquaculture,* 230: 181–187. https://doi.org/10.1016/S0044-8486(03)00429-0.

Hervant F., Garin D., Mathieu J., Freminet A. 1999. Lactate metabolism and glucose turnover in the subterranean crustacean *Niphargus virei* during post-hypoxic recovery. *The Journal of Experimental Biology,* 202. 579–592.

Hill C. 1988. Life cycle and spatial distribution of the amphipod *Pallasea quadrispinosa* in a lake in northern Sweden. *Holarctic Ecology,* 11: 298–304.

Hill C., Furst M., Hammer J. 1990. Introduction of the amphipods *Pallasea quadrispinosa* and *Gammaracanthus lacustris* into lakes in northern Sweden. *Annales Zoologici Fennici,* 27: 241–244.

Holdich D. M. 2002. Present distribution of crayfish in Europe and some adjoining countries. In: Souty-Grosset C., Grandjean F. (eds.) *Knowledge-based management of European native crayfish. Crayfish special.* 4. Bulletin Français de la Pêche et de la Pisciculture, 367 (4): 611–650.

Holeton G. 1974. Metabolic cold adaptation of Polar fish: fact or artefact? *Physiological Zoology,* 47 (3): 137–152. http://www.jstor.org/stable/30157851.

Hudson L. N., Isaac N. J. B., Reuman D. C. 2013. The relationship between body mass and field metabolic rate among individual birds and mammals. *Journal of Animal Ecology,* 82: 1009–1020. https://doi.org/10.1111/1365-2656.12086.

Kalinkina N. M., Sidorova A. I., Polyakova T. N., Belkina N. A., Berezina N. A., Litvinova I. A. 2016. Reduction of the abundance of deep-water macrozoobenthos of Lake Onego under multifactorial impact. *The Principles of Ecology,* 2: 43–61. https://doi.org/10.15393/j1.art.2016.5182 [in Russian].

Kalinkina N. M. 2015. Distribution of relict crustaceans in deep-water lakes of Karelia in connection with the geological features of the region. *The Principles of Ecology,* 2: 38–54. https://doi.org/10.15393/j1.art.2015.4124 [in Russian].

Karaman G. S., Pinkster S. 1977. Freshwater *Gammarus* species from Europe, North Africa and adjacent regions of Asia (Crustacea-Amphipoda), 1. *Gammarus pulex* group and related species. *Bijdragen tot de Dierkunde,* 47 (1): 1–97.

Kernan M., Battarbee R. W., Moss B. (eds.) 2010. *Climate change impacts on freshwater ecosystems.* Blackwell Publishing Ltd.

Kouba A., Petrusek A., Kozák P. 2014. Continental-wide distribution of crayfish species in Europe: update and maps. *Knowledge and Management of Aquatic Ecosystems,* 413: 5. https://doi.org/10.1051/kmae/2014007.

Krogh A., 1914. On the influence of the temperature on the rate of embryonic development. *Zeitschrift fur Physik (Journal for Physics).* 16: 163–177.

Kurashov E. A., Barbashova M. A., Dudakova D. S., Kapustina L. L, Mitrukova G. G., Rusanov A. G., Aleshina D. G., Iofina I. V., Protopopova E. V., Rodionova N. V., Trifonova M. S. 2018. The ecosystem of Lake Ladoga: current state and trends of its change in the late XX-early XXI century. *Biosphere,* 10 (2): 65−121. [In Russian].

MacNeil C., Dick J. T. A., Elwood R. W. 1997. The trophic ecology of freshwater *Gammarus* spp (Crustacea: Amphipoda): problems and perspectives concerning the functional feeding group concept. *Biological Reviews,* 72: 349–364. https://doi.org/10.1111/j.1469-185X.1997.tb00017.x.

Maximov A. A., Berezina N. A., Maximova O. B. 2021. Interannual changes in benthic biomass under climate-induced variations in productivity of a small northern lake. *Fundamental and Applied Limnology/Archiv für Hydrobiologie,* 194 (3): 187–199. https://doi.org/10.1127/fal/2020/1291.

Moore J. W. 1977. The importance of algae in the diet of *Gammarus lacustris* and *Pontoporeia affinis. Canadian Journal of Zoology,* 55: 637−641. https://doi.org/10.1139/z77-083.

Økland F., Økland J., Økland K., Nordseth F., Nordby C. 2011. The unexpected discovery of a brackish water amphipod, *Gammarus zaddachi* Sexton, 1912, found isolated at 150 m depth in an inland freshwater lake in Norway. *Crustaceana,* 84 (5/6): 701−706.

Østbye K., Østbye E., Lien A. M., Lee L. R., Lauritzen S. E., Carlini D. B. 2018. Morphology and life history divergence in cave and surface populations of *Gammarus lacustris* (L.). *PLoS ONE,* 13 (10): e0205556. https://doi.org/10.1371/journal.pone.0205556.

Pankova E. S., Berezina N. A. 2007. Predation rate and size selectivity of the invasive amphipod *Gmelinoides fasciatus* preying upon native isopod *Asellus aquaticus. Acta Zoologica Lituanica.* 17 (2): 144−150. https://doi.org/10.1080/13921657.2007.10512826.

Panov V. E., Berezina N. A. 2002. Invasion history, biology and impacts of the Baikalian amphipod *Gmelinoides fasciatus* (Stebb.). In: Leppäkoski E., Olenin S., Gollasch S. (eds.), *Invasive Aquatic Species of Europe.* Kluwer Publisher, Dordrecht. P. 96-103.

Penk M., Donohue I., Récoules V., Irvine K. 2015. Elevated temperatures interact with habitat quality to undermine survival of ectotherms in climatic refugia. *Diversity and Distributions* 21: 200−210. https://doi.org/10.1111/ddi.12259.

Pöckl M., Webb B. W., Sutcliffe D. W. 2003. Life history and reproductive capacity of *Gammarus fossarum* and *G. roeseli* (Crustacea: Amphipoda) under naturally fluctuating water temperatures: a simulation study. *Freshwater Biology,* 48: 53-66. https://doi.org/10.1046/j.1365-2427.2003.00967.x.

Popov I. 2020. The expansion of freshwater crayfish range to the center of Eurasia. *Nauplius,* 28, e2020016. https://doi.org/10.1590/2358-2936e2020016.

Pörtner H. O. 2006. *Climate-dependent evolution of Antarctic ectotherms: An integrative analysis Deep Sea Research Part II: Topical Studies in Oceanography,* 53 (8–10): 1071–1104. https://doi.org/10.1016/j.dsr2.2006.02.015.

Pörtner H. O., Knust R. 2007. Climate change affects marine fishes through the oxygen limitation of thermal tolerance. *Science.* 315 (5808): 95–97. https://doi.org/10.1126/science.1135471.

Pörtner H. O. 2010. Oxygen- and capacity-limitation of thermal tolerance: a matrix for integrating climate-related stressor effects in marine ecosystems *Journal of Experimental Biology* 2010 213: 881–893. https://doi.org/ 10.1242/jeb.037523.

Salonen J. K., Hiltunen M., Figueiredo K., Paavilainen P., Sinisalo T., Strandberg U., Kankaala P., Taskinen J. 2019. Population structure, life cycle, and trophic niche of the glacial relict amphipod, *Gammaracanthus lacustris*, in a large boreal lake. *Freshwater Biology.* 64: 2176–2188. https://doi.org/ 10.1111/fwb.13404.

Schmidt-Nielsen K. (ed.) 1995. *Animal Physiology, Adaptation and Environment*, 4th ed., Cambridge University Press, Cambridge.

Scholander P. F., Flagg W., Walters V., Irving L. 1953. Climatic adaptation in arctic and tropical poikilotherms. *Physiological Zoology* 26(1):67–92.

Segerstråle S. G. 1957. On the immigration of the glacial relicts of Northern Europe, with remarks on their prehistory. *Comm. Biol. Soc. Sci. Fenn.* 16: 1–117.

Sharov A. N., Berezina N. A., Nazarova L. E., Poliakova T. N., Chekryzheva T. A. 2014. Links between biota and climate-related variables in the Baltic region using Lake Onega as an example. *Oceanologia,* 56 (2): 291–306. https://doi.org/10.5697/oc.56-2.291.

Skadsheim A. 1984. Life Cycles of *Gammarus oceanicus* and *G. salinus* (Amphipoda) in the Oslofjord, Norway. *Holarctic Ecology,* 7 (3): 262–270. https://doi.org/10.1111/j.1600-0587.1984.tb01130.x.

Skurdal J., Taugbol T., 2001. Crayfish of commercial importance – astacus. In: Holdish D. M. (ed.) Biolody of freshwater crayfish. Part 2. *Crayfish of commercial importance.* Blackwell Science. Oxford: 467–510.

Souty-Grosset C., Holdich D., Noël P. Y., Reynolds J. D., Haffner P. (eds). 2006. *Atlas of crayfish in Europe.* Museum national d'Histoire naturelle, Paris.

Spikkeland I., Kinsten B., Kjellberg G., Nilssen J. P, Väinölä R. 2016. The aquatic glacial relict fauna of Norway – an update of distribution and conservation status. *Fauna Norvegica*, 36: 51–65.

Stepanova V. B., Stepanov S. I., Vylezhinsky A. V. 2011. Long-term studies of macrozoobenthos of the Ob Bay. *Vestnik ekologii, lesovedeniya i landshaftovedeniya* [*Bulletin of ecology, forestry and landscape science*], 11: 110–117 [in Russian].

Sushchenya L. M., Semenchenko V. P., Vezhnovets V. V. 1986. *Biology and production of glacial relict crustaceans*. Nauka and technika, Minsk [in Russian].

Sutcliffe D. W. 1992. Reproduction in *Gammarus* (Crustacea, Amphipoda): basic processes. *Freshwater Forum,* 2: 102-128.

Takhteev V. V., Berezina N. A., Sidorov D. A. 2015. *Check-list of Amphipoda (Crustacea) from continental waters of Russia, with data on alien species Arthropoda Selecta*, 24 (3): 335-370. https://doi.org/10.15298/arthsel.24.3.09.

Timofeyev M. A., Kirichenko K. A., Rokhin A. V., Bedulina D. S., Chernyshova K. P., Pobezhimova T. P. 2006. Induction of anaerobic processes in Baikal endemics *Eulimnogammarus vittatus* (Dyb.) and *E. verrucosus* (Dyb.) (Amphipoda, Crustacea). *J. of Stress Physiology Biochemistry,* 2 (1): 56−61.

Uitto A., Sarvala J. 1991. Seasonal growth pattern of *Pontoporeia affinis* and *P. femorata* (Crustacea, Amphipoda) in a northern Baltic Sea archipelago. *Marine Biology,* 111: 237−246. https://doi.org/10.1007/BF01319705.

Vainio J. K., Väinölä R. 2003. Refugial races and postglacial colonization history of the freshwater amphipod *Gammarus lacustris* in Northern Europe. *Biological Journal of the Linnean Society,* 79: 523–542. https://doi:10.1046/j.1095-8312.2003.00207.x.

Węsławski J. M., Draganska-Deja K., Legeżynska J., Walczowski W. 2018. Range extension of a boreal amphipod *Gammarus oceanicus* in the warming Arctic. *Ecology and Evolution,* 8 (15): 7624−7632. https:// doi:10.1046/10.1002/ece3.4281.

Wilhelm F. M., Schindler D. W. 2000. Reproductive strategies of *Gammarus lacustris* (Crustacea: Amphipoda) along an elevation gradient. *Functional Ecology,* 14: 413–422. https://doi.org/10.1046/j.1365-2435.2000.00426.x.

Wohlschlag D. E. 2013. Respiratory Metabolism and ecological characteristics of some fishes in McMurdo Sound, Antarctica. In: B*iology of the Antarctic Seas,* M. O. Lee (Ed.). https://doi:10.1029/AR001p0033.

Woolway R., Weyhenmeyer G. A., Schmid M., Dokulil M. T., de Eyto E., Maberly S. C., May L., Merchant C. J. 2019: Substantial increase in minimum lake surface temperatures under climate change. *Climatic Change,* 155 (1): 81–94. https://doi.org/10.1007/s10584-019-02465-y.

Zadelenov V. A., Dubovskaya O. P., Bazhina L. V., Glushchenko L. A., Isaeva I. G., Kleush V. O., Semenchenko K. A., Matasov V. V., Shadrin E. N. 2017. New Data on Biota of Some Lakes in the Western Part of the Putorana Plateau. *Journal of Siberian Federal University. Biology,* 10: 87–105. https://doi.org/10.17516/1997-1389-0010 [In Russian].

In: Lake Water
Editors: Oleg S. Pokrovsky, Yulia Bespalaya et al.
ISBN: 978-1-53619-275-9

Chapter 10

Freshwater Mollusks in Lakes of the Solovetsky Islands (The White Sea)

Yulia V. Bespalaya[1,3,*], Olga V. Aksenova[1,3], Ivan N. Bolotov[1,2] and Andrey S. Aksenov[2]

[1]N. Laverov Federal Center for Integrated Arctic Research, the Ural Branch of the Russian Academy of Sciences, Arkhangelsk, Russian Federation
[2]Northern (Arctic) Federal University Named after M.V. Lomonosov, Arkhangelsk, Russian Federation
[3]Saint Petersburg State University, Saint Petersburg, Russian Federation

Abstract

Islands present a model system for addressing underlying ecological, evolutionary biology and biogeographical questions. This work presents the results of studies of 9 lakes on Bolshoy Solovetsky Island (the Solovetsky Archipelago, the White Sea, Northern Russia). The purpose of this study was to investigate how the diversity and distribution of freshwater mollusks are related to environmental factors. Twenty species of freshwater mollusks belonging to nine genera and five families were identified during this study. The fauna of mollusks in lakes of Bolshoy Solovetsky Island is taxonomically poor in comparison with the continental lakes of the taiga zone. We found that substrate type, abundance of macrophytes and hydrochemical parameters play an important role in the distribution of the mollusks in the lakes of Bolshoy Solovetsky Island.

Keywords: the Solovetsky Archipelago, freshwater mollusks, biodiversity

Introduction

The study of the species diversity of communities formed in azonal conditions – on islands, along the banks of streams, in zones of geothermal anomalies, in springs, and caves, etc., is of

* Corresponding Author's Email: jbespalaya@yandex.ru.

great interest (Macarthur & Wilson, 1967; Chernov & Matveeva, 2002). It is well known that islands are a model system for studying fundamental questions in ecology, evolutionary biology and biogeography (Graham et al., 2017). In general, the island communities are characterized by a low species diversity (Macarthur & Wilson, 1967; Chernov, 2008; Bespalaya et al., 2011). The impoverishment of the island faunas is due to a number of reasons, among which two groups are the most important. The first group consists of paleogeographic factors associated with the genesis of the islands, including glaciation, marine transgressions, and regressions (Graham et al., 2017). The second group of factors includes the current natural landscape conditions of the islands – area, range of habitats, degree of isolation from the mainland, climate, etc. (Bolotov, 2006).

We have chosen the Solovetsky Islands of the White Sea as a model area for research. The specificity of the region is related to the fact that it is located in the zone of the Pleistocene continental glaciations, including the last one – the Valdai glaciation, which determined the age and migratory character of the biota (Shvartsman & Bolotov, 2008). The formation of the terrestrial biocenoses began only at the termination of the allerod (about 10,000 to 11,000 years ago), after the degradation of the late Valday glaciation and the drainage of periglacial lakes with depths of more than 60 m (Bespalaya et al., 2009). The main goal of this work was to study the effect of environmental factors on the patterns of diversity and distribution of freshwater mollusks in the lakes of Bolshoy Solovetsky Island.

Material and Methods

Study Area

The Solovetsky Archipelago is located in the southern, most shallow part of the White Sea, at the entrance to Onega Bay. The distance of the islands of the archipelago from the mainland is about 20-30 km. The largest of the islands of the archipelago is Bolshoy Solovetsky Island, which has an area of 246.5 km^2 (Figure 1).

The islands of the archipelago have a complicated recent geological history. They are formed by boulder-pebble-sandy, glacial, water-glacial, and lacustrine-glacial sediments of the last Upper Quaternary glaciation (Kolosova, 1999). According to the results of the morphostructural zoning of the White Sea and adjacent continental territories, the Solovetsky Islands are included in the Onega morphostructural region as a separate Solovetsky morphostructural area (Avenarius, 2004). Thus, the Solovetsky Archipelago is the northwestern end of the Onega ridge clearly demarcated as being different from the structures of the Baltic shield lying to the west.

The archipelago is located within the Kandalaksha rift zone of the Kola-Belomorian rift system of the Baltic Shield on its northeastern border with the Russian Plate. The Solovetsky Islands are located near the intersections of a number of deep faults separating the largest structures of the East European Platform. The basement of the islands is Archean-Lower Proterozoic, and in the western part of the archipelago it is partially overlain only by Cenozoic moraine deposits (Shvartsman et al., 2002). The crystalline basement of the Solovetsky Islands is Archean-Lower Proterozoic. The upper part of the basement is composed of granite gneisses, amphibolites and breccias.

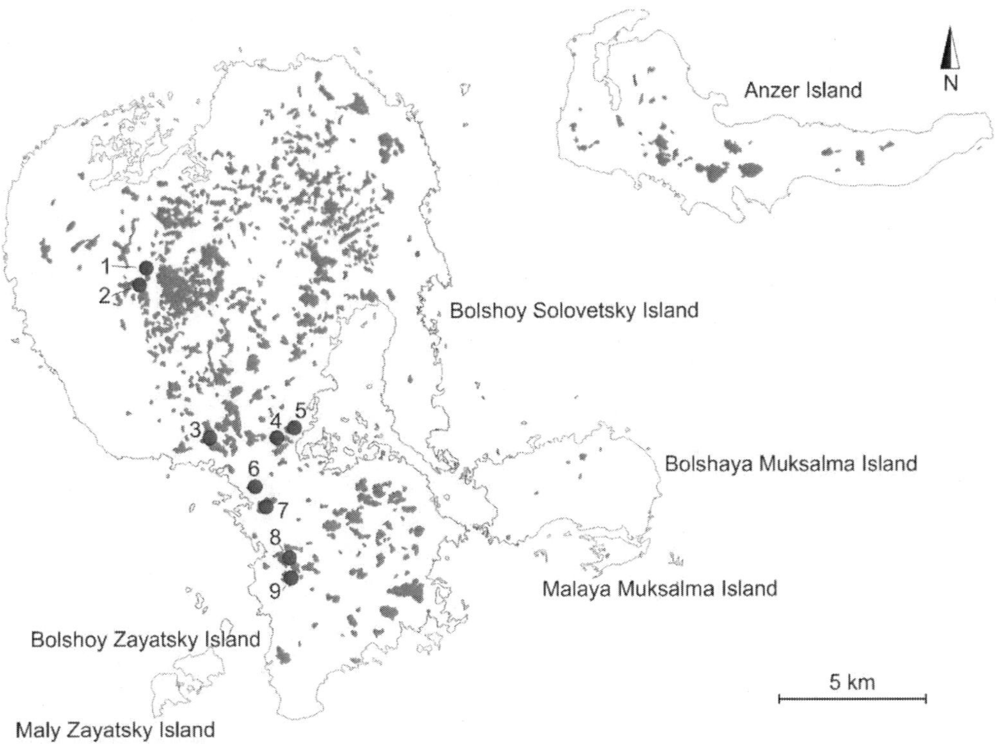

Figure 1. Overview map of the study area with designation of sampling sites on Bolshoi Solovetsky Island (red points): 1 – Karasevoye Lake; 2 – Lesnoye Lake; 3 – Nizhny Pert Lake; 4 – Krasnoarmeyskoye Lake; 5 – Varvarinskoye Lake; 6 – Bannoye Lake; 7 – Svyatoe Lake; 8 – Bolshoye Lobskoye Lake; 9 – Zapadnoye Lobskoye Lake.

The climate of the Solovetsky Archipelago is transitional from maritime to continental. Its formation is influenced by its proximity to the Arctic Circle, the specificity of atmospheric circulation and the White Sea (Gritsevskaya et al., 1972). Winters on the islands are milder and summers are cooler. The average temperature in July is 13.3 °C and in January is 9.6 °C. The coldest month is February, and the warmest month is July (Shvartsman et al., 2005).

The influence of the White Sea, as one of the determining factors, manifests itself differently in different seasons of the year. In the spring and summer months, during the period of ice melting and the period of open water, the water masses of the White Sea absorb most of the heat energy which then moves from warmer to cooler latitudes and to deeper levels and also evaporates. This explains the delay in spring on the islands in comparison with the mainland stations located in the same latitude (Boguslavsky, 1971; Gritsevskaya et al., 1972).

The Solovetsky Archipelago is located within the northern taiga biome. A feature of the Solovetsky Archipelago is the presence on its territory of natural complexes that are atypical for this natural zone. There are both coastal tundra and spruce, mixed and small-leaf forests of the herb type.

The soils on the Solovetsky Islands are podzolic, boggy and peaty, with a predominantly sandy and stony base having a low nutrient content (Kolosova, 1999).

The hydrographic network of Bolshoy Solovetsky Island is devoid of rivers and is represented by lakes connected by short channels or canals and sometimes these lakes have no runoff at all. The area of lakes on Bolshoy Solovetsky Island is large and reaches 12%. Ninety percent (90%) of all lakes of the archipelago are concentrated here (Shvartsman et al., 2005). According to the genesis of the basins, glacial and glacial-tectonic lakes prevail on the island. The significant vertical dissection of the basin promotes the development of thermal zones in lakes, which excludes the possibility of heat accumulation by silts. The maximum depth of the lakes reaches 33.5 m. The Solovetsky lakes are recharged by groundwater and to a lesser extent by atmospheric precipitation. Bottom sediments in lakes are represented by silts, boulders, sands, and clays. Silty soils prevail and occupy up to 90-98% of the entire bottom area (Gritsevskaya et al., 1972).

Figure 2. Lakes of Bolshoy Solovetsky Island: a – Karasevoye Lake; b – Lesnoye Lake; c – Krasnoarmeyskoye Lake; d – Nizhny Pert Lake; e – Varvarinskoye Lake; f – Bannoye Lake; g – Svyatoe Lake; h – Bolshoye Lobskoye Lake; i – Zapadnoye Lobskoye Lake (Photos by O.V. Aksenova).

During the research, the hydrological and hydrochemical characteristics of nine lakes on Bolshoy Solovetsky Island were studied: Varvarinskoye Lake, Svyatoe Lake, Krasnoarmeyskoye Lake, Nizhny Pert Lake, Bannoye Lake, Bolshoye Lobskoye Lake, Zapadnoye Lobskoye Lake, Lesnoye Lake, and Karasevoye Lake (Figure 1 and Figure 2).

The chemical composition of the waters of the Solovetsky Lakes is formed under conditions of low thickness of loose sediments, poorly soluble parent rocks, and highly podzolized soils, which causes low mineralization. The waters of the lakes are mainly sodium bicarbonate, with a slightly acidic or neutral reaction of the environment, and are poor in nutrients (Gritsevskaya et al., 1972). The mineral composition of lake waters reflects the peculiarities of their formation under the conditions of an island location surrounded by seawater, since the main source of nutrition for the lakes is atmospheric precipitation saturated with salts of marine origin (Gritsevskaya et al., 1972).

Field Sampling

Mollusks were collected in June and July 2007–2009 from nine lakes on Bolshoy Solovetsky Island. A total of 285 benthic samples with 13,798 mollusk specimens were collected from 16 stations. In almost all lakes a two-station sampling was undertaken except for two lakes where samples were collected from one station. From 5 to 7 replicates were sampled at each station using a rectangular hand net (dimensions 0.28 m × 0.5 m, mesh size 200 μm, sampling area of 0.14 m^2). Samples were washed using a sieve (mesh size 0.56 mm) and fixed in 96% alcohol. At each sampling station, depth, substrate type, composition of macrophytes, mosses and coarse detritus with plant remains were recorded. The composition of vegetation remains was estimated as a percentage of mosses. The dominant substrate types were recorded following Bespalaya (2015) and Bespalaya et al. (2019). Mollusks were identified according to species and abundances for each species at the 16 sampling stations and were recorded as the average number of individuals per m^2. The confidence interval was calculated using the standard error (SE).

Water samples for chemical analysis were obtained at intermediate depths using a 0.5-l polyethylene bottle that was kept cool and dark until being returned to the laboratory. Samples were analyzed for macro-ions (SO_4^{2-}, Cl^-, HCO_3^-, Mg^{2+}, K^+, Na^+, and Ca^{2+}) using instrumentation of Core Facility Center 'Arktika' of Northern (Arctic) Federal University (Arkhangelsk). The pH and water temperature were determined in the field using liquid analyzer «Expert-001» series (Russia). The conductivity was analyzed at 20 °C by using conductivity meters (HANNA Instruments, Germany).

Taxonomic identifications followed taxonomic keys (Korniushin, 1996, 2001; Glöer & Meier-Brook, 2003; Opredelitel' ..., 2016). Voucher specimens collected during this study were deposited in the collection of the Russian Museum of Biodiversity Hotspots (RMBH) of the N. Laverov Federal Center for Integrated Arctic Research, the Ural Branch of the Russian Academy of Sciences (FCIARctic).

Statistical Analysis

Mollusk communities were classified by cluster analysis using the Bray-Curtis coefficient, with the paired-group cluster analysis depicting the Bray-Curtis similarity. Before the analysis, the initial data (abundance data) were transformed as log (x + 1) using the PAST ver. 3.0 software (Hammer et al., 2001). The significance of differences was estimated based on the Kruskal-Wallis (multiple comparisons) test using the PAST ver. 3.0 software.

Principal component analysis (PCA) was carried out for the ordination of the stations based on the abiotic factors measured during the study period. Abiotic factors were transformed as log (x + 1) before analysis using PAST ver. 3.0 software.

Multifactor estimation of the relationship of freshwater mollusks with environmental parameters was performed by canonical correspondence analysis (CCA) using CANOCO ver. 4.56 software, (ter Braak & Śmilauer, 2002). Mollusk densities in each sample were transformed as log (x + 1) (ter Braak & Śmilauer, 2002). The analysis was focused on interspecies distances using Hill's scaling. The down-weighting of rare species was not corrected. The significance of the canonical axes was assessed by using the Monte Carlo permutation test.

RESULTS

Hydrochemical Characteristic of the Studied Lakes

The studied lakes of Bolshoy Solovetsky Island are characterized by significant differences in both the composition and values of mineral components (conductivity, major ions concentrations), and organic matter (COD / BOD). So, Varvarinskoe Lake is characterized by a relatively high mineralization with a predominance of sodium ions, hydrocarbonates and chlorides and a sufficiently low content of organic components. At the same time, the surface waters of the lakes Bolshoye Lobskoye and Zapadnoye Lobskoye are poor in mineral composition and have high levels of oxidizable organic matter. The main cation in all lakes was Na^+. Cl^- predominates among the anions in several lakes. However, HCO_3^- is the major anion in Karasevoye, Nizhniy Pert, Bannoye, and Varvarinskoye lakes. The results of studying the hydrochemical parameters of lakes of Bolshoy Solovetsky Island are presented in Table 1 and Table 2.

Species Diversity

We found that the molluscan fauna of the studied lakes of Bolshoy Solovetsky Island includes 20 species of mollusks belonging to 5 families of 9 genera. The Sphaeriidae family includes 10 species and the Planorbidae family is represented by 6 species of mollusks (Table 3). The families Acroloxidae, Lymnaeidae, and Valvatidae each contained one to two species (Figure 3).

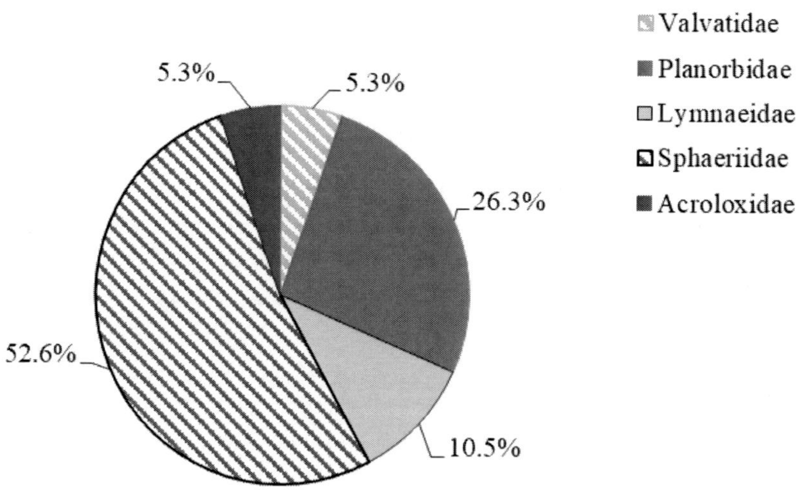

Figure 3. Taxonomic structure of molluscan fauna from lakes on Bolshoy Solovetsky Island (by the share of species for each family of the total number of registered species).

Table 1. Integral hydrochemical characteristic of the studied lakes of Bolshoy Solovetsky Island

Lake	Temperature, °C	pH	Conductivity, µsm/cm	COD	BOD_5
				mg O_2/L	
Lesnoye Lake	14.6 ± 0.5	7.4 ± 0.2	147 ± 1	75.7 ± 1.2	2.0 ± 0.1
Karasevoye Lake	14.4 ± 0.3	9.2 ± 0.1	165 ± 2	60.1 ± 1.0	2.5 ± 0.2
Nizhny Pert Lake	-	7.2 ± 0.1	-	27.7 ± 0.9	0.5 ± 0.1
Zapadnoye Lobskoe Lake	12.4 ± 0.1	6.2 ± 0.1	67 ± 1	85.4 ± 1.0	2.9 ± 0.1
Bolshoye Lobskoe Lake	10.2 ± 0.1	6.2 ± 0.1	59 ± 1	91.0 ± 1.5	1.0 ± 0.2
Varvarinskoye Lake	12.1 ± 0.1	7.7 ± 0.2	156 ± 4	32.5 ± 0.9	2.0 ± 0.3
Krasnoarmeyskoe Lake	13.2 ± 0.1	6.6 ± 0.1	56 ± 1	70.2 ± 0.9	0.9 ± 0.1
Bannoye Lake	12.3 ± 0.1	9.7 ± 0.1	97 ± 1	47.0 ± 1.3	2.0 ± 0.2
Svyatoye Lake	12.7 ± 0.1	7.2 ± 0.1	51 ± 1	43.3 ± 0.5	0.7 ± 0.1

Table 2. General ions concentration of the studied lakes of Bolshoy Solovetsky Island

Lake	Cl^-	SO_4^{2-}	HCO_3^-	K^+	Na^+	Mg^{2+}	Ca^{2+}
	ppm						
Lesnoye Lake	40.8 ± 5.0	6.42 ± 0.50	24.4 ± 0.5	1.85 ± 0.15	22.3 ± 2.1	3.63 ± 1.00	5.36 ± 0.55
Karasevoye Lake	30.7 ± 4.1	6.32 ± 0.55	48.8 ± 1.8	1.69 ± 0.10	24.5 ± 2.2	3.64 ± 1.15	8.05 ± 0.75
Nizhny Pert Lake	8.43 ± 2.2	4.75 ± 0.48	18.6 ± 0.5	1.22 ± 0.15	7.94 ± 1.5	2.03 ± 0.75	3.05 ± 0.35
Zapadnoye Lobskoe Lake	14.8 ± 1.8	2.94 ± 0.35	6.1 ± 0.5	1.02 ± 0.05	11.1 ± 1.5	1.39 ± 0.40	2.04 ± 0.30
Bolshoye Lobskoe Lake	7.28 ± 0.9	4.04 ± 0.66	6.0 ± 0.4	1.13 ± 0.15	7.3 ± 1.0	1.32 ± 0.20	2.29 ± 0.30
Varvarinskoye Lake	35.6 ± 5.1	3.69 ± 0.40	42.7 ± 1.5	1.73 ± 0.10	28.6 ± 4.3	2.71 ± 0.75	3.95 ± 0.40
Krasnoarmeyskoe Lake	9.51 ± 1.5	4.17 ± 0.35	6.1 ± 0.4	1.2 ± 0.15	9.05 ± 0.9	1.88 ± 0.80	2.91 ± 0.30
Bannoye Lake	10.2 ± 2.0	9.07 ± 0.42	30.5 ± 0.6	2.97 ± 0.25	9.6 ± 1.0	3.25 ± 1.20	7.02 ± 0.55
Svyatoye Lake	7.97 ± 1.7	5.76 ± 0.22	6.1 ± 0.7	1.39 ± 0.20	6.57 ± 1.5	1.45 ± 0.55	2.94 ± 0.35

Table 3. Species diversity, average density (ind./m^2), ± SD, and number of mollusk specimens (N) in the studied lakes of Bolshoy Solovetsky Island (the White Sea)

Species name	Bannoye Lake	Bolshoye Lobskoe Lake	Zapadnoye Lobskoe Lake	Varvarinskoe Lake	Nizhny Pert Lake	Krasnoarmeyskoe Lake	Lesnoye Lake	Svyatoye Lake	Karasevoye Lake
Sphaerium corneum (Linnaeus, 1758)	43.4±46.9 (85)	7.14±0 (1)	3.57±0 (1)	0.00	25.00±19.6 (98)	56.90±53.3 (239)	7.14±0 (1)	21.43±24.7 (9)	0.00
Euglesa casertana (Poli, 1791)	30.16±46.5 (38)	0.00	34.52±39.7 (116)	466.1±692.5 (2479)	113.52±187.6 (413)	98.3±181.2 (247)	19.39±14.7 (19)	28.6±20.6 (28)	10.39±9.8 (16)
E. globularis (Clessin in Westerlund, 1873)	11.90±4.1 (5)	30.52±34.4 (47)	10.71±4.12 (6)	45.71±39.0 (64)	23.81±16.1 (20)	14.29±0 (2)	14.29±0 (2)	40.00±25.1 (28)	7.14±0 (1)
E. hibernica (Westerlund, 1894)	15.71±9.3 (11)	17.46±11.1 (44)	14.29±10.8 (16)	0.00	19.84±25.2 (25)	217.86±260.2 (549)	7.14±0 (1)	40.00±36.6 (28)	17.86±7.1 (10)
E. lilljeborgii (Clessin in Esmark et Hoyer, 1886)	0.00	7.14±0 (1)	20.00±15.5 (14)	0.00	39.29±22.2 (22)	12.50±10.7 (7)	17.86±5.1 (5)	42.86±12.4 (18)	0.00
E. milia (Held, 1836)	0.00	7.14±0 (1)	0.00	30.52±29.9 (94)	15.00±7.9 (21)	0.00	0.00	14.29±7.1 (6)	0.00
E. nitida (Jenyns, 1832)	11.90±20.5 (15)	27.38±26.0 (92)	0.00	1084.92±1617.2 (5771)	49.23±84.4 (179)	122.14±251.2 (376)	0.00	0.00	0.00
E. obtusalis (Lamarck, 1818)	0.00	16.67±14.04 (14)	10.7±10.3 (18)	0.00	7.14±0 (2)	24.23±24.1 (95)	0.00	10.71±5.1 (3)	29.3±28.3 (41)
E. subtruncata (Malm, 1855)	43.88±38.6 (43)	11.90±5.8 (10)	18.9±20.1 (37)	75.65±92.9 (233)	81.58±100.9 (217)	35.71±31.8 (75)	0.00	47.62±55.5 (40)	21.43±19.3 (12)
E. supina (Schmidt, 1850)	0.00	0.00	0.00	0.00	0.00	0.00	0.00	497.62±86.1 (209)	0.00
Acroloxus lacustris (Linnaeus, 1758)	0.00	0.00	0.00	0.00	0.00	0.00	7.14±0 (1)	0.00	10.71±3.9 (6)
Valvata sibirica Middendorff, 1851	10.00±3.9 (7)	8.93±3.3 (10)	7.14±3.3 (3)	31.8±34.8 (116)	0.00	25.40±19.3 (32)	21.43±0 (3)	20.24±12.3 (17)	64.84±56.4 (118)
Lymnaea stagnalis (Linnaeus,1758)	8.33±2.9 (7)	0.00	0.00	0.00	7.14±0 (2)	0.00	7.14±0 (3)	0.00	23.8±37.8 (20)
Myxas glutinosa (Müller, 1774)	0.00	0.00	0.00	0.00	8.57±3.2 (6)	0.00	0.00	0.00	0.00
Bathyomphalus contortus (Linnaeus, 1758)	10.71±5.1 (3)	0.00	0.00	0.00	7.14±0 (3)	7.14±0 (1)	16.96±10.8 (19)	0.00	42.86±39.2 (78)

Species name	Bannoye Lake	Bolshoye Lobskoe Lake	Zapadnoye Lobskoe Lake	Varvarinskoe Lake	Nizhny Pert Lake	Krasnoarmeyskoe Lake	Lesnoye Lake	Svyatoye Lake	Karasevoye Lake
Armiger crista (Linnaeus, 1758)	0.00	0.00	0.00	11.90±22.6 (5)	0.00	0.00	0.00	7.14±0 (1)	28.6±22.6 (16)
Gyraulus albus (Müller, 1774)	14.29±0 (2)	0.00	0.00	21.43±15.1 (15)	25.5±22.2 (25)	17.86±15.2 (5)	7.14±0 (1)	16.7±16.5 (7)	0.00
G. acronicus (Férussac, 1807)	47.90±58.4 (114)	11.90±4.1 (5)	11.90±8.3 (5)	36.51±22.7 (138)	13.27±8.8 (26)	20.00±14.1 (42)	15.48±14.9 (26)	30.00±23.5 (42)	23.21±19.4 (26)
G. stroemi (Westerlund, 1873)	23.81±22.9 (10)	0.00	0.00	22.86±16.3 (16)	16.07±6.8 (9)	22.45±21.9 (22)	7.14±0 (1)	0.00	7.14±0 (2)
G. laevis (Alder, 1838)	43.3±72.0 (57)	21.43±0 (3)	11.90±8.2 (4)	47.83±40.8 (154)	17.35±9.0 (17)	35.20±37.7 (106)	25.42±27.7 (57)	40.82±29.7 (40)	28.57±42.4 (24)

Classification of Sampling Stations Based on Mollusk Communities

Based on cluster analysis, four classes of mollusk groupings have been identified (Figure 4). Each of them is confined to certain habitats.

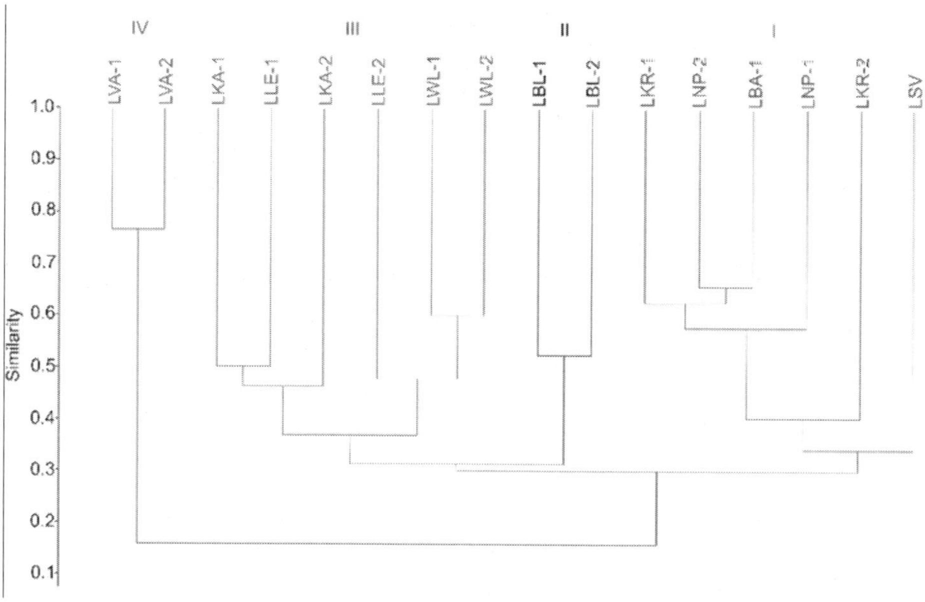

Figure 4. Dendrogram of cluster analysis of species composition and population density for mollusks from lakes of Bolshoy Solovetsky Island by stations (Bray-Curtis coefficient, paired group). Station names: LVA – Varvarinskoye Lake; LKA – Karasevoye Lake; LLE – Lesnoye Lake; LWL – Zapadnoye Lobskoye Lake; LBL – Bolshoye Lobskoye Lake; LKR – Krasnoarmeyskoye Lake; LNP – Nizhny Pert Lake; LBA – Bannoye Lake; LSV – Svyatoe Lake.

The first class includes samples of mollusks from Svyatoe, Krasnoarmeyskoye, Nizhny Pert, and Bannoye lakes. The most abundant species were *Sphaerium corneum*, *Euglesa casertana*, *E. nitida*, *E. subtruncata*, *E. supina*, *E. hibernica*, *E. lilljeborgii*, and *Gyraulus acronicus*. Mollusks inhabit silty-sandy, stony-sandy, and silty soils with plant remains, spatterdock (yellow pond-lily), water shamrock, filamentous algae, and fouling of aquatic mosses. The average density of mollusks was 52.1 ± 110.04 ind./m^2 (from 4.3 to 1171.4 ind./m^2).

Euglesa nitida, *E. globularis*, and *E. hibernica* dominate the groupings of the second class (Bolshoye Lobskoe Lake) (Figure 4). Mollusks are confined to silty-sandy and stony-sandy soils with the presence of plant remains (depths 0.2–0.8 m). The vegetation is represented by spatterdock, water shamrock, filamentous algae, and fouling of aquatic mosses. The average density of mollusks was 20.4 ± 21.14 ind./m^2 (from 7.14 to 114.3 ind./m^2).

The third class is represented by samples from Karasevoye, Lesnoye, and Zapadnoye Lobskoe lakes. *Valvata sibirica*, *Gyraulus laevis*, and *Euglesa casertana* prevailed in high numbers. The depth of the habitats here varied from 0.1 m to 0.7 m. Silty and silty-sandy soils with plant remains and abundant thickets of spatterdock, comarum, sedge, pondweed, and arrowhead predominate. The average density of mollusks was 24.2 ± 29.9 ind./m^2 (from 6.7 to 178.5 ind./m^2).

Euglesa nitida and *E. casertana* sharply predominate in numbers in groupings of class IV (Varvarinskoe Lake) (Table 3). Such groups are formed on silty-sandy and silty soils with plant remains at a depth of 0.1 to 1.0 m. The average density of mollusks was 296.4 ± 767.9 ind./m^2 (from 69.4 to 8107.1 ind./m^2) which is higher compared to the groupings of classes I – III (Kruskal-Wallis test: $H = 116.3$, $df = 3$, $P < 0.0001$). Moreover, the species richness is highest in groupings of class I (Figure 5).

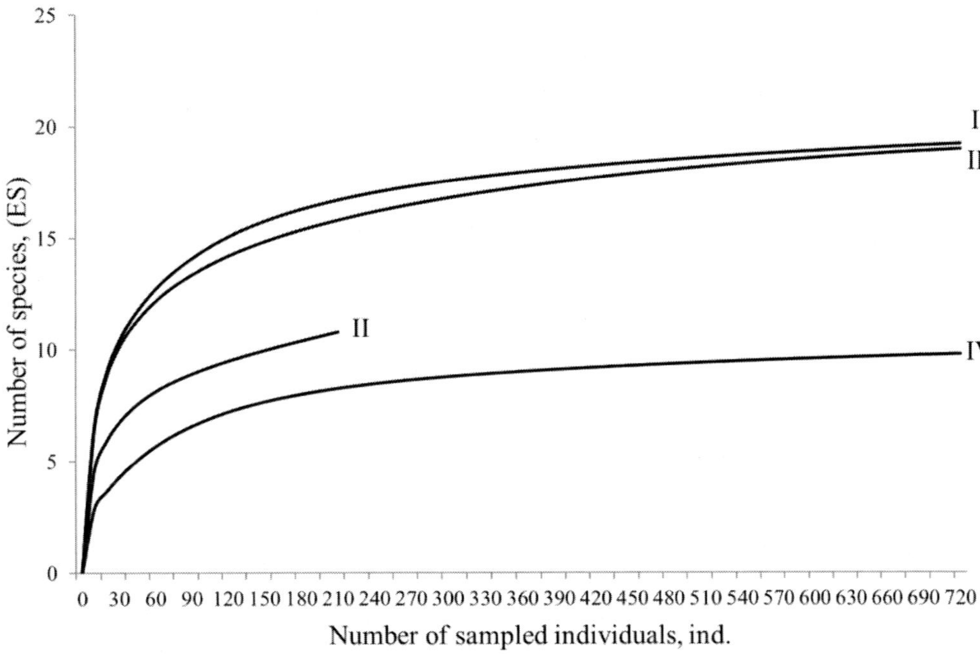

Figure 5. Species accumulation curves for mollusk fauna from lakes of Bolshoy Solovetsky Island. Curves I, II, and IV are truncated at the level of N = 720 specimens to improve the clarity of the graph.

Environmental Variables

Principal component analysis (PCA) of environmental factors revealed differences between stations based mainly on depth values, soil type, and hydrochemical parameters (Figure 6). The first axis describes 38.6% dispersion, which correlates positively with the concentration of macro-ions (Cl^-, Mg^{2+}, K^+, Na^+, and Ca^{2+}), as well as the pH value, and correlates negatively with the depth and amount of plant remains. The second axis describes 17.9% dispersion, which correlates positively with the substrate type composed mainly of sandy and stony-sandy soils, and negatively with the substrate represented by silty soil with plant remains.

Canonical correlation analysis (CCA) was performed to determine the relationship between the density of mollusks and 17 environmental variables (hydrochemical parameters, soil type, depth, presence of plant remains and composition of aquatic vegetation at sampling sites) (Figure 7). The analysis focused on interspecies distances with Hill scaling; the contribution of rare species was not corrected. The eigenvalues (lambda) are 0.167, 0.121, 0.109, and 0.072 for the first (horizontal), second (vertical), third and fourth axes, respectively. The first two axes explain 42.8% of the variance, and all canonical axes together explain 69.8% of the variance.

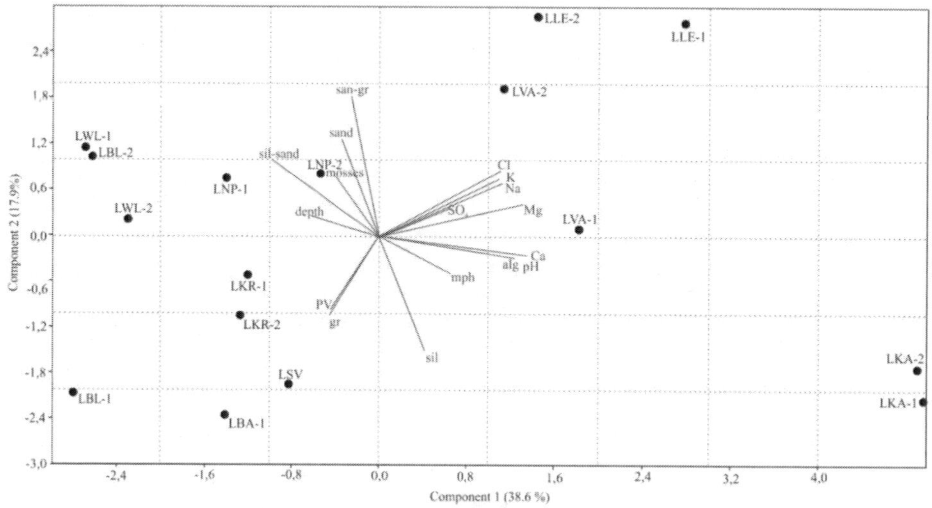

Figure 6. Distribution of sampling stations depending on environmental factors based on principal component analysis (PCA). For station designations, see legend in Figure 4. Designation of factors: sil-san – silty-sandy; san-gr – sandy-stony; sand – sandy; gr – stony; sil – silty; mph – macrophytes; mosses – moss; PV – plant remains; alg – algae; depth – depth; SO_4^{2-}, Mg^{2+}, K^+, Cl^-, Na^+, Ca^{2+} – macroions.

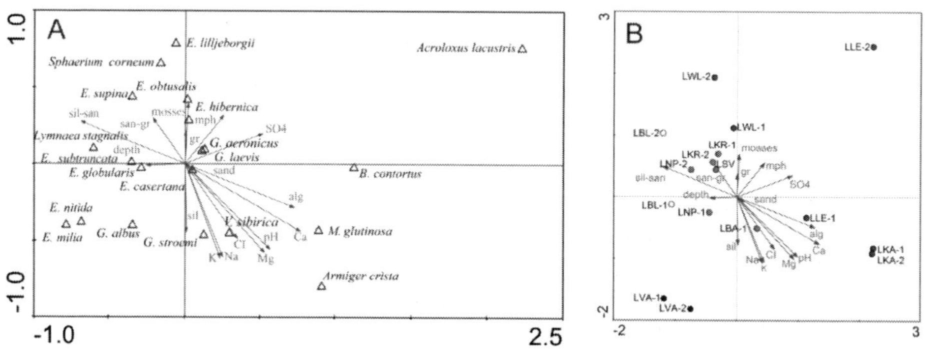

Figure 7. Ordinal diagram of the canonical analysis of relationships between the abundance of mollusks (A) and sampling stations (B) with environmental variables in lakes of Bolshoy Solovetsky Island. For station designations, see caption to Figure 4.

The canonical correlations between the abundance of freshwater mollusks and environmental variables for axes 1 and 2 are 1.0 and 0.99, respectively. The significance of the axes was determined based on the Monte Carlo test (the first axis was $F = 0.331$, $P = 0.002$, all axes were $F = 3.560$, $P = 0.002$). The strongest influence among the analyzed factors was exerted by the type of substrate and the hydrochemical parameters of water bodies. The first axis reflects the distribution of mollusks and sampling stations depending on the concentration of macro-ions, mainly SO_4^{2-}, Mg^{2+}, Ca^{2+}, pH level and the presence of algae (Figures 7A and

7B), the projection of the second axis was positively correlated with the presence of mosses and negatively with a silted type of substrate, concentration of macro-ions Mg^{2+}, Ca^{2+}, K^+, Cl^-, Na^+, and pH level. In general, the CCA results correspond to the four distinguished classes of mollusk groupings (Figure 7). Stony-sandy and silty-sandy soils with thickets of moss are the optimal habitat for bivalve mollusks *Sphaerium corneum, Euglesa supina, E. subtruncata, E. obtusalis, E. lilljeborgii, E. hibernica,* and *Lymnaea stagnalis*. *E. casertana, E. globularis, E. nitida*, and *E. milia* inhabit deeper areas of silted bottom. Thickets of aquatic vegetation and filamentous algae, with a high content of macro-ions and pH levels, are mainly inhabited by gastropods – *Acroloxus lacustris, Valvata sibirica, Bathyomphalus contortus, Armiger crista, Gyraulus albus, G. acronicus, G. stroemi, G. laevis,* and *Lymnaea stagnalis* (Figure 8).

Figure 8. Freshwater mollusks on various substrates from the lakes of Bolshoy Solovetsky Island: a – *Lymnaea stagnalis* from Karasevoye Lake; b – *Valvata sibirica* on the back of the yellow pond-lily leaf from Krasnoarmeyskoye Lake; c – *Myxas glutinosa* from Nizhny Pert Lake (Photos by O.V. Aksenova).

Mollusk Diversity and Abundance in Lakes

In general, the molluscan fauna of the studied lakes is taxonomically poor in comparison with the continental lakes of the taiga zone (Bespalaya et al., 2011; Sokolova et al., 2010). Island communities are formed under certain conditions (isolation) in comparison with the mainland; therefore, they are characterized by certain structural features (Bespalaya et al., 2011; Sokolova et al., 2010). On the one hand, this is the specificity of the island habitat which includes cenotic relationships in a limited area, and on the other, the initial biological and ecological characteristics of the species (Chernov, 2008).

The mollusk communities of lakes of Bolshoy Solovetsky Island depending on the specific of the habitat were dominated by *Valvata sibirica, Gyraulus acronicus, G. laevis, Euglesa casertana, E. globularis, E. hibernica, E. nitida,* and *E. subtruncata* (Figure 9). These species distinguish a wide Holarctic, Palearctic, and European-Siberian range. *E. casertana* is cosmopolitan species (Graf, 2013). According to some authors, *E. casertana* is the most ecologically tolerant of all sphaeriids (Guralnick, 2005; Bespalaya et al., 2020).

We found that the type of substrate, abundance of macrophytes and hydrochemical parameters all play important roles in the distribution of the mollusks in lakes. It is well known that the fauna of lakes is normally overly sensitive to physical and chemical disturbances (Raddum & Fjellheim, 2002). Many researchers have stressed the importance of water chemistry in determining the structure of freshwater mollusk communities (Moore, 1978; Dillon, 2000; Horsák & Hajek, 2003; Bespalaya, 2015; Bespalaya et al., 2019, 2020).

The average density of mollusks varied significantly between lakes (Table 3). The highest density of mollusk settlements was found in Varvarinskoye Lake where the maximum density of *Euglesa casertana* was 3,476 ind./m^2 and *E. nitida* 8,107 ind./m^2. The waters of Varvarinskoye Lake are characterized by an increased content of macro-ions, in particular Mg^{2+}, Ca^{2+}, K^+, Cl^-, Na^+, and a high level of conductivity (Table 2, Figure 7B).

Figure 9. Freshwater mollusk diversity from lakes of Bolshoy Solovetsky Island: a – *Sphaerium corneum*; b – *Euglesa globularis*; c – *E. hibernica*; d – *E. lilljeborgii*; e – *E. milia*; f – *E. nitida*; g – *E. obtusalis*; h – *E. supinum*; i – *Acroloxus lacustris*; j – *Valvata sibirica*; k – *Lymnaea stagnalis* (juv.); l – *Myxas glutinosa*; m – *Bathyomphalus contortus*; n – *Armiger crista*; o – *Gyraulus stroemi*; p – *G. laevis* (Photos by O.V. Aksenova).

Another especially important environmental factor affecting the benthic community structure is the substrate type (Moore, 1978; Beaty et al., 2006). In agreement with previous data, we found that substrate type and sediment characteristics are two of the most important factors affecting the distribution of benthic species, including mollusks (Beaty et al., 2006; Bódis et al., 2011; Bespalaya, 2015; Bespalaya et al., 2019, 2020).

According to some authors the vegetation composition is the main factor with which to explain the distributions of mollusks in freshwater ecosystems (Horsák & Hajek, 2003; Bolotov et al., 2014; Bespalaya, 2015). Thus, we found that the gastropods *Acroloxus lacustris, Valvata sibirica, Bathyomphalus contortus, Armiger crista, Gyraulus albus, G. acronicus, G. stroemi, G. laevis,* and *Lymnaea stagnalis* prefer habitats with high densities of aquatic vegetation

(Figure 7 and Figure 8). Horsák & Hajek (2003) note that the vegetation can reliably predict the composition of the mollusk community.

CONCLUSION

The fauna of mollusks in lakes of Bolshoy Solovetsky Island is taxonomically poor in comparison with the continental lakes of the taiga zone. Twenty species of freshwater mollusks belonging to 9 genera and 5 families were identified during this study. Low taxonomic diversity is due to a combination of paleogeographic and modern ecological factors. We found that substrate type, abundance of macrophytes and hydrochemical parameters play important roles in the distribution of the mollusks in the lakes of Bolshoy Solovetsky Island.

These results expand our understanding of the environmental requirements of freshwater mollusks in island lakes ecosystems. The spatial distribution of freshwater mollusk communities along environmental gradients is based on the ecological preferences of specific species. At the same time, a complex set of various environmental factors has a significant impact.

ACKNOWLEDGMENTS

The Ministry of Science and Higher Education of the Russian Federation supported the study of biodiversity of freshwater bivalves (project no. AAAA-A17-117033010132-2 and 0793-2020-0005). The Russian Science Foundation supported the study of environmental parameters (no. 19-14-00066). The Russian Foundation for Basic Research supported the study of the distribution of freshwater mollusks depending on environmental factors (no. 20-04-00361).

REFERENCES

Avenarius I. G. 2004. Morphostructure of the White Sea. *Geomorphology,* 3: 48–56 [in Russian].

Beaty S. R., Fortino K., Hershey A. E. 2006. Distribution and growth of benthic macroinvertebrates among different patch types of the littoral zones of two arctic lakes. *Freshwater Biology,* 51: 2347–2361.

Bespalaya Y. 2015. Molluscan fauna of an Arctic lake is dominated by a cosmopolitan *Pisidium* species. *Journal of Molluscan Studies,* 81: 294–298.

Bespalaya Y., Joyner-Matos J., Bolotov I., Aksenova O., Gofarov M., Sokolova S., Shevchenko A., Travina O., Zubry N., Aksenov A., Kosheleva A., Ovchinnikov D. 2019. Reproductive ecology of Pisidium casertanum (Poli, 1791) (Bivalvia: Sphaeriidae) in Arctic lakes. *Journal of Molluscan Studies,* 85: 11–23. https://doi.org/10.1093/mollus/eyy050.

Bespalaya Y. V., Aksenova O. V., Sokolova S. E., Shevchenko A. R., Tomilova A. A., Zubrii N. A. 2020. Biodiversity and distributions of freshwater mollusks in relation to chemical and physical factors in the thermokarst lakes of the Gydan Peninsula, Russia. *Hydrobiologia.* doi: 10.1007/s10750-020-04227-9.

Bespalaya Y. V., Bolotov I. N., Usacheva O. V. 2011. Structure and species diversity of topical groups of mollusks in lakes of the Solovetsky Islands and Onega Peninsula, northwestern Russia. *Russian Journal of Ecology* 42(2): 143–150. doi: 10.1134/S1067413611010048.

Bespalaya Y. V., Bolotov I. N., Zubry N. A. 2009. Topical groups of mollusks in the lakes of Bol'shoy Solovetskiy Island (Solovetskiy Archipelago, White Sea, Northwestern Russian). *Inland Water Biology,* 2(2): 177–186. doi: 10.1134/S1995082909020114.

Bódis E., Nosek J., Oertel N., Tóth B., Hornung E., Sousa R. 2011. Spatial distribution of bivalves in relation to environmental conditions (middle Danube catchment, Hungary). *Community Ecology,* 12: 210–219.

Boguslavsky G. A. 1971. *Solovetsky Islands*. Arkhangelsk: North-West. book publishing house [In Russian].

Bolotov I. N. 2006. Diurnal lepidopterans (Lepidoptera, Diurna) from the Solovetskiye Islands (Northwestern Russia, the White Sea). *Entomological Review,* 85(8): 943–949. doi: 10. 1134/S0013873806050034.

Bolotov I. N., Bespalaya J. V., Aksenova O. V., Gofarov M. Y., Sokolova, S. E. 2014. Mollusks in the Zoobenthos of Relict Lakes with abnormally high biological production in the eastern European Subarctic. *Inland Water Biology,* 7: 61–71.

Chernov Yu. I. Species Diversity and Compensatory Phenomena in Communities and Biological Systems, in *Ekologiya i biogeografiya (Ecology and Biogeography),* Moscow, 2008, 529–556 p.

Chernov Yu. I., Matveeva N. V. 2002. Landscape-Zonal Distribution of Species in the Arctic Biota, *Uspekhi Sovremennoy Biologii*, 122(1): 26–45 (in Russian).

Dillon R. T. 2000. *The ecology of freshwater molluscs*. Cambridge University Press, Cambridge.

Gloër P., Meier-Brook C. 2003. Süsswassermollusken. *Ein Bestimmungschlüssel für die Bundesrepublik Deutschland* [*A determination key for the Federal Republic of Germany*]. 13 Auflage. Deutscher Jugendbund für Naturbeobachtun, Hamburg.

Graf D. L. 2013. Patterns of freshwater bivalve global diversity and the state of phylogenetic studies on the Unionoidae, Sphaeriidae, and Cyrenidae. *American Malacological Bulletin,* 31:135–153.

Graham N. R., Gruner D. S., Lim J. Y, Gillespie R. G. 2017 *Island ecology and evolution: challenges in the Anthropocene Environmental Conservation*, 44 (4): 323–335.

Gritsevskaya G. L., Kyabileva G. K., Nikolaeva L. A., Semenova V. I. 1972. Hydrology and hydrochemistry of the Solovetsky Lakes. *Materials for the integrated study of the Solovetsky lakes*. Petrozavodsk: Karelia. P. 6–43.

Guralnick R. P. 2005. Combined molecular and morphological approaches to documenting regional biodiversity and ecological patterns in problematic taxa: a case study in the bivalve group *Cyclocalyx* (Sphaeriidae, Bivalvia) from western North America. *Zoologica Scripta* 3: 469–482.

Hammer O., Harper D. A. T., Ryan P. D. 2001. PAST: paleontological statistics software package for education and data analysis. *Palaeontologia Electronica,* 4: art. 4.

Horsák M., Hájek M. 2003. Composition and species richness of mollusc communities in relation to vegetation and water chemistry in the Western Carpathian spring fens: the poor-rich gradient. *Journal of Molluscan Studies,* 69: 349–357.

Korniushin A. V. 2001. Taxonomic revision of the genus *Sphaerium* sensu lato in the Palaearctic Region, with some notes on the North American species. *Arch Molluskenkd,* 129: 77–122.

Korniushin A. V. 1996. *Bivalve molluscs of the superfamily Pisidioideain the Palaearctic region: fauna, systematics, phylogeny.* Schmalhausen Institute of Zoology, Kiev [In Russian].

MacArthur R. H., Wilson E. O. 1967. *The Theory of Island Biogeography*, Princeton, NJ: Princeton Univ. Press,

Moore J. S. 1978. Some factors influencing the diversity and speciescomposition of benthic invertebrate communities in twenty Arctic and Subarctic Lakes. *International Review of Hydrobiology,* 63: 757–771.

Opredelitel' zooplanktona i zoobentosa presnykh vod Yevropeyskoy Rossii (*Keys to zooplankton and zoobenthos of fresh waters in European Russia*). 2016. Vol. 2. Zoobentos/Eds. V. R. Alekseyev, S. Ya. Tsalolikhin. – M-SPb.: Tovarishchestvo nauchnykh izdaniy KMK, 457 p. [In Russian].

Raddum G. G., Fjellheim A. 2002. Species composition of freshwater invertebrates in relation to chemical and physical factors in high mountains in southwestern Norway. *Water Air and Soil Pollution*. Focus, 2: 311–328.

Shvartsman Yu. G., Bolotova G. N., Bolotov I. N., Iglovsky S. A. 2005. Spatial distribution of the components of natural complexes of the Solovetsky Archipelago. *Arctic Evironmental Research,* 7(1): 4–17 [in Russian].

Shvartsman Yu. G., Bolotov I. N. 2008. *Spatial and temporal heterogeneity of the taiga biome in the area of Pleistocene continental glaciations*. Ekateriburg, URO RAS. 302 p.

Sokolova S. E., Bespalaya Yu. V. 2010. Local fauna of mollusks of the European North of Russia: Kenozero and Kozhozerskaya group of lakes. *Arctic Evironmental Research*, 1: 63–71 [in Russian].

ter Braak C. J. F., Šmilauer P. 2002. CANOCO reference manual and CanoDraw for Windows user's guide: software for canonical community ordination (version 4.5). *Microcomputer Power,* Ithaca, NY.

In: Lake Water
Editors: Oleg S. Pokrovsky, Yulia Bespalaya et al.

ISBN: 978-1-53619-275-9

Chapter 11

FRESHWATER MOLLUSKS IN LAKES OF THE BOLSHEZEMELSKAYA TUNDRA (THE VASHUTKINY LAKES)

Yulia V. Bespalaya*[1,*], *Olga V. Aksenova*[1,3], *Andrey S. Aksenov*[2], *Ivan N. Bolotov*[1], *Svetlana E. Sokolova*[1] and *Alexander R. Shevchenko*[1]

[1]N. Laverov Federal Center for Integrated Arctic Research, the Ural Branch of the Russian Academy of Sciences, Arkhangelsk, Russian Federation

[2]Northern (Arctic) Federal University named after M. V. Lomonosov, Arkhangelsk, Russian Federation

[3]Saint Petersburg State University, Saint Petersburg, Russian Federation

ABSTRACT

We studied the effect of environmental factors on the patterns of diversity and distribution of freshwater mollusks in the relict Vashutkiny Lakes System. Twenty-four species of freshwater mollusks were identified during a survey of the Vashutkiny Lakes System. The Vashutkiny Lakes are comparable to those in the north of the boreal zone in terms of the level of species diversity and the density of mollusk settlements. The high productivity of lakes in the east of the Bolshezemelskaya tundra is associated with the history of their genesis and the action of zonal and intrazonal factors. We found that the type of substrate, abundance of macrophytes, and depth play an important role in the distribution of the mollusks in lakes.

Keywords: freshwater mollusks, species diversity, envinronmental factors, relict ecosystems, the Vashutkiny Lakes System

* Corresponding Author's Email: jbespalaya@yandex.ru.

INTRODUCTION

The problem of relics can be considered one of the main problems of biogeography (Birshtein, 1947). In addition to a purely taxonomic interpretation of this term (relict species or taxa of a different rank), they often speak of relict ecosystems that were widespread in the past and are now preserved in the form of small isolated areas. The European Subarctic in the Pleistocene was partially affected by several continental glaciations, and many ecosystems formed here relatively recently, in the last 10-15 thousand years (Velichko, 2002; Markova, 2008; Hubberten et al., 2004; Mangerud al., 2004). Therefore, only a few examples of relict ecosystems are given for this territory. Among them is the Vashutkiny Lakes system in the east of the Bolshezemelskaya tundra, which belong to the basin of the Adzva River (a tributary of the Usa River) (Goldina, 1972; Zvereva et al., 1966). Hydrobiological research in 1960–1961 showed that life in the Vashutkiny Lakes is very rich and diverse despite unfavorable climatic and trophic conditions (Getsen, 1964; Zvereva et al., 1970; Zvereva et al., 1964). Aquatic plants are developed abundantly there, algae, zooplankton, and benthos are numerous and varied. "... Indicators of the composition, abundance and biomass of plankton and benthos for the lakes of the Far North are completely exceptional, they do not fit into the existing schemes and ideas about the poverty of life and low biological productivity of polar lakes and even the entire north of our country from 60 parallel and higher ..." (Zvereva et al., 1964). Later it turned out that anomalously high productivity is also characteristic of a number of other large lacustrine systems in the east of the Bolshezemelskaya tundra (Padimeysky and Kharbeysky lakes) (Vlasova et al., 1976a, b; Zvereva, 1970; Zvereva et al. 1970; Sidorov, 1974). The Vashutkiny Lakes represent the largest water system in the East of the Bolshezemelskaya Tundra, and they have attracted the attention of hydrobiologists for many years (Zvereva et al., 1964, 1966; Leshko, 2002). The study of freshwater fauna of the Vashutkiny Lakes was started by A.V. Zhuravsky (1904). More than half a century later, complex studies were carried out by the researchers of the Komi Branch of the USSR Academy of Sciences in European northeast Russia (Zvereva, 1966; Zvereva et al., 1964, 1966; Popova, 1966). These lakes are comparable to northern boreal lakes in terms of mollusk species diversity and density. The hypotheses proposed by Zvereva (1966) on the relict origin of the ecosystem of these lakes and on the leading role of intrazonal factors in the formation of their abnormally high level of production have been analyzed by Bolotov et al. (2014). Both hypotheses are supported by modern data.

The aim of this work was to study the effect of environmental factors on the species diversity and distribution of freshwater mollusks in the relict Vashutkiny Lakes System.

METHODS

Study Area

The east of the Bolshezemelskaya tundra is a huge territory of the Nenets Autonomous Okrug (Figure 1), adjacent to the western macroslope of the Polar and Subpolar Urals (Pochvennogeologicheskie ..., 1984). The study area is located within the tundra zone and is characterized by long cold winters with stable snow cover, a short growing season and a large amount of precipitation (Goldina, 1972; Khokhlova, 2002). A feature of the natural conditions of the area is the presence of permafrost sections (Rebristaya, 1977). The average temperature of the warmest month (July) ranges from +11.6 to +12.3°C, and the coldest month (February) ranges from –18.5 to –19.6°C (Goldina, 1972).

Pre-Quaternary and Quaternary deposits take part in the geological structure of the region (Goldina, 1972). The question of the history of the Quaternary period of the southeastern part of the Bolshezemelskaya tundra has long remained controversial, in particular, the question of the origin of deposits of this period (Zarkhidze, 1972; Zarkhidze et al., 2010).

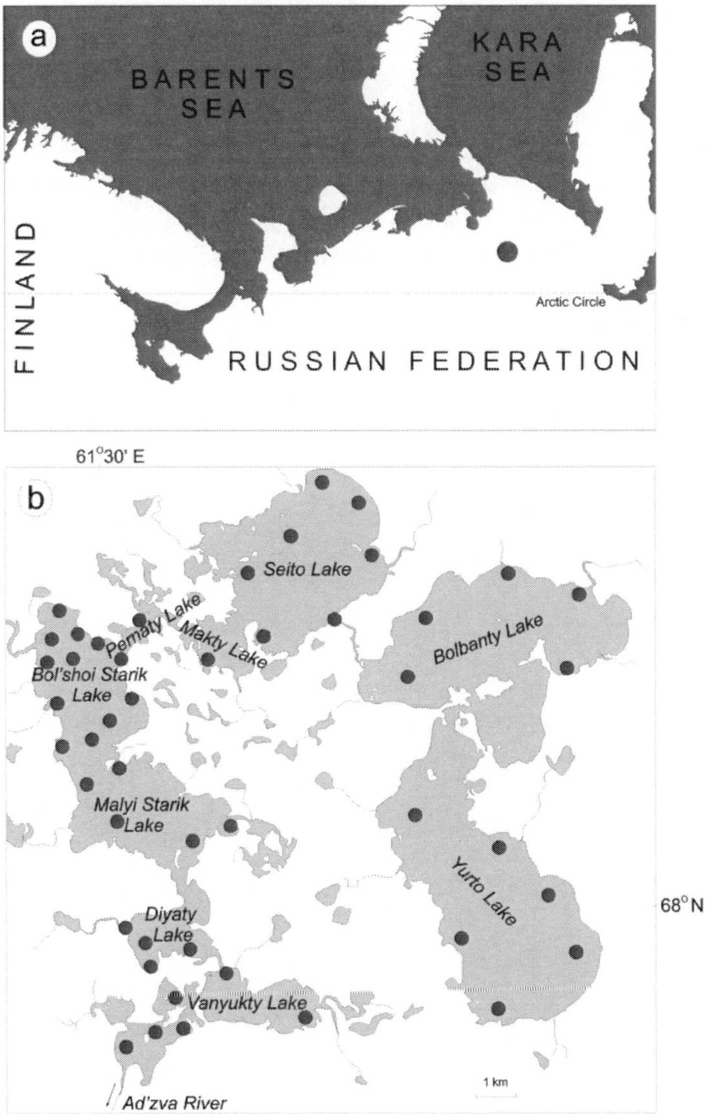

Figure 1. Overview map of the study area (a) of the Vashutkiny Lakes with the designation of sampling points (b).

The results of modern paleogeographic studies prove that there was no cover glaciation in the east of the Bolshezemelskaya tundra in the Late Pleistocene (Maslenikova & Mangerud,

2001; Mangerud et al., 2004). Reconstructions of the contours of periglacial lakes show that in northeastern Europe there was a large reservoir with an area of 76 thousand km^2, known as Lake Komi, which reached its maximum development about 80-100 thousand years ago (Hubberten et al., 2004; Maslenikova & Mangerud, 2001; Mangerud et al., 2004; Svendsen et al., 2004). The Vashutkiny Lakes region, located at relatively high hypsometric levels, was not closed by the waters of the "Komi Lake," although it entered its basin (Maslenikova & Mangerud, 2001). In the penultimate (60-50 thousand years ago) and the last (20-18 thousand years ago) glacial maximums, the eastern part of the Bolshezemelskaya tundra was also located outside the glaciers; apparently, during these periods there were no dammed lakes (Mangerud et al., 2004). At the base of the Quaternary deposits, there is a moraine, represented by loams, clays, enriched with sand, gravel, and boulders (Goldina, 1972).

A characteristic feature of the hydrographic network of the eastern part of the Bolshezemelskaya tundra is many lakes (Goldina, 1972). The Vashutkiny Lakes are the largest lake system of the Bolshezemelskaya tundra belonging to the basin of the Adzva River (a tributary of the Usa River) with a total area of about 85 km^2 (Zvereva et al., 1966). The system includes nine lakes: Vanyukty, Diyaty, Bolshoi Staryk, Maly Staryk, Pernaty, Makty, Bolbanty, Seyto, and Yurto (Figure 1).

The ice formation in the Vashutkinsky Lakes system begins in October (Goldina, 1972). The maximum ice thickness is 1.5–1.8 m (Goldina, 1972). Ice break-up of water bodies occurs in late June and early July (Zvereva et al., 1966; Goldina, 1972).

The Vashutkiny Lakes are characterized by low mineralization, low content of dissolved organic matter, pH values are close to neutral. The differences between the lakes in terms of hydrochemical parameters are insignificant.

Field Sampling

Mollusks were collected in July 2009 and August 2016 from nine lakes on the Vashutkiny Lakes system. A total of 206 benthic samples with 4519 mollusk specimens were collected from 49 stations. From 5 to 7 replicates were collected at each station using a rectangular hand net (dimensions 0.28 m × 0.5 m, mesh size 200 μm, sampling area of 0.14 m^2) or with a Peterson dredge (0.024 m^2) in deeper waters. Samples were washed using a sieve (mesh size 0.56 mm) and fixed in 96% alcohol. At each sampling station, depth, substrate type, composition of macrophytes, mosses and coarse detritus with plant remains were recorded. The dominant substrate types were recorded following Bespalaya (2015) and Bespalaya et al. (2019). Mollusks were identified to species and abundances for each species at the 49 sampling stations were recorded as the average number of individuals per m^2. The confidence interval was calculated using the standard error (SE).

Hydrochemical characteristics of water from lake's surface (pH, conductivity, dissolved O_2, Na^+, K^+, Ca^{2+}, Mg^{2+}, HCO_3^-, SO_4^{2-}, Cl^- concentrations) were obtained according to the methodology (Shevchenko et al., 2016).

Taxonomic identifications followed taxonomic keys (Korniushin, 1996, 2001; Glöer & Meier-Brook, 2003; Opredelitel' ..., 2016). Voucher specimens collected during this study were deposited in the collection of the Russian Museum of Biodiversity Hotspots (RMBH) of the N. Laverov Federal Center for Integrated Arctic Research, the Ural Branch of the Russian Academy of Sciences (FCIARctic).

Statistical Analysis

Mollusk communities were classified by cluster analysis using the Ward method. Before the analysis, the initial data (abundance data) were transformed as log (x + 1) using the PAST ver. 3.0 software (Hammer et al., 2001). The significance of differences was estimated based on the Kruskal-Wallis (multiple comparisons) test using the PAST ver. 3.0 software.

Principal component analysis (PCA) was carried out for the ordination of the stations based on the abiotic factors measured during the study period. Abiotic factors were transformed as log (x + 1) before analysis using PAST ver. 3.0 software.

Multifactor estimation of the relationship of freshwater mollusks with environmental parameters was performed by canonical correspondence analysis (CCA) in the CANOCO ver. 4.56 software, (ter Braak & Śmilauer, 2002). Mollusk densities in each sample were transformed as log (x + 1) (ter Braak & Śmilauer, 2002). The analysis was focused on interspecies distances using Hill's scaling. The down-weighting of rare species was not corrected. The significance of the canonical axes was assessed by using the Monte Carlo permutation test.

RESULTS

Hydrochemical Characteristic of the Studied Lakes

The results of studying the hydrochemical parameters of the Vashutkiny Lakes are presented in Table 1. All of 9 lakes had high concentration of dissolved oxygen, from 9 to 11 mg/L. Value of pH varied from neutral (7.0 in Seito Lake) to slightly alkaline (8.9 in Pernaty Lake). Conductivity measured at 14…17°C change from minimum in Yurto Lake (47.7 μS/cm) to maximum in Diyaty Lake (66.1 μS/cm).

Main cation in Vashutkiny water system is Ca^{2+}, which average concentration ranged 4.9…9.3 ppm, while main anion is HCO_3^-, with concentration varying 29.4…50.3 ppm. Other anions are SO_4^{2-} and Cl^-. Distribution of other cations followed the $Na^+ > Mg^{2+} > K^+$.

Species Diversity

In lakes Vanyukty, Bolbanty, Pernaty, Maly Starik, Diyaty, Makty, Seyto, and Yurto, 24 species, 9 genera and 5 families of mollusks were identified (Table 2). Such species as *Euglesa casertana* (17.6%), *Ampullaceana lagotis* (13.6%), *Stagnicola palustris* (8.2%), *Gyraulus acronicus* (7.6%), and *E. dilatata* (7.0%) had the largest number of individuals in the total sample (N = 4519 specimens). The largest number of species is recorded in Vanyukty Lake (22 species of freshwater molluscs), and the smallest in Pernaty Lake (10 species of freshwater mollusks).

Table 1. Hydrological and hydrochemical characteristics of the Vashutkiny Lakes system (Bolshezemelskaya tundra) (SD)

Lakes	pH	Conductivity, (μS/cm)	O_2, %	HCO_3^-	SO_4^{2-}	Cl^-	Na^+	Mg^{2+}	K^+	Ca^{2+}
				ppm						
Bolbanty Lake	7.5 ± 0.7	54.0 ± 3.8	92.5 ± 1.6	30.9 ± 7.8	2.3 ± 0.2	2.3 ± 0.2	2.6 ± 1.0	1.7 ± 0.1	0.5 ± 0.1	6.2 ± 0.7
Bolshoy Starik Lake	7.8 ± 0.7	62.6 ± 7.9	ND	34.8 ± 6.5	3.6 ± 0.4	2.7 ± 0.3	7.3 ± 1.5	1.6 ± 0.2	1.7 ± 0.4	5.6 ± 0.7
Diyaty Lake	7.6 ± 0.5	66.1 ± 9.6	95.6 ± 0.2	50.3 ± 9.9	3.8 ± 0.5	2.1 ± 0.2	1.9 ± 0.8	2.4 ± 0.3	0.4 ± 0.1	9.3 ± 1.2
Makty Lake	8.7 ± 0.5	57.4 ± 1.1	ND	ND	1.8 ± 0.2	1.0 ± 0.2	1.6 ± 0.5	1.9 ± 0.2	1.2 ± 0.3	9.0 ± 1.0
Malyi Starik Lake	7.4 ± 0.4	60.7 ± 0.5	91.6 ± 1.3	34.4 ± 1.7	2.6 ± 0.2	2.4 ± 0.3	2.1 ± 0.5	1.8 ± 0.4	0.5 ± 0.1	6.9 ± 0.7
Pernaty Lake	8.9 ± 0.9	57.6 ± 0.6	ND	ND	2.5 ± 0.2	2.2 ± 0.2	1.7 ± 0.3	1.8 ± 0.4	0.4 ± 0.1	7.1 ± 0.5
Seyto Lake	7.0 ± 0.2	62.9 ± 12.3	93.9 ± 5.9	33.9 ± 2.7	4.3 ± 2.90	2.3 ± 0.3	2.0 ± 0.3	2.2 ± 0.7	0.4 ± 0.1	8.7 ± 2.8
Vanyukty Lake	8.0 ± 0.7	62.3 ± 1.0	98.4 ± 1.8	34.9 ± 3.9	2. 5 ± 0.49	2.5 ± 0.4	2.1 ± 0.2	1.9 ± 0.2	0.5 ± 0.1	7.0 ± 1.6
Yurto Lake	7.7 ± 0.9	47.7 ± 1.4	98.7 ± 1.9	29.4 ± 6.4	1.8 ± 0.1	2.2 ± 0.1	2.1 ± 0.1	1.6 ± 0.2	0.5 ± 0.1	4.9 ± 0.4

ND – no data.

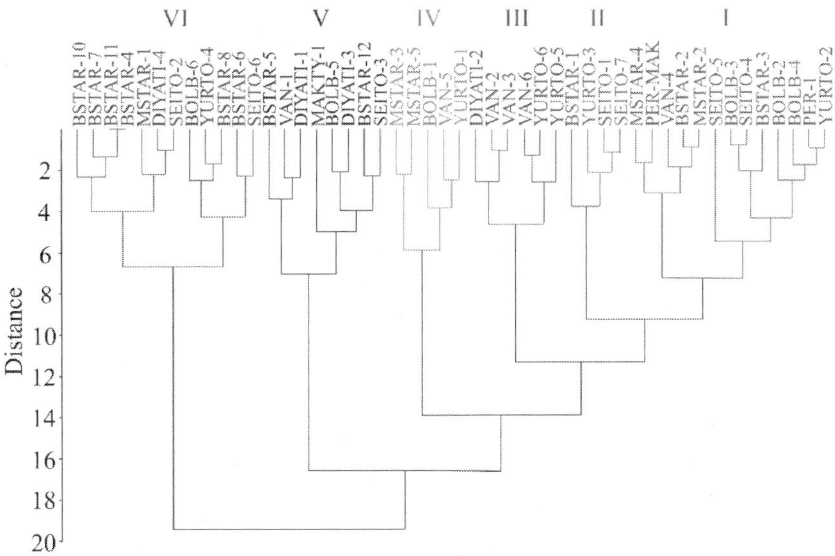

Figure 2. Dendrogram of cluster analysis of species composition and population density of mollusks at stations in the Vashutkiny Lakes (Bolshezemelskaya tundra) (Ward method).
Station designations: BSTAR – Bolshoy Starik Lake; MSTAR – Malyi Starik Lake; BOLB – Bolbanty Lake; VAN – Vanyukty Lake; PER – Pernaty Lake; PER-MAK – Pernaty-Makty lakes; YURTO – Yurto Lake; SEYTO – Seyto Lake; DIYATY – Diyaty Lake.

Based on cluster analysis, six classes of mollusk groupings were identified (Figure 2). Each of them is characterized by certain habitats. The depth of habitat of groups of mollusks of classes I, III, and IV is less than that of classes II, V, and VI (Kruskal-Wallis test: H (df = 5, N = 206) = 159.1, P <0.0001). The density of mollusk settlements is minimal in habitats of classes I, II, II, VI and maximum in class IV and V Kruskal-Wallis test: H (df = 5, N = 206) = 52.6, P <0.0001).

In groupings of class I, *E. casertana* (19.3%), *A. lagotis* (12.2%), *S. palustris* (9.8%), and *E. dilatata* (9.0%) prevailed in numbers. The settlements of mollusks are mainly confined to the littoral zone, with an average depth of 0.97 ± 0.13 m (from 0.1 to 3.0 m). Such groups are formed on silty-sandy and stony-sandy soils in thickets of arctophila (*Arctophila fulva*), horsetail and algae (Figure 3). The average density of mollusks here was 36.7 ± 3.1 ind./m^2 (from 0 to 333.3 ind./m^2).

In groupings of class II, *E. casertana* (28.5%) and *E. dilatata* (16.9%) predominate in numbers. Here mollusks inhabit deep-water areas (depths up to 7.0 m), the average depth of habitats is 2.9 ± 2.3 m. Silty, silty-sandy soils with the presence of plant remains predominated. The average density of mollusks here was 81.9 ± 11.9 ind./m^2 (from 0 to 250.0 ind./m^2).

The gastropods *A. lagotis* (21.0%) and *G. stroemi* (14.8%) were dominated in groupings of class III. Mollusks inhabit the littoral zone (depths 0.2–1.0), predominantly stony-sandy soils with a high content of plant remains. Aquatic plants are represented by dense thickets of pondweed, sedge, arctophila, filamentous algae and mosses (Figure 3). The average depth in this class was 0.6 ± 0.03 m. The average density of mollusks was 51.6 ± 6.1 ind./m^2 (from 0 to 333.3 ind./m^2).

The settlements of mollusks from the fourth class are confined to the coastal zone, silty-sandy and silty soils with plant remains. They differ from those of the previous class by a lower level of siltation and an abundance of plant remains. The depth varied from 0.1 to 1.5 m, with an average value of 0.9 ± 0.2 m. Aquatic plants are represented by thickets of arctophila, horsetail, sedge, and moss (Figure 3). *E. casertana* (15.3%), *E. hibernica* (14.8%), and *A. lagotis* (13.2%) prevailed here in numbers. The average density of mollusks here was 53.3 ± 9.9 ind./m^2 (from 0 to 1166.7 ind./m^2).

In groupings of class V, mollusks inhabit mainly deep-water areas of lakes. Mollusks are confined to clayey, silty-clayey, silty soils with the presence of plant remains. Mollusks have also been observed in deep areas in algae and mosses. The species *E. casertana* (19.5%) and *P. dilatatum* (10.8%) dominated in abundance in deep-water areas (1.5–34 m); *A. lagotis* (11.8%) dominated in the coastal zone up to 1.5 m. The average density of mollusks here was 64.7 ± 6.0 ind./m^2 (from 0 to 541.6 ind./m^2). The depth varied from 0.5 to 34.0 m and averaged 2.9 ± 0.3 m.

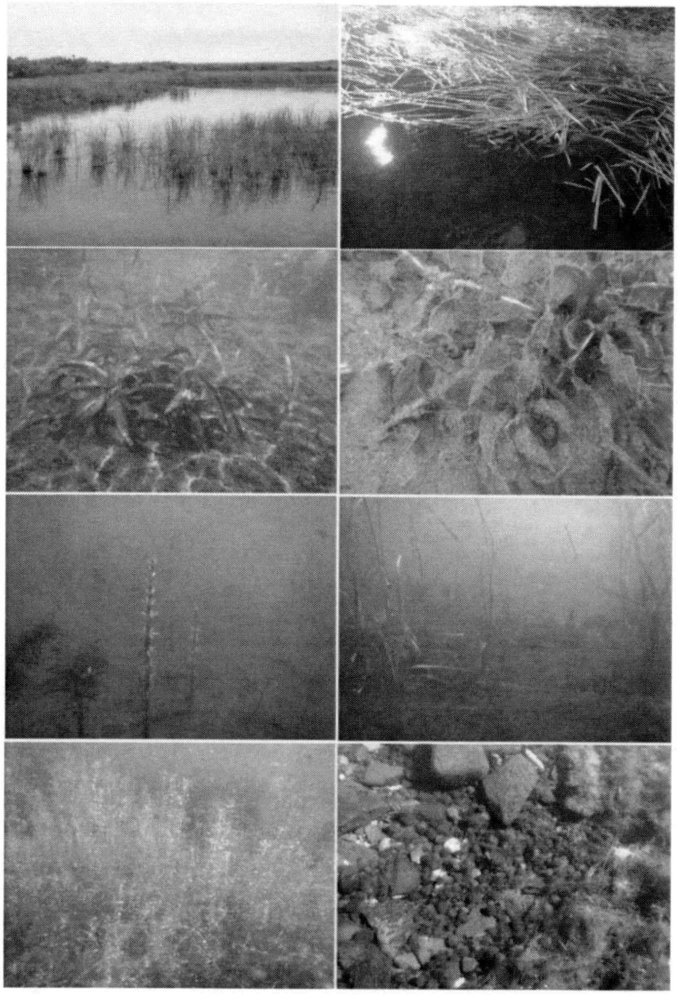

Figure 3. Aquatic plants and algae inhabit the shallow water or littoral zone in the Vashutkiny Lakes (Photos by O.V. Aksenova).

Table 2. Species diversity, average density (ind./m^2), ± standard error of the mean (SEM) and number of mollusks specimens (N) from the Vashutkiny Lakes

Species	Bolbanty Lake	Bolshoy Starik Lake	Vanyukty Lake	Diyaty Lake	Makty Lake	Malyi Starik Lake	Pernaty Lake	Pernaty-Makty Lakes	Seyto Lake	Yurto Lake
Sphaerium corneum (Linnaeus, 1758)	-	175 ± 46,4 (21)	19,5 ± 6,1 (17)	17,1 ± 6,2 (12)	125 ± 0 (9)	7,1 ± 0 (2)	-	-	52,1 ± 10,4 (5)	208,3 ± 48,1 (15)
Euglesa casertana (Poli, 1791)	106,8 ± 40,7 (88)	141,8 ± 42,3 (142)	46 ± 12,1 (59)	95,7 ± 37,3 (105)	61,1 ± 19,9 (16)	74,2 ± 33,8 (94)	97,6 ± 60,5 (20)	-	109,2 ± 20,2 (110)	123,5 ± 29,7 (165)
E. globularis (Clessin in Westerlund, 1873)	68,3 ± 31,3 (38)	134,5 ± 120,3 (13)	50,7 ± 14,3 (21)	138,1 ± 93,4 (29)	28 ± 13,7 (3)	34,4 ± 10,8 (24)	45,2 ± 38,1 (3)	35,7 ± 7,1 (10)	126,2 ± 33,2 (19)	56,8 ± 29,6 (27)
E. hibernica (Westerlund, 1894)	71,4 ± 0 (101)	-	10,7 ± 3,6 (3)	-	41,7 ± 0 (1)	142,9 ± 0 (20)	-	-	-	107,1 ± 0 (15)
E. lilljeborgii (Clessin in Esmark et Hoyer, 1886)	78,2 ± 48,6 (84)	90,2 ± 33,4 (43)	43,3 ± 13,3 (45)	50 ± 23,2 (42)	61,5 ± 34,5 (21)	12,5 ± 5,4 (7)	125 ± 0 (3)	-	39,3 ± 14,7 (22)	74,3 ± 32,7 (46)
E. milia (Held, 1836)	70,8 ± 29,2 (15)	7,1 (1)	14,3 ± 0 (2)	10 ± 1,7 (7)	24,4 ± 17,3 (2)	7,1 ± 0 (1)	-	7,1 ± 0 (1)	-	-
E. obtusalis (Lamarck, 1818)	-	-	32,1 ± 10,7 (18)	11,9 ± 4,8 (5)	10,7 ± 3,6 (3)	7,1 ± 0 (1)	-	7,1 ± 0 (1)	7,1 ± 0 (1)	17,9 ± 3,6 (5)
E. personata (Malm, 1853)	-	125,0 (3)	-	-	-	-	-	-	125 ± 0 (3)	-
E. subtruncata (Malm, 1855)	83,3 ± 0 (2)	42.0 ± 18,1 (17)	18,7 ± 11,5 (3)	39,3 ± 10,7 (22)	41,7 ± 0 (2)	10 ± 2,9 (7)	-	28,6 ± 0 (4)	83,3 ± 0 (2)	-
E. supina (Schmidt, 1850)	-	59,5 ± 23,8 (7)	157,1 ± 0 (22)	-	-	468,7 ± 349 (47)	-	-	-	64,3 ± 42,9 (18)
Pisidium amnicum (Müller, 1774)	44,8 ± 22,7 (14)	121,4 ± 78,6 (34)	7,1 ± 0 (1)	-	62,5 ± 20,8 (3)	-	-	-	84,9 ± 43,3 (26)	-
P. dilatatum Westerlund, 1897	194,4 ± 152,8 (14)	21 ± 10,5 (4)	67,1 ± 25,2 (47)	66,1 ± 36,5 (37)	85,9 ± 36,1 (19)	39,8 ± 5,8 (42)	80,4 ± 44,6 (8)	53,6 ± 32,1 (15)	146,2 ± 22 (91)	43,5 ± 15,6 (5)
Valvata depressa C. Pfeiffer, 1821	28 ± 13,7 (3)	16,4 ± 6,7 (11)	20 ± 6,9 (28)	69 ± 54,8 (29)	68,7 ± 37,6 (24)	21,4 ± 7,1 (7)	35,4 ± 4,0 (15)	-	60,9 ± 23,7 (15)	64,9 ± 35,2 (40)
V. sibirica Middendorff, 1851	30,2 ± 11,5 (3)	33,3 ± 10,9 (32)	34,1 ± 15,9 (38)	47,4 ± 30,8 (21)	79,3 ± 21,2 (12)	44,1 ± 17,1 (43)	14,3 ± 7,1 (4)	-	83,3 ±24,1 (6)	208,3 ± 125 (10)
V. piscinalis O.F Müller, 1774	21,4 ± 0 (3)	15,6 ± 8,6 (4)	14,3 ± 3,5 (14)	7,1 ± 0 (1)	35,7 ± 0 (5)	7,1 ± 0 (1)	7,1 ± 0 (1)	7,1 ± 0 (1)	10,7 ± 3,6 (3)	72,3 ± 30,0 (26)

Table 2. (Continued)

Species	Bolbanty Lake	Bolshoy Starik Lake	Vanyukty Lake	Diyaty Lake	Makty Lake	Malyi Starik Lake	Pernaty Lake	Pernaty-Makty Lakes	Seyto Lake	Yurto Lake
Ampullaceana lagotis (Schrank, 1803)	28,3 ± 6,9 (54)	34,2 ± 7,3 (98)	42,8 ± 11,4 (202)	20,4 ± 6,1 (68)	41,7 ± 0 (4)	22 ± 6,1 (95)	17.9± 3.4 (11)	8,6 ± 1,4 (6)	22,4 ± 7,8 (48)	41,4 ± 27,2 (29)
Radix auricularia (Linnaeus, 1758)	50 ± 0 (7)	25 ± 17,9 (8)	17,9 ± 6,8 (10)	83,3 ± 35,1 (51)	64,3 ± 0 (9)	14,3 ± 0 (15)	-	-	-	45,2 ± 38,1 (3)
Stagnicola palustris (O.F. Müller, 1774)	7,1 ± 0 (14)	35,9 ± 13,1 (46)	16,3 ± 2,4 (42)	31,3 ± 5,3 (72)	54,6 ± 29,8 (46)	23,8 ± 5,8 (60)	-	19 ± 6,3 (8)	25,9 ± 6,7 (82)	-
Bathyomphalus contortus (Linnaeus, 1758)	-	7,1 ± 0 (4)	16,1 ± 6,5 (27)	17,9 ± 8,7 (20)	17,9 ± 10,7 (5)	27,6 ± 15,9 (27)	-	58,3 ± 25,4 (49)	-	-
Gyraulus acronicus (Férussac, 1807)	41,7 ± 0 (2)	41,2 ± 12,3 (49)	48,5 ± 21,4 (65)	51,4 ± 33,1 (148)	52,7 ± 11 (15)	37,8 ± 9,4 (46)	-	11,4 ± 2,9 (8)	18,7 ± 11,5 (4)	28 ± 18,5 (6)
G. albus (Müller, 1774)	-	42,9 ± 0 (6)	-	14,3 ± 0 (2)	-	-	-	-	-	7,1 ± 0 (1)
G. laevis (Alder, 1838)	31,5 ± 10,1 (4)	21,4 ± 0 (3)	7,1 ± 0 (3)	10,7 ± 3,6 (3)	-	132,1 ± 75 (37)	-	-	21 ± 10,5 (4)	-
G. stelmachoetius (Bourguignat, 1860)	56,5 ± 23,3 (24)	51,8 ± 10,7 (29)	22,9 ± 6,9 (16)	25 ± 3,6 (7)	14,3 ± 0 (2)	142,9 ± 30 (31)	7,1 ± 0 (1)	-	7,1 ± 0 (6)	7,1 ± 0 (3)
G. stroemi (Westerlund, 1873)	24,4 ± 17,3 (4)	50 ± 32,6 (35)	38,5 ± 9,8 (103)	17,7 ± 4,7 (38)	30,2 ± 11,5 (3)	7,1 ± 0 (1)	7,1 ± 0 (1)	-	32 ± 10,2 (18)	29,5 ± 9,9 (22)

Finally, *A. lagotis* (31.1%) prevails in the groupings of class VI in terms of numbers. In general, the average density of mollusks here was 31.4 ± 5.1 ind./m^2 (from 0 to 200.0 ind./m^2). Settlements of mollusks are mainly confined to the littoral zone of water bodies, the depth varied from 0.1 to 2.0 m, the average depth was 1.9 ± 0.4 m. Mollusks were found on stony-sandy and silty-sandy soils. The plants contain thickets of pondweed.

Environmental Variables

The distribution of sampling stations depending on environmental factors was analyzed by the principal component analysis (PCA) (Figure 4). Differences between stations are mainly due to the nature of the substrate, composition of aquatic plants, depth and hydrochemical parameters. The projection of the first axis positively correlated with the concentration of macro-ions (K^+, Ca^{2+}, Mg^{2+}) and electrical conductivity on the positive side of the axis and with the nature of the substrate (silty-sandy soil) and the concentration of Cl^- located on the other side of the axis. Along the second axis, the difference between stations is mainly related to the nature of the substrate, the composition of aquatic plants and hydrochemical parameters (SO_4^{2-}, conductivity) on the positive side of the axis and the depth value located on the other side of the axis.

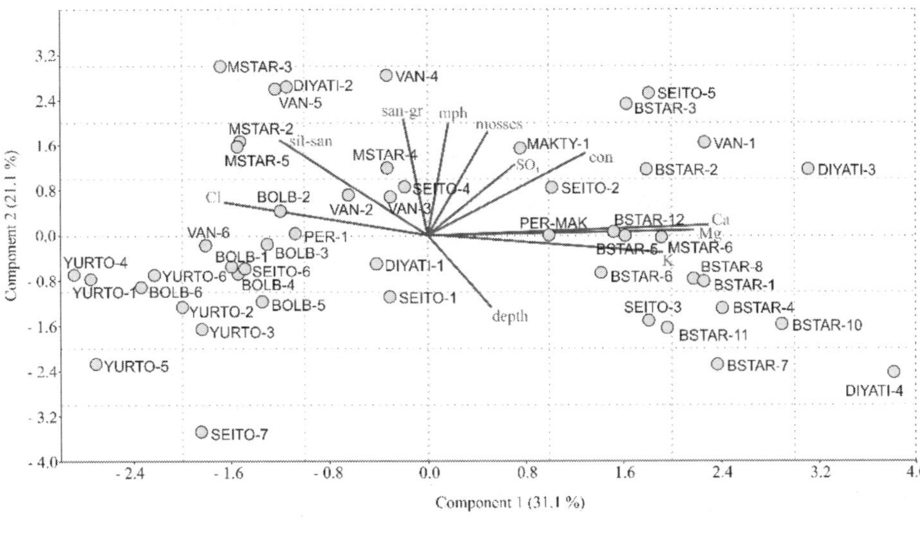

Figure 4. Distribution of sampling stations depending on environmental factors based on principal component analysis (PCA). For station designations, see Figure 2. Abbreviation of factors: sil-san – silted-sandy; san-gr – stony-sandy; mph – macrophytes; mosses – moss; SO_4^{2-}, Mg^{2+}, K^+, Cl^-, Na^+, Ca^{2+} – macroions; con – electrical conductivity.

As a result of the canonical correspondence analysis (CCA), it was found that out of 19 environmental factors, only 4 have a significant impact on the distribution and abundance of mollusks (Figure 5). The analysis focused on interspecies distances with Hill scaling, the contribution of rare species was not corrected. The eigenvalues (lambda) are 0.078, 0.032, 0.021 and 0.014 for the first (horizontal), second (vertical), third and fourth axes, respectively.

Figure 5. Ordination diagram of canonical correlation analysis (CCA) showing relationships between the abundance of mollusks with environmental variables in the Vashutkiny Lakes.

Figure 6. Freshwater gastropods on various substrate type in the Vashutkiny Lakes. (Photos by O. V. Aksenova).

The canonical correlations between the abundance of freshwater mollusks and environmental variables for axes 1 and 2 are 0.740 and 0.635, respectively. The significance of the axes was determined based on the Monte Carlo test (the first axis was $F = 2.417$, $P = 0.04$, all axes were $F = 1.185$, $p = 0.04$). The projection of the first axis was positively correlated with depth and negatively correlated with the nature of the substrate and aquatic plants. Along the second axis, the difference between stations is mainly associated with the presence of higher aquatic plants and depth (Figure 5).

Deep-water areas are optimal habitat, mainly for bivalve mollusks of the Sphaeriidae family (Figure 5). Lymnaeidae and Planorbidae predominated in shallow waters, in thickets of higher aquatic plants (Figure 6).

Mollusk Diversity and Abundance in Relict Lakes: The Role of Zonal and Intrazonal Environmental Factors

The species richness of mollusks in the Vashutkiny Lakes is practically the same as compared to some lakes in the northern taiga subzone (Bolotov et al., 2014). Considering that the lakes are located at 68° N in the subzone of typical tundra, among the "almost ubiquitous" distribution of frozen rocks (Pochvennogeologicheskie ..., 1984), high indicators of the species diversity of mollusks should be recognized as anomalous, and a similar limnological setting was also found in the Padimeysky and Kharbeysky lakes, located approximately 60 km southeast of the Vashutkiny Lakes (Zvereva, 1970; Bolotov et al., 2014). At the same time, the waters of the lakes in the east of the Bolshezemelskaya tundra are characterized by low mineralization, low content of dissolved organic matter, pH values are close to neutral.

According to the literature data, the anomalous bioproductivity of the lakes in the eastern part of the Bolshezemelskaya tundra is due to several reasons (Zvereva, Bolotov et al., 2014). Thus, it has been established that the lakes of this region are relict. In addition, the territory of the east of the Bolshezemelskaya tundra was not subjected to continental glaciations in the late Pleistocene (Zarkhidze et al., 2010; Mangerud et al., 2004; Maslenikova & Mangerud, 2001). Probably, lake ecosystems could have existed here for at least 80–100 thousand years.

An analysis of the distribution paths of freshwater animals in Northern Europe showed a high probability of the presence of a glacial refugium in the Pechora basin (Makhrov & Bolotov, 2006).

If the high species richness of lake biota in the east of the Bolshezemelskaya tundra can be explained by the peculiarities of their genesis and antiquity, then the stable existence in the Subarctic of limnic ecosystems with anomalous biological productivity should be associated with the action of some azonal environmental factors. Designated by O.S. Zvereva (1966) and Zvereva et al. (1964), the role of higher aquatic plants, and first of all arctophiles, in the provision of lake ecosystems with organic matter and biogenic substances is beyond doubt. Thus, the distribution of mollusks by habitat is clearly linked to the development of macrophytes (Figure 3). However, such an intensive development of plants fits into the general picture of anomalous productivity of lakes and is not so much a factor as a consequence of a favorable combination of environmental conditions.

The hypothesis of O.S. Zvereva (1970) on the leading role of intrazonal factors, namely the additional feeding of lakes with groundwater discharged from karst massifs, is of great interest. Despite the low salinity of water in the Vashutkin lakes, in some areas there is a local influence of saline waters of the groundwater runoff, apparently, primarily in winter (Zvereva et al., 1966; 1970).

At the same time, during our research, we did not establish the effect of the hydrochemical parameters of the lakes on the distribution of mollusks. Our results are in accordance with the observation of Bódis et al. (2011), who found that influence of hydrochemical parameters on the distribution of bivalves are small. This is probably due to the fact that the differences between the lakes in terms of hydrochemical parameters within one system are usually low (Zvereva et al., 1966; Zvereva et al., 1970; Goldina, 1972).

In addition, the increased levels of dissolved salts as a result of local influences of mineralized subterranean waters was observed mostly in winter, whereas our research was conducted during the summer period (Zvereva et al. 1966, 1970; Goldina, 1972).

Finally, the basis of the anomalous productivity of the lakes in the east of the Bolshezemelskaya tundra is their flow-through regime, which provides an intense convective heat exchange of water masses with suprapermafrost taliks (Pochvennogeologicheskie usloviya, 1984) and leads to the accumulation of summer heat in loose Quaternary sediments widespread in lake basins. This is facilitated by the huge catchment areas, intense wind mixing of water masses, and the predominance of shallow littoral and sublittoral areas in the bottom relief. Intensive water exchange and wind mixing contribute to the leveling and stability of the gas and hydrochemical regime, and the high transparency of the water dramatically expands the photic layer, therefore, such groups as higher plants, aquatic mosses and mollusks successfully colonize lakes to significant depths (Bolotov et al., 2014).

CONCLUSION

The Vashutkiny Lakes are comparable to lakes in the north of the boreal zone in terms of the level of species diversity and density of mollusks. The high productivity of the lakes in the east of the Bolshezemelskaya tundra is associated with their genesis and the action of zonal and intrazonal factors. We found that the type of substrate, abundance of macrophytes, and depth play an important role in the distribution of the mollusks in lakes. The deep-water areas are optimal habitat, mainly for bivalve mollusks of the Sphaeriidae family. Lymnaeidae and Planorbidae dominate mainly in shallow waters in thickets of higher aquatic vegetation. Further work is needed to undertake the biogeography, phylogeny and taxonomic revision of fauna of freshwater mollusks in the Arctic region based on molecular genetic data.

ACKNOWLEDGMENTS

The Ministry of Science and Higher Education of the Russian Federation supported the study of biodiversity of freshwater bivalves (project no. AAAA-A17-117033010132-2), the study of environmental parameters was supported by the Russian Science Foundation (no. 19-14-00066). The Russian Foundation for Basic Research supported the study of the distribution of freshwater mollusks depending on environmental factors (no. 20-04-00361).

REFERENCES

Bespalaya Y. 2015. Molluscan fauna of an Arctic lake is dominated by a cosmopolitan *Pisidium* species. *Journal of Molluscan Studies,* 81: 294–298.

Bespalaya Yu., Joyner-Matos J., Bolotov I., Aksenova O., Gofarov M., Sokolova S., Shevchenko A., Travina O., Zubriy N., Aksenov A., Kosheleva A., Ovchinnikov D. 2019. Reproductive ecology of *Pisidium casertanum* (Poli, 1791) (Bivalvia: Sphaeriidae) in Arctic lakes. *Journal of Molluscan Studies,* 85(1): 24–34. doi:10.1093/mollus/eyy050.

Birshtein Ya. A. 1947. The term "relic" in biology. *Zoologicheskiĭ Zhurnal,* 26(4): 313–330. (in Russian).

Bódis E., Nosek J., Oertel N., Tóth B., Hornung E., Sousa R. 2011. Spatial distribution of bivalves in relation to environmental conditions (middle Danube catchment, Hungary). *Community Ecology,* 12: 210–219.

Bolotov I. N., Bespalaya Y. V., Aksenova O. V., Gofarov M. Y., Sokolova S. E. 2014. Mollusks in the zoobenthos of relict lakes with abnormally high biological production in the eastern European Subarctic. *Inland Water Biology,* 7(1): 61–71. https://doi.org/10.1134/S1995082914010040.

Getsen M. V. 1964. Higher plants of Vashutkiny Lakes of the Bol'shaya Zemlya tundra (Usa River basin). *Botanical zhurnal,* 49(4): 587–589. (in Russian).

Gloër P., Meier-Brook C. 2003. *Süsswassermollusken. Ein Bestimmungschlüssel für die Bundesrepublik Deutschland* [*Freshwater mollusks. A determination key for the Federal Republic of Germany*]. 13 Auflage. Deutscher Jugendbund für Naturbeobachtun, Hamburg.

Goldina L. P. 1972. *Geografiya ozer Bol'shezemel'skoi tundry* (*Geography of Lakes of the Bol'shaya Zemlya Tundra*), Leningrad: Nauka [in Russian].

Hammer O., Harper D. A. T., Ryan P. D. 2001. PAST: paleontological statistics software package for education and data analysis. *Palaeontologia Electronica,* 4: art. 4.

Hubberten H. W., Andreev A., Astakhov V. I. et al. 2004. The periglacial climate and environment in Eurasia during the last glaciation. *Quaternary Science Reviews,* 23: 1333–1357.

Khokhlova L. G. State of hydrochemical knowledge of surface waters of the Bolshezemelskaya Tundra. In: *Vozobnovimye resursy vodoemov Bolshezemelskoy tundry (Renewable resources of reservoirs of the Bolshezemelskaya Tundra),* (A. I. Taskaev, ed.) Syktyvkar, Tr. Komi Nauch. Tsentra UrO RAN, 2002, pp. 5–14 [in Russian].

Korniushin A. V. 2001. Taxonomic revision of the genus *Sphaerium* sensu lato in the Palaearctic Region, with some notes on the North American species. *Arch Molluskenkd,* 129: 77–122.

Korniushin A. V. 1996. *Bivalve molluscs of the superfamily Pisidioideain the Palaearctic region: fauna, systematics, phylogeny*. Schmalhausen Institute of Zoology, Kiev [In Russian].

Leshko Yu. V. 2002. Molluscan fauna of tundra water bodies In: *Vozobnovimye resursy vodoemov Bolshezemelskoy tundry (Renewable resources of reservoirs of the Bolshezemelskaya Tundra),* (A. I. Taskaev, ed.). Syktyvkar, Tr. Komi Nauch. Tsentra UrO RAN, 2002, pp. 63–71 [in Russian].

Makhrov A. A., Bolotov, I. N. 2006. Dispersal routes and species identification of freshwater animals in Northern Europe: a review of molecular evidence. *Russian Journal of Genetics,* 42(10):1101–1116.

Mangerud J., Jakobsson M., Alexanderson H., et al. 2004. Icedammed lakes and rerouting of the drainage of northern Eurasia during the last glaciation, *Quaternary Science Reviews,* 23: 1313–1332.

Markova A., van Kolfschoten T. *Evolution of the European ecosystems during Pleistocene Holocene transition* (24-8 кyr BP). Moscow, KMK Scientific Press Ltd Editors, 2008.

Maslenikova O., Mangerud J. 2001. Where was the outlet of the icedammed lake Komi, northern Russia? *Global and Planetary Change,* 31: 337–345.

Opredelitel' zooplanktona i zoobentosa presnykh vod Yevropeyskoy Rossii (*Keys to zooplankton and zoobenthos of fresh waters in European Russia*). 2016. Vol. 2. Zoobentos/Eds. V. R. Alekseyev, S. Ya. Tsalolikhin. – M-SPb.: Tovarishchestvo nauchnykh izdaniy KMK, 457 p. [In Russian].

Pochvennogeologicheskie usloviya Nechernozem'ya (*Soil and Geological Conditions of Nechernozemie*), Moscow: Izd. Mosk. Gos. Univ., 1984. [In Russian].

Rebristaya O. V. *Flora vostoka Bolshezemelskoy tundry* (*Flora of the east of the Bolshezemelskaya Tundra*), Leningrad, 1977, pp. 22–27.

Shevchenko A. R., Aksenov A. S., Aksenova O. V., Bespalaya Y. V., Gusakova E. V. 2016. *Surface distribution of general ions in subarctic fresh water systems* (*Vashutkiny Lakes, Russia*). Conference Proceedings "16th International Multidisciplinary Scientific GeoConference - SGEM 2016, GREEN Extended Scientific Sessions," 2–5 November, 2016, Vienna. Book 3, 3: 275–280. doi: 10.5593/SGEM2016/HB33/S02.035.

Sidorov G. P. 1974. *Rybnye resursy Bol'shezemel'skoi tundry* (*Fish Resources of the Bol'shaya Zemlya Tundra*), Leningrad: Nauka [in Russian].

Svendsen J. I., Alexanderson H., Astakhov V. I., Demidov I., Dowdeswell J. A., Funder S., Gataullin V., Henriksen M., Hjort C., Houmark-Nielsen M., Hubberten H. W., Ingólfson O., Jakobsson M., Kjær K., Larsen E., Lokrantz H., Lunkka J. P., Lyså A., Mangerud J., Matiouchkov A., Murray A., Möller P., Niessen F., Nikolskaya O., Polyak P., Saarnisto M., Siegert C., Siegert M., Spielhagen R. & Stein R. 2004. Late Quaternary ice sheet history of Northern Eurasia. *Quaternary Science Reviews,* 23: 1229–1271.

ter Braak C. J. F., Šmilauer P. 2002. CANOCO reference manual and CanoDraw for Windows user's guide: software for canonical community ordination (version 4.5). *Microcomputer Power*, Ithaca, NY.

Velichko A. A. Dinamika landshaftnykh komponentov vnutrennikh morskikh basseinov Severnoi Evrazii za poslednie 130 000 let. Atlas monografiya "Razvitie landshaftov i klimata Severnoi Evrazii. Pozdnii pleistotsen–golotsen–element prognoza" (Dynamics of Landscape Components of Inland Marine Basins of Northern Eurasia in the Past 130000 Years. Monograph Atlas "The Development of Landscapes and Climate of Northern Eurasia: Late Pleistocene–Holocene—The Elements of the Fore cast"), Vol. 2: *Obshchaya paleogeografiya* (*General Paleogeography*), Moscow, GEOS, 2002.

Vlasova T. A., Getsen M. V. 1976. Basic patterns of production processes in tundra lakes, in *Produktivnost' ozer vostochnoi chasti Bol'shezemel'skoi tundry* (*Productivity of Lakes of the Eastern Part of the Bol'shaya Zemlya Tundra*), Leningrad: Nauka, 131–136 [in Russian].

Zarkhidze D. V., Gusev E. A., Anikina N. Yu., et al. 2010. [New data on the stratigraphy of PlioceneQuaternary sediments of the MoreYu River basin (Bolshezemelskaya Tundra)] Novye dannye po stratigrafii pliotsen chetvertichnykh otlozhenii basseina reki MoreYu (Bol'shaya Zemlya tundra), *Geologo Geofizicheskie kharakteristiki litosfery Arkticheskogo regiona*, 210(7): 96–110. (in Russian).

Zarkhidze V. S. 1972. Vashutkinskay svita Timano Uralskoi oblasti. *Voprosi stratigrafii i korrelacii pliocenovich i pleistocenivich otlozhenii severnoi i uzhnoi chastei Preduraliya* (*Vashutkinskaya Suite Timan-Urals region. Stratigraphy and correlation of Pliocene and Pleistocene deposits in the Northern and southern parts of the Urals*), 1:78-82.

Zhuravskiy A. V. About the Mainland west. Topographic appearance and fauna of the tundra. Results of the 1903 summer trip. Trudy Imperatorskogo St.-Peterburgskogo obshchestva estestvoispytateley. *Otdelenie zoologii fiziologii,* 1904, v. 35, no. 1, pp. 81–92; no. 2, pp. 65–95 [in Russian].

Zvereva O. S. 1966. *Benthos and general issues of hydrobiology of Vashutkiny Lakes, in Gidrobiologicheskoe izuchenie i rybokhozyaistvennoe osvoenie ozer Krainego Severa SSSR* (*Hydrobiological Study and Fishery Development of Lakes of the Far North of the USSR*), Moscow: Nauka, 112–136. (in Russian).

Zvereva O. S., 1970. Problems of zonality and intrazonality in limnology of the Far North, in *Biologicheskie osnovy ispol'zovaniya prirody Severa* (*Biological Basics of Nature Management in the North*), Syktyvkar: Komi Knizh. Izd., 244–248 [in Russian].

Zvereva O. S., Getsen M. V., Iz'yurova V. K., 1964. The system of relict lakes in the Bol'shaya Zemlya tundra, *Dokl. Akad. Nauk SSSR*. 155(3): 677–679. (in Russian).

Zvereva O. S., Vlasova T. A., Goldina L. P., Iz'yurova V. K. 1970. Results of limnological research of the Bol'shaya Zemlya tundra, in *Biologicheskie osnovy ispol'zovaniya prirody Severa* (*Biological Basics of Nature Management in the North*), Syktyvkar: Komi Knizh. Izd., 248–253 [in Russian].

Zvereva O. S., Vlasova T. A., Goldina, L. P. 1966. Vashutkiny Lakes and the history of their study, in *Gidrobiologicheskoe izuchenie i rybokhozyaistvennoe osvoenie ozer Krainego Severa SSSR* (*Hydrobiological Study and Fishery Development of Lakes of the Far North of the USSR*), Moscow: Nauka, 4–21. [In Russian].

In: Lake Water
Editors: Oleg S. Pokrovsky, Yulia Bespalaya et al.
ISBN: 978-1-53619-275-9

Chapter 12

ECOSYSTEMS OF LAKES OF THE ARKHANGELSK REGION

Alexander P. Novoselov*, Gennady A. Dvoryankin, Elena V. Pribytkova and Ekaterina N. Imant

Laverov Federal Center for Integrated Arctic Research, the Ural Branch of the Russian Academy of Sciences, Arkhangelsk, Russian Federation

ABSTRACT

The chapter represents data about the lakes in the Arkhangelsk region. The lake area of the Arkhangelsk region is relatively small. Only few significant lake-river systems can be mentioned. They are the upper part of the Onega river with lakes Vozhe and Lacha, the Siya river, which flows through a chain of flowing lakes, as well as the lake systems of the Ken river, the Ilex river and the Kozha river, located in the Western part of the Arkhangelsk region. Below there is the data on several parameters and characteristicts which were analyzed: chemical and physical lake characteristics, plankton, benthos, fish food stocks, vegetation, ichtyofauna, fish stocks.

Keywords: Arctic, lakes, the Arkhangelsk region

INTRODUCTION

In the last half-century, changes in aquatic ecosystems are determined not only by natural conditions, but also by the influence of anthropogenic factors on them. At the same time, their impact is similar to the climate impact, but it appears in a shorter period and causes more abrupt changes. Environmental consequences caused by both direct and indirect effects of technogenic factors lead to the reduction of hydrobiont species` number, changes in their ranges and population, changes in a population structure of commercial fish species, decrease in biological

* Corresponding Author's Email: alexander.novoselov@rambler.ru.

diversity, and the overall transformation of aquatic ecosystems (Reshetnikov, 1995; Kudersky, 1995; Tsepkin, 1995, 1996, etc.).

Lakes, as the habitat of hydrobionts, are very diverse according to their natural qualities. Lakes are primarily diverse due to their size, depth, structure of the coastline and bed, nature of the soil, mineral composition and gas regime of water, temperature regime, flowage and other parameters which depend on the characteristics (soil, vegetation, population) of the catch basin and climate. Geological factors played a major role in the formation of the ichthyofauna composition, as well as time period: the glacial epoch, and then the dry, warm and cooler epoch, together with the presence of water connections between water bodies and river basins at these times. Finally, anthropogenic activities, such as forest devastations, plowing land, draining swamps, building dams and canals, creating settlements and industrial enterprises, has a profound impact on the regime of waterbodies and the ichthyofauna composition in them (Kozmin, 2011).

In contrast to the Republic of Karelia, the lake area of the Arkhangelsk region is relatively small. Only few more or less significant lake-river systems can be mentioned. They are the upper part of the Onega river with lakes Vozhe and Lacha, the Siya river, which flows through a chain of flowing lakes, as well as the lake systems of the Ken river, the Ilex river and the Kozha river, located in the Western part of the Arkhangelsk region. The exact number of lakes in the region has not yet been determined. An approximate number is 224.252 lakes with a water mirror area of 11.359 km^2, which is about 2% of the region area. The vast majority of lakes are small reservoirs with an area of less than 50 hectares. Among the relatively large ones are Lacha (345 km^2), Kozhozero (97 km^2), Kenozero (66.3 km^2), Lekshmozero (54 km^2) (Dvoryankin, 2010; Kozmin, 2011).

By origin, the lakes of the Arkhangelsk region are divided into the following groups: glacial, valley coastal and karst.

Glacial lakes are the largest water bodies, and the largest part of the region's lake stock belongs to them. These are mostly shallow oblong lakes. A number of lakes located in terminal-moranial areas. These lakes are characterized by a multi-lobed shape and have a rather large depth. The largest and most complex is Kenozero in the basin of the Onega river. Glacial lakes, which are located on the mainland of the study region, are gradually changing their appearance, they are covered by precipitations brought by rivers. An example of such lake is Lacha, where the Onega river originates from.

Valley coastal lakes belong to rivers and are an isolated part of the main riverbed. Their dimensions range from hundreds of meters to kilometers with a width of 50-100 m and a depth of up to 3 m. During the spring flood, they are filled with water, and connect with the river. There are a lot of lakes and starits in the floodplain of the Northern Dvina river, there are a few less in the lower reaches of the Kuloy and Mezen rivers.

Karst lakes are deep reservoirs filled with cold clear water. There are especially many of them in the area of gypsum karst that covers the upper and middle flows of the Mekhrenga river, the lower flow of the Pukshenga river and the South-Eastern part of the White Sea-Kuloy plateau. Karst lakes are fed by ground water and often have underground water connections with each other. There are sometimes disappearing lakes in the karst zone. They are Syamgozero and lake Oboksha, as well as a number of small water bodies which lose water once every 4-10 years and temporarily dry up. There are three karst lakes filled with salt water in the basin of the Kuloy river (Kozmin, 2011).

The following proceeding presents general characteristics (physiography, food supply and ichthyofauna) of the largest commercial lake Lacha, located in the zone of the flat taiga, as well as different types` lakes in the territories of specially protected natural areas of the Arkhangelsk region: large lakes - Kenozero and Lekshmozero (Kenozersky National Park) and, small taiga lakes - Sychevo and Pershkovskoe (Pinezhsky state reserve).

Lakes of the Lowland Taiga

Lake Lacha

Lake Lacha, located in the South-Western part of the Arkhangelsk region, is one of the largest inland water body in the North of the European part of Russia. The basin of the lake is of a pre-glacial origin, being appeared on the site of extensive subglacial lakes within the South-Eastern slope of the Baltic shield and has undergone subsequent changes as a result of the glacier impact. Its shape is close to oval and elongated from North to the South. The shoreline of the lake has smooth outlines, broken only in the Southwestern part of it (Figure 1) (Kozmin, Shatova, 1979).

The water area is 334 km^2, the drainage area is 12600 km^2, the length of the lake is 33 km, and the width in the southern part is 13-14 km, in the Northern part - 8.5 km. The lake is shallow, the average depth is 1.3 m, and the maximum does not exceed 5.3 m. The lake is water flowing. The Svid river is the largest of the tributaries, which flows into the southern part of the lake and gives almost half of the surface flow. In the Northern part of the lake the Onega river flows out, which is one of the largest rivers in the European North-East of Russia (Lake Hydrology..., 1979; Pekisheva et al., 2007).

Currently, the lake is located in the area of anthropogenic activity. Widespread deforestation in the catchment area, the use of adjacent land for agriculture, the use of mineral fertilizers and pesticides contribute to the accumulation of organic matter in the lake. There is a so-called "cultural eutrophication" (Buyanovskaya, 1977), which is common characteristic for almost all large freshwater waterbodies. At the same time, successional changes in abiotic environmental factors occur in lakes, which affect the structure of relationships between the main biological parameters of lake ecosystems, as well as all levels of organization of water ecosystems. This leads to a restructuring of the trophic structure in the lake, ranging from phyto- and zooplankton to representatives of zoobenthos and fish (Reshetnikov, 1979, 1982).

In recent decades, changes that indicate the acceleration of the eutrophication of Lacha lake ecosystem became more visible. The water area is often overgrown with higher water vegetation and it is difficult to get access not only by motor ships and boats, but also by small-sized fleets. Because of the intensive development of macrophytes, industrial fishing on the lake with seines has become significantly difficult (Novoselov and Kozmin, 1990; Novoselov, 2006). In the current situation, it is possible to resolve the issue of regulatory measures aimed at increasing its fish productivity only after a detailed analysis of the current state of its ecological components, especially plankton and bottom communities, which play an important role as a food base for fish living in the lake.

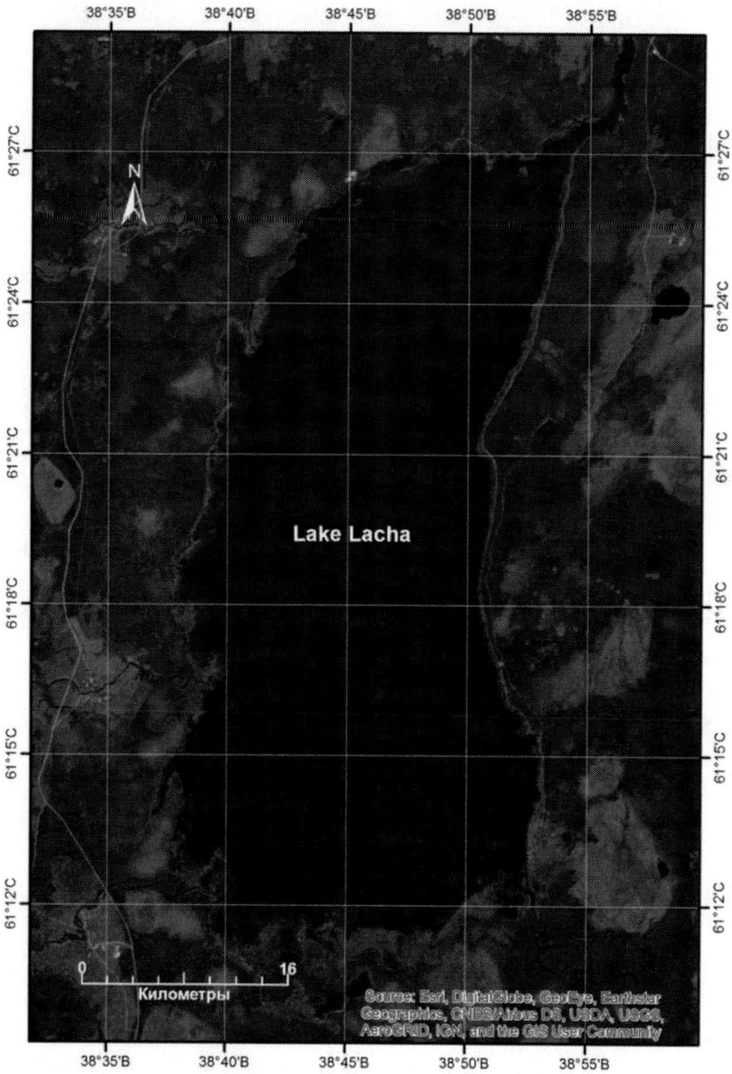

Figure 1. Schematic map of Lake Lacha.

Hydrobiological studies of lake Lacha were previously conducted by the Vologda-Arkhangelsk expedition in the period from 1972 to 1975. The studies were focused on the processes occurring in large shallow water bodies, and were part of the project for transferring the waters of Northern rivers. Zooplankton and zoobenthos were studied as a part of the research. The species composition of hydrobiological communities was determined, as well as the ecological and fisheries status of the water body. Based on research, the lake was defined as a eutrophic water body with an average level of zooplankton and zoobenthos development (Hydrobiology of lakes..., 1978). In recent years, work on the study of the fish food base of lake Lacha were part of the state monitoring conducted by the Northern Department of Polar Russian Institute of Fisheries and Oceanography (Novoselov et al., 2017 a,b)..

This article is based on materials collected in the period from 2003 to 2015. Hydrobiological materials were selected from a grid of 12 monitoring stations located evenly across the lake's water area. Collection and processing of hydrobiological samples were carried out according to generally accepted methods (Zhadin, 1990; Kutikova, 1970; Manuilova, 1964; Determinant..., 1977; Rylov, 1930, 1948; Khazov, 2000). Zooplankton samples were taken by straining 100 liters of water through a small Djedi net. Zoobenthos samples were taken by a Petersen dredger with a capture area of 1/40 m^2. Cameral processing was also carried out using traditional methods (see links).

Zooplankton

Zooplankton plays an important role in aquatic ecosystems, serving as the main food resource for planktophage fish and juveniles of almost all fish species, regardless of their trophic specialization. In addition, planktonic organisms react quickly enough to changing environmental conditions, and its adaptive responses contribute to maintaining the ecological balance of water systems. Zooplankton actively participates in the process of self-purification of water bodies, since its filtration activity is inextricably linked to the feeding process. As a result, the amount of suspended organic matter in the water decreases, which includes the main components of zooplankton food-phytoplankton, bacterioplankton and detritus (Andronnikova, 1976).

The zooplankton of Lacha lake is represented by the species common to the North-Western zone, widely distributed in the Northern part of the Palearctic. According to the results of state monitoring, 58 taxa of planktonic invertebrates live in the lake. Their taxonomic list is also presented (Table 1).

The dominant groups in zooplankton are Copepoda (21 species) and Cladocera (19-21 species), which are represented in approximately equal numbers. Rotifers of Rotatoria (=Rotifera) were less presented in the general species spectrum. So, Lacha lake is “Cladocera-Copepoda” lake due to quality content of zooplankton communities.

Analyzing the level of zooplankton development in the lake, there are two biotopes identified. They differ in both the qualitative and quantitative composition of zooplankton communities.

In the pelagic zone of the lake, the major species during the research period were *Kellicotia longispina, Conochilus unicornis, Asplanchna priodonta, Bosmina (Eubosmina) coregoni, Daphnia cuculata, Limnosida frontosa, Leptodora kindtii, Mesocyclops leuckarti, Thermocyclops oithonoides, Eudiaptomus gracilis*. The biomass of open water zooplankton varied in different years from 0.55 to 1.91 g/m^3, which made it possible to classify this zone as a mesotrophic type and a medium-sized class of water bodies for fish. *Sida crystallina, Eurycercus lamellatus, Acroperus harpae, Peracantha truncata, Pleuroxus uncinatus,* and *Eucyclops serrulatus* dominated in the zone of higher aquatic vegetation. The zooplankton biomass in this zone reached 5.8 g/m^3, The lake was assigned as the eutrophic type and high-feed class (Mikhailov et al., 1988).

In general, in terms of the zooplankton number and biomass, Lacha lake is a mesotrophic medium-sized fish-feeding water system. Taking into account the areas of thickets of individual plant communities, the average weighted biomass was 3.3 g/m^3. At the same time, the most productive associations were pondweed (Potamogeton) and parrots’ feather (Myriophyllum), occupying 84% of the total size of the thickets, where the average zooplankton biomass was 6.05 and 9.1 g/m^3 respectively (Mikhailov et al., 1988).

Table 1. Systematic list of zooplankton of Lake Lacha (according to: Novoselov et al., 2017a)

Taxons	2005 y.	2015 y.
ROTIFERA		
1. Asplanchna priodonta Gosse, 1850	+	+
2. A. herricki Guerne, 1888	+	+
3. Conochilus unicornis Rousselet, 1892	+	+
4. Conochilus sp.	+	+
5. Euchlanis sp.	-	+
6. Kellicottia longispina (Kellicott, 1879)	+	+
7. Trichocerca sp.	-	+
8. Keratella quadrata (Müller, 1786)	-	+
9. Ploesoma hudsoni (Imhof, 1891)	+	+
10. Platyias quadricornis (Ehrenberg, 1832)	+	-
Totally Rotifera:	7	9
CLADOCERA		
11. Acroperus harpae (Baird, 1834)	+	+
12. Alonopsis elongatus G.O. Sars, 1862	-	+
13. Bosmina (Eubosmina) coregoni Baird, 1857	+	+
14. B. (B.) longirostris (O.F. Müller, 1785)	+	+
15. Bosmina sp.	-	+
16. Camptocercus rectirostris Schödler, 1862	+	-
17. Ceriodaphnia dubia Richard, 1894	+	+
18. Ceriodaphnia sp.	+	-
19. Chydorus sphaericus (O.F. Müller, 1776)	+	+
20. Ch. ovalis Kurz, 1875	+	+
21. Daphnia (Daphnia) cucullata G.O. Sars, 1862	+	+
22. D. (D.) longispina (O.F. Müller, 1776)	-	+
23. D. (D.) cristata G.O. Sars, 1862	-	+
24. D. (D.) longiremis G.O. Sars, 1862	+	+
25. D. (D.) galeata G.O. Sars, 1864	+	+
26. Daphnia sp.	+	+
27. Leptodora kindtii (Focke, 1844)	+	+
28. Limnosida frontosa G.O. Sars, 1862	+	+
29. Limnosida sp.	+	+
30. Polyphemus pediculus (Linnaeus, 1761)	+	+
31. Scapholeberis microcephala G.O. Sars, 1890	+	+
32. Sida crystallina (O.F. Müller, 1776)	+	+
33. Nauplii	+	+
Totally Cladocera:	19	21
COPEPODA		
Calanoida		
34. Calanoida juv.	+	+
35. Nauplii	+	+
36. Acanthodiaptomus denticornis (Wierzejski, 1887)	+	+
37. Diaptomus sp.	-	+
38. Eudiaptomus gracilis (Sars G.O., 1863)	+	+
39. Eudiaptomus sp.	-	+
40. Heterocope appendiculata Sars G.O., 1863	+	+
41. Heterocope sp.	+	-
Totally Calanoida:	6	7
Cyclopoida		
42. Nauplii	+	+
43. Cyclopoida juv.	+	+
44. Acanthocyclops venustus venustus (Norman & Scott T., 1906)	+	+

Taxons	2005 y.	2015 y.
COPEPODA		
45. A. vernalis vernalis (Fischer, 1853)	+	-
46. Cyclops scutifer scutifer Sars G.O., 1863	+	+
47. C. strenuus strenuus Fischer, 1851	-	+
48. Cyclops sp.	+	+
49. Diacyclops languidoides languidoides (Lilljeborg, 1901)	+	+
50. D. languidus languidus (Sars G.O., 1863)	+	-
51. Eucyclops macrurus macrurus (Sars G.O., 1863)	+	+
52. Macrocyclops albidus albidus (Jurine, 1820)	+	+
53. Macrocyclops distinctus (Richard, 1887)	+	-
54. Megacyclops gigas gigas (Claus, 1857)	-	+
55. M. viridis viridis (Jurine, 1820)	+	+
56. Mesocyclops leuckarti leuckarti (Claus, 1857)	+	+
57. Thermocyclops crassus crassus (Fischer, 1853)	+	+
58. Th. oithonoides (Sars G.O., 1863)	+	+
Totally Cyclopoida:	15	14
Totally Copepoda:	21	21

In the summer period of 2015, cladocerans prevailed (ind./m^3), accounting for 53.9% of the total zooplankton species composition. The basis of this plankters' group were *Daphnia cristata, Daphnia cucullata*, and species of the *Bosmina* genus. The role of Copepoda was also high; the juvenile and nauplial forms of Cyclopoida had high quantities. *Mesocyclops leuckarti*, species of the genera *Eudiaptomus* and *Heterocope* can be noted as well. Overall, copepods were 40.6% of the total population of the plankton community. Large Asplanchna and oligosaprobe *Kellicottia longispina* dominated among rotifers (Figure 2).

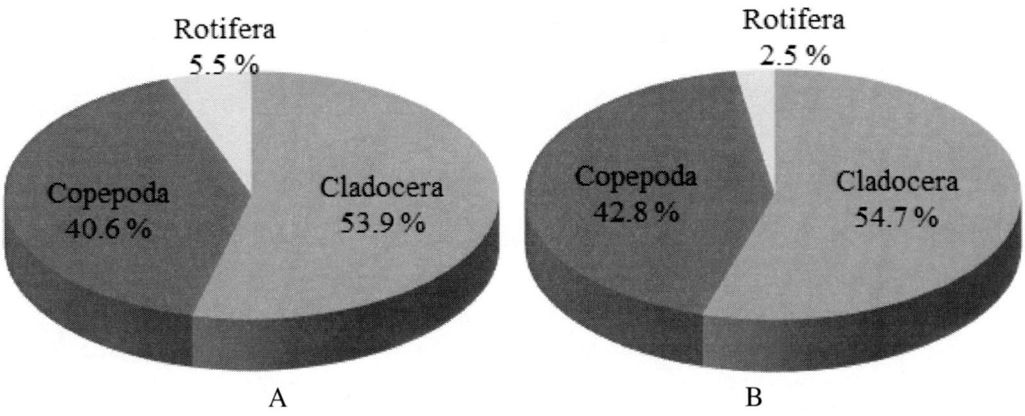

Figure 2. Zooplankton population (A) and biomass (B) rates in Lacha lake in August 2015 (according to: Novoselov et al., 2017a).

The contribution of different zooplankton groups to its formation was different in terms of biomass. At the same time, cladocerans (especially species of the genera *Daphnia* and *Bosmina*) dominated, which accounted for more than half (54.7%) of the total zooplankton biomass. The value of Copepoda biomass, mainly juvenile and nauplial forms of Cyclopoida, was also significant (42.8%) (see Figure 2). The role of rotifers (one of the most sensitive to eutrophication groups of zooplankton) in the formation of biomass was insignificant and was formed by the large rotifers of the Asplanchna genus. On average, the zooplankton biomass of the lake in summer was 0.902 g/m^3 with a population of 29947 ind./m^3. The structure of the zooplankton community in 2015 in Lacha lake allows to conclude that all typical development peculiarities, the spatial and temporal distribution of zooplankton did not go beyond the previously estimated parameters.

While analyzing the distribution of the zooplankton number and biomass, practically the same ratio (both in number and in biomass) of crustaceans and rotifers was observed in the analyzed samples at almost all stations. Their highest values (from 35 to 60 thousand ind./m^3 in population and from 1.3 to 1.6 g/m^3 in biomass) were observed in the South-Western part of the lake at stations # 1, 4 and 16, as well as in the Eastern part of the water body at station # 18. The lowest rates (from 5 to 22 thousand ind./m^3, and from 0.1 to 0.6 g/m^3) were recorded in the North-Western, Central and coastal zones of the water area at stations # 2, 13 and 17, as well as #14.

Comparative analysis of zooplankton species diversity in Lake Lacha showed that its qualitative composition did not change in almost a fifty-year period (from 1968 to 2015) and was represented by the main taxonomic groups: rotifers, as well as cladocerans and copepods. At the same time, their quantitative characteristics changed significantly (Table 2).

Table 2. Changes in the zooplankton community 1968-2015 years, quantity (N, ind./m^3) and biomass (B, g/m^3) (according to: Novoselov et al., 2017a)

Group of organisms	1968		1986		2005		2015	
	N	B	N	B	N	B	N	B
Rotifera	14.3	1.3	33.1	0.8	16.4	11.8	5.5	2.5
Cladocera	26.8	10.7	32.0	84.2	53.1	53.3	53.9	54.7
Copepoda	58.9	88.0	34.9	15.0	30.5	34.9	40.6	42.8
Total	100.0	100.0	100.0	100.0	100.0	100.0	100.0	100.0

Table 2 shows that there were changes in zooplankton composition, affected both the structure of the zooplankton community and its groupquantity. Rotifers, which are considered as indicator test objects in the growing eutrophication of water systems (*Keratella cochlearis hispida, K. cochlearis testa, Trichocerca capucina, T. cylindrica, Filinia longiseta)* appeared and gradually increased among zooplankton communities. Structural changes in zooplankton composition were also observed. Copepods in the total population decreased from 59 to 35% and in biomass - from 88 to 15%. At the same time, the situation with cladocerans had the opposite picture, their population increased from 27 to 32%, in weight - from 11 to 84%. In other words, there was a situation in the lake ecosystem when the appearance and development of small forms of copepods and rotifers resulted in an increase in the zooplankton population without increasing its biomass (Figure 3).

At the beginning of this century (from 2005 to 2015 years), the number of rotifers decreased by more than 3 times (from 16.4 to 5.5%) and biomass – by more than 5 times (from 11.4 to 2.5%). At the same time, the number of large zooplankton copepods increased from 30.5 to 40.6% in number and from 34.9 to 42.8% in biomass, respectively. Quantitative criteria of cladocerans have not changed significantly over the past 10 years (see Table 2, Figure 3).

Figure 3. Dynamics of long-term changes in the population (A) and biomass (B) of the main zooplankton groups in Lake Lacha (according to: Novoselov et al., 2017a).

Zoobenthos

Zoobenthos is one of the main organisms' group, used as biological indicators in water quality assessment during eutrophication process. The representatives of the benthic fauna are characterized by a wide range of ecological features, quite large size, specific habitats, as well as a significant life span that allows them to accumulate pollutants. All this makes them a very convenient object for monitoring freshwater ecosystems (Bakanov, 1997).

According to G. V. Fadeeva and co-authors (Hydrobiology of lakes..., 1978), there are about 200 taxa of benthic and phytophilic invertebrates in lake Lacha. Their taxonomic list is presented below (Table 3).

In general, the benthic fauna of lake Lacha includes representatives of 5 large taxonomic units (types). Freshwater poriferan of Porifera and Coelenterata (including 1 species or 0.5% of the total benthic faunal composition) were isolated. Annelida (24 species or 12.5%) and Mollusca (43 species or 22.3%) were characterized by a slightly larger number of species. More than half of all the detected species (124 or 64.2%) belonged to the arthropod type Arthropoda (Figure 4A). Arachnida (1 species or 0.8%) were isolated among them, more than a third of the species (43 or 37.7%) were Crustacea, and more than half (80 species or 64.5%) of the considered group were Insecta (see Figure 4B). Finally, the last dominant group included 1 species (1.3%) of Lepidoptera water larvae and 21 species (26.3%) of Trichoptera caddisfly larvae. The main part of it (58 species or 72.4%) was represented by the larvae of Diptera (see Figure 4C), consisting exclusively of representatives of the family Chironomidae.

Table 3. Systematic list of invertebrates of zoobenthos and phytophilic fauna (according to: Novoselov et al., 2017b)

Phylum Porifera-freshwater poriferan
1. Spongilla sp. - spongilla
Phylum Coelenterata - coelenterates
Class Hydrozoa - hydroid polyps
1. Hydra sp.
Phylum Annelida – segmented worms
Class Oligochaeta - roundworms
1. *Stylaria lacustris* (L.)
2. *Arcteonais lomondi* (Martin)
3. *Ripistes parasita* (Schmidt)
4. *Nais pseudobtusa* Piguet
5. *N. barbata* Mull.
6. *N. simplex* Piguet
7. *Ophidonais serpentina* (Mull.)
8. *Uncinais uncinata* (Oersted)
9. *Chaetogaster diaphanous* (Gruith)
10. *Limnodrilus udekemianus* Claparede
11. *L. hoffmeisteri* Claparede
12. *Potamothrix hammoniensis* (Mich.)
13. *Isochaetides newaensis* (M ich.)
14. *Psammoryctides albicola* (Mich.)
15. *P. barbatus* (Grube)
16. *Tubificidae* gen. sp. juv.
17. *Peloscolex ferox* (Eisen)
18. *Enchytreidae* gen. sp.
19. *Lumbriculus variegatus* (Muller
Class Hirudinea - leeches
1. *Glossiphonia complanata* L.
2. *Helobdella stagnalis* Blanch
3. *Piscicola geometra* L.
4. *Herpobdella octoculata* (L.)
5. *H. testacea* Savigny
Phylum Mollusca - molluscs
1. *Limnaea stagnalis* (L.)
2. *Radix auricularia* (L.)
3. *R. lagotis* (Schrank)
4. *R. ovata* (Draparnaud)
5. *R. pereger* (Muller)
6. *Galba glabra* (Muller)
7. *Amphipeplea glutinosa* (Muller)
8. *Physa fontinalis* (L.)
9. *Planorbis carinatus* Muller
10. *P. planorbis* (L.)
11. *Coretus corneus* (L.)
12. *Anisus vorticulus* (Troschel)
13. *A. contortus* (L.)
14. *Guraulus albus* (Muller)
15. *G. gredleri* (Bielz) Gredler
16. *G. laevis* (Alder)
17. *Armiger crista* (L.)
18. *Acroloxus lacustris* (L.)
19. *Valvata piscinalis* (Muller)

20. *V. pulchella* Studer 21. *V. cristata* Muller 22. *Viviparus viviparus* (L.) 23. *Bithynia tentaculata* (L.) 24. *B. leachi* (Sheppard) 25. *Unio pictorum* (L.) 26. *Anodonta anatina* (L.) 27. *Sphaerium solidum* (Normand) 28. *Sph. subsolidum* Clessin 29. *Sph. corneum* (L.) 30. *Sph. nitidum* Clessin 31. *Sph. lacustre* (Muller) 32. *Pisidium amnicum* (Muller) 33. *P. supinum* A. Schmidt 34. *P. henslowanum* (Sheppard) 35. *P. casertanum* (Poli) 36. *P. ponderosum* Stelfox 37. *P. nitidum* Jenyns 38. *P. subtruncatum* Malm 39. *P. pusilium* (Gmelin) Jenyns 40. *P. pulchellum* Jenyns 41. *P. obtusale* (Lam.?) Jenyns 42. *P. milium* Held. 43. *P. moitesserianum* Paladilhe
Phylum Arthropoda - arthropods Class Crustacea - crustaceans Order Isopoda - isopods 1. *Asellus aquaticus* L. Order Amphipoda - amphipods 1. *Gammarus lacustris* Sars 2. *Pallasea quadrispinosa* Sars Order Decapoda - decapods 1. *Astacus leptodactylus* L. Order Ostracoda - shells 1. *Nyocypris decipiens* Masi 2. *Cypridopsis vidum* O.F. Muller 3. *Cyclocypris ovum* (J.) 4. *C. serena* (Koch) - + 5. *Cypria ophthalmica* (J.) - + 6. *Candona candida* 7. *C. protzi* Hartwig 8. *C. caudata* Kaufmann 9. *C. pratensis* Hartwig 10. *Darvinula stevensoni* Br. et Rob. 11. *Cytherissa lacustris* G.O. Sars 12. *Limnocythere sancti pafticii* Br. et Rob Class Arachnida – arachnids Order Araneina - the spiders *1. Argyroneta aquatica* L. Class Insecta - insects Order Trichoptera - caddisfly 1. *Rhayacophila* sp. 2. *Agapetus comatus* Piet. 3. *Agraylea multipunctata* Curt. 4. *Hydroptila femoralls* Eaton 5. *Hydroptila* sp. 6. *Orthotrichia tetensii* Kolbe 7. *Oxyethira flavicornis* Pictet 8. *Oxyethira* sp. 9. *Holocentropus picicornis* Steph.

Table 3. (Continued)

10. *Cyrnus flavidus* Mc. L.
11. *Phryganea bipunctata* Retz.
12. *Ph. grandis* L.
13. *Phryganea* sp.
14. *Molanna angustata* Curt
15. *M. albicans* Zett.
16. *Molanna* sp.
17. *A. cinereus* Curt.
18. *Mystacides nigra* L.
19. *Oecetis ochracea* Curt.
20. *Oecetis* sp.
21. *Limnephilus* sp.

Order Lepidoptera - butterflies

1. *Acentropus niveus* Ollv.

Order Diptera - Diptera

1. *Micropsectra gr. praecox* Mg.
2. *Stempelina* gr. *bausei* Kieff.
3. *Tanytarsus* gr. *lobatifrons* Kieff.
4. *T. gr. gregarius* Kieff.
5. *T. gr. mancus v.d.* Wulp
6. *T. pedice lliferus* Birula
7. *Cryptochironomus gr. fuscimanus* Kieff.
8. *C. gr. defektus* Kieff.
9. *C. gr. vulneratus* Zett.
10. *C. gr. conjugens* Kieff.
11. *C. gr. camptolabis* Kieff.
12. *C. gr. viridulus* P.
13. *C. gr. anomalus* Kieff.
14. *C. gr. pararostratus* Lenz
15. *Cryptochironomus* sp. Lipina
16. *Einfeldla gr. carbonaria* Mg.
17. *Pseudochironomus gr. prasinatus* Staeg.
18. *Endochironomus gr. dispar* Mg.
19. *E. albipennis* Mg.
20. *Grlyptotendipes gr. gr ipekoveni* Kieff.
21. *Lauterborniella brachylabis* Edv.
22. *Limnochironomus gr. nervosus* Staeg.
23. *L. gr. tritomus* Kieff.
24. *Microtendipes gr. chloris* Mg.
25. *Paratendipes gr. albimanus* Mg.
26. *P. intermedius* Tschern.
27. *Pentapedilum executrn* Kieff.
28. *Polypedilum breviantennatum* Tschern.
29. *P. gr. convictum* Walk.
30. *P. gr. nubeculosum* Mg.
31. *P. gr. scalaenum* Schr.
32. *Stictochironomus gr. histrio* P.
33. *S. psammophilus* Tschern.
34. *Stictochironomus* sp.
35. *Chironomus f.1. bathophilus* Kieff.
36. *Ch. f.1.plumosus* (L.)
37. *Ch. f.l. thummi* Kieff.
38. *Tendipedini gen.?*
39. *Pagastiella orophilla* Edw.
40. *Allochlranomus* Kieff.
41. *Cricotopus gr. silvestris* (Fabr.)
42. *C. algarum* Kieff.

43. *C. versidentatus* Tschern.
44. *C. biformis* Edw.
45. *Psectrocladius gr. psilopterus* Kieff.
46. *P. medius* Tschern.
47. *Microcricotopus bicolor* Zett.
48. *Trichocladius gr. lucidus* Staeg.
49. *Diamesa campestris* Edw.
50. *D. gaedi* Mg.
51. *Prodiamesa oltvacea* Mg.
52. *Syndiamesa gr. nivosa* Geotgh.
53. *Corynoneura* sp.
54. *Thienemaniella* sp.
55. *Ablabesmyia gr. lentiginosa* Pries.
56. *A. gr. monilis* L.
57. *Clinotanypus nervosus* Mg.
58. *Procladius* Skuze

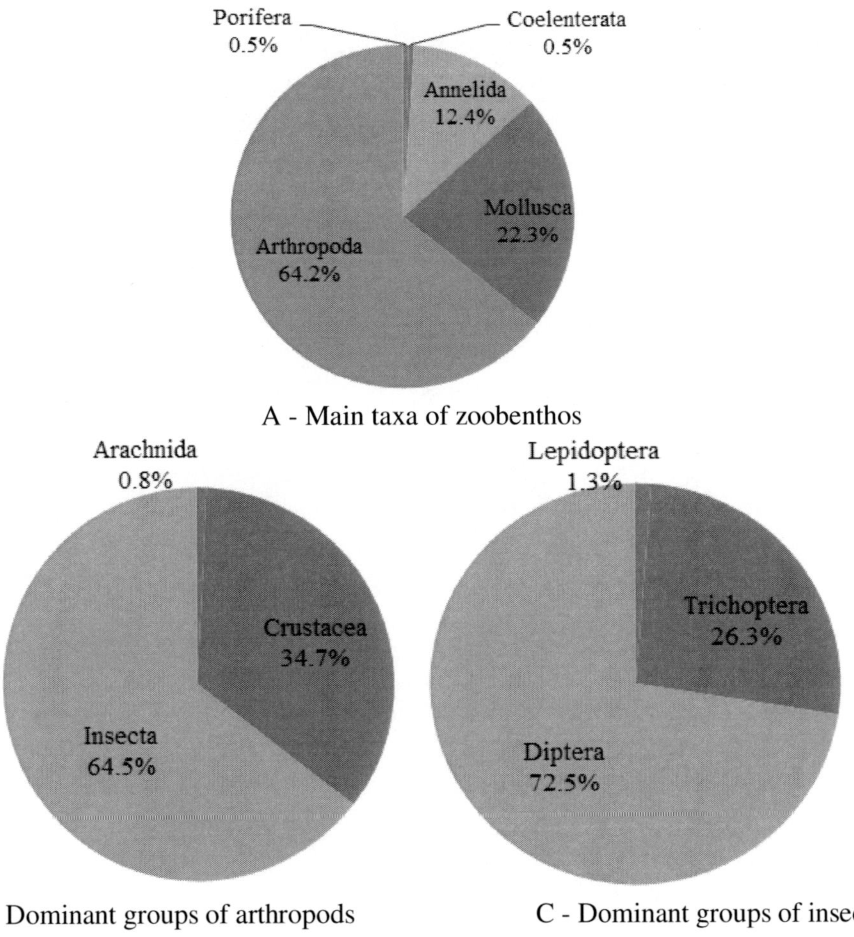

Figure 4. Composition of the benthic fauna of Lake Lacha: large taxa-types (A), representatives of the dominant groups-arthropods (B) and insects (C) (according to: Novoselov et al., 2017b).

Data for the period since the beginning of regular observations in 2003, allow to assess the lake as a water body with an average level of benthic fauna development, characterized by a fairly large species diversity.

Totally 16 systematic benthic groups were observed in the benthic fauna of the lake during this period (Figure 5). These are representatives of worms (Nematoda, Oligochaeta, Hirudinea), water crustaceans (Isopoda, Phillopoda, Cladocera, Ostracoda), mollusks (Bivalvia, Gastropoda), water mites (Acari) and water larvae and pupae of insects (Ephemeroptera, Trichoptera, Heleidae, Chironomidae, Simuliidae and other Diptera).

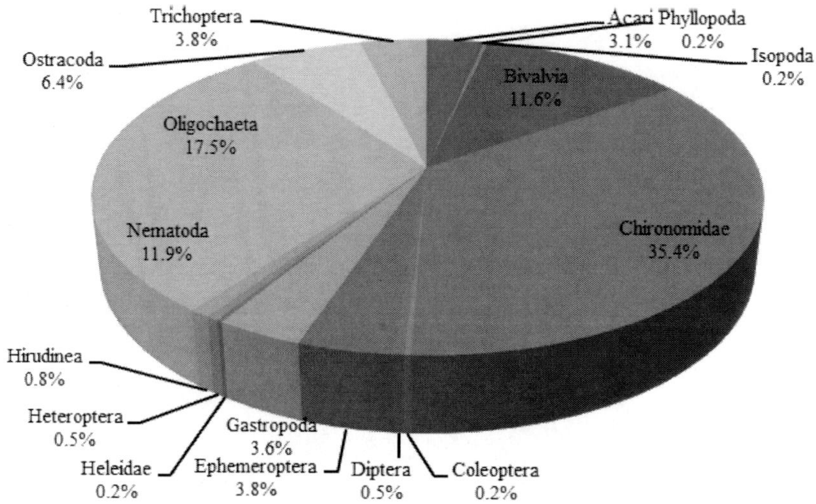

A

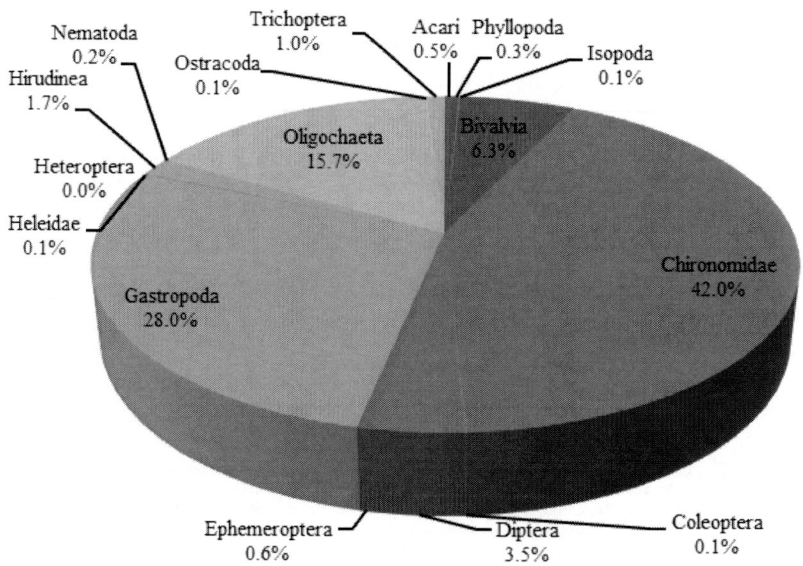

B

Figure 5. Groups' Ratio of benthic fauna in lake Lacha based on multi-year (2003-2015) research (A-number, B-biomass) (according to: Novoselov et al., 2017b).

In general, the chironomid larvae (Chironomidae lv.) presented in all samples, and dominated in population (35.4%) and biomass (42.0%). Besides them, there were oligochaetes (17.5%), nematodes (11.9%) and bivalves (11.6%). In terms of biomass, gastropods (28.0%)

and oligochaetes (15.7%) played a significant role in the formation of benthic biomass additionally to chironomid larvae.

Chironomid larvae (*Chironomus plumosus, Procladius, Tanitarsus gr. mascuc, T. gr. lauterborni, Polypedium breviantenattum*) and mollusks from the pea family Pisidiidae (*Euglessa casertana, E. nitida, E. supina, E. milium, E. ponderosa*) are mass species of the open part of the lake. Benthic biomass was 10.44 g/m^3 in average during the research period. They were dominated by Phytophilic chironomid larvae (*Endochironomus albipennis, Gliptotendipes gr. gripekoveni, Gricotopus gr. silvestris, Criptochironomus gr. pararostratus*) and oligochaetes (*Stylaria lacustris, Ripistes parasita, Nais barbata*) dominated in macrophyte thickets. The average biomass did not exceed 5.7 g/m^2.

The sands, being a productive biotope, are widely populated by chironomid larvae (*Tanytarsus gr. mancus, T. gr. gregarius, Polypedium breviantennatum, P. gr. scalaenum*), mollusks (Pisidiidae) and oligochaetes (*Stylaria lacustris, Nais barbata*). Psammophilic invertebrates inhabit the silted sand biotope, where larvae of chironomids of the genera Tanytarsus and Procladius, molluscs of the family Pisidiidae and Valvata pischinalis, and oligochaetes are dominating. The zoobenthos population on the sand was 3666 ind./m^2, its biomass was 5.8 g/m^2, and on silted sands there was 2450 ind./m^2 and 4.8 g/m^2, respectively. The surf zone fauna of rocks is mainly represented by molluscs of the pond snails genus Radix ovata and a few chironomid larvae. Zoobenthos population was relatively low here (150 ind./m^2) and biomass was 0.46 g/m^2 during the sampling period. In the safe zone in rocks overgrown with cladophora, mainly oligochaetes, nematodes and chironomid larvae live. Their average number was 9720 ind./m^3, and their biomass was 3.18 g/m^2.

In general, long-time annual average rates for the benhic fauna in the lake were 3620 ind./m^2 in population, and 10.6 g/m^2 in biomass.

Material for hydrobiological analysis was collected throughout the lake's water area at 12 stations, mainly on muddy biotopes during the growing season (August 2015). When analyzing the distribution of the zoobenthos population and biomass in the water area, almost all stations showed the same ratio (both in number and in biomass) of insects, namely aquatic chironomid larvae. Their greatest population numbers (from 600 to 900 ind./m^2) were also found in the South-Western part of the lake, and biomass (from 2.4 to 3.7 g/m^2) – in the southern part of the water area and in the Western part. The lowest values (from 1 to 300 ind./m^2 and the biomass from 0.001 to 1.2 g/m^2) were observed at all other stations along the Northern and Southern coasts of lake Lacha.

The benthic quantity during the sampling period in 2015 was significantly lower than the long-term annual average. Only 4 groups of benthic hydrobionts were observed, where chironomid larvae was dominant, and occupied by 49.5% of the total number and 77.3% of the total biomass (Figure 6), oligochaetes were the subdominant group, with 28.4 and 18.5% respectively.

The average benthos population was 423 ind./m^2, with an average biomass of 1.26 g/m^2. Low quantity of the benthic fauna in the 2015 and the depletion of the species composition were likely due to poor hydrological conditions of the year, as well as massive departure of amphibians during that period.

Benthic invertebrates were numerous and produced significant biomass. However, quantity of the benthic fauna fluctuated significantly in different research years (Figure 7).

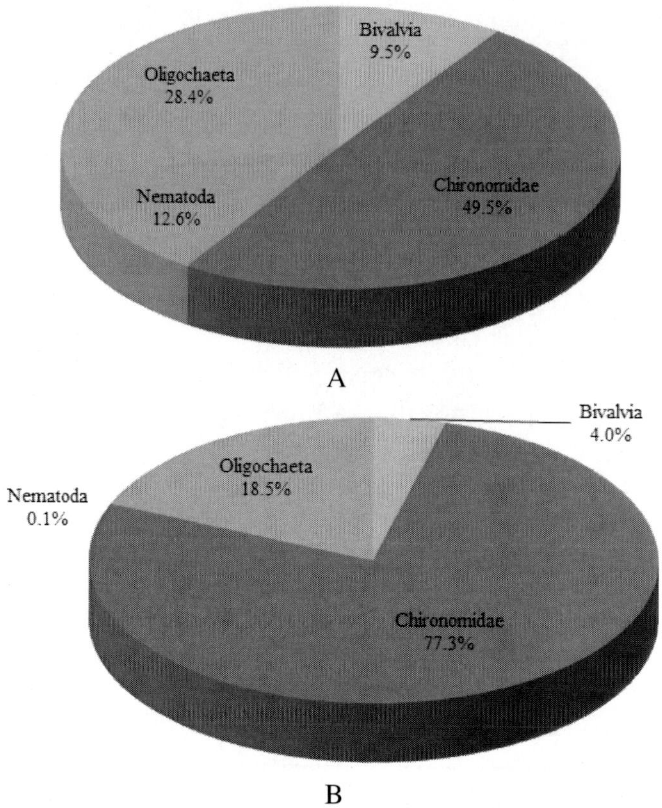

Figure 6. Groups' ratio of benthic fauna in Lake Lacha, August 2015 (A-population, B-biomass) (according to: Novoselov et al., 2017b).

The average benthic population ranged from 423 ind./m^2 (2015) to 3.538 ind./m^2 (2005), while the average biomass fluctuated from 1.48 g/m^2(2003) to 4.53 g/m^2(2014). There was low population at the beginning of the research period (2003-2004) and in some seasons in the in 2013 and 2015. The population was made of few Diptera, oligochaetes and bivalves. Single chironomid larvae performed Diptera, whose share in the biomass formation was insignificant. Biomass was mainly formed by the limited amount of molluscs of the genus Pisidium. In some years (2005, 2007, 2010), population of Diptera, oligochaetes, and molluscs was significantly higher. In addition, the benthic composition was expanding qualitatively due to representatives of caddislfies, dayflies, water mites and beetles, as well as benthic plankton crustaceans. However, the biomass of the benthic fauna was represented mainly by bivalves and chironomid larvae. Since 2005, there has been an increase in the number of oligochaetes, whose number was consistently high until 2012. Oligochaetes also formed significant biomass during this period. But later, their quantity decreased. Fluctuations in benthic fauna population were probably related to the hydrological conditions of specific years. In general, according to the fish-farming classification (Pidgayko et al., 1968), lake Lacha can be classified as a water body with medium feeding capacity for benthic fish.

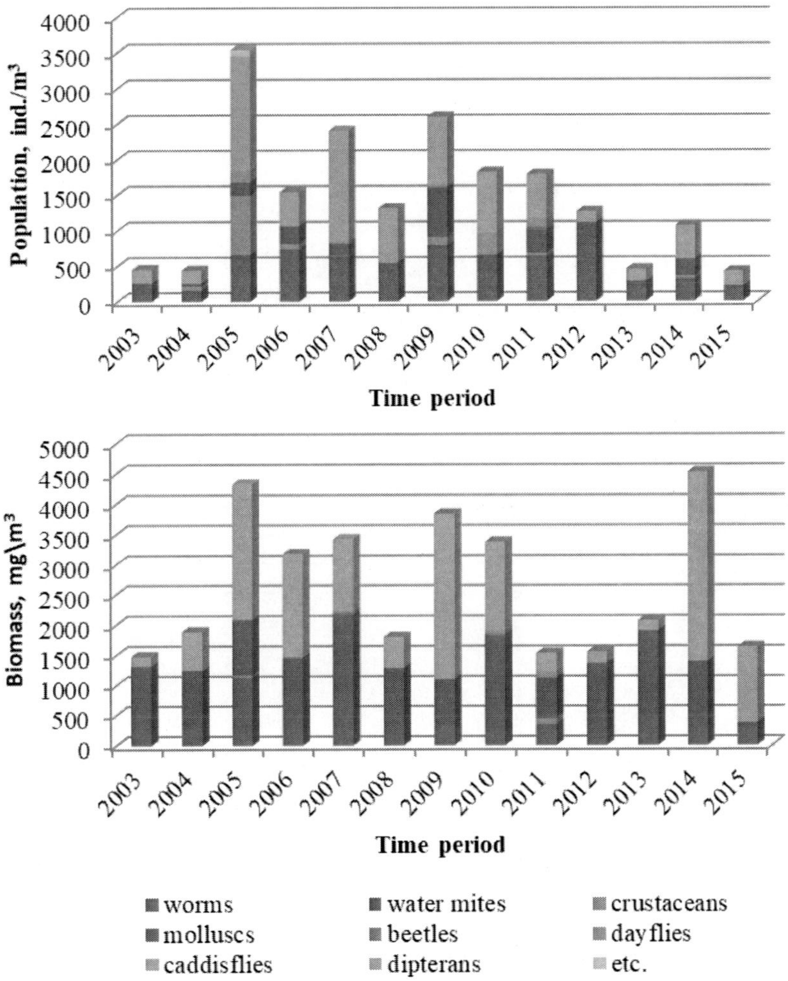

Figure 7. Quantity dynamics of the main benthic groups in Lake Lacha (according to: Novoselov et al., 2017b).

Ichthyofauna

For a long time there was fishery on lake Lacha. Even N. Ya. Danilevsky (1876) in his report about fishing in Russia noted that "due to long-term fishing with close mesh nets there is a few fish in lake Lacha." From the information collected in 1861,there were up to 300 close mesh nets every year in the lake, each of which caught from 30 to 50 poods of fish. The total annual catch was approximately 200 tons. Basically, ruff was caught, which used for the dried fish production. Danilevsky insisted on banning fishing with close mesh nets, hoping that this would help increase the number of large fish.

After the October revolution, fishing statistics remained unestimated for a long time. The approximate number of fish caught in the lake before 1914 was 12.0 thousand poods per year, or 200 tons. In 1928, the catch was 500 tons, while the main catch was in the summer and autumn season – 400 tons, and in the spring and winter - only 100 tons. In 1932, caught fish ratio was as follows: pike – 11.3%, ide – 1.7%, ruff – 44.7%, smelt – 4.1%, burbot – 1.5%, roach – 25.2%, perch – 11.4%, whitefish – 0.1%. From the Table 4, ruff, roach and pike were main fishing species. In the late 30s-early 40s of the 20th century, fish catches in lake Lacha were 527 t in 1938, 421 t - in 1939, 300 t - in 1940, 280 t - in 1941. Fish diversity was small in catches, main species were ruff, perch and roach (Kuznetsov, 1951). Analysis of the fish catch dynamics over a long period shows that there are significant changes in the fish community in the lake. In the pre-war years, fishing was based on early-matured groupings of perch, roach and ruff. Nowadays, bream dominates, whose age reaches 20 years. Analysis of catch dynamics over a long period shows that there were significant changes in the ichthyocenosis of the lake; bream, pike, ide and walleye appeared instead of small ordinary fish. Whitefish were extremely rare, and they were not included in fishing statistics (Table 4).

Table 4. Fishing in the Lake Lacha, t (according to: Kozmin, Shatova, 1997)

Year	Bream	Pike	Burbot	Ide	White bream	Perch	Roach	Ruff	Pike perch	Other kinds	In total
1961-1970	56.4	21.5	3.0	8.3	6.0	11.6	4.3	8.8	-	14.5	135.6
1971-1980	67.8	23.0	7.8	3.5	1.6	6.6	11.7	7.1	-	8.0	137.5
1981-1990	111.2	12.7	4.3	2.8	2.0	8.2	14.1	3.4	-	1.4	161.4
1991-1995	155.5	2.8	2.0	2.2	2.6	7.6	4.6	-	-	-	180.0
1996	71.0	0.8	0.1	1.8	0.5	1.3	0.5	0.2	-	-	76.2
1997	60.9	0.2	0.1	0.7	0.1	1.3	0.3	0.1	-	-	63.7
1998	64.2	0.1	0.4	-	-	1.1	0.1	0.1	-	-	66.0
1999	60.8	0.3	0.3	0.8	-	0.8	0.1	-	-	-	63.1
2000	52.4	2.0	0.5	1.8	-	2.6	0.8	0.8	-	4.2	65.1
2001	43.0	6.5	2.5	2.5	-	5.5	2.5	3.4	4.5	-	70.4
2002	44.9	7.9	2.4	2.4	-	5.9	3.4	1.9	6.7	-	75.5
2003	52.1	15.4	3.5	6.6	0.1	8.6	4.4	2.1	15.6	-	108.4
2004	12.1	2.7	1.0	1.2	-	2.6	1.1	-	2.4	0.1	23.2
2005	11.0	4.0	0.6	1.1	-	1.3	0.6	-	1.0	0.9	20.5
2006	9.0	4.0	0.9	0.9	-	1.0	0.5	0.7	0.8	-	17.8

Bream is the main commercial fish in lake Lacha. Its fishing is regulated by setting a limit in order to maintain bream stocks at an optimal level since 1990.

The average catch per fishing effort is an important parameter, characterized stocks' state of a particular type of fish. In the summer of 2005, small-sized bream, which is not popular among the bream population, predominated in non-water catches in lake Lacha.. In this regard, fishermen were forced to release young bream (less than 25 cm) from catches back into the lake. A large-scale trawl with a length of 1000 m worked on lake Lacha until 2003. The average catch varied from 456 to 639 kg.

The analysis of the current situation in the fishing complex on lake Lacha showed the irrational use of the fish resources. Only bream is caught by fishing, and small ordinary fish have not been caught for the past ten years. The lake is heavily overgrown with aquatic vegetation, there is a gradual siltation of bream and pike spawning grounds, which reduces the efficiency of natural fish reproduction. In addition, siltation of the lake reduces the availability of benthic organisms and is the reason for the late feed-switching of bream, resulting in a formation of a slow-growing bream population in the lake. A similar phenomenon was noted earlier during the eutrophication of Syamozero in 1973-1977 (Dgebuadze, 1979). At the same time, the deterioration of the size and age structure of the bream population is confound by the fishery of just large fish.

Lake Lacha is a eutrophic water body. Its commercial fish productivity is 10 kg/ha on average. In the early 90-ies of the last century, the recorded annual catch of fish here reached 180 tons. By including perch, roach and ruff in the fishery, the current annual catch can be raised to 200 tons, including 100 tons of bream (Kozmin, 2011).

Kenozero Lake (Kenozersky National Park)

Lake Kenozero is the third largest lake in the territory of the Arkhangelsk region. Its total area is 74.1 km^2, and the water mirror area is 66.3 km^2. Kenozero is part of the Onega river basin. The river Kena flows from the Kenozero. The water body is located on the territory of Kenozersky national park.

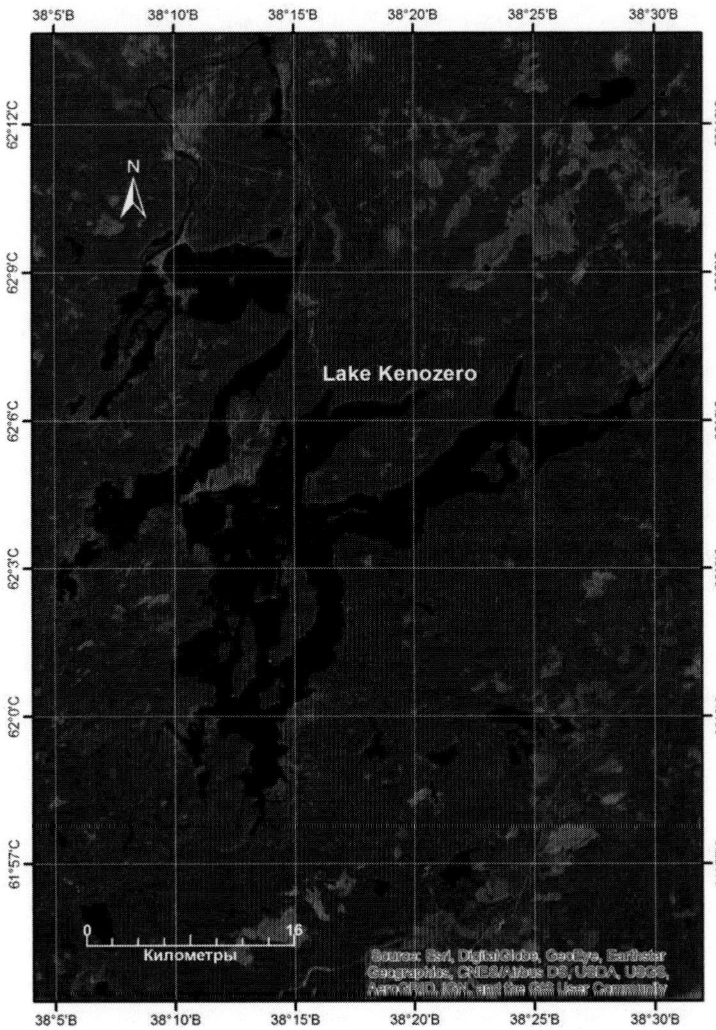

Figure 8. Lake Kenozero.

The lake is of a tectonic origin. The topography of the coastline is very rugged. Numerous islands and peninsulas divide the lake into separate stretches and bays (Figure 8). The lake has a very complex bottom relief and close proximity of large depths with banks and shoals. Its reliably established maximum depth is 77 m. The littoral zone is poorly developed; shallow areas with a depth of up to 5 m occupy a narrow coastal strip and shoals in the central part of the lake.

The water body has a good flow rate, its conditional water exchange rate equals to 1,1. Kenozero is characterized by transverse water flow, which leads to the formation of local hydrological and hydrochemical conditions in bays and reaches of the Northern and Southern parts of the lake. The water mass capacity of Kenozero lake is 1.343 km^3.

In general, the water system has a lake-river level regime, mainly due to the annual flow of the main tributary of the Kenozero – the Pocha river. There are four cleary expressed seasons in the annual fluctuations of the level: autumn flood, winter low-water period, spring flood and summer low water. According to the average multi-year data, autumn flood caused by autumn rains associated with increased cyclonic activity is observed in October-November. The average drop in water levels continues until the end of March and the beginning of April. The highest water levels are observed during the spring flood – from mid-April to mid-May. The average annual level amplitude in the Kenozero is 208 cm.

Kenozero water mass is characterized by stable thermal stratification of the hydrological regime. Homothermy is observed here only in the spring and autumn periods, when convective mixing covers the entire water of the lake from its surface to the bottom in all parts, both shallow-water and deep-water (Dvoryankin, 2010).

Hydrochemistry

Institute of Ecological problems of the North conducted hydrochemical studies of the lake in 1999-2001 (Kalinkina, 1999; Lukin, 2001), while Northern Department of Polar Russian Institute of Fisheries and Oceanography - in 2007-2008 (Dvoryankin, 2010). Kenozero waters are slightly mineralized (Table 5). During the research period, the overall level of mineralization did not change. The waters of Kenozero lake contains a relatively large amount of organic matter of allochthonous origin, primarily of a humus nature. Due to that fact water colour ranged from 77 to 109 degrees in summer, reached up to 150 degrees in the upper layer in winter at some stations (Kalinkina, 1999; Lukin, 2001). The water transparency was therefore very low – 1.5 m. The color of Kenozero waters gradually decreased with increasing depth.

The content of total nitrogen in water of Kenozero lake in the winter of 1999 was small (not more than 0.5 mg/L). It remained at about the same level in 2001, 2007 and 2008 (Table 5). Comparison with data from 1999-2001 shows that the amount of mineral forms of nitrogen ($N\text{-}NH_4$ and $N\text{-}NO_3$) in the water also remained at the same level. The waters of Kenozero lake had low concentrations of total phosphorus. The maximum content of total phosphorus was observed in the surface. In recent years, the amount of this biogenic element has not changed much. Phenols' content in the water has fallen by more than three times, which was obviously due to the ban on wood rafting on the lake, introduced in the 90s of the XX century. The content of petroleum products was also low – 0.035 mg/L.

Table 5. Chemical water composition of lake Kenozero, mg/L (according to: Kalinkina, 1999; Lukin, 2001; Dvoryankin, 2010)

Parameters	1999	2001	2007-2008	
			autumn	spring-summer
Common nitrogen, mg/L	0.500	0.360	0.300	0.220
Ammonium nitrogen	0.010	0.005	0.020	-
nitrate, mg/L	0.050	0.020	0.060	0.050
nitrite, mg/L	-	-	0.008	0.006
Common phosphorus	0.020	0.006	0.030	0.044
mineral, mg/L	-	-	0.010	0.013
organic, mg/L	-	-	-	0.031
pH	-	7.20	-	-
Mineralization, mg/L	44.30	-	50.9	43.5
Phenols, mg/L	0.007	-	0.0002	-
Mineral oils, mg/L	-	-	0.035	-
Fe, mcg/L	313	-	-	-
Al, mcg / L	128	-	-	-
Si, mcg/L	-	-	190	193

The catchment area is the main source of organic substances, biogenic elements and microelements. There are relatively small volumes of wastewaters entering the Kenozero from coastal settlements. They do not have a significant impact on the main water quality characteristics neither in winter nor in summer. There are no industrial facilities on the coast of the lake. Thus, the waters of Kenozero lake retain their natural quality (Dvoryankin, 2010).

Higher Water Vegetation

Higher aquatic vegetation of Kenozero lake is slightly diverse in its species composition. The most common are common reeds and lake reeds. Wide thickets of these two species are located almost along the entire shoreline of the lake. In some areas there are spatterdock, clasping-leaved pondweed, sometimes pondweed forms short areas of thickets in the open part of the lake.

Figure 9. Kenozero Lake. The highest water vegetation near the village of Vershinino.

The largest thickets of common reed and lake reed are located in the shallow waters of the Northern part of the lake, opposite the village Kenozersky pogost, Vershinino, in the gulf of Persluchta (Figure 9). Weak development of macrophytes (their area is less than 3% of the total area of the water body) is caused by the complex relief of the lake basin, the presence of a sharp and steep coastal slope that begins immediately from the water's edge (Dvoryankin, 2010).

Phytoplankton

First phytoplankton studies on Kenozero lake were conducted at the end of august 2001 by the expedition of Northern Water Problems' Institute Karelian Research Centre, Russian Academy of Sciences (Lukin, 2001). Samples were taken at 8 stations across the entire water area, which made it possible to obtain an integral characteristic of the entire water body. There were 85 algae taxons identified in the plankton of the Kenozero, including: Cyanophyta – 16, Chrysophyta – 5; Cryptophyta – 4, Bacillariophyta – 23, Dynophyta – 3; Chlorophyta – 34 (Table 6). In average, there were 37.1 taxons in the samples. The Shannon index (Sh) of species diversity reached relatively high values and was 3.78 in average. The taxons population equalization index $((Sh\text{-}Sh_{min}) / (Sh_{max} - Sh_{min}))$ was 42%. The Simpson index was equal to 0.14. High values of the index concern to the high diversity of phytoplankton community in the lake. The analysis of the species similarity showed that the phytoplankton distribution was uniform.

Table 6. The integral quantitative phytoplankton characteristics of Kenozero lake, August 2001 (according to: Lukin, 2001)

Taxon	N	EN	N, %	B	EB	B, %	f, %
CYANOPHYTA							
Anabaena sp.	0.26	0.12	0.79	3.89	1.85	0.14	50.0
Aphanothece sp.	0.21	0.11	0.65	1.07	0.53	0.04	62.5
Aphanocapsa sp.	0.01	0.01	0.04	0.03	0.03	0.00	12.5
Aphanocapsa elachistica W. et G.S.West	0.46	0.39	1.41	1.65	1.29	0.06	25.0
Aphanocapsa delicatissima W. et G.S.West	0.09	0.04	0.29	0.45	0.21	0.02	50.0
Chroococcus minutus (Kutz.) Nageli	0.11	0.08	0.34	0.70	0.47	0.02	25.0
Coelosphaerium kuetzingianum Nag.	0.57	0.57	1.75	2.29	2.29	0.08	12.5
Limnothrix planctonika (Wolocz.) Meffert	0.22	0.07	0.66	2.17	0.72	0.08	87.5
Merismopedia tenuissima Lemm.	0.30	0.30	0.90	0.03	0.03	0.00	12.5
Merismopedia warmingiana Lagerch.	0.81	0.50	2.47	0.08	0.05	0.00	37.5
Phormidium tenuis (Menegh.) Gom.	0.01	0.01	0.02	0.23	0.23	0.01	12.5
Phormidium sp.	0.07	0.03	0.20	2.36	1.21	0.08	50.0
Snowella atomus Kom. at Hindak	0.65	0.29	2.00	0.81	0.35	0.03	75.0
Snowella lacustris (Chod.) Kom. et Hindak	0.55	0.31	1.68	1.67	1.13	0.06	37.5
Snowella sp.	0.13	0.07	0.39	0.28	0.16	0.01	37.5
Woronichinia sp.	0.22	0.22	0.68	0.67	0.67	0.02	12.5
CHRYSOPHYTA							
Dinobryon divergens Imh.	0.15	0.03	0.47	4.62	1.02	0.16	100.0
Dinobryon bavaricum Imh.	0.12	0.04	0.36	3.58	1.07	0.13	100.0
Dinobryon acuminatum Ruttn.	0.01	0.01	0.03	0.29	0.29	0.01	12.5
Mallomonopsis sp.	0.06	0.02	0.18	7.41	3.09	0.26	62.5
Pseudostaurastrum hastatum (Reins.) Chod.	0.01	0.01	0.04	3.40	3.40	0.12	12.5
BACILLARIOPHYTA							
Acanthoceros zachariasii (Brun) Simons.	0.15	0.05	0.45	74.67	25.39	2.65	62.5
Asterionella formosa Hass.	8.74	0.79	26.73	305.96	27.58	10.85	100.0
Taxon	N	EN	N, %	B	EB	B, %	f, %

BACILLARIOPHYTA							
Aulacoseira italica (Ehr.) Sim.	0.27	0.10	0.82	15.34	5.60	0.54	62.5
Aulacoseira italica var. tenuissima (O.Mull.) Sim	0.09	0.09	0.26	4.29	4.29	0.15	12.5
Aulacoseira distans (Ehr.) Sim.	0.05	0.04	0.16	3.91	3.05	0.14	25.0
Aulacoseira alpigena (Grun.) Krammer	1.77	0.35	5.43	113.91	26.36	4.04	100.0
Aulacoseira islandica (O.Mull.) Sim.	0.17	0.05	0.52	13.76	3.90	0.49	75.0
Aulacoseira granulata (Ehr.) Sim.	0.55	0.15	1.69	331.07	90.09	11.74	100.0
Cocconeis sp.	0.08	0.03	0.24	2.24	0.81	0.08	62.5
Cyclotella sp.	0.24	0.07	0.72	46.35	18.49	1.64	87.5
Cyclotella comta (Ehr.) Kutz.	0.03	0.02	0.10	4.93	3.26	0.17	25.0
Cyclotella glomerata Bachm.	0.03	0.03	0.09	1.48	1.48	0.05	12.5
Fragilaria crotonensis Kitt.	0.56	0.37	1.71	47.25	32.65	1.68	62.5
Fragilaria capucina Desm.	0.05	0.05	0.14	5.57	5.57	0.20	12.5
Rhizosolenia longiseta Zach.	0.72	0.16	2.21	211.90	54.31	7.52	100.0
Stephanodiscus sp.	0.08	0.03	0.24	7.57	3.05	0.27	62.5
Synedra ulna (Nitzsch.) Ehr.	0.03	0.03	0.08	6.81	6.81	0.24	12.5
Synedra acus Kutz.	0.15	0.06	0.47	50.74	18.78	1.80	75.0
Synedra berolinensis Lemm.	0.03	0.03	0.09	0.74	0.74	0.03	12.5
Synedra nana Meister	0.54	0.10	1.64	4.83	0.89	0.17	100.0
Tabellaria fenestrata (Lyngb.) Kutz.	0.61	0.39	1.85	146.86	97.27	5.21	50.0
Tabellaria fenestrata var.asterionelloides Grun.	5.67	1.32	17.35	1138.39	177.58	40.37	100.0
Tabellaria flocculosa (Roth.) Kutz.	0.01	0.01	0.03	2.84	2.84	0.10	12.5
CRYPTHOPHYTA							
Croomonas acuta Uterm.	0.04	0.02	0.13	0.49	0.27	0.02	37.5
Chroomonas nordstedtii Hans.	0.01	0.01	0.02	0.11	0.11	0.00	12.5
Cryptomonas sp.	0.02	0.01	0.06	3.07	2.11	0.11	25.0
Chilomonas minor Ehr.	0.01	0.01	0.04	0.04	0.04	0.00	12.5
DYNOPHYTA							
Ceratium hirundinella (O.F.M.) Bergh.	0.05	0.02	0.14	31.39	12.10	1.11	62.5
Peridinium sp.	0.14	0.06	0.43	44.21	18.84	1.57	75.0
Peridinium pusillum (Penard) Lemm.	0.03	0.03	0.09	8.18	8.18	0.29	12.5
CHLOROPHYTA							
Staurodesmus sp.	0.01	0.01	0.02	0.20	0.20	0.01	12.5
Staurodesmus mammillatus (Nord.) Teiling	0.03	0.02	0.10	1.80	1.21	0.06	37.5
*Staurodesmus megacanthus (Lund.)*Teiling	0.01	0.01	0.04	3.26	3.26	0.12	12.5
Cosmarium sp.	0.02	0.02	0.07	2.77	1.82	0.10	25.0
Cosmarium phaseolum Breb.	0.02	0.02	0.06	1.91	1.91	0.07	12.5
Staurastrum sp.	0.03	0.02	0.09	1.46	0.92	0.05	37.5
Staurastrum anatinum Cooke et Wiills	0.01	0.01	0.05	5.95	5.95	0.21	12.5
Staurastrum sebaldi Reinsch	0.01	0.01	0.04	1.30	1.30	0.05	12.5
Staurastrum petsamoense (Boldt) Jarnef.	0.01	0.01	0.02	6.43	6.43	0.23	12.5
Staurastrum paradoxum Meyen	0.01	0.01	0.03	1.91	1.91	0.07	12.5
Spondylosium planum (Wolle)	0.35	0.09	1.07	26.05	9.69	0.92	87.5
Closterium sp.	0.16	0.11	0.49	26.62	17.81	0.94	25.0
Closterium dianae Ehr.	0.01	0.01	0.02	2.32	2.32	0.08	12.5
Closterium ceratium Perty	0.29	0.09	0.88	10.07	3.57	0.36	62.5
Closterium pronum Breb.	0.02	0.01	0.06	7.81	5.48	0.28	25.0
Botryococcus protuberans W. et G.S.West	0.07	0.07	0.23	0.37	0.37	0.01	12.5
Crucigenia tetrapedia (Kirchn.) W. et W.	0.21	0.07	0.65	3.73	1.24	0.13	87.5
Elakatotrix gelatinosa Wille	0.59	0.21	1.80	5.89	2.05	0.21	100.0
Eutetramorus fottii (Hindak) Kom.	0.27	0.27	0.83	1.36	1.36	0.05	12.5
Dictyosphaerium subsolitarium Goor	0.26	0.26	0.80	0.86	0.86	0.03	12.5
Didymocystis inconspicua Korch.	0.59	0.17	1.79	2.07	0.70	0.07	75.0
Kirchneriella contorta (Schmidle) Bohl.	0.17	0.15	0.52	2.08	1.43	0.07	25.0
Monoraphidium contortum (Thur.) Kom.-Legn.	0.04	0.02	0.12	0.24	0.13	0.01	37.5
Monoraphidium komarkovae Nugard	0.05	0.02	0.14	0.59	0.25	0.02	50.0

Table 6. (Continued)

Taxon	N	EN	N, %	B	EB	B, %	f, %
CHLOROPHYTA							
Oocystis borgei Snow.	0.03	0.03	0.08	0.65	0.65	0.02	12.5
Oocystis sp.	0.18	0.09	0.56	2.45	1.32	0.09	75.0
Scenedesmus quadricauda (Turp.) Breb.	1.42	0.27	4.35	13.10	2.77	0.46	100.0
Sphaerocystis schroeteri Chod.	1.06	0.45	3.23	8.96	3.60	0.32	62.5
Tetraedron minimum Delf.	0.04	0.02	0.12	0.84	0.44	0.03	37.5
Pediastrum tetras (Ehr.) Ralfs.	0.17	0.10	0.53	0.53	0.28	0.02	37.5
Pediastrum boryanum (Turpin) Meg.	0.17	0.16	0.52	3.43	3.22	0.12	25.0
Eudorina elegans Ehr.	0.14	0.10	0.41	2.75	2.30	0.10	25.0
Chlamydomonas sp.	0.22	0.12	0.68	2.89	1.86	0.10	62.5
Pandorina morum (Mull.) Bory	0.10	0.10	0.29	0.48	0.48	0.02	12.5
Total	32.70	2.91	100.00	2819.69	309.24	100.00	100.0

Note: N - average number, thousand cell/l; EN – average number error; N - relative number; %; B - average biomass, mg/m^3; EB - average biomass error; B - relative biomass, %; f - occurrence, %.

The diatoms *Asterionella formosa* and *Tabellaria fenestrata var. asterionelloides, Aulacoseira granulata, A. alpigena* and *Rhizosolenia longiseta* dominated in population and biomass. In addition, *Dinobryon divergens, D. bavaricum, Synedra nana, Elakatotrix gelatinosa* and *Scenedesmus quadricauda* were found in all samples. The number of Kenozero phytoplankton in the summer 2001 was $(32.7\pm2.9)*10^3$ cells/L, and the biomass was 2.8 ± 0.3 g/m^3. Maximum values of phytoplankton biomass (4.1 and 3.6 g/m^3) were observed in the Northern part of the water body.

Due to analysis, most of the detected phytoplankton taxons belong to species widely distributed in continental waters. In terms of biogeography, algoflora is represented mainly by cosmopolitan forms, with few North-Alpine and boreal species. More than half of the species were indifferent to pH, while the others were approximately equal alkaliphilic and acidophilic. Saprobic indicator species were up to 60% of the total algae composition. Most of them were oligo-b- and b-mesosaprobes. Thus, during summer period 2001, phytoplankton of Kenozero lake was taxonomically diverse and had a uniform structure throughout the water area. The lake had mesotrophic level due to lake typification scale (Lukin, 2001).

In October 2009, there were 53 forms of algae in Kenozero lake, including 36 diatoms, 9 green algae, 3 blue – green algae, 1 golden algae and 4 peridinium algae. Diatoms, blue-green and green algae, such as *Melosira granulate, Tabellaria fenestrate, Fragillaria capucina, Asterionella Formosa*, and *Achnantes linearis* were basically in systematic diversity.

Psychrophilic diatoms dominated in the autumn phytoplankton of Kenozero lake, both in species diversity (64.3%) and in population number (82…95%). Green algae were also relatively abundant. Blue-green algae were less diverse and their biomass was small (table 6). Peridinium algae, constantly presented in plankton in small quantities, were represented by the genus Gymnodium, which characterized Kenozero as an oligotrophic water body.

Hydrobiological studies of the central part of Kenozero lake were conducted in June 2018 (Figure 10).

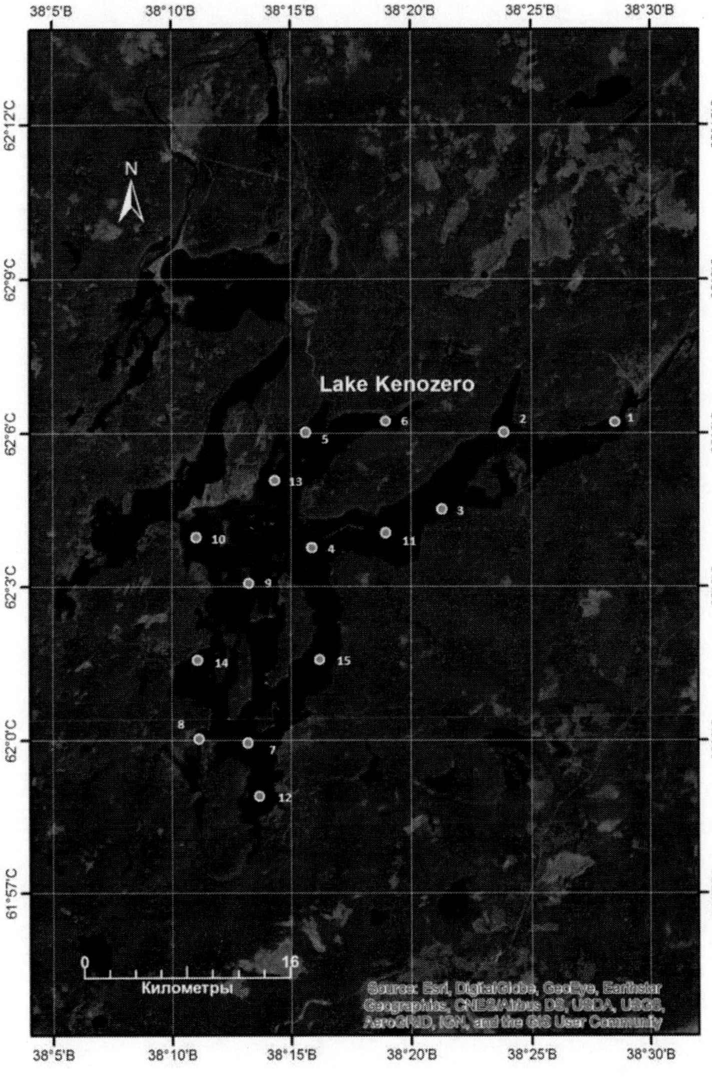

Figure 10. Sampling map of Kenozero phytoplankton in June 2018 (according to: Otchenash et al., 2019).

The phytoplankton community of Kenozero during this period was represented by microalgae belonging to seven divisions: Bacillariophyta, Dinophyta, Chlorophyta, Cyanophyta, Chrysophyta, Xanthophyta and Euglenophyta. Totally there were 70 species and supraspecific taxons of microalgae. Diatoms were the most diverse in species composition. There were 50 diatomic species, it was 72% of the total number of discovered and identified taxons. Green and blue-green algae were in a minority – 7 and 4 species, respectively. There were 3 species of dinophytic algae and 3 species of yellow-green algae, 2 species of golden algae and 1 species of euglenic algae (Figure 11). Such phytoplankton community structure with diatoms dominating is typical for Northern water systems which were not subjected to anthropogenic impact (Otchenash et al., 2019).

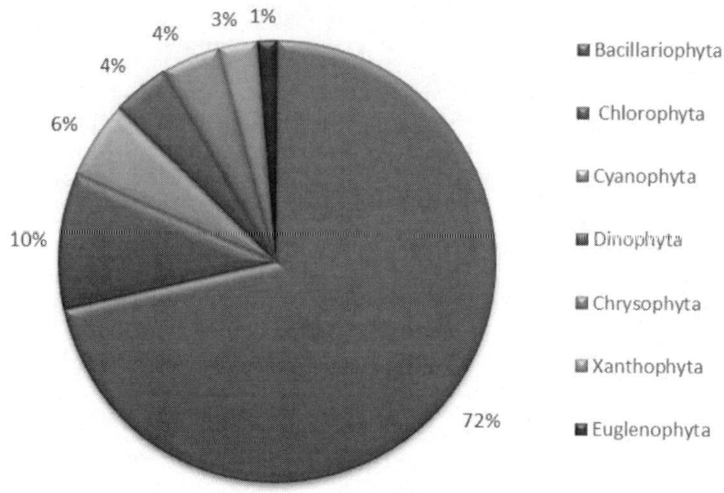

Figure 11. Phytoplankton taxonomic composition of Kenozero lake in June 2018 (according to: Otchenash et al., 2019).

The most common species of planktonic algae were Melosira granulata and Fragilaria crotonensis (present at all stations), and less common Tabellaria fenestrata, Gymnodinium sp., Asterionella formosa, Closterium acutum and Protoperidinium bipes.

Golden algae of the genus Dinobryon inhabited water with a minimum content of mineral phosphorus, which is an indicator of oligotrophic conditions (Barinova et al., 2006). Most of the identified microalgae have the status of widespread species or cosmopolitan and are common representatives of the lake algoflora of the North (Otchenash et al., 2019).

The total amount of planktonic microalgae in June 2018 varied from 3.52 to 50.96 thousand cells/L. The largest number was registered at station 3, the smallest at station 15. The average microalgae abundance was 23.75 thousand cells/L (Figure 12). Based on obtained data, Kenozero lake in the spring-summer period is a water body with a very low number of lower algae (Otchenash et al., 2019).

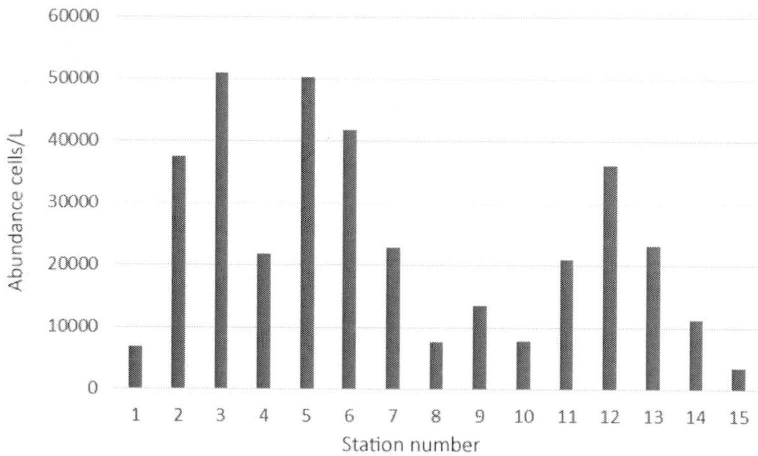

Figure 12. Phytoplankton abundance in Lake Kenozero (June, 2018) (according to: Otchenash et al., 2019).

The total phytoplankton biomass varied from 8.01 to 138.56 mg/L. The maximum of total biomass was registered at station 3, located in the northern part of the lake, whereas the smallest value was at station 15. The average phytoplankton biomass was 47.61 mg/L (Figure 13).

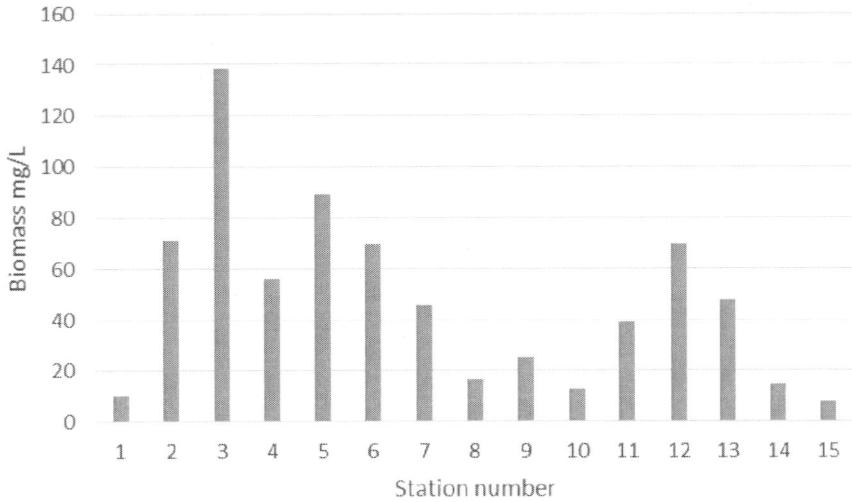

Figure 13. Biomass of phytoplankton in Lake Kenozero (June, 2018) (according to: Otchenash et al., 2019).

Representatives of the Bacillariophyta division - *Melosira granulate, Asterionella formosa* and euglenic algae (which could not be identified before) had the highest biomass at all stations. There were members of the genus Fragilaria, Tabellaria, Nitzschia and Dinophyta identified. Representatives of Euglenophyta dominated at stations 1, 2, 3 and 11. This type algae inhabits areas affected by organic pollution (Komulainen et al., 2006).

The values of the Shannon-Weaver diversity indices for phytoplankton abundance ranged from 1.9 at station 3 to 3.6 at station 12. Biomass varied from 1.9 at station 3 and 14 to 3.5 at station 6. In average, diversity indices were 2.9 for abundance and 2.8 for phytoplankton biomass. Low index values were relevant to low species diversity.

The saprobity index of V. Sladechek varied from 1.57 to 1.8 (station 1) and it was 1.6 in average. Due to saprobity index values waters of Kenozero correspond to the oligo-β-meso-test zone (saprobity index 1.5 – 2.5), or II-III class of water quality with a moderate content of organic substances (Assessment Methods..., 2015).

Phytoplankton composition of Kenozero is typical for most fresh water systems in our region. There are microalgae, widely distributed in the North of Russia, and cosmopolitan species. Kenozero is a floristically depleted oligotrophic water body with a significant predominance of diatoms and low quantitative indicators of abundance and biomass at the beginning of the growing season. There was a small areal contamination of the Kenozero in the Pershlakhta Bay, where there were euglenic algae – indicators of organic water pollution. This could be due to the fact that the Bay is actually the mouth of the lake and collects almost all the water from the Kenozero water area along with anthropogenic pollutants. Additional research is required to assess the degree of pollution and to determine its causes (natural or anthropogenic).

Zooplankton

In 1968, the biomass of zooplankton in the Kenozero was 0.15 g/m^3 in average, and the population was 14.1 thousand ind./m^3 (Anukhina, 1969). Plankton had a high feeding value, keeping in mind plankton variety.

Table 7. Zooplankton species composition of Kenozero lake, August 2001 (according to: Lukin, 2001)

Taxon	Northern part	Southern part	Bays
ROTIFERA			
Trichocerca sp.	+	+	+
Synchaeta sp.	+	–	–
Polyarthra major Burckhardt	+	+	+
Bipalpus hudsoni (Imhof)	+	+	+
Asplanchna herricki Guerne	+	+	+
A. priodonta Gosse	+	+	+
Trichotria sp.	+	–	–
Euchlanis dilatata Ehrenberg	+	+	+
Keratella cochlearis (Gosse)	+	+	+
Kellicottia longispina (Kellicott)	+	+	+
Conochilus unicornis Rousselet	+	+	+
CLADOCERA			
Limnosida frontosa Sars	+	+	+
Diaphanosoma brachyurum (Lievin)	+	+	+
Daphnia longispina hyalina (Leydig)	+	+	+
Daphnia cristata Sars	+	+	+
Ceriodaphnia sp.	+	–	+
Rhynchotalona falcata (Sars)	+	–	–
Chydorus sphaericus (O. F. Muller)	+	+	+
Grabtoleberis testudinaria (Fischer)	+	–	–
Bosmina longirostris (O. F. Muller)	+	+	+
B. obt. obtusirostris Sars	+	+	+
B. obt. lacustris Sars	+	+	+
B. coregoni coregoni (Baird)	+	+	+
B. coregoni gibbera (Schoedler)	+	+	+
B. kessleri (Uljanin)	+	+	+
Leptodora kindtii (Focke)	+	+	+
COPEPODA			
Calanoida			
Limnocalanus macrurus Sars	+	+	+
Eudiaptomus gracilis (Sars)	+	+	+
Eurytemora lacustris (Poppe)	+	+	+
Heterocope appendiculata Sars	+	+	+
Cyclopoida			
Cyclops strenuus Fischer	+	+	+
C. kolensis Lilljeborg	+	+	+
Acanthcyclops sp.	–	+	–
Mesocyclops leuckarti (Claus)	+	+	+
Thermocyclops oithonoides	+	+	+

Note: + present in samples, - missing in samples.

In the summer of 2001, the zooplankton community of Kenozero included 35 taxons: Calanoida – 4, Cyclopoida – 5, Cladocera – 15, and Rotifera – 11 (Table 7). Due to species diversity and its amount, Kenozero is a b-oligotrophic water system. The average abundance of zooplankton in August 2001 was 23.5 thousand ind./m^3, and the biomass was 0.51 g/m^3. In the upper water layer (0-2 m), these values increased, to 44.0 thousand ind./m^3 and 0.9 g/m^3, respectively. (Lukin, 2001).

In June 2018, there were 30 species of micro- and mesozooplankton in the zooplankton fauna of lake Kenozero, including Rotifera – 7, Cladocera – 11 and Copepoda – 12 species.

According to zoogeographic characteristics, the majority of zooplankton species consisted of cosmopolitan and Palearctic organisms. In the cold-water systems *Daphnia cristata* is the most common cladoceran, the Holarctic *Kellicottia longispina* is common rotifer, a relic *Cyclops scutifer* - copepod. Most planktonic organisms are eurythermic (Anukhina, 1969). The zooplankton included water quality indicator species – *Kellicottia longispina, Limnocalanus macrurus, Cyclops scutifer*. The presence of these species indicates a low level of biogens in the water and its purity (Determinant of zooplankton and zoobenthos..., 2010).

In samples from 2018 copepods (mainly Calanoida) had the biggest contribution to the zooplankton core in both numbers and biomass. Two species of bosmin – eurybiont cosmopolite *Bosmina longirostris* and eurybiont species of the European part of Russia *Bosmina coregoni* - predominated among cladocerans, providing up to 93% of the biomass among Cladocera. The majority of rotifers exist under the ice. Immediately after opening, in the first half of May-June, they had their maximum development. Due to this fact, the biomass of rotifer plankton had quite high values – 19.8%, because of large rotifers-cosmopolitans *Asplanchna priodonta* and *Asplanchna herricki*. The average number and biomass of Kenozero zooplankton in June 2018 was 1.517 ind./m^3 and 36.2 mg/m^3, respectively. Crustaceans had a leading place in planktonic fauna. On average, they occupied about 80% of the total population and biomass (Table 8). In general, copepods were dominating in June. In early summer there were plenty of nauplial and copepodite stages of Copepoda, eggs and young Cladocera.

Table 8. Abundance and biomass of zooplankton in lake Kenozero, June 2018 (according to: Otchenash et al., 2019)

Taxon	N		B	
	ind./m^3	%	mg/m^3	%
Rotifera	313	20.6	7.211	19.8
Cladocera	455	30.0	11.680	32.4
Copepoda	749	49.4	17.330	47.8
Total	1517	100.0	36.221	100.0

The zooplankton composition and the ratio of systematic groups remained common. Copepods dominated at most stations in abundance and biomass. The exception were stations 1-3, where two *Bosmina* species dominated (Table 9). Calanoida is more sensitive to the process of eutrophication among copepods, their number decreases (Vezhnovets, 1981). At the same time, the presence of *Bosmina longirostris*, which is an indicator of the α-β-mesosaprobe zone, may indicate a slight contamination of the waters the lake (Andronnikova, 1980, 1996, 2007).

Due to the data obtained on zooplankton, Kenozero lake is a poor water system for fish-plantation. However, despite the low average numbers and biomass, it has certain reserves of food for plankton-eating fish, taking into account the vast area of the lake. At the same time, the crustacean nature of zooplankton, where copepods and cladocerans outnumber rotifers, and determines its value as a food for fish.

Table 9. Abundance, biomass and dominant zooplankton species of lake Kenozero, June 2018 (according to: Otchenash et al., 2019)

Station number	Total number, ind./m³	Total biomass, mg/m³	Dominant Species
1	5110	146.53	*Bosmina longirostris, Bosmina coregoni*
2	1130	21.92	*Bosmina longirostris, Bosmina coregoni*
3	2140	39.63	*Bosmina longirostris, Bosmina coregoni*
4	1120	24.32	*Thermocyclops oithonoides, Asplanchna herricki*
5	420	10.2	*Thermocyclops oithonoides, Limnocalanus macrurus*
6	650	12.8	Cyclopoida juv., Nauplii Copepoda *Bosmina coregoni*
7	980	19.9	Cyclopoida juv., *Asplanchna priodonta*
8	1070	27.6	Cyclopoida juv., *Asplanchna priodonta, Eudiaptomus sp.*
9	2370	50.36	Cyclopoida juv., *Limnocalanus macrurus*
10	1430	35.43	Cyclopoida juv., *Asplanchna herricki*
11	550	13.59	Cyclopoida juv., *Eudiaptomus* sp.
12	1860	44.29	*Thermocyclops oithonoides, Eudiaptomus* sp.
13	750	27.22	*Limnocalanus macrurus*
14	1800	38.98	*Thermocyclops oithonoides, Limnocalanus macrurus*
15	1370	31.14	*Cyclops* sp., *Asplanchna priodonta*
Average	1517	36.2	

In June 2018, zooplankton of Kenozero lake included indicator species such as *Kellicottia longispina, Limnocalanus macrurus, Cyclops scutifer*, etc. Their presence indicated a low level of organic pollution and purity of the most area of the lake.

The presence in the zooplankton community of indicator species of α-β-mesosaprobic zone (*Bosmina longirostris)* shows a slight pollution of the waters in the Gulf of Perslucht (V. 1, 2, 3). This fact was confirmed by the phytoplankton studies in the area, where species-indicators of water pollution were also identified. Additional research is required to assess the degree of pollution and determine its causes (natural or anthropogenic).

Zoobenthos

In 1968, there was first assessment of the fishing potential of the Kenozero: macrozoobenthos species composition of the lake, its structure, distribution and degree of quantitative development. According to the obtained data, there were 17 systematic groups of benthic invertebrates observed. Among aquatic insects, there were larvae of insects-chironomidae (29 species), trichoptera, ephemeroptera, ceratopogonidae, chaoborus sp. Oligochaeta (14 species) and Bivalvia. It was noted that a complex of relict crustaceans, such as *Monoporeia (Pontoporeia) affinis, Pallasea Quadrispinosa, Mysis relicta, Gammaracantus lacustris* inhabited Kenozero lake. Areas with benthic biomass up to 1.0 g/m² occupied 81,5% of the total area of the lake, with biomass from 1 to 3 g/m² – 18.1%, and over 3 g/m² – only 0.4%. The highest biomass was observed in the littoral zone, where it reached 6.0 g/m² and in the ore-bearing sandy areas of the open part of the lake – up to 4.0 g/m². Total average benthic biomassranged from 0.5 to 1.6 g/m^2. Thus, according to the qualitative composition and degree of quantitative development of the benthic fauna, Kenozero was oligotrophic type lake (Anukhina, 1969).

Research conducted in 1999 and 2001 (Kalinkina, 1999; Lukin, 2001) showed that there was poor biological diversity of macrobenthic invertebrates in Kenozero. The Shannon species diversity index varied from 0.81 to 2.42 in March 1999, and it was 2.91 in the summer of 2001. There were totally 18 taxons from 7 systematic groups observed: Nematodes, Oligochaetes, Bivalves, Amphipods (Crustaceans), Trichopters, Chironomids, Chaoborids (Insects) (Table 10).

Table 10. Taxonomic composition of macrozoobenthos Kenozero (according to: Kalinkina, 1999; Lukin, 2001)

Taxon	N	N, %	f, %	N	N, %	f, %
	March 1999			August 2001		
NEMATODA	23.81	2.87	57.1	–	–	–
OLIGOCHAETA	80.95	9.77	100.0	57.02	23.64	73.7
BIVALVIA	204.8	24.71	85.7	70.18	29.09	94.7
CRUSTACEA						
Monoporeia affinis	28.57	3.45	14.3	22.81	9.45	21.1
TRICHOPTERA						
Mystacides nigra	–	–	–	0.88	0.36	5.3
CHIRONOMIDAE						
Tanytarsus sp.	157.1	18.97	42.9	27.19	11.27	15.8
Chironomus sp.	–	–	–	12.28	5.09	15.8
Demicryptochironomus vulneratus	–	–	–	0.88	0.36	5.3
Paracladopelma camptolabis	4.76	0.57	14.3			
Limnochironomus nervosus	–	–	–	0.88	0.36	5.3
Sergentia coracina	109.52	13.22	42.9	0.88	0.36	5.3
Polypedilum scalaenum	19.05	2.30	28.6	0.88	0.36	5.3
Stictochironomus crassiforceps	19.05	2.30	57.1	8.77	3.64	31.6
Protanypus sp.	–	–	–	0.88	0.36	5.3
Prodiamesa bathyphila	9.52	1.15	14.3	7.89	3.27	31.6
Eukiefferiella longicalcar	–	–	–	0.88	0.36	5.3
Orthocladius saxicola	–	–	–	0.88	0.36	5.3
Procladius sp.	95.24	11.49	57.1	22.81	9.45	63.2
CHAOBORIDAE						
Chaoborus sp.	76.19	9.20	71.4	5.26	2.18	21.1
Total	828.6	100.00	100.0	241.2	100.00	100.0

Note: N - average number, ind./m^2; N - relative number, %; f – occurrence, %.

In March 1999 the number of organisms varied from 233 to 1867 ind./m^2, biomass - from 0.26 to 3.59 g/m^2. In August 2001, zoobenthos abundance was from 67 to 917 ind./m^2 and biomass - from 0.035 to 1.922 g/m^2. The average quantitative values of macrobenthic communities in that period are shown in Table 11.

Chironomid larvae (about 50%) and bivalve mollusks of the family Sphaeridae (about 25%) were dominating in number and in biomass. At depths of 4.5 to 9.0 m, species of Tanytarsus and Procladius genera were prevailing. On the gray-green silts of the profundal zone

(20.0-50.0 m), benthic fauna was represented by *Prodiamesa bathyphila, Sergentia coracina, Stictochironomus crassiforceps, Procladius* sp., Bivalvia, Oligochaeta and larvae of *Chaoborus* sp.

Table 11. Major quantitative characteristics of macrozoobenthos in Kenozero lake (according to: Kalinkina, 1999; Lukin, 2001)

Taxon	N	EN	N, %	Ib	B	EB	f, %
March 1999							
Nematoda	23.81	9.52	2.87	0.80	0.002	0.001	57.1
Oligochaeta	80.95	21.65	9.77	1.22	0.064	0.022	100.0
Crustacea	28.57	28.57	3.45	6.00	0.090	0.090	14.3
Bivalvia	204.76	57.01	24.71	3.33	0.396	0.146	85.7
Chaoboridae	76.19	44.07	9.20	5.35	0.303	0.172	71.4
Chironomidae	414.29	123.41	50.00	7.72	0.786	0.279	85.7
Totally	828.57	202.29	100.0	10.37	1.641	0.474	100.0
August 2001							
Oligochaeta	57.02	22.45	23.64	10.08	0.118	0.061	73.7
Crustacea	22.81	14.49	9.45	10.49	0.009	0.053	21.1
Bivalvia	70.18	19.10	29.09	5.93	0.193	0.055	94.7
Trichoptera	0.88	0.88	0.36	1.00	0.004	0.004	5.3
Chaoboridae	5.26	2.57	2.18	1.43	0.008	0.006	21.1
Chironomidae	85.09	27.77	35.27	10.34	0.145	0.033	89.5
Totally	241.23	57.03	100.00	15.37	0.552	0.128	100.0

Note: N - average number, ind./m^2; EN - its error, ind./m^2; N - relative number, %; IB-Blackman index (distribution: IB=1-random, >1-aggregated, <1-uniform); B - average biomass, g/m^2; EB - its error (g/m^2); f – occurrence, %.

In the spring of 2008, there were 5 groups of benthic hydrobionts identified in Kenozero lake. Oligochaetes (40.4% in number and 75.7% in biomass) and chironomid larvae (38.3% in number and 20.5% in biomass) predominated. Nematodes (Nematoda), water mites (Hydracarina), and Brooks (Trichoptera) were widely distributed as well. The number and biomass of the benthic population remained low and was 412 ind./m^2 and 0.39 g/m^2, respectively (Dvoryankin, 2010). The data is presented in Table 12.

Table 12. Benthic fauna of Kenozero, may 2008 (according to: Dvoryankin, 2010)

Taxon	Frequency incidence, %	N		B	
		ind./m^2	%	g/m^2	%
Nematoda	65.9	61	14.9	0.001	0.2
Oligochaeta	87.9	167	40.4	0.29	75.7
Hydracarina	65.9	9	2.1	0.003	0.7
Trichoptera	65.9	18	4.3	0.01	2.9
Chironomidae lv.	98.4	158	38.3	0.08	20.5
Totally		412	100	0.39	100

In general, zoobenthos in Kenozero lake was relatively poor in taxonomic composition. Due to the structure and degree of quantitative development the lake is common to oligotrophic Northern lakes (Kitaev, 1984). However, the food supply for benthic fish species is not low, since the lake has productive areas with high benthic biomass. Taking into account the large area of the water body, these zones with high benthic biomass provide a fairly high fish productivity of the Kenozero.

Ichthyofauna

Modern fish fauna of Kenozero lake has two species of lampreys and 23 species of fish, which belong to 12 families and 8 orders (Figure 14) (Dvoryankin, 2016).

Over the past thirty years, fish community structure of the lake has undergone serious changes caused by both natural and anthropogenic factors. In comparison with 2000, the list of fish was fulfilled with 4 new species - *Sander lucioperca, Abramis ballerus, Scardinius erythrophthalmus* and *Oncorhynchus gorbuscha*. There was the first data about the rudd existing in the basin of the Onega river (Kenozero). *Sander lucioperca* appeared in the Kena river on the territory of the park, it was migrating along the river Onega from lake Lacha. The number of *Abramis ballerus*, which was not recorded in catches until 2000, increased sharply in Kenozero lake. (Dvoryankin, 2016). Two species of Kenozero fish *Stenodus leucichthys nelma* and *Cottus gobio* were included in the Red books of the Arkhangelsk region (2008), the Nenets Autonomous district (2006) and Russia (2001).

The ichthyofauna of Kenozero lake is heterogeneous regarding zoogeography. The genesis of the fauna, the nature of the species area, and its ecological characteristics are the main criteria for assigning a species to a particular faunal complex. (Nikolsky, 1974, 1980; Reshetnikov, 1981). Due to the data obtained, there are five freshwater faunal complexes: Arctic freshwater, boreal plain, boreal submountain, Ponto-Caspian freshwater, and ancient late Tertiary.

1) The Arctic freshwater faunal complex includes whitefish *Stenodus leucichthys nelma, Coregonus albula, Osmerus eperlanus, Lota lota* and lamprey (Table 13). Some of them have circumpolar distribution and live in cold water with a high oxygen content. These species are characterized by a cylindrical body, pelagic (*Stenodus leucichthys nelma*, and whitefish) or benthic (*Lota lota*) color, rather complex species structure and wide variability (whitefish). They spawn mainly in autumn or winter, laying low-adhesive or non-adhesive benthal eggs with a long incubation period (except smelt and lampreys, which have spring spawning and fairly rapid development of eggs). The types of food are diverse – there are euryphages, planktophages, and predators (minogues have a predatory-parasitic type of feeding).
2) Cyprinid fish (*Rutilus rutilus, Leuciscus idus, Leuciscus leuciscus*) mainly represent the boreal plain faunal complex, adapted to life in the stagnant and slow-flowing waters of the plains of the boreal zone. It also includes *Perca fluviatilis, Gimnocephalus cernuus* and *Esox lucius* (Table 13). Most representatives of the boreal plain complex are less oxyphilous, and among them there are species that have adapted to life in water with low oxygen content. The fish of this faunal complex are quite diverse in body shape, color, and reproduction patterns. But most of them are phytophiles, which lay eggs on underwater vegetation. Spawning in these species occurs in spring or early summer. According to the nature of their diet, they are euryphages, except for pike, which is an ambush predator.

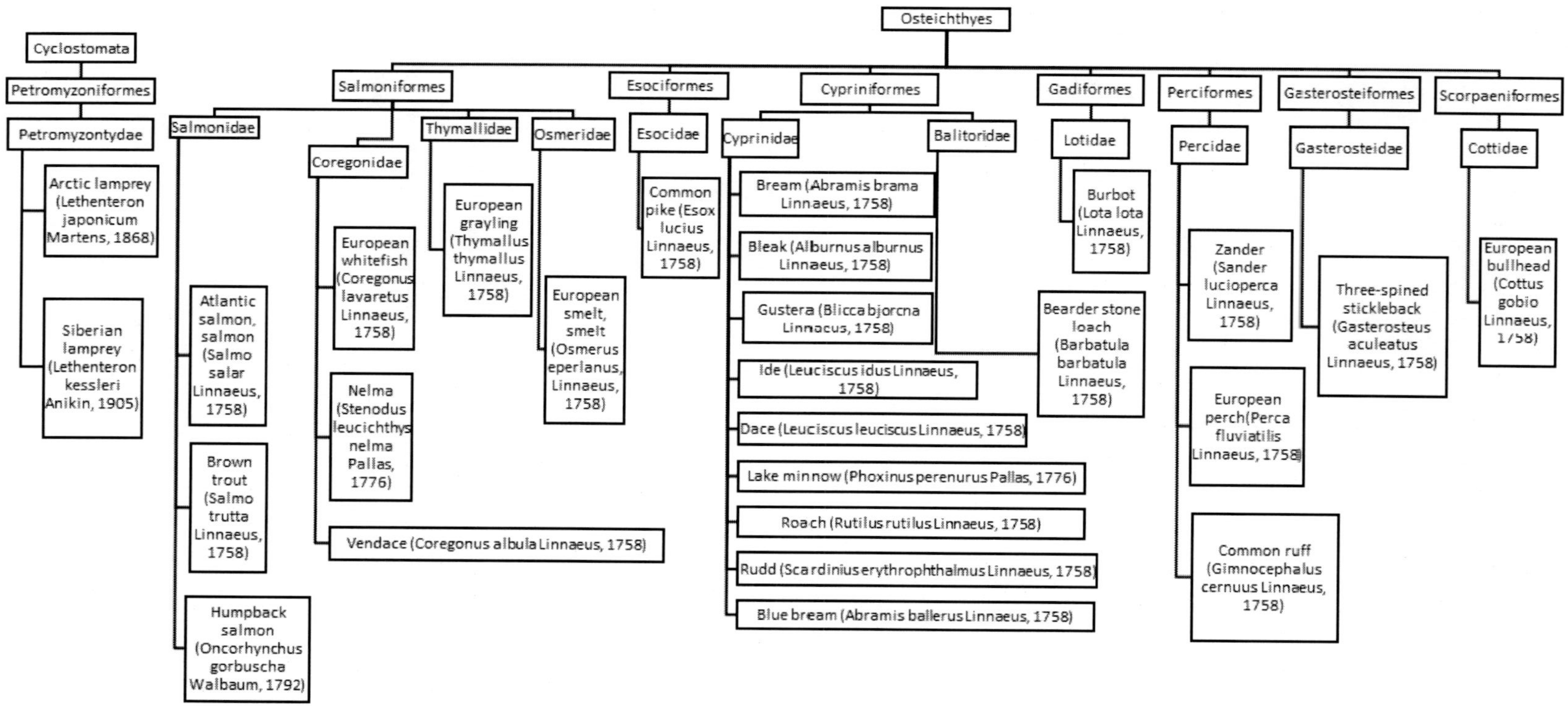

Figure 14. Taxonomy of fishes of Kenozero lake.

3) The boreal submountain faunal complex includes *Salmo trutta, Oncorhynchus gorbuscha, Thymallus thymallus, Phoxinus perenurus, Cottus gobio* (Table 13). All of them are oxyphilic species, demanding the content of oxygen in water. They are characterized by a spindle shaped body covered with small scales (except for *Cottus gobio*) and having a near bottom color. The intraspecific structure is relatively simple, and differentiation is mainly determined not by the nature of nutrition, but by the timing and places of reproduction. Spawning occurs both in spring (*Thymallus thymallus, Phoxinus perenurus* etc.) and in autumn (Salmonidae), Eggs are laid on the ground or buried in it. By the nature of their diet, they are mainly predators and benthics.
4) The Ponto-Caspian freshwater faunal complex includes the remaining part of cyprinid fish: *Abramis brama, Blicca bjorcna, Abramis ballerus, Scardinius erythrophthalmus, Alburnus alburnus* and *Gasterosteus aculeatus* (Table 13). Most of its representatives prefer slow-flowing and stagnant water bodies that are sufficiently saturated with oxygen. There are species with pelagic coloration and a flattened from the sides body. By the nature of reproduction, they are mainly spring-spawning phytophiles. Male *Gasterosteus aculeatus* actively guard nests with laid eggs. They are mainly bentophages due to feeding character.
5) The Ancient upper-tertiary faunal complex is represented by *Sander lucioperca*, which only recently appeared in the Kenozero basin (Table 13). It is an active predator that is extremely demanding of oxygen and water quality. It is characterized by early spawning, while laying eggs on stony ground, arranging primitive nests.

Table 13. Fish species diversity of Kenozero lake (according to: Dvoryankin, 2016)

Faunal complexes				
Arctic freshwater	Boreal plain	Boreal submountain	Ponto-Caspian freshwater	Ancient upper-tertiary
Lethenteron japonicum	*Esox lucius*	*Salmo salar*	*Abramis brama*	*Sander lucioperca*
Lethenteron kessleri	*Leuciscus idus*	*Salmo trutta*	*Abramis ballerus*	
Coregonus albula	*Leuciscus leuciscus*	*Oncorhynchus gorbuscha*	*Alburnus alburnus*	
Osmerus eperlanus	*Gimnocephalus cernuus*	*Barbatula barbatula*	*Scardinius erythrophthalmus*	
Coregonus lavaretus	*Perca fluviatilis*	*Cottus gobio*	*Gasterosteus aculeatus*	
Lota lota	*Rutilus rutilus*	*Thymallus thymallus*	*Blicca bjorcna*	
Stenodus leucichthys nelma		*Phoxinus perenurus*		

Currently, there are classifications that divide animals into ecological or biological groups due to their environment and habitat. Fish are most often classified in relation to salinity and habitat into the following ecological groups: marine, freshwater, anadromous and brackish (Nikolsky, 1974, 1980). In Kenozero lake there are two ecological groups of fishes: freshwater and anadromous.

Anadromous fish include *Oncorhynchus gorbuscha, Salmo trutta* and *Lethenteron japonicum*. These are generative freshwater species that feed in the sea and migrate to fresh water to reproduce. These species are found singly, and a lake is part of a migratory route to spawning for them.

The majority of the fish of Kenozero lake are freshwater fish (except for *Stenodus leucichthys nelma*) and they live in the lake constantly. They can be divided into three ecological forms: rheophilic (river), limnophilic (lake) and limno-rheophilic (mixed habitat). The most numerous is the limno-rheophilic group (Table 14).

Table 14. Ecological groups of fish of Kenozero lake (according to: Dvoryankin, 2016)

Fish species by ecological group			
Freshwater			Anadromous
river	lake	mixed habitat	
Lethenteron kessleri	*Coregonus albula*	*Cottus gobio*	*Lethenteron japonicum*
Stenodus leucichthys nelma	*Coregonus lavaretus*	*Abramis brama*	*Salmo salar*
Thymallus thymallus	*Osmerus eperlanus*	*Alburnus alburnus*	*Salmo trutta*
Leuciscus leuciscus	*Phoxinus perenurus*	*Blicca bjorcna*	*Oncorhynchus gorbuscha*
Barbatula barbatula	*Scardinius erythrophthalmus*	*Leuciscus idus*	
	Gasterosteus aculeatus	*Rutilus rutilus*	
		Esox lucius	
		Lota lota	
		Gimnocephalus cernuus	
		Perca fluviatilis	
		Sander lucioperca	
		Abramis ballerus	

According to the feeding, all fish living in the Kenozero are planktophages, benthophages, predators (including the predatory parasitic type of feeding for lamprey) and euryphages. There are no purely herbivore and detritivore fish, although plant residues and detritus are found in the digestive tracts of many fish. The only obligate planktophagus among fish of Kenozero lake is *Coregonus albula*. Five species of benthophages are represented – *Coregonus, Abramis brama, Blicca bjorcna, Gimnocephalus cernuus, Cottus gobio*. There are 9 species among predators – lampreys, *Salmo salar, Salmo trutta, Oncorhynchus gorbuscha, Stenodus leucichthys nelma, Esox lucius, Lota lota, Sander lucioperca*. Euryphages include the 10 remaining species. The division of fish into these groups is quite conditional. In real natural conditions, fish feed on the type of food that is most available and its quantity is sufficient to cover the energy costs of searching and consumption (Dvoryankin, 2016).

Almost all fish of Kenozero lake use for reproduction the spawning substrate of water. About half of them spawn on the stony-pebble (lithophile) and sandy (psammophila) soils. Whitefish (*Coregonus lavaretis, Coregonus albula, Stenodus leucichthys nelma, Thymallus thymallus Linnaeus, Barbatula barbatula, Gimnocephalus cernuus* and *Cottus gobio*) are lithophile. Lampreys, *Osmerus eperlanus, Lota lota* and some species of cyprinid fish (*Leuciscus leuciscus*) are psammophiles. Many fish lay their sticky eggs on the vegetable substrate. These are mainly spring-spawning cyprinid fish species (*Abramis brama, Abramis ballerus, Blicca bjorcna, Alburnus alburnus, Leuciscus idus, Phoxinus perenurus, Rutilus rutilus, Scardinius erythrophthalmus* as well as *Esox lucius* and *Perca fluviatilis*. *Salmo salar,*

Salmo trutta, Oncorhynchus gorbuscha, Gasterosteus aculeatus, Sander lucioperca build nests (Dvoryankin, 2016).

In Kenozero, the number of commercial fish includes 13 species of which the most valuable and main are *Coregonus lavaretus, Coregonus albula, Esox lucius, Abramis brama, Leuciscus idus, Esox lucius* and *Lota lota*, less valuable are – *Rutilus rutilus, Perca fluviatilis, Abramis ballerus, Osmerus eperlanus, Leuciscus leuciscus, Gimnocephalus cernuus* and *Alburnus alburnus*.

Coregonus Lavaretus

It is a valuable, but small species that require control over a condition of stocks. Its annual catches since 2012 are fairly stable, but insignificant and, according to the fishing statistics, do not exceed 200…300 kg. *Coregonus lavaretus* become sexually mature at the age of 3+. At this age, it has a length of 25…30 cm (here and further indicates the fishing length – from the beginning of the snout to the end of the scale cover) and a weight of 250…350 g. Whitefish spawning begins in early November on sandy-shingle and stony soils at a depth of 3…4 m and continues until the end of December. The water temperature during this period is close to 0°C.

Coregonus Albula

In the Kenozero vendace is represented by the small lake form. It is the second largest population of vendace Kenozersky Park (after Lekshmozero). In the catches, there are representatives of four age groups: from fingerlings to grouse at the age of 3+ years. Vendace production in this water body is carried out both by minnow seines and by set nets. The catches are dominated by fish aged 1+, with an average length of 11…12 cm. Kenozero vendace matures in the second year of life, spawns in October-November. The average fecundity of one female, depending on age, ranges from 2.5 to 4.5 thousand eggs.

Taking into account the high number of species in the water body, the stability of size and age characteristics and the stability of the fishery (which tends to increase, according to the fishing statistics of the catch, from 5 tons in 2014 to 9.5 tons in 2018), the state of vendace population in the Kenozero is estimated as very prosperous.

Esox Lucius

In the catches on the Kenozero there was a common pike aged from 3 to 11 years with length from 31 to 90 cm and weighing from 232 to 6300 g. There are mature fish 4…6 years old, 40…60 cm long and weighing 700…2000 g dominating in the catches. The pike reaches sexual maturity in the water bodies of the Kenozersky national park at the age of 3…5 (males at 3, females at 4…5) with a body length of 35…40 cm. It spawns from late April to mid-May. The beginning of spawning coincides with the melting of ice on spills, in small, well-warmed places. The eggs are large, yellowish in color, are deposited on the flooded vegetation. The incubation period lasts 8…10 days. The basis of the spawning stock is made up of fish aged 4…6 years – that is, mostly first-time spawners and come to spawn a second time. The average absolute fecundity increases with age from 13.4 thousand eggs in females at 4 years to 90.8 thousand in 7-year-old individuals.

Common pike predominately feeds by the most numerous species of fish – European perch and roach. Smelt plays a significant role in its feeding. There were cases of feeding with frogs. After two years pike becomes predator, before that the diet consists of benthic organisms (usually insect larvae).

Pike in Kenozero lake is of a great importance as an object of amateur and sport fishing. The high and stable population of the Kenozero pike is indicated by the statistics of its catch. The annual production of pike in Kenozero lake in 2012 - 2018 has varied slightly, according to statistics it was 14…16 tons.

Abramis Brama

In Kenozero lake there is one of the most numerous in the region populations of bream. It is represented by 14 age groups (from 1 to 14 years). Linear and weight growth of Kenozerskiy bream is characterized by relatively low parameters for all age groups. Kenozero bream matures late - females become sexually mature only by 10…11 years with a length of more than 33 cm and a weight of more than 800 g. The absolute fecundity of Kenozero bream at this age is 52.5 thousand eggs. The stability of Kenozero lake bream withstands to fishing and helps to keep the nature of the population variable in age. Fishing statistics also confirm the relatively high number of the species and its stability. Over the past 6 years, the catch of bream in the Kenozero has not changed much and, according to statistics, amounted to 10-11 tons annually.

Abramis Ballerus

The catches of blue bream in Kenozero lake are represented by individuals aged 4 to 10 years. Kenozersky blue bream is a slow-growing fish. Its average size ranges from 13.7 cm in length and 32 g in 4-year-olds to 25 cm and 229 g in fish aged 10 years.

In Kenozero, the blue bream reaches sexual maturity at the age of 5…6 years (females a year later) with a body length of 16…18 cm and a weight of 60…80 g. It spawns from the end of May to mid-June at a water temperature of 12…14 ^{0}C. Eggs are small, greenish in color, deposited on the flooded vegetation. The basis of the spawning population consists of fish aged 7-8 years. The ratio of males to females is approximately 1,5:1, but from the age of 9, females begin to predominate. Average fertility increases with age from 6.5 to 14 thousand eggs.

Commercial statistics for 2012 - 2018 represent relatively high and stable population of this species in Kenozero lake. Annual catches of blue bream for 6 years are relatively large and vary slightly within the range of 4.9…5.8 tons.

Leuciscus Idus

In the Kenozero, ide has a high population. In amateur and control catches, this species is represented by individuals aged from 2 to 13 years. The average size of the Kenozero ide varies from 14,2 cm in length and 55 g in weight in 2-year-olds to 42 cm and 1670 g in fish aged 13 years.

Ide reaches its sexual maturity in Kenozero at the age of 7 years with a body length of 32…34 cm and a body weight of 700…800 g. Spawning begins in early May at a water temperature of 5…7°C on spills. Small sticky eggs are deposited on the roots of coastal trees, snags and vegetation at a depth of 0.5…0.7 m. the Absolute fecundity of females at the age of 12 years ranges from 100 to 110 thousand eggs.

The relatively high and stable number of ide is evidenced by the data of its catch statistics for 2012 - 2018. The annual production in the Kenozero lake in the last 6 years is sufficiently large and remains stable, averaging 4.6…4.8 tons.

Lota lota of Kenozero lake has a relatively low linear-weight characteristics. Its average size ranges from 22.4 cm and 91 g in 4-year-olds to 69 cm and 2628 g in burbot aged 12 years. Burbot reaches sexual maturity at the age of 4…6 years. There is no information on its fecundity in the lake. According to literature data, the absolute fecundity of burbot ranges from 50 thousand to 5 million eggs (on average from 300 to 400 thousand). Spawning is portioned, stretched, occurs from December to March. Spawning grounds are located at the confluence of streams, where there is good aeration, the water is clear and the temperature is lower than in the riverbed. Eggs are semi-pelagic, with a fat drop, non-sticky, with a diameter of 0.75…0.92 mm in the ovarian membrane and 1.05…1.15 mm already in the water after spawning. Spawning occurs on sandy or pebbly ground at depths of 0.5…3.0 m. With a weak movement of water, the eggs are kept suspended on the spawning grounds or settles to the bottom and got hammered under the stones, and in the spring, with an increase in the flow rate in the rivers, they gradually drifts down. Eggs develop at a water temperature close to 0°C, but hatching occurs at a water temperature of + 1.5 to +3…4°C 60…90 days after fertilization.

The satisfactory state of the Kenozero burbot population is indicated by the steady trend of increasing its production from 0.9 tons in 2013 to 5 tons in 2018, according to fishing statistics.

Perca Fluviatilis

In Kenozero lake European perch forms two environmental groups. There is a slow-growing European perch, which inhabits thickets of higher water vegetation in shallow waters and feeds on benthic fish. Open and deeper places are the ecological niche of the large predator perch. Both groups make up the same initial population, in which there is a gradual differentiation of perch size during the first 4…5 years. Then the larger perch switch to eating fish and dramatically increase in the growth. Smaller individuals remain benthic and, over time, become more and more behind in size from predators. Elder perches of all groups have large fluctuations in length and weight in each age range. Slow growing perch is represented by individuals aged 1 to 8 years. Its average size varies from 8 to 23 cm in length and from 8 to 219 g of weight, due to age. It becomes fast-growing perch after 4 years of life. Its average size ranges from 20 cm in length and 143 g in 5-year-olds, to 29 cm and 464 g in fish aged 8 years.

In the Kenozero, perch reaches sexual maturity at 3…4 years. In all populations, spawning is very elongated – it begins in mid-May at a water temperature of 8…10°C and can last until the end of June. Eggs are small, white-yellow in color, deposited on underwater vegetation in the form of ribbons. The ratio of males and females is close to 3:1, but from the age of 8, females begin to predominate in the catches. Absolute fertility increases with age. In tall individuals, it increases from 6.5 thousand eggs in females at the age of 4 years to 21.5 thousand in 7-year old fish. A female fast-growing form of perch at the age of 8 years had a fecundity of 35 thousand eggs.

The high and stable population of this species in the Kenozero is evidenced by the multi-age structure of the commercial part of the perch population and data on its catch statistics for 2012-2018. Its annual catches in the Kenozero is quite large and has varied only slightly, averaging 10.5…11.5 tons.

Rutilus rutilus of Kenozero lake has a relatively low linear-weight characteristics. The catches are dominated by fish aged 5…7 years with an average length of 13…18 cm and a weight of 40…100 g. Sexual maturity roach reaches at 3…4 years (females a year later). Spawning takes place quite rapidly in the second half of May in the coastal zone of the lake and on spills with last year's vegetation. Eggs are sticky, pale yellow, develop within 10…14 days. Hatched larvae after 2…3 days switch to active feeding of zooplankton. Fertility of roach of Kenozero lake ranged from 4.7 to 12.3 thousand eggs.

The high and stable population of this species in the Kenozero is evidenced by the multi-age structure of the commercial part of the roach population and data on its catch statistics for 2012 - 2018. Annual catches of roach in Kenozero lake is quite large and amount to an average of 6.8…8.8 tons.

Osmerus Eperlanus

Kenozero smelt is a small fish with an average length of 8…11 cm and a weight of 5…8 g. The body is thin, elongated, translucent. The color is light with a greenish tinge, the fins are light transparent. The scales are also transparent, easily falling off, without silvery pigment. Kenozero smelt reaches sexual maturity at the age of 2 years. It usually spawns in the first decade of May at a water temperature of 6…8°C. The spawning season lasts for 8…10 days. During this period, huge amount of smelt enter the river Pocha, which flows into the Kenozero in its northern part. The eggs are small and yellow in color, are deposited on a sandy-pebbly soils.

There is a high population of smelt in Kenozero, but it is not the main commercial species. This is due to the peculiarity of its biology. Smelt fishing is conducted for several days during its entry into the river Pocha on spawning. At the same time, smelt spawning occurs at the beginning of May during the prohibition of fish production by commercial fishing tools. Thus, local residents do not have the opportunity to catch it in large quantities. Permitted smelt fishing by dip nets can not affect the number of this species in the water body. In this regard, the state of the populations of smelt in the Kenozero lake is assessed as good and stable.

Leuciscus leuciscus, Alburnus alburnus and *Gimnocephalus cernuus* are secondary commercial species in Kenozero lake. There is no targeted production of these species, and they are caught as accidental by-catch. The prevalence of these species throughout the Kenozero water area indicates their abundance, and the volume of annual catch, according to fishing statistics, is insignificant: dace - less than 0.5 t, common ruff - less than 0.2 t, bleak catch statistics are not taken into account due to small volumes. In this regard, it can be argued that the current intensity of fishing in the water body does not affect the number and stability of populations of these species (Dvoryankin, 2016).

Lake Lekshmozero (Kenozersky National Park)

Lekshmozero is the fourth largest water body in the Arkhangelsk region. Its water surface area is 54.4 km^2. There are no islands. The lake has an oval shape. The coastline, which is 31 km long, is very poorly indented. The coefficient of its meandering is 1.6. The length of the lake is 12.6 km the maximum width is 6 km (Figure 15).

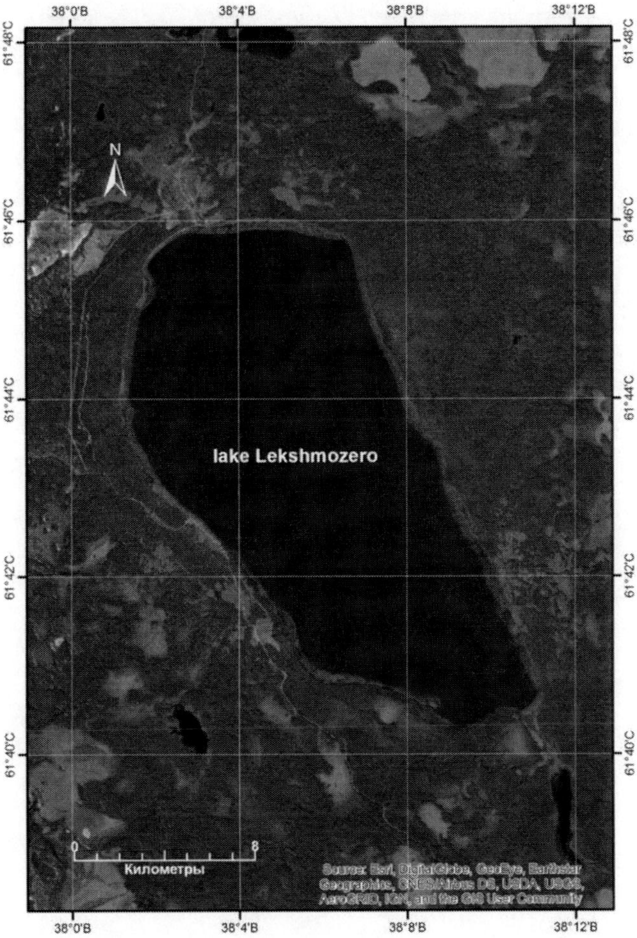

A

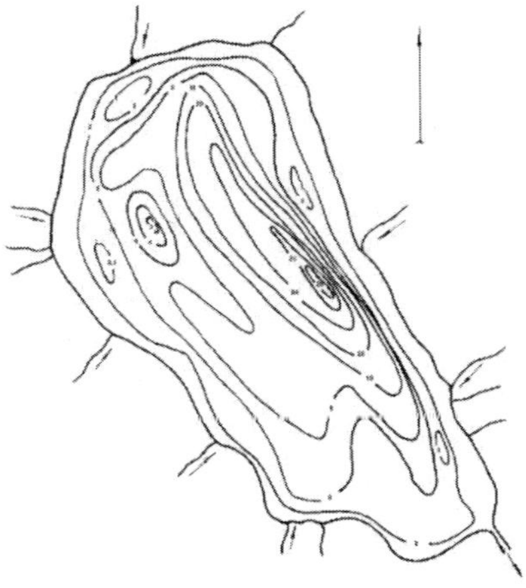

B

Figure 15. Configuration (A) and bathymetric depth map (B) of the lake Lekshmozero.

The lake bottom is flat, without sharp changes in depth. There are two shoals in the Northwestern part of the lake. The area of great depths is drawn out in the form of a furrow, shifted to the Eastern shore. Its area is relatively small (4%). The maximum depth of the lake is 23.8 m, the prevailing depths range from 5 to 7.5 m, the average depth is 5.9 m. The water volume is 0.316 km3 in the lake. Most of the bottom area is covered with silt. The coastal part is usually dominated by sand-and-shingle deposits (Figure 16). Gradual silting of the sands begins at a depth of 2-3 m, and at 8-9 m they turn into soft silts. The shores of the lake have different heights: Western and Southern-low, Northern and Eastern-high, steep.

Figure 16. Coastal areas of the lake Lekshmozero (photo by G. A. Dvoryankin).

Hydrology and Hydrography

Lekshmozero belongs to the basin of the Onega river and to the White sea. The water body is located close to the White Sea-Baltic watershed, so the catchment area is small and is 197 km^2. The nature of water exchange of the lake is low-flow, the ratio of its conditional water exchange is 0.2. Shallow streams are mainly the inflows of the lake, evenly distributed along the coast. Runoff from the lake is carried out by the river Lekshme in the lake Lacha. According to the thermal regime, Lekshmozero belongs to the moderate type. In general, the pattern of water temperature of surface horizons in the summer period repeats the pattern of air temperatures. In winter, the surface water temperature is 0.1-0.2°C. In late April-early May, the water temperature in the lake passes through 0.2°C. The maximum temperature is recorded in July, it decreases in August and, usually, in November when the ice is frozen, its values again pass through 0.2°C. Despite the relatively large depth of the lake, its water temperature changes slightly with depth. Temperature stratification is established only in summer. Homothermy or similar to it state is established as a result of convective and dynamic mixing, which develops due to the open winds of the water surface.

The lake is located in an area of excessive moisture. However, due to the small area of the catchment, there are no pronounced periods during changes in its levels. The predominant role in feeding the water body belongs to meltwater, so the maximum water levels on the Lekshmozero are observed in spring – at the end of May (up to 0.1 – 0.3 m). The duration of standing at maximum levels ranges from 5 to 20 days. The duration of the water level decline can reach 2.5 months. Low water levels are in the summer-autumn period, usually for about 2 months. Rain impact in water levels is insignificant. Small ups and downs of water levels are associated with wind setdowns and setups. The water of the Lekshmozero is very clean, the transparency of the Secchi disk is 4.8 m (Dvoryankin, 2010).

Hydrochemistry

The major role in the formation of hydrochemical regime is its hollow. The absence of large tributaries determines a high degree of autochthonous processes in the water body. Due to the high dynamics of water masses, the lake is characterized by uniformity of the chemical composition of water, both in the water area and in depth.

Institute of Ecological problems of the North conducted hydrochemical studies in 1987-1988 (Vorobyova, 2008) and Northern Department of Polar Russian Institute of Fisheries and Oceanography - in 2007-2008 (Dvoryankin, 2010). The obtained data provide information on changes in the main hydrochemical parameters of the water body over these twenty years and the ecological state of the Lekshmozero.

Waters of Lekshmozero lake are low-mineralized. The value of their total mineralization in the 80's did not exceed 105.7 mg/L, and in 2007 it was equal to an average of 78 mg/L in May-June and 107 mg/L in November. According to the degree of hardness, the water is very soft, the total hardness very slightly - from 0.979 to 1.319 mg/L. Vertical profile of the water temperature determines dissolved oxygen content . There is clearly expressed anaerobic layer at deep-sea bottom stations of Lekshmozero lake. The content of dissolved oxygen decreased vertically from 90 to 11% in winter and from 117 to 56% in summer. In the shallow part of the lake, the content of dissolved oxygen is equal both in the surface and bottom layers and is 80-92% in winter and 63-91% in summer (Vorobyova, 2008).

The content of the organic matter for COC (chemical oxygen consumption) is within permitted values, prescribed by fisheries regulations. Biological oxygen consumption for five days (BOC5) in 1987 ranged from 1.619 to 6.519 mgO/L with an average annual rate of 3.787 mgO/L. In 1988, BOC5 values declined to 2.874 mgO/L. The amount of suspended substances in water varied on average from 4.624 to 7.649 mg/L. There was small amount of nutrients in the water. In the 80's, the average annual content of ammonium nitrogen was 0.082 mg/L, nitrates - 0.026 mg/L. The nitrite content was less than 0.01 mg/L. The highest value of total phosphorus was 0.046 mg/L (average annual value was 0.026 mg/L). In 2008, the situation with nutrients remained almost unchanged: ammonium nitrogen and phenols – in traces, nitrates - 0.025 mg/l, nitrites – in traces, total phosphorus - 0.041 mg/L. Compared to 1987, the content of petroleum products decreased by 3-4 times (Table 15). This is obviously due to a general reduction of pollution sources (significant reduction of agricultural land and enterprises in the catchment area, outflow of local population), as well as stricter measures for the protection and control of water bodies by the national park administration.

Table 15. Chemical composition of Lekshmozero water (according to: Vorobyova, 2008; Dvoryankin, 2010)

Parameters	Average annual, mg/L			
	1987	1988	2007-2008	
			spring-summer	autumn
Oxygen	11.63	12.84	-	-
BPK_5	3.787	2.874	-	-
COC	19.52	19.92	-	-
Ammonium nitrogen	0.052	0.112	0.000	0..001
nitrate	0.024	0.028	0.014	0.037
nitrite	0.000	0.002	0.001	0.000
Common phosphorus	0.024	0.027	0.053	0.029
Suspended substances	4.624	7.649	-	-
pH	5.812	7.000	-	-
Mineralization	105.7		78.1	106.7
Phenols	traces	0.001	traces	traces
Mineral oils	0.089	0.019	0.014	0.030

Microbiological studies in 2008 confirmed the stability of the Lekshmozero ecosystem and the water of the lake was established as clean. However, potentially negative is a steady lack of oxygen in its bottom layers. It was the lack of oxygen in the summer of 1996 that led to the mass extinction of vendace and other fish species. Ecological disaster could happen if unfavorable external conditions coincide in the lake (Vorobyova, 2008).

Higher Aquatic Vegetation

Higher aquatic vegetation in Lekshmozero is poorly developed and grows only in coastal areas. Thickets of reeds and rushes up to 100-200 m in width grow in locus along the Western, Northern and Eastern shores of the water body. Occasionally there are thickets of water smartweed and spatterdock. For twenty years (by 2008), the overgrowth area of the water body has increased by about 30 per cent, mainly due to siltation of shallow waters. However, the total area occupied by aquatic flora does not currently exceed 2-3% of the entire surface of the lake (Dvoryankin, 2010).

Phytoplankton

In October 2009, the composition of the phytoplankton consists of 40 species of algae. There were 24 diatoms, 9 green, 3 blue-green, 1 golden, 2 peridinium and 1 euglene species. Diatoms dominated in autumn. This group prevailed both in species diversity (61.5%) and in quantity (97-99%). At the same time, the diatom *Fragillaria capucina* at all stations accounted for 79-82% of all species. Blue-green algae were less diverse and their biomass was low. Green algae, despite their relative species diversity, were in a depressed state (Table 16).

Table 16. The number of phytoplankton in lake Lekshmozero, autumn 2009, thousand cells/L (according to: Dvoryankin, 2010)

Systematic groups	Sampling location			Average
	northern part of the lake	western part of the lake	central part of the lake	
diatom	188.64	93.92	499.68	260.75
blue-green	0.78	2.72	2.08	1.86
golden	-	-	0.48	0.16
green	2.1	0.88	0.56	1.18
peridinium	0.84	0.8	0.4	0.68
euglenid	-	0.08	-	0.03
Total	192.36	98.4	503.2	264.65

Zooplankton

In 2008 the spring-summer zooplankton of the Lekshmozero was represented by 24 species and forms (Copepoda – 8, Cladocera – 11, Rotifera – 5 species). In terms of population and biomass, copepods (46.9 and 43.2% of the total population and biomass, respectively) and rotifers (42.1 and 41.2%, respectively) dominated. The most numerous in plankton were juveniles (21.4%) and nauplial stages (22.9%) of Copepoda, as well as rotifer *Asplanchna priodonta* (27.1%). The biomass of the zooplankton community was represented by young copepods (27.5%) and *Asplanchna priodonta* (40.9%). The average number of zooplankton Lekshmozero was 103 thousand ind./m^3 with an average biomass of 1.54 g/m^3 (Table 17). Due to obtained data Lekshmozero is a medium-sized water body for planktophage fish (Dvoryankin, 2010).

Table 17. Species composition of zooplankton in lake Lekshmozero, May-June 2008 (according to: Dvoryankin, 2010)

Taxon	Parameters					
	P, %	N	EN	Ib	B	EB
ROTIFERA						
Polyarthra sp.	62.6	8.49	8.49	0.40	0.01	0.00
Asplanchna priodonta	82.5	29072.3	17643.8	504.60	632.31	383.76
Kellicottia longispina	87.9	7639.44	3889.81	93.33	1.91	0.97
Conochilus sp.	65.9	6705.73	6203.41	270.43	2.01	1.86
Filinia sp.	62.6	21.22	21.22	1.00	0.01	0.00
CLADOCERA						
Diaphanosoma brachyurum	62.6	8.49	8.49	0.40	0.38	0.38
Holopedium gibberum	62.6	16.98	16.98	0.80	1.70	1.70
Daphnia sp.	87.9	38.20	19.45	0.47	0.14	0.09
Daphnia cucullata	62.6	8.49	8.49	0.40	0.38	0.38
Daphnia cristata	63.4	2147.53	2109.45	97.64	92.39	90.68
Ceriodaphnia sp.	88.4	84.88	42.44	1.00	0.77	0.39
Chydorus sp.	62.6	212.21	212.21	10.00	3.40	3.40
Chydorus sphaericus	65.1	2206.95	2080.92	92.46	35.31	33.29
Bosmina obtusirostris	71.2	4838.31	3953.20	152.21	57.21	41.44
Bosmina coregonii	98.5	1821.98	443.10	5.07	49.27	11.96
Leptodora kindtii	62.6	8.49	8.49	0.40	0.48	0.48
COPEPODA						
Calanoida juv.	90.3	1655.21	767.47	16.77	29.79	13.81
Cyclopoida juv.	98.6	23554.9	5658.40	64.05	423.99	101.85
Nauplii	96.5	19310.8	6153.99	92.42	77.24	24.62

Table 17. (Continued)

Taxon	Parameters					
	P, %	N	EN	Ib	B	EB
COPEPODA						
Heterocope appendiculata	62.6	42.44	42.44	2.00	2.76	2.76
Eudiaptomus sp.	73.5	2504.04	1931.43	70.20	107.67	83.05
Cyclops strenuus	98.4	101.86	25.46	0.30	3.97	1.12
Mesocyclops leuckarti	90.0	1315.68	616.50	13.61	21.05	9.86
Thermocyclops oithonoides	62.6	8.49	8.49	0.40	0.14	0.14
Total	92.5	103336	43118.0	847.82	1544.31	698.94

Note: P - probability that the average number differs from 0, %; N - average number, ind./m^3; EN - its error; IB - Blackman index (distribution: IB=1-random, >1-aggregated, <1-uniform); B - average biomass, mg/m^3; EB - its error; f – occurrence, %.

Zoobenthos

The benthic fauna of Lekshmosero in 2008 included 12 groups of benthic hydrobionts. It had representatives of worms (oligochaetes, nematodes and leeches), water mites, bivalves, benthic copepods (Cyclopoida) and water larvae of insects – mayflies (Ephemeroptera), caddis flies (Trichoptera), heleidae (Heleidae), larvae and pupae of chironomids (Chironomidae), as well as adult beetles (Coleoptera). In terms of population and biomass chironomid larvae were 54.0 and 44.0%, respectively, and oligochaetes were 24.1 and 44.4%, respectively. Among others, there was a high number of nematodes, up 13.3% of the total number. In general, the benthic fauna of the Lekshmozero produced an average biomass of 3.71 g/m^2 with an average population of 8311 ind./m^2 (Table 18). Due to obtained data Lekshmozero is a medium-sized water body for benthic fish (Dvoryankin, 2010).

Table 18. Lekshmozero benthic fauna, May 2008 (according to: Dvoryankin, 2010)

Taxon	Frequency incidence	N		B	
		ind./m^2	%	g/m^2	%
Nematoda	66.8	1105	13.3	0.04	1.0
Oligochaeta	64.3	2000	24.1	1.65	44.4
Hirudinea	62.6	18	0.2	0.18	4.9
Hydracarina	62.6	18	0.2	0.01	0.4
Cyclopoida	62.6	18	0.2	0.001	0.02
Bivalvia	95.5	263	3.2	0.13	3.7
Epemeroptera	76.2	88	1.1	0.02	0.4
Trichoptera	62.6	35	0.4	0.02	0.4
Coleoptera im.	62.6	18	0.2	0.01	0.3
Chironomidae lv.	79.3	4491	54.0	1.63	44.0
Chironomidae pp.	77.5	240	2.9	0.01	0.4
Heleidae	62.6	18	0.2	0.002	0.1
Total	-	8311	100	3.71	100

Ichthyofauna

In the structure of There are 11 species of fish that belong to 5 families and 5 orders in the modern fish fauna of the lake (Figure 17) (Dvoryankin, 2016):

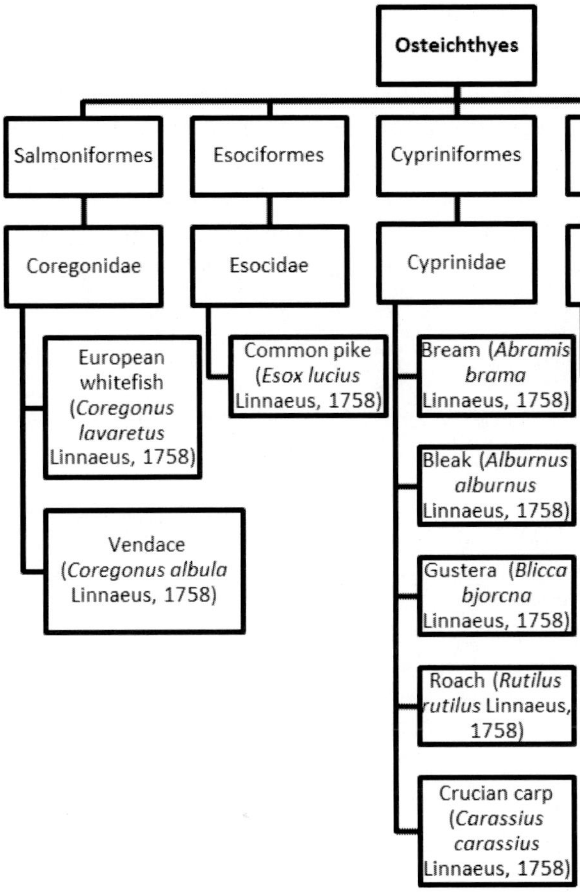

Figure 17. Taxonomy of fish in Lekshmozero lake (according to: Dvoryankin, 2016).

There are representatives of three freshwater faunal complexes in the modern ichthyofauna of the water body: Arctic freshwater, Boreal plain, and Ponto-Caspian freshwater (Nikolsky, 1980).

1) The Arctic freshwater faunal complex combines coregonids (*Coregonus lavaretus* and *Coregonus albula*) and *Lota lota* (Table 19). These fish have a circumpolar distribution and live in cold water with a high oxygen content. These species are characterized by a rolling body, pelagic (*Coregonus lavaretus*) or bottom (*Lota lota*) color, rather complex species structure and wide variability (*Coregonus lavaretus*). They spawn in autumn or winter (*Lota lota* - until March), laying low-adhesive or non-adhesive bottom eggs with a long period of incubation. The type of food is diverse – benthophages (*Coregonus lavaretus*), planktophages (*Coregonus albula*), predators (*Lota lota*).
2) The Boreal plain faunal complex is represented by *Rutilus rutilus, Perca fluviatilis, Gimnocephalus cernuus* and *Esox lucius* (Table 19). Most representatives of the Boreal plain complex are less oxyphilic. The fish that make up this faunal complex are quite diverse in body shape, color, and breeding patterns. But most of them are phytophiles, which lay eggs on underwater vegetation. Spawning in these species occurs in spring

or early summer. By feeding, they are benthophages and euryphagous, except for *Esox lucius*, it is an ambush predator.

3) The Ponto-Caspian freshwater faunal complex includes the remaining part of the cyprinid fish – *Abramis brama, Blicca bjorcna, Alburnus alburnus, Carassius carassius* (Table 19). Most of its representatives prefer slow-flowing and stagnant water bodies that are sufficiently saturated with oxygen (with the exception of *Carassius carassius*, which has adapted to a lack of oxygen). There are predominately species with pelagic coloration and a flattened body from the sides. Fish of this faunal complex are spring-spawning phytophiles. According to feeding, they are mainly benthophages.

Table 19. The distribution of fish Lekshmozero for faunistic complexes (according to: Dvoryankin, 2016)

Arctic freshwater	Boreal plain	Ponto-Caspian freshwater
Coregonus lavaretus	*Esox lucius*	*Abramis brama*
Coregonus albula	*Gimnocephalus cernuus*	*Alburnus alburnus*
Lota lota	*Perca fluviatilis*	*Blicca bjorcna*
	Rutilus rutilus	*Carassius carassius*

Fish in Lekshmozero lake can be divided into two ecological groups: limnophilic (lake) and limno-rheophilic (mixed lake-river habitat) (Nikolsky, 1974). The first group includes *Coregonus albula* and *Coregonus lavaretus*, the second – all other types of fish.

All fish that live in Lekshmozero can be classified as planktophages, benthophages, euryphages and predators. The only obligate planktophage among the fish of Lekshmozero is *Coregonus albula*. Benthophages are *Coregonus lavaretus, Abramis brama, Blicca bjorcna, Gimnocephalus cernuus, Carassius carassius*. There are two predator species: *Esox lucius* and *Lota lota*. Euryphages are *Perca fluviatilis, Rutilus rutilus,* and *Alburnus alburnus*. It should be noted that the division of fish into these groups is quite conditional. In natural conditions, fish feed on the type of food that is most available and its quantity is sufficient to cover the energy costs of food searching and consumption.

Almost all fish use bottom of the pond and its vegetation for breeding. Some of them lay their eggs on stony-pebbly (lithophiles) and sandy (psammophiles) soils. Whitefish (*Coregonus albula, Coregonus lavaretus*), and *Gimnocephalus cernuus* are among lithophiles. *Lota lota* is psammophile. Other fish species lay their sticky eggs on the plant substrate. Cyprinidae fish (*Abramis brama, Blicca bjorcna, Alburnus alburnus, Rutilus rutilus, Carassius carassius*), as well as *Esox lucius* and *Perca fluviatilis* are spring-spawning (Dvoryankin, 2016).

There are 7 commercial fish species in Lekshmozero lake. *Coregonus lavaretus, Coregonus albula, Esox lucius, Abramis brama* and *Lota lota* are most valuable and major commercial species, *Rutilus rutilus, Perca fluviatilis* and *Blicca bjorcna* are secondary. *Gimnocephalus cernuus* and *Alburnus alburnus* are not actually commercial species, they fall into nets accidentially and its statistics is not taken into account, as well as *Carassius carassius*, which is still found in the catches only occasionally.

Coregonus Albula

Vendace is represented by the small lake form in the Lekshmozero. In amateur and scientific catches, there are individuals of four age groups: from 1+ (fish that have lived for 1.5 years) to vendace at the age of 4+ years (who have lived for 4.5 years). Fish with a length of 12 to 14 cm predominate (here and further the fishing length is indicated – from the beginning of the snout to the end of the scale cover) at the age of 1+ … 2+ years, which may account for more than 90% of the total catch of the species. However, in some years, the age structure of the population of *Coregonus albula* varies considerably. So, in 2004 more than 50% of the total catch was fish aged 1+ with a length of less than 11 cm, and in 2009 – 87% of the extracted vendace was 3+ years old and up to 16 cm long. The reason for such fluctuations is obviously the natural cyclical features of the life span of this population.

Vendace spawning is annual, one-time in autumn-winter. Breeding periods are determined by the temperature regime of water bodies. The beginning of spawning begins in the second half of October at a water temperature of +3…+4°C. The main approach of liquid producers of vendace is usually observed in late October – early November at water temperature +1…+2°C. Spawning ends under the ice in mid-December. Spawning grounds are located on sandy-shingle and stony soils at depths of 2…5 m. Vendace of the Lekshmozero matures in the second year of life. Its absolute fecundity is relatively small and ranges on average from 2425 eggs for two-year-olds to 6431 eggs for fish aged 4+.

In the Lekshmozero *Coregonus albula* is the main commercial species and plays an important role in the life of the local population. Its share in the total annual volume of catch on the water body ranges from 50 to 80%. Taking into account the high number of species in the water body, the stability of size and age characteristics and the stability of the fishery (in 2014 - 2018, according to the fishing statistics of the Kenozerskiy national park, the catches of vendace were 20.6…22.7 t), the population status of the Lekshmozero vendace is very good.

Coregonus lavaretus is represented in catches by individuals of 5 age groups (from 3+ to 7+ years). The average size of the studied fish ranged from 23.6 cm and 179 g in 4-year-olds to 35.5 cm and 555 g in fish aged 7+. The modal group in the samples is expressed quite well – individuals aged 4+…5+ years made up about 82% of the total fish studied. Whitefish becomes sexually mature at the age of four. Its spawning begins in early November on sandy-shingle and stony soils at a depth of 3.4 m and continues until the end of December. Thus, most of the spawning period of whitefish is during the period of ice formation and in the absence of amateur fishing, which allows fish to spawn safely.

Coregonus lavaretus is one of the most valuable fish in the Lekshmozero. The amount of its production over the past six years (2013 - 2018) was insignificant. According to the data of commercial catch statistics of the Kenozerskiy national park, the production changed quite strongly. It is 0.18…0.30 t, which indicates the unstable status of the population. The low number of *Coregonus lavaretus* requires constant monitoring of its stock consistency.

Esox Lucius

In catches on Lekshmozero lake, this species is mainly represented by individuals aged from 2 to 7 years. In fishing nets, *Esox lucius* appeared in huge amounts starting from the age of 4. The average size of the extracted pike ranges from 45.7 cm in length and 944 g in weight for 4-year-olds to 70 cm and 3200 g for 7 year- old fish. There are individuals up to 14…15 kg, according to fishermen reports. The modal group in the catches is expressed quite well – about

80% of the studied fish are mature individuals aged 5-6 years, 55…60 cm long and weighing 1.5…2.5 kg.

The pike reaches sexual maturity at the age of 3…5 years (males at 3, females at 4…5) with a body length of 35…40 cm. It spawns from late April to mid-May. The beginning of spawning coincides with the melting of ice on spills in small, well-warmed places. Eggs are large, yellowish, and are deposited on flooded vegetation. The incubation period lasts 8...10 days. The basis of the spawning stock is made up of fish at the age of 4…6 years – that is, mostly first-time spawners - and came to spawn for a second time. The average absolute fecundity increases from 13,4 thousand eggs in females at the age of 4 years to 90.8 thousand eggs in 7-year-old individuals. The pike mostly feeds on the most numerous types of fish – *Perca fluviatilis* and *Rutilus rutilus*. In autumn pike changes to the feeding on *Coregonus albula*.

The number of pike in the Lekshmozero is relatively low, and catches tend to decrease (according to the fishing statistics of the Kenozerskiy national park, from 4 t - in 2014 to 1.9 t - in 2018). The multi-age nature of the Lekshmozero pike population maintain its stability. The presence of multiple age groups largely determines the success of reproduction – a decrease in the number of one or two generations due to adverse conditions (for example, fishing pressure) is compensated by the participation in reproduction of other age groups. However, like whitefish, pike requires constant control over its stock consistency.

Abramis Brama

It is established that in Lekshmozero *Abramis brama* appeared only in the late-seventies of the last century. During the comprehensive studies in 1969, it was noted the absence of representatives of *Abramis brama* in the Lekshmozero (Anukhina, 1969). Perhaps, there was a natural introduction of *Abramis brama* from the neighboring lake Lacha at a high level of the river Lekshma, which connects these two water bodies. The deliberate introduction of the species by amateur fishermen cannot be excluded, living conditions for bream were very favorable. In 2007, there was an individual with a commercial length of 75 cm and a weight of 6.1 kg caught in the lake. This is the largest studied representative of this species, caught on the territory of the Arkhangelsk region. A characteristic feature of the Lekshmozero bream is the multi-age structure of its population. There are fish aged from 3 years to 21 years. Bream sizes range from 16,4 cm in length and 93 g in weight to 75 cm and 6100 g, respectively. The modal group is well expressed in the catches. The major part of the fish caught is made up of individuals aged 7 years with an average length of 34 cm and a weight of 800…900 g. This is the age of bulky maturation of bream and it is about 60% of the total catch of the species. The fecundity of bream females at the age of 7 years reaches 60…70 thousand eggs. Spawning takes place in late May – mid-June at a water temperature of at least 14°C. Small sticky eggs are deposited on the flooded last year vegetation. The incubation period lasts for two weeks.

The multi-age nature of the commercial part of the bream population indicates its stability. The catches are quite consistent. According to the data of commercial statistics of the Kenozerskiy national park, catches are 2…2.5 tons annually. Consequently, the population status of *Abramis brama* in the Lekshmozero is satisfactory.

Lota Lota

The population of Lekshmozero burbot is also characterized by a multi-age structure. In catches there are individuals aged from 2 to 9 years. The average size of burbot from the catch varies from 24 cm in length and 109 g in weight for 2 year-olds, to 64 cm and 2.35 kg for 9 year-old burbots. However, according to fishermen, there were fish up to 5…6 kg found in the catches. *Lota lota* reaches sexual maturity in the lake at the age of 4…6.

The total number of burbot in the Lekshmozero is relatively low, catches are not stable and tend to decrease (according to the fishing statistics of the catch of the Kenozerskiy national park, from 2.7 t - in 2014 to 1.4 t - in 2018). Thus, the burbot of the Lekshmozero requires constant monitoring of its stock.

Perca Fluviatilis

In Lekshmozero perch forms two environmental groups. There is a slow-growing perch in the shallows overgrown with higher water vegetation. It feeds on benthos. Open and deeper places are the ecological niche of the large perch-predator. Probably, both groups make up the same initial population, in which during the first 4…5 years there is a gradual differentiation of perch size. Then the larger *Perca fluviatilis* switches to fish food and dramatically increase the growth rate. Smaller individuals remain benthophages and, over time, become far behind in size from predators. At the same time, in older age groups, fluctuations in the length and weight of perch in each age group are very large.

In the catches, perch is represented by individuals of 9 age groups: from 3+ to 11+ years, with an average length of 11…12 to 40 cm and a weight of up to 1…1.5 kg. According to fishermen, there are perches up to 2…2.5 kg caught in the lake. The modal group is stretched: individuals at the age of 3+ to 7+ years dominate, which make up more than 90% of the total perch produced. In the Lekshmozero perch reaches sexual maturity at the age of 3…4. Spawning is very stretched – it begins in mid-May at a water temperature of 8…10°C and can last until the end of June. Eggs are small, white-yellow, and are deposited on underwater vegetation in the form of ribbons. The ratio of males and females, in general, is close to 3:1, but from the age of 8, females begin to predominate in the catches. In the Lekshmozero, the absolute fecundity of *Perca fluviatilis* ranges from 4…5 thousand eggs in 4…5 year-old females to 27 thousand eggs in eight-year-old fish. In fast-growing individuals, the absolute fecundity is much higher and also increases with age from 13.6 to 166.4 thousand eggs.

Perca fluviatilis is one of the most numerous species of fish living in the water body (in particular, it accounts for more than 80% of all fish caught by fishing rods in winter). At the same time, the annual catches of perch on the water body are small (it is not a commercially valuable fish) and range from 2 to 4 tons. The multi-age characteristics of the perch and its big amount indicate the well-being of this species in the water body.

Rutilus Rutilus

In catches, roach is found from 3 to 10 years old, its size ranges from 9 to 28 cm in length and from 13 to 380 g of weight, respectively. Individuals at the age of 8+…9+ are dominanting, they are more than 70% of all caught roach. Roach reaches sexual maturity at 3…4 years (females a year later). Spawning takes place quite rapidly in the second half of May in the coastal zone of lakes and on spills with last year vegetation. Eggs are sticky, pale yellow, develop within 10…14 days. Hatched larvae switch to active nutrition in 2…3 days. The absolute fecundity of Lekshmozero *Rutilus rutilus* ranges from 6.6 thousand eggs in 5-year-olds to 21 thousand eggs in fish at the age of 8.

Along with *Perca fluviatilis*, roach is one of the most numerous fish of the water body, although its annual catches on the water body are also small – from 1.5 to 3 tons. The multi-age characteristics of the harvested roach, the predominance of older age groups in the catches and its amount also indicate the prosperity of this species in the Lekshmozero.

Alburnus Alburnus

In catches, this species is represented by 4 to 5 year-old individuals. The average size of caught bleak is 11…13 cm long and 20…27 grams of body weight. Bleak becomes matured at the age of 3. It spawns late – in June at a water temperature of at least 16°C. Spawning is spread and portioned. Sticky yellowish eggs are deposited on coastal vegetation at a depth of 0.1…0.6 m. The average fecundity of first-time spawning individuals is 4…4.5 thousand eggs.

Bleak in the Lekshmozero has a very high population. In experimental catches by small-scale nets in 2018, its number was the highest compared to other fish species. Bleak does not have any commercial value. It is caught while catching spawning aggregations of *Coregonus albula* in September-November. The high number and lack of fishing indicates a very prosperous state of the population.

Blicca bjorcna in the Lekshmozero lake is relatively numerous. In the catches, there are mainly matured individuals 6…7 years old with a length of 14 cm and a weight of 55 g to 17.8 cm and 122 g, respectively. *Blicca bjorcna* matures in the Lekshmozero at the age of 3…4. The absolute fecundity of females at the age of 6…7 ranged from 17 to 36 thousand eggs. The commercial value of gustera is small. Its small size and low taste quality make it a secondary fish for fishing. Gustera usually falls into the fishing nets while fishing for *Abramis brama* and large *Rutilus rutilus*. It is not included in official fishing statistics. High amount of fish and low fishing lead to a good state of the species' population.

Gimnocephalus cernuus is another abundant representative of the Lekshmozero ichthyofauna. In control catches there are individuals from 3 to 9 years old. The size of *Gimnocephalus cernuus* ranges from 7,5 cm in length and 6 g in weight for 3-year-olds to 13 cm and 37 g for fish at the age of 9. The modal group in the samples consists of matured individuals 5…6 years old with an average length of 8…10 cm and a body weight of 10…17 g. The ruff reaches sexual maturity at the age of 3. Spawning is extended and takes place in May-June. Eggs are small, yellow, and they are deposited on sandy or stony ground. The absolute fecundity of the ruff increases with age from 1 thousand eggs in 3-year-old females up to 5 thousand eggs in fish at the age of 10.

Ruff in large numbers is caught while ice fishing on *Perca fluviatilis*. It has no commercial significance and is not taken into account by official statistics. High abundance and weak fisheries provide good conditions for the population in the Lekshmozero (Dvoryankin, 2016).

Carassius carassius appeared in the lake as a result of deliberate fish introduction by local fishermen in the early 2000s. Currently, there is a small population of this species in the Lekshmozero and its number is growing from year to year.

This study was supported by the Ministry of Science and Higher Education of the Russian Federation within a state assignment, no. AAAA-A19-119011690119-9.

REFERENCES

Andronikova, I. (1976) *Zooplankton output.* Biological productivity of lake Krasnoe and conditions of its formation, 160-178. Leningrad.

Andronikova, I. (1980) *Changes in the zooplankton community due to eutrophication process.* Eutrophication of mesotrophic lake, 78-99.

Andronikova, I. (1996) *Structural and functional organization of zooplankton in lake ecosystems*, 190. Saint-Petersburg.

Andronikova, I. (2007) Evaluation of descriptiveness of zooplankton indicators as a bioindicator in monitoring of lake ecosystems. *Bioindication in freshwater ecosystem monitoring*, 212-216.

Anukhina, A. (1969) *Fisheries research of lakes in the Arkhangelsk region*: lakes of the Kenozersk group and Lekshmozero lake, 246. Petrozavodsk.

Bakanov, A. (1997) Using the characteristics of zoobenthos diversity for monitoring the state of freshwater ecosystems. *Monitoring of biodiversity*, 278-282. Moscow.

Barinova, S. Medvedeva L. A., Anisimova O. V. (2006) *Biodiversity of algae-indicators of the environment*, 498. Moscow.

Bayanov, N. (1997) Zooplankton of different types of karst lakes of the Pinezhsky reserved area and their use for monitoring purposes. *Abstract. diss. cand. biol. nauk*, 24. Moscow.

Bayanov, N. (2000) Zooplankton and benthos of the reserve's lakes. Structure and dynamics of natural components of the Pinega nature reserve (Northern taiga of ETR, Arkhangelsk district). *Biodiversity and geo-diversity in karst areas.*, 113-127. Arkhangelsk.

Buyanovskaya, A. (1977) The Problem of anthropogenic eutrophication in The Academy Sciences of the USSR. In the book: *Anthropogenic eutrophication of natural waters*, Vol. 1, 7-12. Chernogolovka.

Danilevsky, N. (1876) Description of fishing in the North-Western lakes. *Research on the state of fishing in Russia*, 155. Saint-Petersburg.

Dgebuadze, Yu. (1979) Bream growth in water bodies of different latitudes. *Variability of fish in pre-modern ecosystems*, 74-92. Moscow.

Dvoryankin, G. (2010) Modern state of ichthyofauna and ways of managing fish resources of lakes of Kenozersky National Park. *Diss. kand. biol. Sciences*, 201. Kaliningrad.

Dvoryankin, G. (2016) *Fish of Kenozersky National Park*, 100. Arkhangelsk.

Hydrobiology of lakes Vozhe and Lacha (*in connection with the forecast of water quality moving to the South*) (1978), 276. Leningrad.

Hydrology of lakes Vozhe and Lacha (*in connection with the forecast of water quality transferred to the South*) (1979), 288. Leningrad.

Freshwater invertebrates' indicator of the European part of the USSR (1977), 510. Leningrad.

Kalinkina, N. (1999) *Ecological assessment of the current state of lakes in the Kenozerskiy National Park*, 117. Petrozavodsk.

Khazov, A. (2000) *Analysis of hydrobiological data and its software implementation*, 154. Petrozavodsk.

Kitaev, S. (1984) *Ecological bases of bioproductivity of lakes of different natural zones*, 207. Moscow.

Komulainen, S., Chekryzheva, T., Vislyanskaya, I. (2006) Algoflora of lakes and rivers of Karelia. *Taxonomic composition and ecology*, 81. Petrozavodsk.

Kozmin, A., Shatov, V. (1997) *Fish management characteristics of lakes in Arkhangelsk region*, 79. Arkhangelsk.

Kozmin, A. (2011) *Fish resources of rivers and lakes of the European North-East of Russia: their conservation and use*, 314. Murmansk.

Kudersky, L. (1995) General trends in changes in the fish population of freshwater systems. *Tez. docl. Vseros. Conf. for Probl. Providence prognosis.*, 76-76. Murmansk.

Kutikova, L. (1970) *Rotifers of the fauna of the USSR*, 744. Leningrad.

Kuznetsov, N. (1951) *Fishing in rivers and lakes of Arkhangelsk region*, 166. Arhangelsk.

Lukin, A. (2001) *Assessment of the fish community current state in lake Kenozero*, 72. Petrozavodsk.

Manuylova, E. (1964) *Branched crustaceans (Cladocera) of the fauna of the USSR*, 327. Moscow-Leningrad.

Methods of water quality assessment based on hydrobiological indicators: educational and methodological development of the course "Hydrobiology" (comp.: Derevenskaya O. Yu.) (2015), 44. Kazan.

Mikhailov, Yu., Andronikova, I., Vlasov, V., and others. (1988) Assessment of possible changes in the ecosystem of lake Lacha during the transfer of runoff from Onega Bay of White sea. *VINITI RAN 26.02.88. 2052-V*, 26.

Nikolsky, G. (1974) *Ecology of fish*, 367. Moscow.

Nikolsky, G. (1980) *Structure of the species and regularities of fish variability*, 183. Moscow.

Novoselov, A., Kozmin, A. (1990) On the need for fisheries reclamation of lake Lacha. *Environmental problems of rational use and protection of water resources in the North-West of the European part of the RSFSR.*, 48-49. Vologda.

Novoselov, A. (2006) Reasonability of the Lacha Lake technical melioration (Arkhangelsk Region, Russia). *European Large Lakes Symposium*, S. 98. Estonia.

Otchenash, N., Dvoryankin, G., Imant, E. (2019) State of spring phytoplankton and quality of the Kenozero waters in 2018. *Arctic Environmental Research*, № 19 (1), 43-48.

Pekisheva, V., Onishchuk, R., Tuchina, O. (2007) Lake Lacha. Pomorskaya encyclopedia, vol. 2. *Nature of the Arkhangelsk North.*, 276. Arkhangelsk

Pidgayko, M., Alexandrov, B., Ioffe, T., and others. (1968) Brief biological and production characteristics of water bodies of the North-West USSR. *Izv. GOSNIORH*, Vol. 67, 205-225. Leningrad.

Red book of Arkhangelsk region (1995), 330. Arkhangelsk.

Red book of Russian Federation (2001), 860. Moscow.

Red book of Nenets Autonomous district (2006), 490. Naryan-Mar.

Reshetnikov Yu. (1979) Fish variability and ecological forecasting. In the book: "*Fish variability of freshwater ecosystems,*" 5-12. Moscow,

Reshetnikov, Yu. (1981) G. V. Nikolsky's ideas about faunal complexes and their modern development. *Modern problems of ichthyology*, 75-96. Moscow.

Reshetnikov, Yu. (1982) Change of lake ecosystems. In the book: "*Changing the structure of the fish population of the eutrophied reservoir,*" 5-16. Moscow.

Reshetnikov, Yu. (1995) Modern problems of studying whitefish. *Questions of ichthyology*, Vol. 35, No. 2, 156-174.

Rylov, V. (1930) Freshwater Cladocera of the USSR. *Indicator of freshwater organisms of USSR*, 288. Leningrad.

Rylov, V. (1948) Fauna of USSR. Crustacea. *Cyclopoida of fresh water.* Publishing house of the USSR Academy of Sciences, vol. 3, issue 3, 318. Moscow-Leningrad.

Tsepkin, E. (1966) Influence of climatic and anthropogenic factors on the freshwater fish fauna of Northern Eurasia (based on archaeological materials). *Vestnik MGU*, Ser. 16, No. 1, 38-43.

Tsepkin, E. (1995) Changes in the commercial fish fauna of continental water bodies of Eastern Europe and Northern Asia in the Quaternary period. *Vopr. Ichthyology*, issue 35, No. 1, 3-17.

Vezhnovets, V. (1981) Biological characteristics of the population of relict planktonic crustacean *Limnocalanus macrurus* Sars in connection with eutrophication of the lake. Circulation of matter and energy in water body. *Elements of the biotic cycle*, Release 2, 67. Irkutsk.

Vorobyova, T. (2008) *Results of complex studies of lakes in the southern part of the Kenozersky National Park*, 18. Arkhangelsk.

Zhadin, V. (1960) *Methods of hydrobiological research*, 191. Moscow.

Zhakov, L. (1984) *Formation and structure of the lakes' fish population in the North-West of USSR*, 144. Moscow.

Zooplankton and zoobenthos indicator of fresh waters of European Russia, Vol. 1: Zooplankton (edited by V. R. Alekseev, S. Ya. Tsalolikhin) (2010), 495. Moscow-SPb.

ANNEX 1. THE PHYTOPLANKTON OF LAKE KENOZERO (OCCURRENCE BY YEAR)

Taxon	Years	
	2001	2018
CYANOPHYTA		
Anabaena sp.	+	+
Aphanizomenon flos-aquae	-	+
Aphanothece **sp.**	+	-
Aphanocapsa sp.	+	-
Aphanocapsa elachistica	+	-
Aphanocapsa delicatissima	+	-
Chroococcus minutus	+	-
Coelosphaerium kuetzingianum	+	-
Gloeocapsa sp.	-	+
Limnothrix planctonika	+	-
Microcystis sp.	-	+
Merismopedia tenuissima	+	-
Merismopedia warmingiana	+	-
Phormidium tenuis	+	-
Phormidium sp.	+	-
Snowella atomus	+	-
Snowella lacustris	+	-
Snowella sp.	+	-
Woronichinia sp.	+	-
CHRYSOPHYTA		
Dinobryon divergens	+	+
Dinobryon bavaricum	+	-
Dinobryon spirale	-	+

(Continued)

Taxon	Years	
	2001	2018
CHRYSOPHYTA		
Dinobryon acuminatum	+	-
Mallomonopsis sp.	+	+
Pseudostaurastrum hastatum	+	-
BACILLARIOPHYTA		
Acanthoceros zachariasii	+	-
Achnanthes sp.	-	+
Asterionella formosa	+	-
Amphora coffeaeformis	-	+
Amphora exigua	-	+
Amphora ovalis	-	+
Aulacoseira italica	+	-
Aulacoseira italica var. tenuissima	+	-
Aulacoseira distans	+	-
Aulacoseira alpigena	+	-
Aulacoseira islandica	+	-
Aulacoseira granulata	+	-
Asterionella formosa	-	+
Caloneis bacillum	-	+
Caloneis silicula	-	+
Caloneis sp.	-	+
Cocconeis sp.	+	+
Cyclotella bodanica	-	+
Cyclotella sp.	+	+
Cyclotella comta	+	+
Cyclotella planctonica	-	+
Cyclotella glomerata	+	-
Cymbella ventricosa	-	+
Diatoma hiemale	-	+
Diploneis interrupta	-	+
Diploneis ovalis	-	+
Eunotia pectinalis	-	+
Eunotia praerupta	-	+
Eunotia sp.	-	+
Fragilaria bicapitata	-	+
Fragilaria crotonensis	+	+
Fragilaria capucina	+	+
Fragilaria construens	-	+
Gomphonema acuminatum	-	+
Gomphonema gracile	-	+
Gyrosigma acuminatum	-	+
Melosira granulata	-	+
Meridion circulare	-	+
Navicula gastrum	-	+
Navicula lanceolata	-	+
Navicula mutica	-	+
Navicula placentula	-	+
Navicula sp.	-	+
Navicula tuscula	-	+
Nitzschia acuminata	-	+
Nitzschia gracilis	-	+
Nitzschia holsatica	-	+
Nitzschia linearis	-	+

Taxon	Years	
	2001	2018
BACILLARIOPHYTA		
Nitzschia longissima	-	+
Nitzschia palea	-	+
Nitzschia sigmoidea	-	+
Nitzschia sp.	-	+
Nitzschia subtilis	-	+
Nitzschia tryblionella	-	+
Nitzschia vermicularis	-	+
Rhizosolenia longiseta	+	-
Stephanodiscus sp.	+	-
Stauroneis anceps	-	+
Stephanodiscus hantzschii	-	+
Synedra ulna sensu	-	+
Synedra ulna	+	-
Synedra acus	+	-
Synedra berolinensis	+	-
Synedra nana	+	-
Tabellaria fenestrata	+	-
Tabellaria fenestrata var.asterionelloides	+	+
Tabellaria flocculosa	+	-
CRYPTHOPHYTA		
Croomonas acuta	+	-
Chroomonas nordstedtii	+	-
Cryptomonas sp.	+	-
Chilomonas minor	+	-
DYNOPHYTA		
Ceratium hirundinella	+	-
Gymnodinium sp.	-	+
Peridinium sp.	+	+
Peridinium pusillum	+	-
Protoperidinium bipes	-	+
CHLOROPHYTA		
Ankistrodesmus convolutus	-	+
Ankistrodesmus falcatus	-	+
Staurodesmus sp.	+	-
Staurodesmus mammillatus	+	-
Staurodesmus megacanthus	+	-
Cosmarium sp.	+	-
Cosmarium phaseolum	+	-
Staurastrum sp.	+	-
Staurastrum anatinum	+	-
Staurastrum sebaldi	+	-
Staurastrum petsamoense	+	-
Staurastrum paradoxum	+	-
Spondylosium planum	+	-
Closterium sp.	+	+
Closterium dianae	+	-
Closterium acutum	-	+
Closterium ceratium	+	-
Closterium pronum	+	-
Botryococcus protuberans	+	-
Crucigenia tetrapedia	+	-
Elakatotrix gelatinosa	+	-
Eutetramorus fottii	+	-
Dictyosphaerium subsolitarium	+	-

(Continued)

Taxon	Years	
	2001	2018
CHLOROPHYTA		
Didymocystis inconspicua	+	-
Kirchneriella contorta	+	-
Monoraphidium contortum	I	-
Monoraphidium komarkovae	+	-
Oocystis borgei	+	-
Oocystis sp.	+	-
Scenedesmus quadricauda	+	-
Sphaerocystis schroeteri	+	-
Tetraedron minimum	+	-
Pediastrum tetras	+	-
Pediastrum boryanum	+	-
Eudorina elegans	+	-
Chlamydomonas sp.	+	-
Pandorina morum	+	-
EUGLENOPHYTA		
Euglena sp.	-	+
XANTHOPHYTA		
Centritractus sp.	-	+
Tribonema sp.	-	+

ANNEX 2. THE ZOOPLANKTON OF LAKE KENOZERO (OCCURRENCE BY YEAR)

Taxon	Years		
	2001	2008	2018
ROTIFERA			
Asplanchna sp.	-	+	-
Asplanchna priodonta	-	+	-
Euchlanis sp.	-	+	-
Keratella quadrata	-	+	-
Kellicottia longispina	-	+	-
Filinia sp.	-	+	-
Trichocerca sp.	+	-	+
Synchaeta sp.	+	-	-
Polyarthra major	+	-	-
Bipalpus hudsoni	+	-	+
Asplanchna herricki	+	-	+
Trichotria sp.	+	-	-
Euchlanis dilatata	+	-	-
Keratella cochlearis	+	-	+
Conochilus unicornis	-	-	-
CLADOCERA			
Limnosida frontosa	+	-	+
Diaphanosoma brachyurum	+	-	-
Daphnia longispina hyalina	+	-	+
Daphnia cristata	+	-	+
Daphnia. hyalina	-	-	+
Ceriodaphnia sp.	+	-	+
Rhynchotalona falcata	+	-	-

Taxon	Years		
	2001	2008	2018
	CLADOCERA		
Chydorus sphaericus	+	-	-
Grabtoleberis testudinaria	+	-	-
Bosmina longirostris	+	-	+
Bosmina obt. obtusirostris	+	+	-
Bosmina obt. lacustris	+	-	-
Bosmina coregoni coregoni	+	-	+
Bosmina coregoni gibbera	+	-	-
Bosmina kessleri	+	-	-
Leptodora kindtii	+	-	-
Alona quadrangularis	-	-	+
Alonopsis elongatus	-	-	+
Polyphemus pediculus	-	-	+
	COPEPODA		
Acanthocyclops sp.	-	-	+
Limnocalanus macrurus	+	-	+
Limnocalanus grimaldi macrurus	-	+	-
Eudiaptomus gracilis	+	-	-
Eudiaptomus sp.	-	+	+
Eurytemora lacustris	+	-	+
Heterocope appendiculata	+	-	+
Cyclops strenuus	+	+	+
Cyclops colensis	+	-	-
Cyclops scutifer	-	+	+
Acanthcyclops sp.	+	-	-
Mesocyclops leuckarti	+	-	+
Mesocyclops oithonoides	+	-	-
Diacyclops bicuspidatus	-	-	+
Megacyclops viridis	-	-	+
Thermocyclops oithonoides	-	-	+

In: Lake Water
Editors: Oleg S. Pokrovsky, Yulia Bespalaya et al.
ISBN: 978-1-53619-275-9

Chapter 13

LAKES OF NENETS AUTONOMOUS AREA

Alexander P. Novoselov*, Gennady A. Dvoryankin, Elena V. Pribytkova and Ekaterina N. Imant
Institute for Complex Arctic Studies,
Federal Center for Integrated Arctic Research of the Russian Academy of Sciences,
Arkhangelsk, RF

ABSTRACT

The chapter describes hydrochemistry, hydrobiology and ichthyology of lakes located in the Nenets Autonomous Area, NE European Russia. We present the information regarding three types of lake system: floodplain lakes (Golodnaya Guba lake), Malozemel'skaya tundra lakes (Indiga and Urdyuzhskoe lakes), and the Bolshezemelskaya tundra lakes (Pilney, Prosunduy, Vashutkin and Padimey-Harbey lakes). We analyze and discuss chemical and physical lake characteristics, plankton, benthos, fish food stocks, vegetation, ichtyofauna, and fish stocks.

Keywords: lake system, Arctic, Nenets Autonomous Area

INTRODUCTION

The main part of the Eastern European tundra on the territory of the Nenets Autonomous Area is occupied by the Bolshezemelskaya tundra with an area of more than 100 thousand km^2. The geological structure and climatic conditions led to the development of a dense hydrographic network. It consists of watercourses and lakes that represent characteristic elements of the landscape. A distinctive feature of the region is the presence of soil permafrost at a shallow depth, which prevents precipitation from penetrating into the subsurface layers. Constant excessive moisture contributes to the wide spread of marsh soils. The total length of watercourses with a length of more than 10 km exceeds 6 thousand km. In terms of the lakes number, the Bolshezemelskaya tundra is almost twice as large as the Republic of Karelia, so

* Corresponding Author's Email: alexander.novoselov@rambler.ru.

called “a country of lakes” – (Goldina, 1972). On the territory of the Bolshezemelskaya tundra, the main lake pool is concentrated, there are about 80 thousand lakes with a total area of 5 thousand km^2, which is comparable to the total amount of lakes in the Komi Republic and the Nenets Autonomous district. Most of them have a small water mirror – less than 0.1 km^2. Large lakes and watercourses are of the greatest economic interest. The number of lakes with an area of 2.5 km^2 is about 700 and the largest of them are the Vashutkin, Padimey, Harbey, Kolvin lake and river water systems with a total water area of 150 km^2. These lake-river systems can be used to create commercial farms and sport fishing (Kozmin, 2011).

Chemical composition of the water in the Bolshezemelskaya tundra is due to the origin of its basins. Lakes are divided into glacial (lake Vasutkin, Harbey, Padimeya, Naulto, Vatyarti and others), the lagoon (the lake is Big Tarovei) and thermokarst (lake Maliy Izyati).

Vashutkin lakes (Adzva river basin) is the largest lake system, with a total area of about 85 km^2 (Solovkina, Sidorov, 1966; Sidorov, 1974, 1976). Harbeyskaya and Padimeyskaya lakes are located in the river basins of Korotfiha and Bolshaya Rogovaya. The largest lakes of the Padimey system are Bolshoy Padimeyty (9.6 km^2) and Lola Kuliga (5.0 km^2), connected by a wide strait. They are characterized by a large depth (up to 57 m) and a complex bottom relief. Bolshoy and Maliy Harbeity, Golovka, and Kresty are a part of Harbei system of lakes. The study of the Harbey lakes was carried out from 1968 to 1972 by the Komi USSR Academy of Sciences in the framework of the International biological program. One of the largest lakes of the Bolshezemelskaya tundra is lake Bolshoy Harbeity – has an area of 23.1 km^2 and a maximum depth of 17 m. At higher latitudes, relative to the lakes of the Harbei system, lake Naulto is located, surveyed in 1993 (Khokhlova, 1994). The lake has an area of 2.2 km^2 and serves as the source of the Naulyakha river, which carries its waters to the Khaypudyr Bay. In 1990 lakes of the basins of Kolva – Piseity, Vozeity, Evsyaty, Settaity rivers were studied. These lakes of glacial origin are residual water body from the large lake-glacial water body that existed in the late tertiary period (Kolvinsky lake) and are associated with inter-hill depressions of moraine ridges. The largest of these lakes – lake Vozeyty – has a maximum depth of 47 m and the area of the water mirror of 6.1 km^2 (Sadyrin, 1998).

Malozemelskaya tundra is a hilly plain between rivers Indiga and Pechora, gradually lowering towards the sea. The rivers are fed mainly by bogs and partly due to lakes, which are especially rich in Malozemelskaya tundra. The vast majority of the lakes are of small sizes (0.5 to 3.0 hectares), but there are several large ones. The largest of the lakes in Malozemelskaya tundra are Urdyuzhskoe lake (area of 71 km^2), Aputei lake (49 km^2), Matervisochnoe lake (23 km^2), Ikcha lake (21 km^2), Tail lake (12 km^2). Small lakes are usually located in closed drainless relief depressions. Lakes of secondary origin often appear in the sites of eroded peat bogs. Quite often, bogs or shallows overgrowing the dried bottoms with bushes appear instead of lakes.

Floodplain Lakes

Golodnaya Guba Lake (The Delta of the Pechora River)

Physical and Geographical Features

Golodnaya Guba lake is the largest water body of the Nenets Autonomous district, located on the territory of the Malozemelskaya tundra, 26 km to the North of Nar’yan-Mar, between delta of the Pechora river in the East and the Nenets ridge in the West. The lake is extended in the latitudinal direction, its maximum length is 40 km, and the maximum width is 10 km. There are three sections in the water body – the Large and the Small pools and the Central part (Figure 1). The lake is characterized by small depths - up to 3.5 m. The bed of the water body is a pebble-sand littoral (with some large boulders), which gradually turns into a profundal filled with fine silts. Coastal sands, silted sand zone and silts of pools and bays are the main biotopes (Golodnaya Guba Lake, 2013).

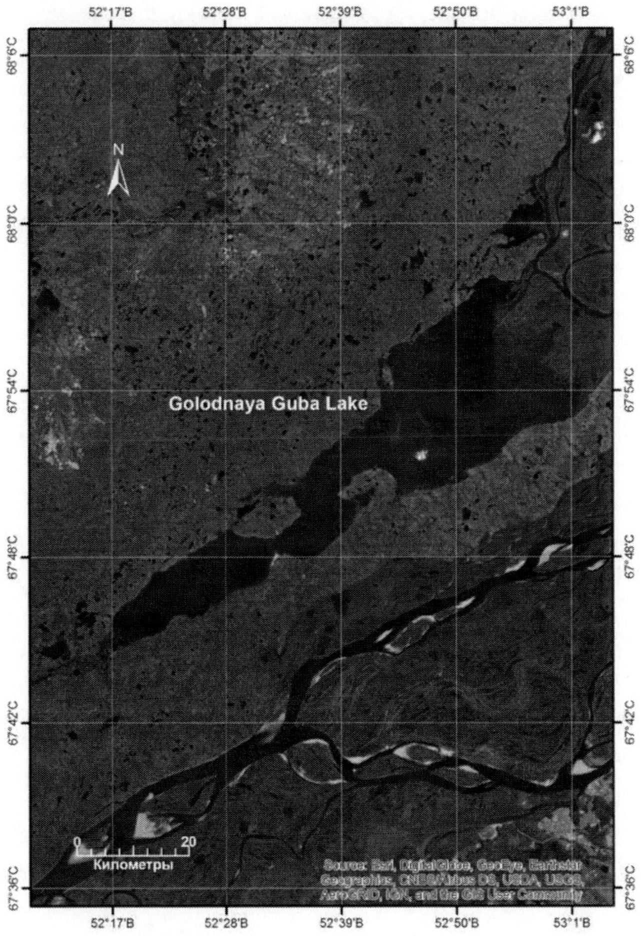

Figure 1. Golodnaya Guba Lake.

The nature of the landscape in the vicinity of Golodnaya Guba lake and the features of geological structure of the basin indicate that the water body, located at the bottom of the inter-row depression, was previously an element of the marine landscape. The lake is connected to the arms of the Pechora delta by three channels. Currently, the lake is in the process of becoming a closed water body due to strong erosion of the banks and the formation of a vast coastal shallow water. In these areas, in low-water years and in cold winters, when the ice thickness reaches 100…120 cm, the lake freezes to the bottom and becomes isolated from the Pechora river for almost six months (Kozmin, 2011).

Golodnaya Guba lake has always had a great fishing significance due to the fact that the dominant commercial fish were valuable species of whitefishes (whitefish, peled, broad whitefish, sardine-cisco). The lake was the main commercial water body in the middle of the last century and produced up to 100 tons of whitefish species annually, which made up about 80% of all production in delta of the Pechora river. Lake Golodnaya Guba has always served as a place for wintering and feeding young fish of all types, as well as fattening of whitefish producers. Golodnaya Guba has favorable conditions for summer feeding, both adult whitefish and juveniles. A rich food base is created due to the good warming and homothermy of the waters, as well as due to the constant replenishment of the lake with the Pechora river waters, which are richer in nutrients than the local runoff. It should be noted that Golodnaya Guba lake is also a feeding place for less valuable fish species – ordinary fish and smelt.

The lake is not isolated in winter, and fish can leave the areas of the lake with an unfavorable regime. In autumn and winter, whitefish are observed to descend from the Golodnaya Guba to the delta and to the Korovinskaya Guba. Due to sufficient isolation and a peculiar hydrological regime, there is a rich food base, in terms of zooplankton and zoobenthos, which significantly exceeds those of the lower reaches of the Pechora river. At the same time, the water body is part of a single system with the Korovinskaya Guba and the Pechora river delta, and during periods of unfavorable hydrological conditions, whitefish migrate from the lake to the delta channels (Kornilova, 1967; Ivanova, Chuksina, 1985).

Back in the sixties of the last century, a number of researchers (hydrographers, hydrochemists, and ichthyologists) noted that the lake entered "the stage of transformation into a closed water body," caused in some years by the disturbance of water exchange between the Golodnaya Guba and the Pechora river channels (Zhila, 1963; Kornilova, 1967; Maccabeeva, 1967). A tendency for a gradual eutrophication of the water body has been observed due to modification of the hydrological regime and a reduction in the intensity of water exchange. Both in winter and in summer, there was deficiency of bottom-level oxygen in some parts of the lake, causing fish death. After the largest accident on the oil pipeline in 1994, a large amount of oil fell into the Pechora river basin. As a result, the content of petroleum hydrocarbons in the water and heavy metals in the lake's soil increased, which became evidence of chronic pollution of Golodnaya Guba by pollutants. The largest impact was made by petroleum products and phosphorus, which stimulates the development of eutrophication processes. The lake's soils have an extremely high content of petroleum hydrocarbons, and the concentration of heavy metals in almost all components exceeded the baseline values. Due to the conceptual law of ecology "everything is connected with everything," changes in abiotic environmental conditions inevitably led to the restructuring of biotic components of the aquatic ecosystem (Novoselov et al., 2005).

In the following years, due to industrial overfishing, poaching and intensive economic activities related to the development of oil and gas fields, the stocks of valuable fish species in the water body significantly decreased. As a result of anthropogenic impact, conditions for feeding young and adult whitefish in the lower reaches of the Pechora river also deteriorated. Reproduction conditions have also deteriorated at many traditional spawning sites. As a result, by the beginning of the XXI century catches of valuable fish species fell tenfold in Golodnaya Guba, and after 2003, fishing was stopped and is still not conducted. At the same time, the Golodnaya Guba continues to be a feeding and wintering place for whitefish species, a place of concentration of their juveniles. The increase of anthropogenic impact on the herds of valuable fish species of the Pechora river delta in recent years increases the importance of the

Golodnaya Guba in the reproduction of whitefish, located to a greater extent in the Delta channels.

Hydrobiology

The study of the forage base in the water body of the lower reaches of the Pechora river (including Golodnaya Guba lake) began in the early sixties (Kornilova, 1967) and continues to the present day. The results of hydrobiological surveys make it possible to monitor changes in the trophic status of the water body. The analysis of changes in the development of invertebrate fauna allows to assess the ecological situation in the water body due to the ongoing technogenic impact.

Zooplankton

Cladocerans and copepods were the main groups that determined both the species and taxonomic diversity of zooplankton of Golodnaya Guba lake in the period from 1980 to 2016. During the entire observation period, 65 zooplankton species were identified in the water body, belonging to 37 genera, 19 families and 5 orders (Table 1).

Table 1. Zooplankton species composition of the Golodnaya Guba Lake (Imant et al., 2018)

Species	Ecological properties	Distribution
ROTIFERA		
Asplanchna priodonta	eurybiont	cosmopolitan
Asplanchna herricki	eurybiont	cosmopolitan
Asplanchna sp.		
Conochilus sp.		
Euchlanis dilatata	lake-pond, littoral	cosmopolitan
Euchlanis sp.		
Filinia sp.		
Kellicottia longispina	lake, pelagic, cold-water	Northern latitudes
Keratella cochlearis	eurybiont	cosmopolitan
Keratella quadrata	eurybiont (mainly warm-water)	cosmopolitan
Notholca labis	eurybiont (mainly cold-water	cosmopolitan
Notholca acuminata	eurybiont, cold-water	Northern hemisphere
Ploesoma hudsoni	eurybiont (mainly cold-water)	cosmopolitan
Ploesoma sp.		
Polyarthra sp.		
Trichocerca capucina	eurybiont, phytophilic	cosmopolitan
Trichocerca sp.		
CLADOCERA		
Alona affinis	eurybiont	Europe, North Africa, Asia
Alona quadrangularis	lake-pond, littoral	Northern hemisphere
Alonopsis elongatus	eurybiont	Palearctic
Bosmina coregoni	-	-
Bosmina longirostris	eurybiont (mainly warm-water)	cosmopolitan
Bosmina kessleri	lake, phytophilic	Palearctic
Bosmina sp.		

Table 1. (Continued)

Species	Ecological properties	Distribution
Bythotrephes cederstroemi	eurybiont	North of the European part of Russia
Ceriodaphnia quadrangula	eurybiont	cosmopolitan
C. reticulata	eurybiont	
Chydorus ovalis	lake, pelagic	cosmopolitan
Ch. sphaericus	eurybiont	cosmopolitan
Daphnia (Daphnia) cristata	lake, pelagic, cold-water	North of Eurasia
D. (D.) galeata	eurybiont	Palearctic
D. (D.) hyalina	eurybiont	Temperate zone of Western and Eastern Europe, Caucasus
D. (D.) cucullata	lake, pelagic, warmwater	North of Eurasia
D. (D.) longiremis	lake, pelagic, cold-water	Northern hemisphere
D. (D.) longispina	eurybiont (mainly warm-water)	cosmopolitan
Diaphanosoma brachyurum	eurybiont (mainly warm-water)	Eurasia, America, North Africa
Eurycercus lamellatus	lake, phytophilic	Northern hemisphere, South America
Leptodora kindtii	lake, pelagic, warmwater	Northern hemisphere
Limnosida frontosa	lake-pond, littoral	Palearctic
Pleuroxus uncinatus	lake, littoral	Holarctic
Monospilus dispar	lake, littoral	Palearctic
Scapholeberis mucronata	eurybiont	Palearctic
Sida crystallina	lake, phytophilic, warmwater	Eurasia, America,
Simocephalus vetulus	lake-pond, littoral	Cosmopolitan, except Australia
COPEPODA		
Cyclopoida		
Acanthocyclops sp.		
Cyclops scutifer	lake, pelagic, cold-water	Palearctic
C. strenuus	lake-pond, littoral-pelagic, planktonic, cold-water	North of Eurasia, North Africa
C. vicinus	lake, pelagic	Palearctic
Diacyclops bicuspidatus	lake, river, intertidal, meiobenthos	cosmopolitan
D. languidoides	common in swamps, meiobenthos	-
D. languidus	common in swamps, coldwater, meiobenthos	-
Eucyclops serrulatus	lake, littoral, meiobenthos	Eurasia
Species	Ecological properties	Distribution
Megacyclops viridis	different types of water bodies, littoral,meiobenthos	European Russia
Mesocyclops leuckarti	eurybiont (mainly warm-water)	cosmopolitan
Paracyclops fimbriatus	lake, river	cosmopolitan
Thermocyclops crassus	lake, littoral-pelagic, planktonic	cosmopolitan
Th. oithonoides	lake, pelagic, planktonic	cosmopolitan
Calanoida		
Diaptomus glacialis	river, small water body	Coast of the Arctic ocean

Diaptomus sp.		
Eudiaptomus graciloides	lake-pond	Europe, Siberia, China
E. gracilis	lake-pond	Europe, Siberia
Eudiaptomus sp.		
Eurytemora affinis	lake, river	Europe, Asia, North America
E. lacustris	lake, river (in the delta)	Europe, Western Siberia
Heterocope appendiculata	lake	North of the European part of Russia

Long-term hydrobiological studies of the lake indicate that its zooplankton community is represented by species typical to northern water body. According to zoogeographic classification, the bulk of zooplankton species in the Golodnaya Guba Lake are organisms that have a cosmopolitan and Palearctic distribution.

Figure 2 shows complex of species, dominating in Golodnaya Guba for several years. Widespread cladocerans was basic zooplankton community. Representatives of the Bosmina genera, as well as Chydorus, are indicators of eutrophic conditions. In 2012, Bosmina accounted for 71.6% of the total zooplankton population. A characteristic feature of summer plankton is the abundance of nauplial and copepodite stages of copepods, eggs and young cladoceras.

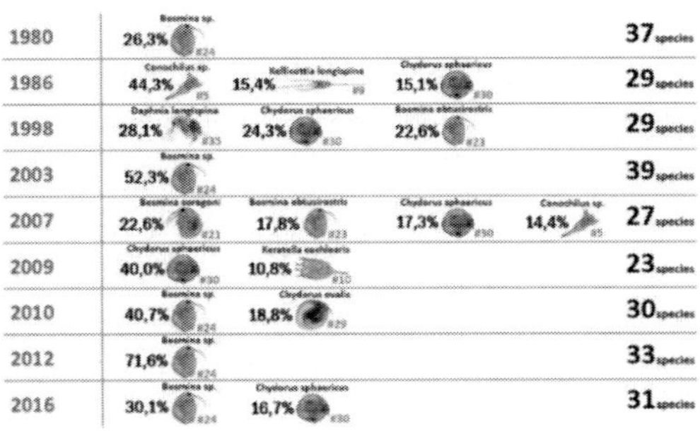

Figure 2. Dominant zooplankton species of the Golodnaya Guba Lake (% of total population) (Imant et al., 2018).

Table 2. Dynamics of changes in zooplankton biomass in the Golodnaya Guba Lake (Novoselov et al., 2005)

Periods	Zooplankton biomass, g/m^3
60'	> 5.00
80's	2.03
Average value 1990-1994	2.12
Post-accident period (15 years)	
Average value 1995 - 1999	1.73
Average value 2000 - 2004	1.69
Average value 2005 - 2009	1.57
Current state	
Average value 2010 - 2017	2.10

According to the level of the food base for plankton-eating fish, the Golodnaya Guba lake was the most productive water body in the lower reaches of the Pechora river. According to the accepted classification (Pidgayko et al., 1968), in the sixties of the last century, it was characterized as a high-feed water body with a zooplankton biomass of more than 5.0 g/m^3. In the eighties, the zooplankton biomass averaged just over 2.0 g/m^3, thus the lake was classified as a medium-feed water body. In the early nineties, the lake's production status was maintained (with a zooplankton biomass of 2.12 g/m^3). In the late nineties, after an emergency oil spill, the average annual biomass decreased to 1.73 g/m^3, but the lake was still a medium-sized water body in terms of production capacity. In the 2000s of the current century, the biomass of planktonic animals continued to remain at the same low level (1.69…1.57 g/m^3), which was caused by the ongoing anthropogenic eutrophication of the water body. In addition, in some years of the decade, there was a strong warming of the water in the lake, which, in turn, at low depths caused phytoplankton blooms. Consequently, it affected the development of invertebrates inhabiting the water column (Novoselov et al., 2005). In the period of 2010 to 2017, the average annual zooplankton biomass returned to the pre-accident level (2.10 g/m^3), and the lake was characterized as a β-mesotrophic water body with an average level of zooplankton biomass development. Currently, its production capacity is defined as above the average yield (Table 2).

Zoobenthos

In the sixties, the ecosystem of the lake has not yet experienced any strong anthropogenic impact, the environmental situation was favorable, and the benthic community was characterized by mass development of chironomids and mollusks. Since the eighties, changes in the benthic fauna caused by pollution of the Pechora basin occured. The dominating species, along with chironomids and mollusks, were oligochaetes – indicators of organic pollution, including oil. Over the past two decades, the number of meiobentic organisms (nematodes, ostracods, and hydracarines) in the samples has increased, which indicates an unfavorable state of the aquatic environment. The biomass of the benthic fauna remained high throughout all the years of observations (since the 60s), and in terms of productivity, Golodnaya Guba retains the status of a high-feeding water body for benthic fish (Pidgayko et al., 1968) (Table 3).

Table 3. Dynamics of changes in zoobenthos biomass in the Golodnaya Guba lake (Novoselov et al., 2005)

Periods	Zoobenthos biomass, g/m^2
60's	Rather eutrophic
80's	55.20
Post-accident period (15 years)	
Average value 1995 - 1999	29.83
Average value 2000 - 2004	36.74
Average value 2005 - 2009	16.64
Current state	
Average value 2010 - 2012	55.16

At the same time, changes in the biomass of benthic communities in the long-term aspect exhibited the same trend as in zooplankton communities. The level of zoobenthos development in the post-accident period decreased from 55.2 g/m^2 in the 80s to 17 to 37 g/m^2 in the late 90s and in the 2000s, and returned to the pre-accident period of 55 g/m^2 in 2010 - 2012. It should be noted that development of both zooplankton and zoobenthos largely depends on the hydrological conditions of each particular year and can vary significantly.

So far, according to the level of the forage base, the lake still continues to be a good feeding water system for benthic fish (whitefish, broad whitefish), as well as for planktophage fish (peled, vendace, smelt, and juveniles of all other species), providing them with trophic conditions above average feeding capacity.

Ichthyofauna

According to literature (Kornilova, 1967) and the results of our research, the ichthyofauna of the Golodnaya Guba Lake is represented by 8 families and 15 taxa of fish, including species, subspecies and ecological forms (Table 4, Figure 12 in Appendix).

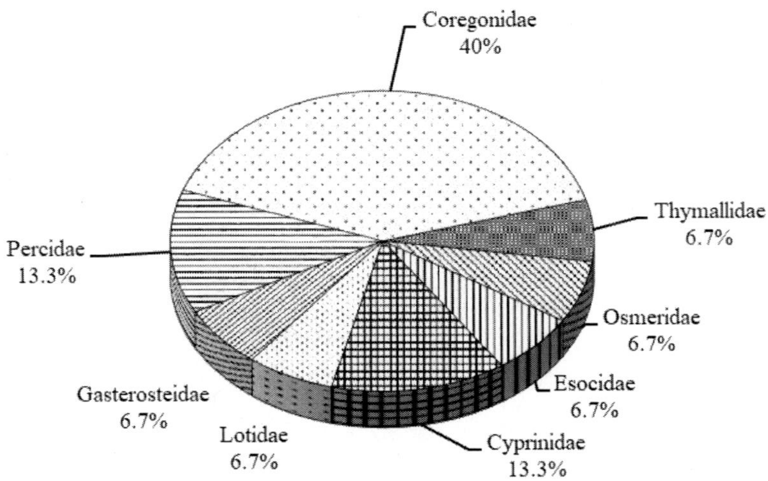

Figure 3. Fish ratio in the ichthyofauna of Golodnaya Guba lake (by: Novoselov, 2013).

Coregonidae is the most abundant family (6 species or 40.0% of the total number of species), represented by Coregonus sardinella marisalbi, Coregonus lavaretus, Coregonus peled, Coregonus nasus, Coregonus autumnalis, Stenodus leucichthys nelma. Two species (13.7% each) are part of Cyprinidae (Leuciscus idus, Rutilus Rutilus) and Percidae (Gymnocephalus cernuus, Perca fluviatilis) families. One species (6.7%) represents the families of Thymallidae (Thymallus thymallus), Osmeridae (Osmerus eperlanus var. spirinchus), Esocidae (Esox lucius), Lotidae (Lota lota), and Gasterosteidae (Pungitius pungitius) (Figure 3).

According to the classification of G. V. Nikolsky (1980), all fish living in Golodnaya Guba lake can be split into 4 faunal complexes (Figure 4). In terms of species, the largest number of freshwater species (8 or 53.3%) belongs to the Arctic freshwater complex. It is presented by vendace, whitefish, peled, broad whitefish, Arctic cisco, nelma, burbot and lake smelt. 5 species (33.3%) represent the boreal plain faunal complex. It includes pike, roach, ide, ruff and perch. One species (6.7%) is included in the boreal submontane (european grayling) and Ponto-Caspian freshwater (nine-spined stickleback) complexes.

Table 4. Ichthyofauna composition of the Golodnaya Guba Lake (by: Novoselov, 2013)

Taxon	Due to Kornilova, 1967	Our data
Coregonidae		
Coregonus sardinella marisalbi	+	+
Coregonus lavaretus	+	+
Coregonus peled	+	+
Coregonus nasus	+	+
Coregonus autumnalis	+	-
Stenodus leucichthys nelma	+	+
Thymallidae		
Thymallus thymallus	+	+
Osmeridae		
Osmerus eperlanus var. spirinchus	+	+
Esocidae		
Esox lucius	+	+
Cyprinidae		
Leuciscus idus	+	+
Rutilus rutilus	-	+
Lotidae		
Lota lota	+	+
Gasterosteidae		
Pungitius pungitius	-	+
Percidae		
Gymnocephalus cernuus	+	+
Perca fluviatilis	+	+
Totally	13	14

Note: + the species; – the species was not found.

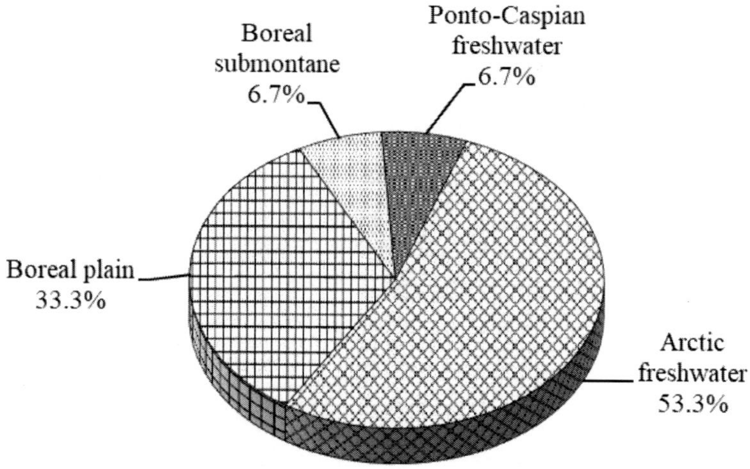

Figure 4. Fish ratio of faunal complexes in the Golodnaya Guba Lake (by: Novoselov, 2013).

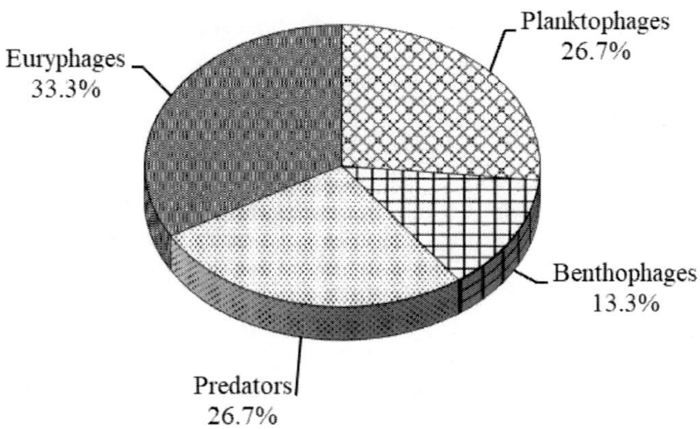

Figure 5. Fish ratio due to different types of food in lake Golodnaya Guba (by: Novoselov, 2013).

Two types of fish are typical for the local environment. These are an ecological form of the Pechora vendace, which has the local name "saurey" (Novoselov, 1991; Sendek, 2000) and a residential lake form of smelt, which has the local name "nagish" (Kirpichnikov, 1935). Two species – the Arctic cisco and the European grayling – are typically river species and are found only once in the Golodnaya Guba (Kornilova, 1967). The grayling are encountered only at the estuaries that flow into the lake of rivers and streams. The remaining 11 species are lake-river and are mainly represented by widespread fish (whitefish, peled, broad whitefish, nelma, pike, roach, ide, burbot, perch, ruff and nine-spined stickleback). There are 4 passable (Arctic cisco) and semi-passable (whitefish, peled, nelma) forms. All other species are aquatic fish whose entire life cycle occurs in certain lake or river biotopes.

All types of fish of the Golodnaya Guba can be classified into four groups: planktophages, benthophages, predators and euryphages (Figure 5). Typically, four types of fish (26.7%) have planktonic food – vendace, peled, Arctic cisco and smelt "nagish." At the same time, small forms of zooplankton are consumed by young fish of almost all species at the early stages of ontogenesis. Whitefish and broad whitefish have predominantly benthic diet (13.3% of the total ichthyofauna). Four fish species are characterized by predatory feeding type (26.7%). These are nelma, pike, burbot and perch. The largest group is euryphages (5 species, or 33.3%). As a rule, euryphages have a wide food range with a predominant consumption of a particular type of feed. In the Golodnaya Guba Lake, these species are represented by European grayling, roach, ide, ruff and nine-spined stickleback.

According to the type of reproduction, all fish in the lake use a certain substrate for laying eggs. The largest number of species (7 or 46.7% of all fish species) are lithophiles and thet lay eggs on rocky areas of the bottom (Figure 6). These are migratory fish (Acrtic cisco) and semi-anadromous (nelma, whitefish, peled and broad whitefish). This group also includes lake vendace "saurey" and European grayling. The psammophilus species that lay eggs on sandy areas of the bottom include 3 species (20.0%): smelt "nagish," burbot and ruff. There are 4 species (26.7%) that prefer to lay eggs on a plant substrate (spring-spawning species): pike, roach, and perch. One species (6.7%) arranges special nests for laying eggs (nine-spined stickleback).

Economically, all fish species living in lake Golodnaya Guba can be classified into the following categories: commercially valuable (7 species or 46.6%), commercially less valuable (7 species or 46.6%) fish, non-commercial species (1 species or 6.8%). Valuable fish include whitefishes, which are in high demand due to high consumer qualities (nelma, vendace, Arctic cisco, whitefish, broad whitefish, peled), as well as grayling. Commercially less valuable fish are mainly spring-spawning ordinary fish that form the basis of spring fishing in the water body (pike, perch, ide, roach), as well as smelt "nagish," burbot and ruff. The category of non-commercial fish includes only nine-spined stickleback (6.7% out of total amount), which is not of interest for fishing due to its low gastronomic value.

Among the ichthyofauna of the lake, there are no fish species that have sharply reduced their numbers and are on the edge of extinction. At the same time, one species is included in the "Red book" list of fish on a purely formal basis, since the presence of species in the Red book of the Russian Federation obliges them to be included in the regional Red books of those subjects of the Federation on the territory of which they live. For this reason, nelma was included in the Red book of the Nenets Autonomous Area in 2006. However, this species is not threatened with extinction and its current state does not cause any concerns in the basin of the Pechora river.

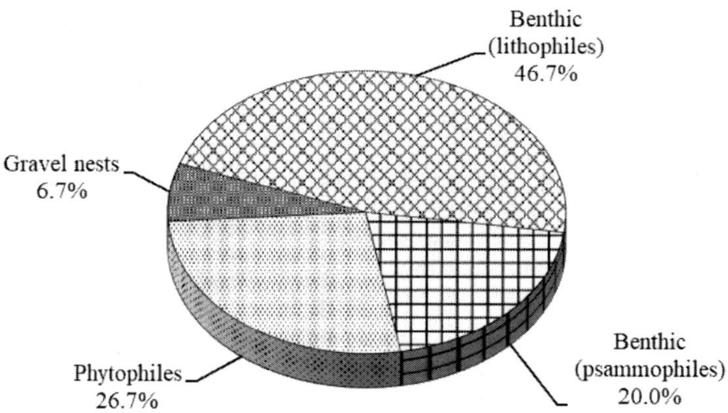

Figure 6. Fish ratio regarding type of reproduction (by preferred substrate) in the Golodnaya Guba Lake (by: Novoselov, 2013).

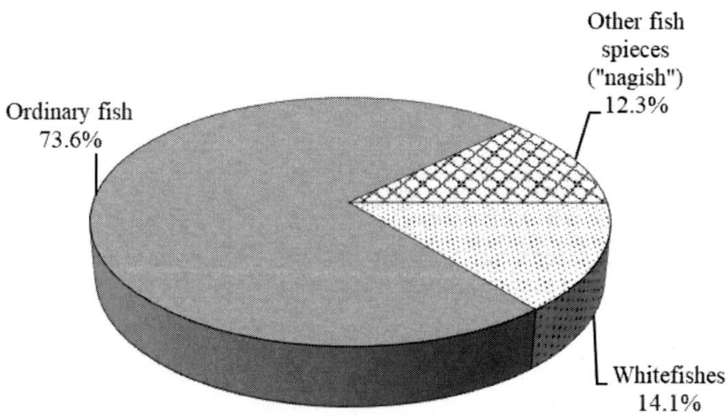

Figure 7. Fish species structure in lake Golodnaya Guba (according to the composition of catches by small-scale net) (by: Novoselov, 2013)

There are successional processes, occurred in the lake: changes in the species structure of the ichthyocenosis, i.e., the ratio of whitefish, ordinary and other fish species (Figure 7). In the

50-80th of the last century, whitefish accounted for more than 80% of the control catches conducted by small-scale nets. The share of ordinary fish species did not exceed 15%, and other fish species (mainly smelt which constituted between 5 and 6%).

In the 90s, after large-scale pollution of the Pechora river, the number of whitefish in catches decreased to 14%, while the share of ordinary fish and smelt "nagish" increased up to 73 and 12%, respectively. In the 2000s, the number of whitefish continued to decrease (up to 6.5%), as well as ordinary fish species - up to 38%. At the same time, the number of smelt "nagish" became more than half (55%) of the total catch. In 2010 – 2012 the revealed trend continues to strengthen, while the share of the whitefish decreased to 1%, and that of the ordinary fish – to 2…3% against the background of an equally significant increase in the number of smelt "nagish"– to 96%.

Smelt and ruff widely distributed in the lake are food competitors to the most valuable species – whitefish, peled and broad whitefish. Juveniles of the last three species, in turn, are objects of nutrition of a fairly large pike. Thus, in order to preserve the Golodnaya Guba Lake as the main feeding water body for valuable fish species, it is advisable to carry out systematic reclamation work in it, aimed at reducing the number of ordinary fish and smelt.

Changes occurring in the biota of the Golodnaya Guba lake ecosystem have also affected its fish population at both the population and organizational levels.

After a catastrophic oil leak into the Pechora river basin in 1994, histological analysis of whitefish in Golodnaya Guba lake revealed serious abnormalities in their organs and tissues. The main pathological changes wereincluded: 1) non-specific inflammation (vascular reactions, alterations and proliferation of tissues) with a predominance of dystrophic and necrotic phenomena; 2) circulatory disorders and vascular changes, namely – congestive venous and capillary fullness, slowing or stopping blood flow, blockage of blood vessels, destruction of vascular walls, hemolysis and parenchymal hemorrhages; 3) edema and violation of protein and lipoprotein metabolism, degeneration of epithelial and cartilage tissues; 4) tumors of several types: in the gills and kidneys – epithelial hyperplasia, sarcoma, undifferentiated cancer type of lymphocytic leukemia; in the liver – round-cell undifferentiated cancer; 5) excessive development of connective tissue. In the late 90's diagnostics of organs and tissues of whitefish showed that the general state of organisms became satisfactory and fairly stable. This was explained as a general temporary reduction in the consequences of the accident in 1994, and quite naturally increased death (elimination) of sick fish. Later studies have shown that the established diagnoses in whitefish for all the organs practically did not change (Lukin et al., 2000).

The current state of the Golodnaya Guba Lake and the ongoing succession changes (transformation of the water ecosystem) are the result of three determining factors. First of all, these are climatc changes, which are expressed in the gradual warming of the region and lead to an increase in the trophic capacity of the lake, its overgrowth and changes in the hydrological regime. The next aspect is hydrographic, which is manifested in the continuous process of soil erosion in the lower course of the river and sedimentation (silting) of the Pechora delta in general and the lake in particular. The last factor is the anthropogenic impact, which is accompanied by the entry of pollutants into the lake and the restructuring of biotic components of the ecosystem under the influence of pollution. The consequences caused by both direct and indirect effects of man-made factors affect the state of biological communities of different trophic levels: phytoplankton – zooplankton and zoobenthos – fish, which is appeared in a reduction in the number of species, changes in their ranges and numbers, and as a result of total biomass, a decrease in biological diversity and, in general, the transformation of aquatic ecosystems. Lake Golodnaya Guba still retains the status of a "forage biotope," being medium-fed for zooplankton and high-fed for zoobenthos, but it begins to lose the value of a feeding reservoir for young whitefish due to changes in the structure of ichthyocenosis.

The dominant role of whitefish in the ichthyofauna of the Pechora basin and their importance in solving the socio-economic problems of the Nenets Autonomous Area requires special attention of scientific, environmental and economic structures to Golodnaya Guba lake. It is necessary to continue monitoring studies for operational control over the state of the lake (monitoring of trophic status), determining ways of rational use of its aquatic biological resources and developing an environmental forecast.

Lakes of the Malozemelskaya Tundra

The largest water body of Malozemelskaya tundra is Indiga lake system and Urdyuzhskoye lake, located in the basins of the Indiga and Sula rivers.

Indigskie Lakes

The lakes of the Indiga group are located in the Northwestern part of Malozemelskaya tundra on the territory of Nenets Autonomous Area and belong to the Indiga river basin. In total, there are about 2 thousand lakes in the Indiga catchment area. However, the vast majority of them belong to small water body, the total area of which is 103 km^2 (only 3% of the area of the river basin). The four largest lakes are located in a close group and are connected by channels. Two lakes (Bolshoe and Korgovoe) have water areas of 12.5 km^2, the Srednee and Nizhniye lakes have 4 to 5 km^2 surface area This system also includes two small water bodies (Maloe Indigskoe and Bezymyannoe) with an area of less than 1 km^2 each (Figure 8).

By their origin, the lakes are relict (residual) and occupy lagoons formed during the post-glacial period. The banks are low and swampy in places. Bottom sediments of all water body are represented by silts, clays, sands and rare boulders. These are flowing water bodies located in a close group and connected by channels. The size of their water areas is from 0.6 to 12.5 km^2. Large lakes (Bolshoe, Korgovoe, Srednee, and Nizhniye) have the greatest depths of 15 to 19 m, the depth does not exceed 7 m in lakes Maloe Indigskoe and Bezymyannoe (Kozmin, 2011). The main limnological parameters of the Indigskie lakes are shown in Table 5.

There were representatives of 44 planktonic invertebrates species recorded in the Indigskie lakes. In Srednee, Nizhniye, Maloe Indigskoe and Korgovoe lakes, the main role in the formation of biomass belongs to copepods, and in lake Bolshoe – to cladocerans. According to the degree of quantitative development of zooplankton, five water bodies were characterized as low-feeding, and only in lake Nizhnye the average zooplankton biomass reached 1.08 g/m^3. The benthic fauna of the lakes included 14 groups of invertebrates. The chironomid group had the most diverse species.

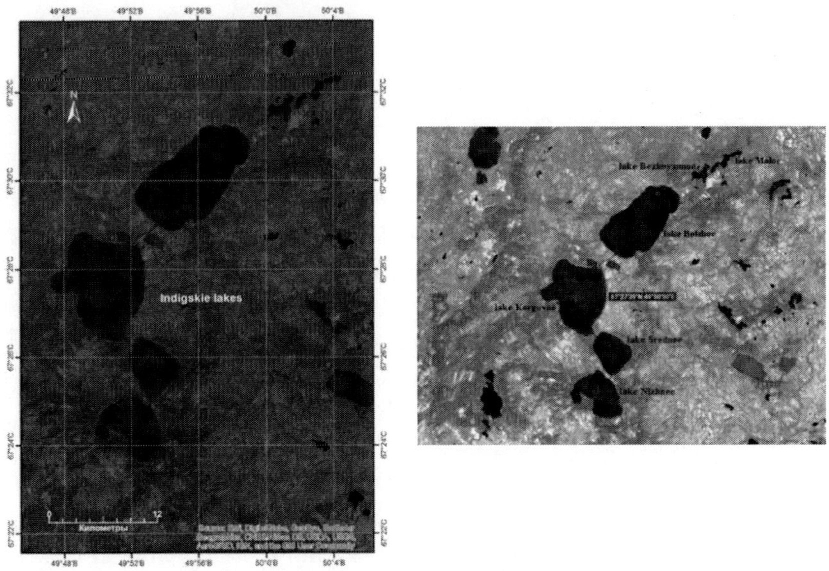

Figure 8. Indigskie lakes.

Table 5. Limnological and hydrobiological characteristics of lakes of the Indiga group (by: Kozmin, 2011)

Lake	Area, km^2	Max. depth, m	Zooplankton		Zoobenthos	
			ind./m^3	g/m^3	ind./m^2	g/m^2
Bolshoe	12.5	18.0	109370	0.64	2006	2.44
Korgovoe	12.5	19.0	104163	0.52	2670	2.59
Srednee	4.0	18.0	118119	0.63	2981	3.73
Nizhniye	5.6	15.0	182748	1.08	1825	2.25
Maloe Indigskoe	0.7	5.0	64135	0.25	4419	9.53
Bezymyannoe	0.6	7.0	63136	0.42	4390	12.70

There was a poor feeding base for fish-benthophages in Bolshoe lakes, Korgovoe, Srednee and Nizhnie. The most highly fed lakes are Maloe Indigskoe and Bezymyannoe (Table 5).

The ichthyofauna of Indigskie lakes was represented by 9 species belonging to 6 families: *Coregonus peled, Coregonus lavaretus, Thymallus thymallus, Esox lucius, Leuciscus idus, Lota lota, Perca fluviatilis, Rutilus rutilus, Gimnocephalus cernuus*. *Thymallus thymallus* was also found in rivers that flow into lakes. There were singular visits of salmon to the lakes.

The most valuable and one of the most numcrous fish species in these lakes is *Coregonus peled*. There are 9 age groups (3+…11+), sizes from 23 to 44 cm (fork length) and weight from 150…200 to 1350 g in fishing catches. Peled becomes sexually matured in the fourth year of life. The catches are dominated by individuals aged 7+…8+ (72% of all production). Peled is characterized by a low growth rate, which indicates a low feeding base of water body and a high density of the population living there. The last assumption is confirmed by the results of the analysis of pike and perch nutrition in Indigskie lakes: peled juveniles was the basis of the diet of predatory fish and was found in 60% of analyzed pike stomachs and 90% of large perch stomachs.

Esox lucius is the main commercial species in Indigskie lakes. The basis of the commercial herd (more than 75%) is fish of the length from 45 to 65 cm and weight from 700…800 to 3000 g, aged 5+…9+. Due to the fact that pike maturation in lakes occurs in the fourth or fifth years of life, its population state is satisfactory.

Rutilus rutilus in catches in Indigskie lakes ranked third after pike and peled for a couple of years. In 1972, the catch was 2.2 tons, or 39% of the total fish caught. In catches, roach reached a length of up to 25 cm and a weight of 323 g. The fishing herd of roach is represented by fish of 11 age groups with a predominance of individuals of six to ten years.

Perca fluviatilis is also widespread in Indigskie lakes. It ranks fourth in catch statistics. Studies have shown that the commercial herd of perch is characterized by a multi-age structure and is represented by fish of twenty age groups. At the same time, the basis of catches (more than 76%) are individuals 14+…18+ years. At this age, the commercial length of perch varies from 22 to 28 cm, and the weight – from 200 to 600 g. The data obtained indicated that the perch population was in good condition and fishing did not have an impact on its size and structure.

Coregonus lavaretus, Thymallus thymallus, and *Lota Lota* were found in catches as a by-catch. Their number in the lake system was not high. Other types of fish in catches were rare.

Indigskie lakes are of particular importance for fishing, since, unlike other similar water bodies, there are mainly valuable fish species in them. Thus, out of 12 tons of all fish caught in 1979, 5.9 tons (or 49% of the total catch) was *Esox lucius* and 4.3 tons (36%) – *Coregonus peled.* According to data, the general state of fish stocks in Indigskie lakes was very satisfactory.

The main object of industrial fishing in the spring is *Esox lucius*, to a lesser extent - *Perca fluviatilis* and *Rutilus rutilus*. In the autumn, the fishermen catch primarily *Coregonus peled.* Fishing is seasonal. During the year, the lakes are fished for no more than three months. The average annual catch is 4 tons. The maximum catch was 12 tons in 1979 (Table 6).

Table 6. Fish yield of Indigskiye lakes (by: Kozmin, 2011)

Year	Fish, tons							
	Coregonus peled	Esox lucius	Rutilus rutilus	Perca fluviatilis	Lota lota	Coregonus lavaretus	Thymallus thymallus	Total
1971	1.13	1.58	1.80	0.58	0.01	-	-	5.10
1972	0.50	2.31	2.21	0.60	0.10	-	0.03	5.75
1973	0.11	0.09	0.70	0.11	0.01	-	-	1.02
1974	0.75	1.07	1.65	0.32	0.08	0.14	-	4.01
1975	0.92	1.05	-	0.08	-	-	-	2.05
1976	0.25	1.28	1.09	0.48	0.02	-	-	3.12
1977	0.30	2.61	0.19	0.21	0.01	-	0.02	3.34
1978	1.02	4.82	0.34	0.53	0.23	-	-	6.94
1979	4.32	5.90	0.94	0.48	0.05	0.18	-	11.87
1980*	0.90	4.70	0.78	0.22	0.04	-	-	6.64
Totally	10.20	25.41	9.70	3.61	0.55	0.32	0.05	49.84

*In 1980 the catch was performed for six months.

A team of 6 to 8 fishermen equipped with non-water boats with mechanical winches and the necessary equipment is sufficient for effective commercial use of the fish stocks of Indigskie lakes. There are three fixed fishing stations on the lake system that are suitable for primary processing of fish and accommodation of fishermen. At least 20 tons of fish can be caught annually, including 5 tons of *Coregonus lavaretus* and *Coregonus peled* with the rational management of fishing on the lakes.

To summarize, Indiga lakes, located in the Northwestern part of the Malozemelskaya tundra in the Nenets Autonomous Area, belong to the Indiga River basin. They are relict (residual) in origin and occupy lagoons formed in the post-glacial period. The sizes of their water areas range from 0.6 to 12.5 km^2, with maximum depths ranging from 7 to 19 m. They are inhabited by representatives of 44 species of planktonic invertebrates. In the Lower, Middle, Small Indiga lakes and lake Korgom, the main role in the formation of biomass belongs to copepods, in Lake Bolshoye - cladocerans. Five water bodies had low level of zooplankton development, and only in Lake Nizhny, the average biomass reached 1.08 g/m^3. The benthofauna of the lakes has 14 groups of invertebrates; the chironomid group had the highest species diversity. Among all the lakes of the group, Maloe Indigskoe and Bezymyannoe are the most nutritious. The ichthyofauna of the lakes of the Indiga group is represented by 9 species belonging to 6 families: peled, whitefish, grayling, pike, roach, dace, perch, ruff, burbot. Grayling is also found in rivers flowing into lakes. Indiga lakes are promising for commercial fishing.

Lake Urdyuzhskoe

Urdyuzhskoe lake is located on the watershed of the Indiga and Sula rivers and belongs to the Pechora river basin. The water body has an irregular rectangular shape, the area of the water mirror is 61.7 km^2. Satellite lakes with an area of 0.8 to 4.0 km^2 are located around Urdyuzhskoe lake. These water bodies are connected to the main lake through small shallow streams, resulting in fish migration throughout the lake system (Figure 9). The water body is flowing – from it flows the Urdyuzhskaya Viska river with a length of 42 km, which flows into the river Soima. In spring and autumn large fishing boats are free to roam at the Urdyuzhskaya Viska river. The total length of the waterway from Lake Urdyuzhskoe to the nearest settlement – the village of Kotkino, whose inhabitants are engaged in fishing, is approximately 100 km.

Lake Urduzhskoye is a shallow water body. Depths of 2…3 m occupy most of its water area, and only in the center of the lake basin there are two small depressions of 4 to 5 m. Bottom sediments are mostly represented by silty soil. The littoral zone is covered with thickets of sedge. The banks are low and swampy.

The qualitative composition of the fish fauna of Lake Urduzhskoye is quite diverse. There are *Coregonus peled, Coregonus nasus, Coregonus lavaretus, Thymallus thymallus, Esox lucius, Leuciscus idus, Lota lota, Perca fluviatilis, Rutilus rutilus, Carassius carassius* and *Gimnocephalus cernuus*. are *Coregonus peled, Esox lucius, Leuciscus idus, Rutilus rutilus* and *Perca fluviatilis* are commercially valueable. *Coregonus peled* is the main commercial fish, with an average annual catch of 4.2 tons, or 46% of the total annual production. However, its catches are fluctuating. There are years when peled does not arrive at the fish receiving point at all. One of the reasons for this are starvation of whitefish, sometimes observed in the Urduzhskoye Lake.

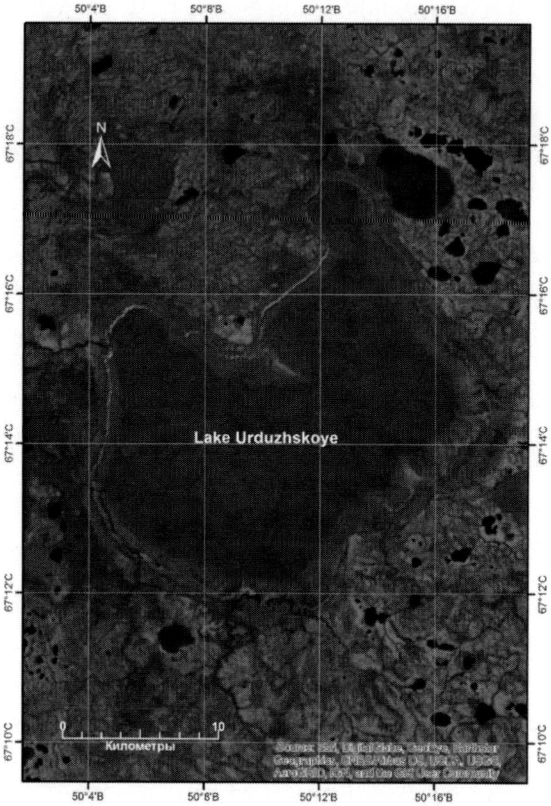

Figure 9. Lake Urduzhskoye.

At the beginning of the development of fisheries on Urduzhskoye lake, the main fishing tools were gill nets, nets and partly stationary traps. In the 50s, fishing on the lakes was carried out with 200 gill nets, 10 ridges, and 1 trawl. Six or seven fishermen participated in the fishing. Fishing was usually seasonal: in the spring, clusters of spring-spawning fish were caught, and in the fall, the fishing was focused on whitefish. Due to the distance of the lake from the settlements, all fish caught in the spring and summer season were salted in barrels, and in winter, they were delivered frozen to the fish receiving point. These methods of storing fish remained for many decades. Fishing on the lake has gradually declined in recent years. Meanwhile, this water body is promising for industrial ice fishing with shore seines. In November and December, shallow areas of the lake freeze to the bottom, and fish are located 3 to 5 meter deep. As a result, it is possible to catch fish with seines with less labor. In one season, one can get 30 tons of fish on lake Urdyuzhskoe, half of which will be valuable species (Kozmin, 2011).

To summarize, Lake Urdyuzhskoe is located on the watershed of the Indiga and Sula rivers and belongs to the Pechora river basin. It is a shallow water body with depths of 2 to 3 m and small lagoons; it has depth of 5 m in the central part of the lake hollows. The level of the food supply in the lake has not been studied. Ichthyofauna is pretty varied and represented by both whitefish (peled, chir, whitefish-pyzhyan) and ordinary (grayling, pike, ide, burbot, perch, roach, crucian carp and ruff) species of fish. Commercial peled, pike, ide, roach and perch are important. Peled was the main commercial fish, the average annual catch of which reached 46%

of the total annual catch. Nowadays fishing on the lake is sporadic. Lake Urdyuzhskoe is promising for the restoration of industrial fishing.

LAKES OF THE BOLSHEZEMELSKAYA TUNDRA

The relief of the Bolshezemelskaya tundra is a slightly hilly plain with single hills up to 50…60 m. Its Eastern part is characterized by the most severe climate in the European North of Russia. Winter lasts for 240…250 days with little snow cover. Temperature fluctuations exceeds 80°, the frost-free period is 50…68 days, the annual precipitation rate increases from 385 to 570 mm from the West and East, and the largest amount falls in summer.

Plain relief, the presence of isolated deeps, close proximity to the earth's surface 'permafrost', excessive moisture, and weak surface evaporation determine the abundance of water bodies. Small water bodies (less than 0.1 km^2) are dominating, they make up to 71% of the investigated area. However, lakes over 1 km^2 are also common, and constitute up to 1.3%. The total lake content of the area reaches 55%. The diversity of water body in shape, size, and other morphometric parameters is a consequence of heterogeneous genesis, which is accepted by most researchers. The origin of big lake systems is still under discussion, since there is no single view of their genesis and history.

According to the theory of cover glaciation, the occurrence of these water bodies is associated with the activity of the glacier. According to the marine hypothesis, these lakes originated in the roughness of the primary relief created by the uneven marine and glacial-marine accumulation of quaternary deposits. Other researchers consider the lakes to be remnants of the once vast water body formed during the Riess-Wurm interglacial epoch after the retreat of the boreal transgression.

The landscape uniformity of the area determines similar limnological characteristics of lake systems. However, there are some differences related to local conditions of formation. The temperature regime and its seasonal dynamics, with the constant influence of winds on water circulation are specific for shallow and deep-water lakes. The shallow lakes are characterized by uniform warming to the bottom, while the deep-water lakes are characterized by thermal stratification. The most intense warming of the waters takes place in July. Relatively high water temperature (12 to 15°C) lasts for two months: July and August.

The maximum depth for most lakes is 15 to 20 m, in some water bodies it reaches to 40 to 57 m. As a rule, the coastal zone of lakes is dominated by stony, pebbly, sandy soils. At depths of more than 2 to 3 m soils are silty. There is an alternation of high and low banks everywhere, while swampy banks are rare.

The average hydrochemical parameters of the water in the open period were: transparency –3 to 5 m, color – 4 to 12°, pH – 7.1, dissolved oxygen in water – 100 to 125% saturation, the concentration of CO_2 –1.7 to 5.0 mg/L, permanganate index – 3.4 to 7.3 mgO_2/L. In the bottom horizons (30 to 57 m deep), the oxygen saturation of water did not fall below 70%, the CO_2 content did not exceed 10.6 mg/L, and the maximum permanganate index was 10.1 mgO_2/L. The most important lakes of the Tundra are Pilninsky group of lakes, which unites 39 water bodies with a total area of 200 km^2, Vashutkini lakes, including 15 water bodies with an area of 94 km^2, and Prosunduyskiy system, consisting of 29 lakes with a total area of 26 km^2 (Kozmin, 2011).

Pilna lakes

The Pilna lake system is located in the Northeastern part of the Bolshezemelskaya tundra, 42 km from the village of Varandey. Hydrographically, the lakes belong to the basin of the Khaypudyr Bay (Pechora sea), at the distance of 12 km. The Pilna lakes are located in the area of residual water bodies of marine origin. Most of them are characterized by small size of the water mirror, with the exception of lake Pilnya (51.6 km^2) and lake Toboy (11.6 km^2). The lakes are shallow, with average depths ranging from 0.5…1.4 m. The maximum depth of 3.2 m was established in lake Pilnya (Table 7). The majority of the lakes are oval with simple outlines. The coastline is weakly developed. Powerful peatlands, which form banks of the lake, fall into the lake due to continuous wave abrasion. As a result, these peatlands replenish the bottom sediments with organic substances in the form of plant residues and, eventually, turn into peat silts. Intensive abrasion occurred in the sandy headlands on lake Pilnya. Products of abrasion fill the basin and smooth the bottom relief of water body. Chemical composition of waters of the Pilnya group is rather special due to the high content of chlorides (up to 19.6 mg/L in autumn in lake Pilnya) and generally low, because of close location to the sea in conditions of excessive moisture (Kozmin, 2011).

There are 63 species and forms of algae belonging to 7 types in the summer. A group of protococcal algae is dominating in diversity and abundance. Their number reaches 20.2 million cells/L. The qualitative composition of the zooplankton community is not rich. Cladocerans dominate both in number and in biomass. A characteristic feature is absence of rotifers in small lakes. Quantitative indicators characterize the zooplankton of Lake Pilnya, Toboy and Lake # 1 are poor for feed due to quantity of zooplankton. The other four lakes are highly foraged for plantophage fishes.

Table 7. Limnological and hydrobiological characteristics of lakes of the Pilnya group (by: Kozmin, 2011)

Lake	Area, km^2	Max. depth, m	Zooplankton	Zoobenthos		
			ind./m^3	g/m^3	ind./m^2	g/m^2
Pilnya	51.60	3.2	27557	0.27	2300	2.89
Toboy	11.60	2.0	53448	0.36	7231	8.36
Lake № 1	0.68	1.6	35809	0.55	3035	2.95
Lake № 2	0.83	1.3	244870	1.44	2044	3.24
Lake № 3	0.15	2.0	212000	1.70	2827	7.32
Lake № 4	0.24	2.1	163977	5.34	2412	3.09

The species composition of the benthic population is uniform. Chironomid larvae is the basis of benthic population. In Lakes # 2 and # 3, and in lake Pilnya, shellfish was dominated. Toboy lake had the highest food capacity, while lake Pilnya had low biomass of zooplankton and benthos (Table 7).

The ichthyofauna of the Pilnya lakes is not very diverse and rather common to any large lakes of the Bolshezemelskaya tundra. *Coregonus albula, Coregonus peled, Coregonus nasus, Coregonus lavaretus, Thymallus thymallus* and *Pungitius pungitius* are dominating. Such widespread fish species as *Perca fluviatilis, Rutilus rutilus* and *Esox lucius* were not detected in catches. The only representative of the boreal-plain complex that lives in the lake Pilnya is a *Gimnocephalus cernuus*. With the exception of *Coregonus nasus*, all the above fish are nonmigratory forms, i.e., their entire life cycle occurs in this freshwater water body. Along with nonmigratory forms there are anadromous fish in Pilnya lake suh as *Salmo salar* and *Oncorhynchus gorbuscha*. Unlike *Salmo salar*, which comes here only once, *Oncorhynchus gorbuscha* in some years is found in the catches of dozens of fishes. In commercial catches, *Stenodus leucichthys nelma* and *Coregonus autumnalis* are occasionally caught, the main feeding places of which are located in the coastal desalinated areas of Pechora and Haipudyr bays. Probably, the visits of these fish species to the Pilnya lake system are random and occur due to strong wind surges. There are no places for spawning and incubation of eggs for *Salmo salar*, *Stenodus leucichthys nelma* and *Coregonus autumnalis* in the lake Pelnya and its tributaries, all these fish are migrants and do not form significant accumulations. The main commercial fish in the lake Pilnya are *Coregonus albula, Coregonus peled, Coregonus nasus*.

With the exception of short-term fishing during the war years, more or less regular fishing on the lake has existed since 1970, and fishing was only in open water. The maximum catch was 5.7 t in 1972. Due to isolated location of lake Pilnya, it is advisable to fish once every three to four years. At the same time, it is possible to catch up to half of the total fish stock without compromising reproduction. In general, every year in the lake Pilnya and the adjacent waters, one can catch up to 20…25 tonnes of whitefish species of fish.

To summarize, the Pilnya lake system is located in the Northeastern part of the Bolshezemelskaya tundra and belongs to the Khaypudyrskaya Bay basin (Pechora sea). For most lakes, there is characteristically small size of water surface, except for lakes Pilnya and Toboy. The lakes are shallow, with average depths of 0.5…1.4 m and a maximum depth of 3.2 m (lake Pilnya). The phytoplankton of lakes in the summer period is represented by 63 species and forms of algae; the group of protococcal algae prevails both in the diversity and in the abundance. The qualitative composition of the zooplankton community is not rich; both in number and in biomass, cladocerans predominate, with the complete absence of rotifers. The species composition of the bottom population is uniform and is based on chironomid larvae. Shellfish are dominating in a number of satellite lakes, as well as in the lake itself. The ichthyofauna of the Pilnya lakes is also not very diverse, with a pronounced dominance of representatives of the Arctic freshwater complex – *Coregonus albula, Coregonus peled, Coregonus nasus, Coregonus lavaretus, Thymallus thymallus* and *Pungitius pungitius*. With the exception of *Coregonus nasus*, all the above fish are aquatic forms, i.e., their entire life cycle takes place in this freshwater water body. The main commercial fish in the lake Pilnya are *Coregonus albula, Coregonus peled, Coregonus nasus*. With proper organization of fishing on the lakes of the Pilnya system, one can catch annually up to 20…25 tons of whitefish species.

Prosunduy Lakes

The Prosunduy lake group is located in the upper reaches of the Kuya river basin, which flows into the Pechora river 20 km below Naryan-Mar. From West to East, the lake system has a length of 23 km, from North to South – 12 km. It is divided into two lake groups – Prosunduy and Pahanduy. In 1978 - 1979, 12 of the largest lakes were studied. The obtained data on limnological and hydrobiological characteristics of water body analysis are shown in Table 8.

According to the size of the water area, the lakes belong to the classes of small and very small water bodies. The area of their water mirror varies from 0.1 to 5.35 km^2. According to the shape of the lake basins, they are divided into oval and oval with lobed separations.

The deepest point is 13 m in lake Krivoe. Many water bodies, ranging from the water's edge to a depth of 1.5 to 2.0 m, have a sandbank, and the deeper sections of the bed are covered with silt of various shades. In summer, the lakes are well warmed up, and direct stratification is observed in deep-water areas.

Table 8. Limnological and hydrobiological characteristics of lakes of the Prosunduy group (by: Kozmin, 2011)

Lake	Area. km^2	Max. depth, m	Zooplankton		Zoobenthos	
			ind./m^3	g/m^3	ind./m^2	g/m^2
Prosunduy	2.11	2.8	11375	0.11	1574	5.12
Krivoe	5.35	11.0	32555	0.66	3908	6.23
Soldatskoe	0.10	6.5	21016	0.69	1100	1.99
Maloe	1.01	2.5	26002	0.59	1427	1.33
Lake № 1	1.08	2.3	23205	0.83	5985	12.80
Ryzhkovo	0.75	6.5	50187	1.00	1099	5.19
Sukhoi Pahanduy	3.84	1.3	17038	1.53	4264	12.56
Kamenka	1.83	13.0	18800	0.16	90	7.80
Sigovoe	2.17	8.0	2224	0.50	1815	6.20
Eazevets	1.74	9.5	20000	0.60	2000	10.30
Utka-Ty	1.52	7.0	18921	0.42	2154	11.13

All water bodies are overflow lakes, the flow is directed to the Kuya river by first-order tributaries or through a whole system of intermediate lakes and channels.

The oxygen regime of the waters of most lakes is favorable. Waters are very soft, low-mineralized and have a diverse chemical composition. Calcium-bicarbonate and sodium-bicarbonate are dominating in waters. There are lakes with mixed ionic composition of water in the autumn season. This diversity, according to Vlasova and Goldina (1967), is due to the specific chemical composition of underground and ground water involved in the nutrition of the water body. The water pH ranges from slightly acidic to slightly alkaline. The content of organic matter is low, with the exception of lakes Utka-Ty, Sukhoi Pakhanduy, Lake № 1, Ryzhkovo and Prosunduy.

The qualitative composition of zooplankton is not rich and is represented by common widespread species. Copepods and rotifers are most often dominant both in number and biomass, less often - cladocerans. The average total number of zooplankton in the summer of 1978 - 1979 varied from 11.375 to 50.187 ind./m^3, the average biomass - from 0.11 to 1.53 g/m^3. Most lakes are low-feeding water bodies for plankton-eating fish.

Chironomid larvae were dominating in the benthic fauna of most water bodies; mollusks were dominating in the lake Pahanduy and Lake #1. Lakes Sukhoi Pahanduy, Yazevets, Utka-Ty and Lake #1 are highly nutritious for benthic fish. Lake Kamenka is a moderately nutritious water body.

There are 13 species of fish belonging to 7 families in the Prosunduy lakes. The entire fish population can be divided into 2 groups – non-migratory and migratory. *Stenodus leucichthys nelma, Coregonus nasus, Coregonus lavaretus, Coregonus peled* and *Leuciscus idus* are migratory fish. *Esox lucius* and *Coregonus peled* provide the basis of commercial catches.

During the spring period, fishing on the lakes of the Prosunduy group was based on catching spawning clusters of spring-spawning fish. There were 200...300 gill nets with a mesh size from 40 to 60 mm and several dozen wing nets placed on fish migration routes to catch *Perca fluviatilis, Rutilus rutilus, Esox lucius* and *Leuciscus idus*. Spring fishing was usually conducted by small teams of fishermen. Each team was fishing in 2...3 lakes. There were fishing stations on the five main water bodies, where up to 15 fishermen could work in the spring and summer season.

In the summer, the most effective fishing is with trawl because blue-green algae are blooming, so the nets are quickly clogged and the fish is poorly caught. For whitefish fishing, trawls with a length of 500 m and a mesh size of 40 mm were most often used. Small ordinary fish were caught with 200-meter small mesh trawls. The water surface of most lakes is small, so in the summer season it is advisable to conduct fishing with a non-water mechanized team of 6...7 people. The team must have vehicles with high cross-country performance, small trawl boats equipped with mechanical cranes, and the necessary set of tools for industrial fishing. If lakes of the Prosunduy group are intensively fished in open water, there are 20 to 30 tons of whitefish caught annually. Due to the shallow water in winter, commercial fishing is not economically efficient on these lakes.

To summarize, the Prosunduy lake group is located in the upper reaches of Kuya river basin, which flows into the Pechora river. According to the size of the water area, the lakes are small and very small water bodies with areas from 0.1 to 5.35 km^2, with the deepest point of 13.0 m (lake Krivoe). The oxygen regime of the waters is favorable, and the content of organic matter is low. Zooplankton is represented by common widespread species, copepods and rotifers predominate both in number and in biomass, less often - cladocerans. Most of the lakes are quantitatively low-nutritious for plankton-eating fish. The benthic fauna of most water body is dominated by chironomid larvae, and some by mollusks. There are 13 species of fish belonging to 7 families in the Prosunduy lakes. The basis of commercial catches are *Esox lucius* and *Coregonus peled*. The lakes are promising for the restoration of local industrial fishing. Potential annual fish production may amount to 20-25 tons.

Vashutkin Lake

Physical-Geographical and Limnological Characteristics

Vashutkin lake is the largest lake system in the Bolshezemelskaya tundra. Its total area is approximately 94 km^2. It is located in the upper reaches of the Adzva river, a right tributary of the Pechora river, about 120 km Northwest of Vorkuta. The system includes 7 large lakes with an area of 6 to 20.4 km^2 and many small water bodies. The main lakes are almost a closed chain of water bodies connected by channels. The Adzva river originates from the terminal lake Vanyukty (Figure 10). The incoming part of water balance in Vashutkin lakes (61 to 68% of total area) consists of the waters arriving from the catchment. The composition of the waters is bicarbonate-calcium. The total water mineralization ranges from 30 to 100 mg/L (Sidorov, 1971).

Vashutkin lakes are located in the region of permafrost. The relief and hydrographic net were mainly formed during the interglacial epoch via erosion processes. Soil and vegetation cover are heterogeneous. The area near by the lakes is the tundra subzone with alternating dwarf birch (yernik) and lichen formations on hills and flat dry spaces with willows and sedges along the banks of water bodies. It is common to see ledum and mossy tundra areas. There are large areas of mixed grass meadows in the floodplains of rivers and lakes.

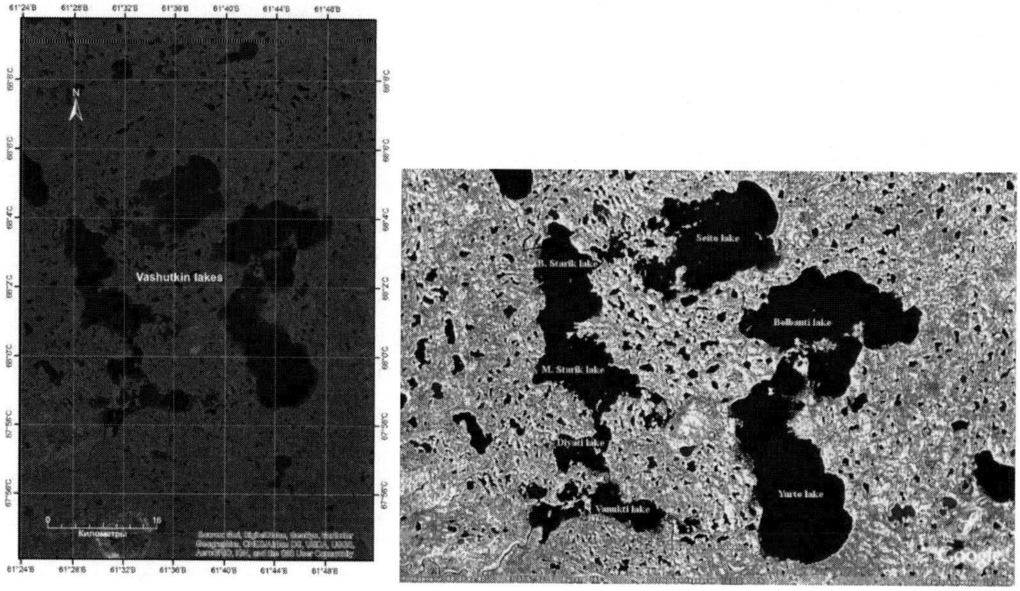

Figure 10. Maps of Vashutkin lakes.

Despite the large number of water bodies, the surface of the tundra in the area of Vashutkin lakes is well drained, and there are few swamps. The banks of water bodies are of the same type, they are determined by the terrain of the tundra: when crossing hills, the height of the banks increases to 8-12 m, in the depression between the hills it decreases to 1-1.5 m. Vashutkin lakes, in addition to the usual floodplain terrace, have an additional extensive branched floodplain system. In this regard, during high water, Vashutkin lakes turn into a single water system, which is of great importance for hydrobionts. By filling small lakes and channels, flood waters help improve the gas regime and facilitate fish migration. When the level decreases, water is enriched with nutrients (Kozmin, 2011).

Vashutkin lakes are divided into four groups based on their morphological features and presumed genesis:

1. The largest head lakes are Yurto, Balbanty, Seito;
2. Lakes that have an elongated shape and a flow-through regime: Bolshoi and Maliy Stariki lakes;
3. Diyaty, Vanukty lakes are the deepest with an average water flow and the most ancient lake, closing system;
4. Makty, Pernaty lakes are an example of small, shallow lakes.

Lakes vary significantly in depth, but all water bodies are characterized by a predominant sublittoral zone of 1…10 m deep, with the exception of lakes Diyaty and Vanyukty, where the depths exceed 10 m. The area and depth of the main lakes of the Vashutkin system are shown in Table 9.

The ice-off usually occurs in mid-June. Spring floods are low (about 2 m), but the water level falls slowly. The highest water temperature (up to 20 °C) is observed in late July. The average water temperature in August is 12 to 14°C. In early September, during the low-water level, the water temperature drops to 2…4°C. Shore ice appears fast in the shallow waters at the end of September. Full freezing of large lakes occurs in the second decade of October.

Table 9. Limnological characteristics of Vashutkin lakes (by: Kozmin, 2011)

Parameter	Lake								
	Yurto	Balbanty	Seito	Makty	Pernaty	Bolshoi Starik	Maliy Starik	Diyaty	Vanukty
Length, km	8.0	7.0	4.5	1.0	1.0	4.0	3.6	1.8	5.0
Width, km	3.5	4.0	3.5	1.0	0.8	2.5	3.0	1.5	2.0
Area, km^2	20.5	18.0	14.7	1.9	2.1	7.1	8.9	2.8	6.3
The length of the coastline, km	27.6	22.8	19.0	6.6	11.0	13.6	12.4	8.4	28.6
Maximum depth, m	17.8	12.5	20.0	5.0	5.4	19.0	13.0	40.0	39.0

During the warm period, a well-defined stratification is observed in the lake pelagial: an epilimnion with a maximum temperature in the surface layer (0…2 m), a stretched layer of a temperature jump, and there is a hypolimnion from 12 to 15 m with a constant temperature of 4 to 5°C (Sidorov, 1974). Vashutkin lakes are located in the "permafrost" region and the snow cover lasts 7 months a year. The ice cover lasts on the lakes for more than 9 months; the open water period is about 3 months. Opening of lakes occurs in mid-June, clearing of ice - at the end of the first decade of July. The first ice appears in shallow waters at the end of September. Full freezing of large lakes usually occurs by mid-October.

Hydrochemistry

The water has a high transparency (2.5 to 4.8 m) and a low color degree – 10-16° of the platinum-cobalt scale. The pH ranges from 6.6…7.6. The water is saturated with oxygen in both spring and summer. In the surface layer, the O_2 content reached 156% saturation. The minimum oxygen levels at a depth of 30 m (lake Diyaty) were 74 to 70%. The water of Vashutkin lakes has a low mineralization and does not exceed 100 mg/L. In general, it belongs to the class of bicarbonate waters of the calcium group.

Higher Aquatic Vegetation

There were 19 species of flowering plants reported in the lakes. Half of them (Potamogeton, Myriophyllum, etc.) are typically aquatic, while the rest are considered amphibious. *Carex aquatilis, Arctophila fulva* and *Equisetum fluviatile* are the most important among the thickets in the lakes. *Carex aquatilis* usually forms dense thickets on the shallows near the water's edge. *Arctophila fulva* spreads to a depth of 1 m, occupying large areas in shallow waters. *Equisetum fluviatile* is often found in isolated heavily silted bays (Kozmin, 2011).

Ichthyofauna

A comprehensive study of the fish population of Vashutkin lakes was conducted by the Komi branch of the USSR Academy of Sciences in 1960 - 1961 and the expedition by the Polar branch of the FSBSI "VNIRO" in 2004-2007. In the course of research, 15 fish species belonging to 8 families were identified in the lake system's ichthyofauna (Table 10) (Dvoryankin and Timofeev, 2007a).

Table 10. Composition of ichthyofauna of Vashutkin lakes (Dvoryankin and Timofeev, 2007a)

Families	Species
Coregonidae	Coregonus peled, Coregonus nasus, Coregonus lavaretus, Stenodus leucichthys nelma, Coregonus albula
Thymallidae	Thymallus thymallus
Cyprinidae	Rutilus rutilus, Leuciscus idus, Phoxinus percnurus
Esocidae	Esox lucius
Percidae	Perca fluviatilis, Gimnocephalus cernuus
Lotidae	Lota lota
Gasterosteidae	Pungitius pungitius
Cottidae	Cottus gobio

Coregonus lavaretus is the main commercial fish of Vashutkin lakes. It inhabits all lakes of the system, from small and the smallest to the largest and deepest. In the autumn period, especially during the spawning period, this fish constitutes up to 85% of the total catch. *Coregonus lavaretus* becomes sexually matured at the age of six when it reaches a body length of 28 to 30 cm and a mass of 300 to 400 g. Its growth rate is quite good and is comparable to representatives of this species in the Pechora river. Local *Coregonus lavaretus* is a typical benthophage. Visual analysis of the stomachs showed the predominance of bivalve mollusks in its diet.

There are very small changes in biological characteristics of *Coregonus lavaretus* over the past decades between the population actively used for fishing 40 years ago and rather virgin population of the present time. There is no big deviations from the size of *Coregonus lavaretus* in samples from 1961 and from 2004 - 2007.

Thymallus thymallus is also one of the most numerous fish in Vashutkin lakes. It is most widely distributed in large and deep lakes. This species completely dominates in the channels between the lakes and at the head of the river Adzva. At the same time, young *Thymallus thymallus* forms clusters in the head of the Adzva river, which is obviously the place of its feeding. Large individuals are more common in lakes. During the greatest warming of the lakes (late July – early August), when the water level decreases, the whitefish goes to deep water areas, the share of *Thymallus thymallus* in commercial catches reaches 40% of all fish species. *Thymallus thymallus* becomes sexually matured at the age of seven. *Thymallus thymallus* is euryphagous.

The growth rate of the local *Thymallus thymallus* is satisfactory and practically does not differ from *Thymallus thymallus* from the Usa river located to the south of Vashutkin Lakes. A comparative analysis of *Thymallus thymallus* growth rates from control catches in 1961 and 2004 - 2007, as well as of the whitefish, did not reveal significant differences.

Coregonus peled is another commercially important fish. In the Vashutkin lakes, *Coregonus peled* has negligible migrations between the major and small water bodies, and its life cycle is essentially within this lake system. According to the data obtained by Komi branch of the Academy of Sciences, the local *Coregonus peled* lives mainly in small lakes connected by channels to the main system. In these settings, this species accounted for up to 90% of all fish caught, while the fish production of *Coregonus peled* reached 40-60 kg/ha.

Males of the local *Coregonus peled* become sexually mature at the age of four, and females – at the age of five. The growth rate of the local *Coregonus peled* is relatively good and exceeds the growth of representatives of this species in other water body of the Bolshezemelskaya tundra. A comparative analysis of the size of the *Coregonus peled* in 1961 and 2004 - 2007 showed an increase in its growth rate.

Esox lucius, along with *Coregonus lavaretus* and *Thymallus thymallus*, are widely distributed in Vashutkin lakes and are found on all horizons and within all biotopes. Only in satellite lakes its number is low. In this lake system, this species is also one of the most important objects of fishing. In the 60s, almost half of its annual catch occurred during two weeks of spawning in late June and early July. During this period, the share of *Esox lucius* in total fish production reached 75%. After spawning, *Esox lucius* feed in the coastal zone, where other fish species accumulate during this period. According to Sidorov (1974), in Vashutkin lakes, the major food for pike are *Coregonus albula*, *Coregonus peled, Coregonus lavaretus*, *Perca fluviatilis* and *Gimnocephalus cernuus*. In October and November, 95% of its food intake was *Coregonus albula*.

Local *Esox lucius* become mature relatively late. Males become sexually mature at the age of 5, females – at the age of 6. Comparison of the growth rates of *Esox lucius* from the collections of 1961 and 2004 - 2007 showed almost complete identity.

Coregonus nasus. In Vashutkin lakes this species is rare. At the age of 7 to 9 years, the commercial length of the local *Coregonus nasus* is 50 to 55 cm and the weight is 2 to 2.4 kg.

Stenodus leucichthys nelma is rarely found in Vashutkin lakes. Representatives of this species occasionally ascend to the water body of the lake system along the Adzva river. At the age of 9, nelma has a commercial length of 70 to 75 cm and a weight of 4 to 5 kg.

Coregonus albula are quite numerous in Vashutkin lakes. It plays an important role as a food source for local predators. Specialized fishing for *Coregonus albula* in water body has never been conducted. This type of fish did not appear in the statistics of commercial catches.

Lota lota, *Rutilus rutilus, Perca fluviatilis* and *Gimnocephalus cernuus* were also excluded from fishing. At the same time, these species are important components of the fish community of Vashutkin lakes. In the control catches of 2004 - 2007, the share of these fish species was 12 to 15% of the total catch weight.

Pungitius pungitius is found in Vashutkin lakes everywhere. The data obtained indicate its importance in the diet of local predatory fish. In particular, this species is one of the main objects in the diet of large perch and *Thymallus thymallus*. In the stomach of one of *Thymallus thymallus* we found 14 individium of *Pungitius pungitius*.

In 50th – 60th of the last century, the fishing was very active. In 1960, 7 teams of 52 people worked on the water body, in 1961 there were 36 fishermen within 6 teams. Gill nets with a mesh size, usually 40 mm, and trawls with a mesh size of 20…22 mm were used in fishing. The recorded catch of fish in 1959 was 44 tons, in 1960 – 48.9 tons, in 1961 – 53.4 tons. According to the Komi branch of the Academy of Sciences, additional 15 to 20 tons of catch were not officially registered. The qualitative composition of annual catches was as follows:

Coregonus lavaretus – 40%, *Coregonus peled* – 15%, *Thymallus thymallus* – 15%, *Esox lucius* – 30%.

The state of populations of the main commercial fish species in water body, the qualitative structure of catches, the catch per effort and the data from the 60s were analyzed to estimate the size of fish stocks in the lake system at the present time. The catch per effort has almost doubled. In 2004, the daily catch per 1 net of 30…36 mm was 7 kg. The catch on a net with a mesh of 40…45 mm was4 kg. Probably, due to the lack of fishing on Vashutkin lakes in recent years, fish fully restored their population. The qualitative composition of valuable fish species in catches has not changed much over the past years. In the control catches of August 2004, it generally corresponds to the same period of the 60s: *Thymallus thymallus* – 47%, *Coregonus lavaretus* – 25%, *Esox lucius* – 28% (Figure 11).

The analysis of the size and age structure of *Coregonus peled, Esox lucius, Coregonus lavaretus,* and *Thymallus thymallus* populations provides quite satisfactory result. The well-being of populations is also indicated by the result of comparing the biological characteristics of the main fish species from the collections at different years. As already mentioned, in the 60s, there was an intensive fishing in the water body, and since the beginning of the 2000s, no fishing was conducted for 15-20 years. However, the growth rate of the main fish species and their ratio in catches have changed very little over the years. This suggests that the fishing load on the lakes in the 60s did not have a significant impact on fish stocks.

Taking into account the data obtained, the possible annual catch of fish on Vashutkin lakes without damage to fish stocks can reach 100 tons, including: *Coregonus lavaretus* – 30 tons, *Thymallus thymallus* – 15 tons, *Coregonus peled* – 15 tons, *Esox lucius* – 30 tons, and 10 tons of *Coregonus albula*. Specialized fishing for perch and roach during the spawning period can produce at least another 10 tons of fish (Dvoryankin, Timofeev, 2007b).

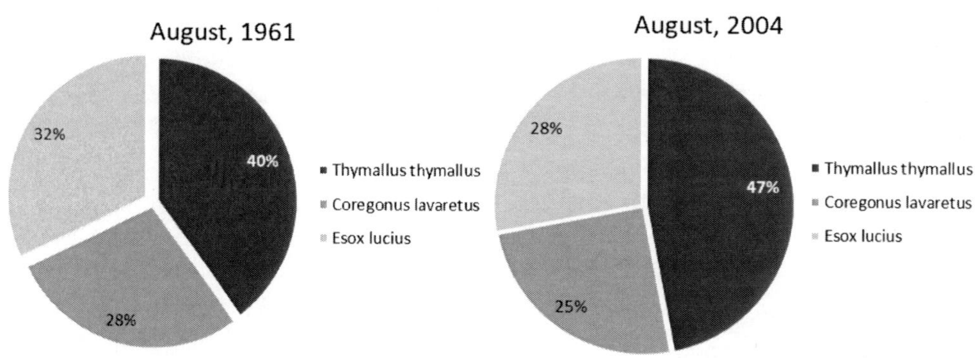

Figure 11. The species composition of the commercial (1961) and scientific (2004) catches on Vashutkin lakes (Dvoryankin and Timofeev, 2007a).

To summarize, Vashutkin lakes is the largest lake system of the Bolshezemelskaya tundra, which includes 7 large lakes and many small water bodies. The major lakes are almost a closed chain of water bodies connected by channels. Lakes vary significantly in depth. Waters is saturated with oxygen, has low mineralization, and belongs to the class of calcium group bicarbonate waters. The highest aquatic vegetation is represented by 19 species of flowering plants. The most important are *Carex aquatilis, Arctophila fulva* and *Equisetum fluviatile*. The lake ichthyofauna includes 15 species of fish belonging to 8 families. The largest number of species are represented by the Coregonidae family (5 species – *Coregonus peled, Coregonus nasus, Coregonus lavaretus, Stenodus leucichthys nelma, Coregonus albula*), three species (*Rutilus rutilus, Leuciscus idus, Phoxinus percnurus*) of the Cyprinidae family, two (*Perca fluviatili* and *Gimnocephalus cernuus*) of the perch family. Each species includes the Thymallidae, Esocidae, Lotidae, Gasterosteidae, and Cottidae families. According to biological parameters, the populations of the main fish species are in a good condition. The possible annual catch of fish on Vashutkin lakes without damage to the fish stock may be up to 100 tons.

Padimey and Charbey Lake System

Physico-Geographical and Limnological Characteristics

The Padimey lakes is a single genetic group of lakes, although they belong to different river basins. Lakes Bolshoy Padimey, Lola-Kuliga, Costy, Yurto are actually one single water body, which originates the Bolshoy Padimeyvis river. There is a lake Maliy Padimey near the lake Bolshoy Padimey, belonging to the basin of the river Karataiha. The area of the main water body ranges from 5.0 km^2 (lake Lola-Kuliga) to 11.2 km^2 (lake Bolshoy Padimey). This group of lakes does not differ significantly from the Vashutkin lake system in the structure and height of the banks, the indentation of the coastline, the nature and location of bottom soils, and some other features. The main difference between them lies in the deep-water nature and low water exchange. The relative area occupied by stony-boulder soils is somewhat increased here, especially in lake Maliy Padimey, which is also characterized by the absence of bays that go deep into the land. The area occupied by depths of 1…5 m is 18…31% in some water bodies of the system. The incoming part of the water balance in the Padimey lakes is 68% due to direct precipitation on the surface of the lake system. The expenditure part of the balance is represented by surface runoff by 63…79%. Depths up to 10 m are prevailing in Padimey lakes. The composition of water in the Padimey group of lakes is hydrocarbonate sodium, and the total water salinity ranged from 30 to 100 mg/L (Sidorov, 1970, 1970a).

The Harbey group of lakes looks like a chain of lakes in the upper reaches of the Harbeyvis river. This system has an intermediate position among the studied lakes in terms of water depth and flow. Its catchment area is mostly swampy. The largest lake in the Harbey system is Bolshoy Harbey (21.3 km^2). Maliy Harbey lake has a water mirror area of 1.7 km^2. The Harbey group of lakes has lower banks comparing to the other lakes (heights up to 2 m prevail).

Golovnoye lake (Golovka lake) is small (3.1 km^2) and has a very rugged coastline. Dense soils are relatively less common here than in other water bodies. The lake has a well-developed subordinate system, which strongly affects its regime. Depths up to 5 m prevail in the Harbey lakes. The depths of 1 to 5 m occupy around 41% of area in lake Golovka, 50% in lake Bolshoy Harbey, and 64% in lake Maliy Harbey. The input part of the balance in the Harbey lakes consists of 61% of water coming from the catchment area, while the expenditure part of the balance is represented by surface runoff (63 to 79%). The composition of water in the Harbey lakes is hydro-carbonate-sodium, and the total water salinity ranged from 30 to 100 mg/L.

The ice-off of the lakes occurs in the second half of June, the ice-on occurs in mid-October. In shallow waters and small lakes, this process begins earlier. During the open water period (3.5 months), there is a continuous decrease in the level by 80…100 cm. A short low-water period occurs at the end of September (Sidorov, 1974).

Species Composition, Biology, and Commercial Significance of Individual Fish Species

The fish species composition of the Harbey and Padimey lake systems is shown in Table 11. It mostly common to the ichthyofauna of the rivers which basins are part of lakes system: the right tributaries of the Usa river and the Korotaiha river. There were totally 13 fish species identified. Unlike Vashutkin lakes there are no *Stenodus leucichthys nelma* and *Pungitius pungitius*. The depletion of ichthyofauna in lake Maliy Padimey was noted due to the loss of Cyprinidae and *Perca fluviatilis*, which was probably due to the different times of fish settlement on the territory of the Bolshezemelskaya tundra and the restructuring of the hydrographic network in the quaternary period.

All fish species are nonmigratory. *Thymallus thymallus* belongs to the lake-river ecological form, other species, as a rule, constantly live in lakes. Places of feeding, spawning and wintering of fish are found in the main and subordinate lakes, as well as in the heads of rivers. *Coregonus lavaretus* dominates in the main lake of the studied system, and *Coregonus peled* dominates in small satellite water bodies. There are commercial fish: *Coregonus peled, Esox lucius, Coregonus lavaretus,* and *Thymallus thymallus*, and *Perca fluviatilis* in the Harbey lakes and lake Bolshoy Padimey (Sidorov, 1970b).

Coregonus albula is found in the main lakes of the Harbey and Padimey systems. Biometric analysis showed that the local species is closer to *Coregonus albula* than to *Coregonus sardinella*. According to growth rates, it does not differ from *Coregonus albula* from other lakes in the Western part of the Bolshezemelskaya tundra (Esipov, 1938). *Coregonus albula* becomes mature at the age of 3 to 4. Its absolute fecundity is, on average, 4.6 thousand eggs (2.4 to 6.3 thousand). *Coregonus albula* is a permanent prey item of *Esox lucius*, and this can serve as an indirect indicator of its abundance in lakes in the absence of specialized fishing. So, according to the frequency of occurrence in the diet of *Esox lucius*, its smallest number is observed in deep-water Padimey lakes (especially in lake Maliy Padimey). The density of *Coregonus albula* is noticeably higher in a well-warmed and shallower lake Bolshoy Harbey.

Coregonus peled lives in Harbey and Padimey lake systems. It is a widespread commercial species. Local *Coregonus peled* grows worse than the floodplain-river peled of the Pechora and the Usa rivers. Mass maturation of local *Coregonus peled* occurs at the age of 6. The absolute fecundity of the Harbey peled is, on average, 32.7 thousand eggs, while the Padimey *Coregonus peled* has significantly more eggs (46.3 thousand).

Coregonus nasus also lives in both lake systems. Its number is small. The growth rate of local *Coregonus nasus* is low and is almost the same in both the Harbey and Padimey lakes.

Coregonus lavaretus is found in most lakes in both systems. It is a commercial species, but its number is slightly lower than *Coregonus peled*. The growth rate of *Coregonus lavaretus* is also low and in terms of size and weight, it is inferior to whitefish from the Usa river. Local *Coregonus lavaretus* become mature at the age of 7 - two years later than *Coregonus lavaretus* of the Pechora and Usa rivers. The absolute individual fecundity of the Harbey *Coregonus lavaretus* is, on average, 8.3 thousand eggs (range of 2.7 to 26.7 thousand), Padimey – 12.2 thousand (range of 6.6 to 23.4 thousand).

Thymallus thymallus is also widely distributed in Harbey and Padimey lake systems. It is a commercial species. Morphologically, *Thymallus thymallus* of both systems is almost identical. The linear and weight growth of local *Thymallus thymallus* is low and lower than that of representatives of this species in most other water bodies of the European North. Mass sexual maturation of *Thymallus thymallus* in both systems occurs at the age of 7…8.

Table 11. The species composition of fish Harbey and Padimey lake systems (Sidorov, 1970)

Fish species			
	Harbey lakes	Bolshoy Padimey lake	Maliy Padimey lake
Coregonus albula	+	+	+
Coregonus peled	+	+	+
Coregonus nasus	+	+	+
Coregonus lavaretus	+	+	+
Thymallus thymallus	+	+	+
Esox lucius	+	+	+
Rutilus rutilus	+	+	–
Leuciscus idus	+	+	–
Lota lota	+	+	+
Perca fluviatilis	+	+	–
Gimnocephalus cernuus	+	+	+
Rhynchocypris percnurus	+	+	–
Cottus gobio	+	+	+

Footnote: + the species was found; – the species was not found.

Esox lucius. It is a mass and commercial species in both lake systems. The local *Esox lucius* grows worse than in the Pechora and Usa rivers. At the same time, *Esox lucius* of the Harbey lakes shows better linear and weight growth than *Esox lucius* from the Padimey lakes. Mass sexual maturation of males in all water bodies occurs at the age of 6, females – at the age of 7.

Rutilus rutilus. It is noted as a rare species in the Harbey lakes and lake Bolshoy Padimey. The growth rate of the local *Rutilus rutilus* is good, its fatness is higher than that of the *Rutilus rutilus* of the Pechora and Usa rivers. There is no *Rutilus rutilus* in Maliy Padimey.

Leuciscus idus. It is very rare in the Harbey lakes and lake Bolshoy Padimey. The growth rate of *Leuciscus idus* in these water bodies is low. There is no *Leuciscus idus* in Maliy Padimey.

Lota lota. It is common for most water bodies of Harbey and Padimey lake systems. It is a by-catch to the main commercial species. There is no specialized *Lota lota* fishing.

Perca fluviatilis is mass species, widely distributed in Harbey lakes and lake Bolshoy Padimey. There is no *Perca fluviatilis* in Maly Padimey. In general, representatives of this species grow worse in the studied water bodies than in the more southern regions of the Pechora river basin. Local *Perca fluviatilis* become mature at the age of 6…7. The absolute fecundity of females reaches 17.5 thousand eggs.

Gimnocephalus cernuus is a common species in the Harbey lakes and in other water bodies. Its growth rate is the same as that of representatives of this species in the upper part of the Pechora River, but lower than that in the Usa River.

Rhynchocypris percnurus and *Cottus gobio*. They live on pebbly and rocky habitats of rivers and large water bodies of Harbey and Padimey lake systems (Smith, 1974). They are non-commercial species.

The Economic Status of the Lakes

In the 50-60s of the last century, a limited set of fishing gear was used for fishing in the Harbey and Padimey lakes: draft seines 80…200 m long and gill nets with a mesh size of

40…45 mm. At the same time, there were two peaks of the fishing season: spring, based on feeding concentrations of all fish species in the coastal part of the lakes, and autumn – on spawning accumulations of *Coregonus lavaretus* and *Coregonus peled*. Winter fishing was very poorly developed due to the lack of proper living conditions and mechanization of labor-intensive processes. Snow of the Northern slopes and glaciers were used in the spring to preserve fresh fish on the lakes, as during summer it was possible to store it for a short time. A significant part of the catches in the summer was preserved via salting, while there was irregular export of fresh fish from the lakes. Transportation of fish to the consumer was by air and off-road vehicles. Currently, there is no commercial fishing in the lakes of both systems.

Based on catch statistics, age structure and biological indicators of commercial fish populations, the possible fish production of Padimey lakes may be up to 6.2 kg/ha, and Harbey lakes – 5.8 kg/ha. Approximately twice the allowable catch, with similar feed resources, is possible on small satellite water bodies. Taking into account the completely unused vendace fishing, as well as the specialized fishing of ordinary fish species, the fishing opportunities of the lakes are very high (Sidorov, 1974).

To summarize, Padimay lakes represent a single genetic group, although they belong to different river basins - the Bolshoy Padimeyvis river basin and the Karataiha river basin. In most physiographic parameters this group of lakes is not significantly different from the Vashutkin lakes system. The Harbey group of lakes is located in the form of a chain in the upper reaches of the Harbeyvis river and occupies an intermediate position among the studied lakes in terms of depth and flow. The input part of the balance in the Padimey lakes is 68% due to precipitation directly falling on the surface of the lake, in the Harbey lakes, 61% is due to water runoff coming from the catchment area. The expenditure part of the balance is a surface runoff by 63…79% in both lake systems. The species composition of fish in the Padimey and Harbey lake systems mainly coincides with the ichthyofauna of the rivers whose basins include water body: the right tributaries of the Usa river and the Korotaiha river. The ichthyofauna includes 13 species of fish, all of which are non-migratory. The natural fish production of the Padimey and Harbey lakes is 5 to 6 kg/ha. The lakes may be promising for the restoration of local fishing. Overall, the lakes of the Nenets Autonomous district can be used for studying, restoring and preserving the biological diversity of lakes in the European Northeast of Russia.

This study was supported by the Ministry of Science and Higher Education of the Russian Federation within a state assignment, no. AAAA-A19-119011690119-9

REFERENCES

Dvoryankin, G., Timofeev, V. (2007a) The ichthyofauna of Vashutkin lakes and the prospects for its rational use. *Materials of the reporting session of the Northern branch of PINRO on the results of research in 2003-2004*, 95-104. Arkhangelsk.

Dvoryankin, G., Timofeev, V. (2007b) Fish stocks of large lake systems in the North of Russia and their rational use (on the example of Vashutkin lakes). *Proceedings of the ICES international conference*, 201. Helsinki.

Esipov, V. (1938) About *Coregonus peled* (Gmelin) from lakes of the Bolshezemelskaya tundra. *Zoological journal,* Volume 17, Issue 2, 303-314.

Goldina, L. (1972) *Geography of lakes of the Bolshezemelskaya tundra* (1972), 103. Leningrad.

Golodnaya Guba lake. (2013) Reserved area "Nizhnepechorsky." *Collection of articles*, 64. Nar'yan-Mar.

Ivanova, N., Chuksina, N. (1985) The role of Golodnaya Guba of the lower reaches of the Pechora river in reproduction of migratory fish. Ecology and reproduction of migratory salmon in the basins of the White and Barents seas. *Collection of scientific papers of PINRO named after N. M. Knipovich*, 129-136. Murmansk.

Imant, E., Zavisha, A., Studenova, M., Novoselov, A., Levitsky, A. (2018) Food base comparative characteristics of different types fish in Northern fisheries basin lakes. III international conference "Actual problems of planktonology." *Conference proceedings*, 86-90. Kaliningrad.

Khokhlova, L. (1994) *Chemical composition of surface waters*. Structural and functional organization of phytocenoses in the Far North, 12-23. Syktyvkar.

Kirpichnikov, V. (1935) *Biological systematic essay of smelt in the White sea*, Cheshskaya Bay and the Pechora river, vol. 21, 103-184. Moscow.

Kornilova, V. (1967) The ichthyofauna of Golodnaya Guba lake in delta of the Pechora river. *Materials of fisheries research in the Northern basin*. Issue IX (works of the Northern branch of PINRO), 32-41. Murmansk.

Kozmin, A. (2011) *Fish stocks of rivers and lakes of the European Northeast of Russia:* preservation and usage, 314. Murmansk.

Lukin, A., Dauvalter, V., Novoselov, A. (2000) *Ecosystem of the Pechora river in modern conditions*, 198. Apatity.

Makkaveeva, P. (1967) Hydrochemical studies of Golodnaya Guba. In the book: "*Materials of fisheries research in the Northern basin.*" Issue 9, 21-26. Murmansk.

Nikolsky, G. (1980) *Species structure and regularities of fish variability*, 182. Moscow.

Novoselov, A. (1991) Distribution of whitefish in the lakes of the Arkhangelsk region. In the collection: *"Biological problems of the North: modern problems of whitefish,"* part 1, 23-37. Vladivostok.

Novoselov, A. (2013) Ecological diversity of the ichthyofauna of Golodnaya Guba lake in the Pechora river delta. Golodnaya Guba lake. *Nizhnepechorsky Nature Reserve*, 28-33. Nar'yan-Mar.

Novoselov, A., Fadeeva, V., Zavisha, A. (2005) Current state of Golodnaya Guba lake ecosystem in the lower reaches of the Pechora river. *Materials of the reporting session of the Northern branch of the Federal state unitary enterprise "PINRO"* (based on the results of research works in 2002-2003), 94-111. Arkhangelsk.

Pidgayko, M., Aleksandrov, B., Ioffe, T, Maksimova, L., Petrov, V., Savateeva, E., Salazkin, A. (1968) Brief biological and production characteristics of the North-West water bodies of the USSR. *Izvestiya GosNIORH*. Vol. 67, 205-225. Moscow.

Red book of the Nenets Autonomous area (2006), 351. Nar'yan-Mar.

Sadyrin, V. (1998) Hydrobiological characteristics of some glacial lakes of the Komi Republic and their productivity. *Hydrobiological journal*, vol. 34, No. 2, 19-30.

Sendek, D. (2000) Phylogenetic analysis of whitefish family Coregonidae by protein electrophoresis. *Abstract. diss. cand. biol. nauk*, 24. Saint-Petersburg.

Sidorov, G. (1976) Biological characteristics and fish production of lake Bolshoy Harbey. *Productivity of lakes in the Eastern part of the Bolshezemelskaya tundra*, 110-130. Leningrad.

Sidorov, G. (1975) Commercial ichthyofauna and prospects of fishery use of large lakes of the Bolshezemelskaya tundra. *Biological fundamentals of the use of the North*, 79-96. Syktyvkar.

Sidorov, G. (1974) *Fish stocks of the Bolshezemelskaya tundra*, 163. Leningrad.

Sidorov, G. (1970) Fish management characteristics of lakes of the Bolshezemelskaya tundra. *Materials of the IV Komi Republican youth scientific conference*, 28-39. Syktyvkar.

Solovkina, L., Sidorov, G. (1966) Commercial fishing importance of lake-river systems of the Bolshezemelskaya tundra. *Hydrobiological study and fish management development of lakes of the Far North of the USSR*, 164-169. Moscow.

Zhila, I. (1963) Golodnaya Guba in delta of the Pechora river. *Observations and research materials of the Arkhangelsk hydrometeorological Observatory,* issue 1, 13-38, Arkhangelsk.

Zvereva, O., Vlasova, T., Goldina, L. (1966) Vashutkin lakes and the history of their research. *Hydrobiological study and fisheries development of lakes in the Far North of the USSR,* 4-22. Moscow.

ABOUT THE EDITORS

Oleg S. Pokrovsky
Research Director, CNRS
CNRS UMR 5563, Observatory Midi Pyrenees, Toulouse France

O.S. Pokrovsky, PhD in Geology and Mineralogy (1994), Research Director (CNRS) possesses both experimental physico-chemical, microbiological and geochemical expertise. He work on biogeochemical cycle of carbon and related nutrients in boreal and Arctic waters with a focus on transport, speciation and bioavailability of dissolved organic matter (DOM) in rivers, lakes, soils and estuaries. Over past decade, he has published more than 100 papers in high rank international journals on the role of DOM and metal cycling in natural waters including lakes and rivers, chemical weathering, metal speciation and interaction with microorganisms, organic matter cycling in surface waters of subarctic and permafrost zone, and colloidal DOM and metal transport and speciation in inland waters of European Russia, Western and Central Siberia. His HI factor (Web of Science) is 52 and total citation number is > 9000.

Yulia V. Bespalaya
Director of Institute
Institute of Biogeography and Genetic Resources, N. Laverov Federal Center for Integrated Arctic Research of Russian Academy of Sciences Federal Agency for Scientific Organizations, Arkhangelsk, Russia.

Y.V. Bespalaya (Director of the Institute of Biogeography and Genetic resources, Head of Laboratory), received her PhD in Biology (2007) on the fauna and ecology of mollusks in island and continental reservoirs of Northern taiga in the west of the Russian plain. She is expert in malacology and limnology, species diversity, freshwater mollusks, ecology, biogeography, adaptation, life cycles, phylogeny, extreme habitats, phylogeny, invasive species, and reproduction. She published 40 peer-reviewed articles in the field of mollusk biology, limnology and lake sciences and participated in numerous field expeditions. Over past years, she led numerous grants of Russian Foundation for Basic Research ("Biological invasions of alien species of mollusks in freshwater ecosystems the watercourses of the basin of the White Sea: ecology, impact on native ecosystems, the prediction of distribution, monitoring", "Phylogeography, ecology and adaptive capacity of freshwater mollusks to extreme climatic conditions of the Arctic", "Mollusks in the zoobenthos of lakes in the European sector of the Arctic: species and genetic diversity, adaptive strategies and life cycles"), coordinated the Program of Federal Agency for Scientific Organizations ("Study of the biological and genetic diversity of invertebrate animals in conditions of climate change in the Arctic with the formation of scientific collections").

Liudmila S. Shirokova

Research Scientist

N. Laverov Federal Center for Integrated Arctic Research, Russian Academy of Science, Arkhangelsk, Russia, and Geoscience and Environment Toulouse UMR 5563 CNRS, Toulouse, France.

L.S. Shirokova (Associate Professor, University of Toulouse), PhD in Biology (2007), is a limnologist who is using biogeochemical and microbiological methods to reveal the functioning of lakes in boreal and subarctic zone. She led numerous grant programs on biogeochemistry and environmental sciences, coordinated collaboration in the field of thermokarst lake research. She demonstrated the importance of lake size control on hydrochemistry of the water column for the case of thermokarst lakes in permafrost peatlands and contributed to studies of carbon emissions from lake surfaces in the Russian subarctic. She has published more than 80 papers in peer -reviewed international journals, 2 books and received more than 1700 citations (HI Web of Science is 26).

Taisia Y. Vorobyeva

Head of Laboratory

N. Laverov Federal Center for Integrated Arctic Research of Russian Academy of Sciences Federal Agency for Scientific Organizations, Arkhangelsk, Russia.

T.Y. Vorobyeva, PhD in Biology (2003), Head of the Laboratory of Freshwater and Marine Ecosystems, N. Laverov Federal Center for Integrated Arctic Research, experts on aquatic microbiology, sewage microbiology, and hydrochemistry of lakes and rivers in the Arctic and Subarctic zone, and hydrology, hydrochemistry, and ecology of the river - sea mixing zone, with a special focus on nitrogen cycle. She created the laboratory on aquatic microbiology at the Federal Center and led numerous national programs and grants on lakes of the Arkhangelsk Region (Kenozersky National Park), organized limnological expeditions in the NW Russian Subarctic and boreal zone. She published more than 60 papers on lakes hydrochemistry, ecosystem assessment and hydrobiology.

INDEX

#

A

B

C

D

E

I

J

K

L

M

N

O

P

R

S

T

U

V

W

Z